From Convexity to Nonconvexity

Nonconvex Optimization and Its Applications

Volume 55

The titles published in this series are listed at the end of this volume.

From Convexity to Nonconvexity

Edited by

R.P. Gilbert
University of Delaware

P.D. Panagiotopoulos

and

P.M. Pardalos
University of Florida

KLUWER ACADEMIC PUBLISHERS
DORDRECHT / BOSTON / LONDON

A C.I.P. Catalogue record for this book is available from the Library of Congress.

ISBN-13:978-1-4613-7979-9 e-ISBN-13:978-1-4613-0287-2
DOI: 10.1007/978-1-4613-0287-2

Published by Kluwer Academic Publishers,
P.O. Box 17, 3300 AA Dordrecht, The Netherlands.

Sold and distributed in North, Central and South America
by Kluwer Academic Publishers,
101 Philip Drive, Norwell, MA 02061, U.S.A.

In all other countries, sold and distributed
by Kluwer Academic Publishers,
P.O. Box 322, 3300 AH Dordrecht, The Netherlands.

Printed on acid-free paper

Contents

Preface xi

1
Frictional contact problems 1
Lars-Erik Andersson, Anders Klarbring
1.1 Introduction 1
1.2 Elementary example of non-uniqueness and non-existence 2
1.3 Classical formulation of the quasistatic frictional contact problem 4
1.4 The static problem 5
1.5 Steady sliding problem 8
1.6 Existence results for quasistatic friction problems 8
1.7 Conclusion 10

References 11

2
Solutions for quasilinear hemivariational inequalities 15
Siegfried Carl
2.1 Introduction 15
2.2 Notations, hypotheses and the main result 16
2.3 Auxiliary results 20
2.4 Proof of the main result 24
2.4.1 Static problem 25
2.4.2 Example 25
2.4.3 Concluding remarks 26

References 27

3
A Survey on Nonsmooth Critical Point Theory 29
Marco Degiovanni
3.1 Introduction 29
3.2 Critical point theory in metric spaces 31
3.3 Subdifferential calculus 33
3.4 Functionals of the calculus of variations 35
3.5 Functionals with quadratic dependence on the gradient 36
3.6 Area-type functionals 38

References 39

4
Exhaustive families of approximations revisited 43
V.F.Demyanov A.M.Rubinov
4.1 Directional derivatives and generalizations 43
4.2 Exhaustive families of upper and lower approximations 45

References 49

5
Optimal shape design 51
Zdzisław Denkowski
5.1 Introduction 51
5.2 Preliminaries 52
5.3 State relations for physical systems 53
5.4 Abstract OSD and direct method 56
5.5 Mapping method and its applications 57
5.6 Some remarks on other methods for OSD problems 59
5.7 Relaxation in OSD problems 61
5.8 Lower semicontinuity of functionals in OSD problems 63

References 63

6
Duality in Nonconvex Finite Deformation Theory 67
David Yang Gao
6.1 Introduction 67
6.2 Framework and Abstract Boundary Value Problem 69
6.3 Conjugate Stress-Strain Tensors and Gap Functions 71
6.4 Potential Extremum Principle 74
6.5 Classical Complementary Energy Principles 75
6.6 Generalized Variational Principles and Triality Theory 77
6.7 Pure Complementary Energy Principles and Minimax Theory 78

References 80

7
Contact Problems in Multibody Dynamics 85
Friedrich Pfeiffer and Christoph Glocker
7.1 Introduction 85
7.2 The Evolution of a Theory 86
7.3 Present Mathematical Formulation 99
7.4 Conclusions 107

References 107

8
Hyperbolic Hemivariational Inequality 111
D. Goeleven[1] and D. Motreanu[2]
8.1 Introduction and formulation of nonsmooth hyperbolic problem 111
8.2 Finite dimensional approximation 113

8.3 Main Results 117

References 121

9
Time–integration algorithms 123
Klaus Hackl

9.1 Introduction 123
9.2 An augmented principle of maximum plastic dissipation 124
9.3 The Evolution Problem 126
9.4 Time–Integration Algorithms and their Stability Properties 126
9.5 Algorithms Involving Operator–Split 130
9.6 Conclusion 135

References 135

10
Contact Stress Optimization 137
J. Haslinger

10.1 Introduction 137
10.2 Formulation of the problem 138

References 145

11
Recent results in contact problems with Coulomb friction 147
J. Jarušek and C. Eck

11.1 Introduction 147
11.2 The static case 148
11.3 Dynamic problem with contact condition in displacement and given friction 152
11.4 Dynamic problem with Coulomb friction and contact condition in velocities 155
11.5 Appendix: Thermal aspects of friction 157
11.6 Conclusion 159

References 159

12
Polarization fields in linear piezoelectricity 161
P. Bisegna and F. Maceri

12.1 Introduction 161
12.2 The linear piezoelectric problem 162
12.3 Weak formulations of the linear piezoelectric problem 165
12.4 Hashin-Shtrikman type variational principles 166
12.5 Conclusions 170

References 170

13
Survey of the methods for nonsmooth optimization 177
M. M. Mäkelä

13.1 Introduction 177

13.2 Convex optimization 178
13.3 Nonconvex optimization 185

References 188

14
Hemivariational inequalities and hysteresis 193
M. Miettinen

References 205

15
Non convex aspects of dynamics with impact 207
L. Paoli and.M. Schatzmann

References 221

16
On Global Properties of D.C.Functions 223
L.N.Polyakova

References 231

17
Variational-Hemivariational Inequalities 233
G.Dinca, G.Pop

References 241

18
Perturbations of Eigenvalue Problems 243
Vicentiu D. Rădulescu

19
Implicit variational inequalities arising in frictional unilateral contact mechanics: analysis and numerical solution of quasistatic problems 255
Marius Cocu, Michel Raous

19.1 Introduction 255
19.2 Quasistatic contact problems with friction 256
19.3 Extension to a model coupling adhesion and friction 259
19.4 Numerical methods 263

References 265

20
Regularity for variational inequalities 269
Rainer Schumann

20.1 Introduction 269
20.2 Variational inequalities and their applications 270
20.3 Regularity 272

References 280

21
A Survey of 1-D Problems of Dynamic Contact and Damage 283

Meir Shillor

21.1 Introduction 283
21.2 Preliminaries 284
21.3 Dynamic Thermoviscoelastic Contact of a Rod 285
21.4 Vibrations of a Beam between Two Stops 287
21.5 A Beam in Frictional Contact 290
21.6 The Elastic Rod with Damage 292

References 294

22
Nonconvexity in plasticity and damage 297
Georgios E. Stavroulakis

22.1 Introduction 297
22.2 Nonsmooth modeling in mechanics 298
22.3 Elastoplasticity 301
22.4 Damage mechanics 305

References 307

23
Augmented Lagrangian Methods for Contact Problems 311
Józef Joachim Telega and Andrzej Gałka

23.1 Introduction 311
23.2 General results 312
23.3 Ito and Kunisch augmented Lagrangian methods 314
23.4 Contact Problems 318
23.5 Parameter Estimation and Optimal Control 322
23.6 Image Restoration 326

References 328

24
Mountain Pass Theorems 333
Stepan A. Tersian

24.1 Introduction 333
24.2 Deformation theorems and (PS) conditions 334
24.3 Mountain pass theorems 339

References 343

25
Proximal Methods for Variational Inequalities with Set-Valued Monotone Operators 345
A. Kaplan and R. Tichatschke

25.1 Introduction 345
25.2 Multi-step Proximal Regularization Scheme 348
25.3 Convergence Analysis 349

References 358

26
Simons' Problem 363

Mirosław Przeworski Dariusz Zagrodny
26.1 Introduction 363
26.2 Some basic facts and definitions 365
26.3 A solution of the generalized Simons' problem 368
26.4 Locating elements of graphs of maximal monotone operators 372

References 378

27
Density estimates of Blake & Zisserman functional 381
Michele Carriero, Antonio Leaci*, Franco Tomarelli***
27.1 Introduction 381
27.2 Notation and preliminary results 383
27.3 Local weak minimizers and essential minimizing triplets. 386
27.4 Density estimates for essential minimizing triplets. 387
27.5 A counterexample 390

References 391

Preface

This collection of papers is dedicated to the memory of Gaetano Fichera, a great mathematician and also a good friend to the editors. Regrettably it took an unusual amount of time to bring this collection out. This was primarily due to the fact that the main editor who had collected all of the materials, for this volume, P. D. Panagiotopoulos, died unexpectedly during the period when we were editing the manuscript. The other two editors in appreciation of Panagiotopoulos' contribution to this field, believe it is therefore fitting that this collection be dedicated to his memory also.

The theme of the collection is centered around the seminal research of G. Fichera on the Signorini problem. Variants on this idea enter in different ways. For example, by bringing in friction the problem is no longer self-adjoint and the minimization formulation is not valid. A large portion of this collection is devoted to survey papers concerning hemivariational methods, with a main point of its application to nonsmooth mechanics. Hemivariational inequalities, which are a generalization of variational inequalities, were pioneered by Panagiotopoulos. There are many applications of this theory to the study of nonconvex energy functionals occurring in many branches of mechanics. An area of concentration concerns contact problems, in particular, quasistatic and dynamic contact problems with friction and damage. Nonsmooth optimization methods which may be divided into the main groups of subgradient methods and bundle methods are also discussed in this collection. There is also some attention given to regularity results for variational inequalities. We believe that the researcher or advanced student wishing to learn more about convex and nonconvex analysis will profit from reading the fine works in this collection.

We would like to take the opportunity to thank the authors of the papers, the anonymous referees, and Kluwer Academic Publishers for making the publication of this volume possible.

R. P. Gilbert and P. M. Pardalos
University of Delaware and University of Florida
January 2001

Preface

This collection of papers is dedicated to the memory of Gaetano Fichera, a great mathematician and also a good friend to the editors. Regrettably it took an unusual amount of time to bring this collection out. This was primarily due to the fact that the main editor who had collected all of the materials for this volume, P. D. Panagiotopoulos, died unexpectedly during the period when we were editing the manuscript. The other two editors in [illegible] of Panagiotopoulos [illegible] [illegible] this collection be [illegible] to both mathematicians.

[illegible]

We would like to take the opportunity to thank the authors of these papers, the anonymous referees and Kluwer Academic Publishers for making the publication of this volume possible.

R. P. Gilbert and P. M. Pardalos
University of Delaware and University of Florida
January 2001

1 A SURVEY OF BASIC MATHEMATICAL RESULTS FOR FRICTIONAL CONTACT PROBLEMS

Lars-Erik Andersson, Anders Klarbring

Linköping University, S-581 83 Linköping, Sweden

leand@mai.liu.se, andkl@ikp.liu.se

Abstract: This paper gives a review of mathematical results of existence and uniqueness of solutions to problems of linear elasticity involving friction. Static, steady sliding and quasistatic problems are discussed. The intention is to describe the state of the art for these problems, as it appeared in 1997 when this manuscript was completed.

1.1 INTRODUCTION

The present paper reviews known results concerning existence and uniqueness of solutions for the problem of a linearly elastic body in frictional contact with a rigid obstacle. The unilateral contact problem without friction was given its basic formulation and was shown to have a unique solution by Signorini; Fichera [8, 9] proved the existence of a solution in the function space of finite energies using a formulation as a quadratic minimization problem. However, when friction is included, the properties of the problem changes dramatically: firstly, any physically reasonable friction model has an evolutionary character, i.e. time enters the formulation; secondly, friction models are non-selfadjoint and the problem is no longer a minimization problem. In Duvaut and Lions [11, 12], where the first mathematical treatment of the friction problem in linear elasticity was given, both of these facts were ignored or suppressed in order to get a tractable mathematical problem. In fact, as described in this paper, the full problem is still only partially solved.

Due to space limitations this review is limited to problems not too far removed from the frictionless Signorini-Fichera problem. That is, we limit ourselves to the quasistatic problem where inertia terms are assumed negligible and to problems derivable by assumptions or approximations from the quasistatic

R.P. Gilbert et al. (eds.), From Convexity to Nonconvexity, 1–14.

problem. Also, we will mostly restrict our attention to so called coercive problems, where some part of the boundary of the elastic body has a prescribed motion, excluding rigid body motions. Most results valid for the coercive case have noncoercive generalizations. How to obtain these generalizations is not always obvious. Section 1.4, however, deals with steady sliding problems, which are usually noncoercive.

Further we do not treat dynamic problems or problems including thermal effects, such as those extensively treated by Shillor and coworkers. Also, we are not attempting to review space discretized problems, although we use such a problem for motivation in the section following this one. For a recent review of discrete problems we refer to [31]. Finally, even concerning problems closely related to the quasistatic problem we are most surely not completely exhaustive. However, we believe that the most important works are covered and that the state of the art is correctly described.

1.2 ELEMENTARY EXAMPLE OF NON-UNIQUENESS AND NON-EXISTENCE

It is known since more than a century, through an example by Paul Painlevé, that problems of rigid body dynamics with Coulomb friction may apparently not have solutions, or, if solutions exist, they may be non-unique. These facts have been debated and investigated up to this day, see e.g. Stewart [38]. What has, on the other hand, seemingly remained unknown until 1980, is that similar difficulties exist in static or quasistatic problems. Such problems may admittedly appear of less fundamental nature than dynamic problems, but have always been of great practical value in engineering practice, in recent years particularly in connection with the finite element method. Janovský [19, 20] was first to discover that a discrete abstract version of the static friction problem of Duvaut and Lions [12] may have non-unique solutions. Klarbring [24, 26] showed that non-uniqueness may also appear in the quasistatic problem, and, furthermore, non-existence of continuous solutions is possible. Similar non-existence is not found in the static problem treated by Janovský. A concrete mechanical example consisting of two connected bars was also given in [24, 26]. Below we will present an example which shows how right hand derivatives of the displacements and contact forces are non-unique and non-existent for particular different loading directions.

Consider a structure, condensed into a single contact node, in two-dimensional contact with a rigid obstacle. The external load and previous evolution of the structure is such that the normal contact force $p_N < 0$ satisfies $p_T = \mu p_N$, where p_T is the tangential contact force and μ is the friction coefficient. That is, the contact force state of the node is on the boundary of the friction cone and two qualitatively different further continuous evolutions are possible:

- the node may evolve into a stick state which means that

$$\dot{p}_T > \mu \dot{p}_N \quad \text{and} \quad \dot{w}_T = 0, \tag{1.1}$$

where $\dot{p}_T$, $\dot{p}_N$ and $\dot{w}_T$ are right hand time derivatives of the contact forces and the tangential displacement;

- the node may evolve into a slip state which means that

$$\dot{p}_T = \mu \dot{p}_N \quad \text{and} \quad \dot{w}_T \geq 0. \tag{1.2}$$

The previously introduced time derivatives $\dot{p}_T$, $\dot{p}_N$ and $\dot{w}_T$, and the time derivative of the normal displacement, $\dot{w}_N$, are linearly coupled through a stiffness equation:

$$\begin{bmatrix} k_{NN} & k_{NT} \\ k_{NT} & k_{TT} \end{bmatrix} \begin{bmatrix} \dot{w}_N \\ \dot{w}_T \end{bmatrix} = \begin{bmatrix} \dot{p}_N \\ \dot{p}_T \end{bmatrix} + \begin{bmatrix} \dot{f}_N \\ \dot{f}_T \end{bmatrix} \tag{1.3}$$

where k_{NN}, etc. are real numbers such that the system matrix is positive definite and $\dot{f}_N$ and $\dot{f}_T$ are time derivatives of the external forces. Since $p_N < 0$ it holds that $\dot{w}_N = 0$ so we may conclude from (1.3), by multiplying the first equation with μ and subtracting, that

$$(-\mu k_{NT} + k_{TT})\dot{w}_T = -\mu \dot{p}_N + \dot{p}_T - \mu \dot{f}_N + \dot{f}_t. \tag{1.4}$$

From (1.1) and (1.4) it follows that if the node goes into a **stick state**, then

$$\mu \dot{f}_N > \dot{f}_t.$$

From (1.2) and (1.4) one finds that if the node goes into a **slip state**, the condition on derivatives of external forces depends on the sign of the "effective stiffness" in (1.4):

$$\begin{array}{llll} \text{if} & -\mu k_{NT} + k_{TT} > 0 & \text{then} & \mu \dot{f}_N \leq \dot{f}_t \\ \text{if} & -\mu k_{NT} + k_{TT} = 0 & \text{then} & \mu \dot{f}_N = \dot{f}_t \\ \text{if} & -\mu k_{NT} + k_{TT} < 0 & \text{then} & \mu \dot{f}_N \geq \dot{f}_t \end{array}$$

We conclude the following for the problem of determining the right hand time derivative of contact forces and displacements when given a rate of change of the external forces:

- if $-\mu k_{NT} + k_{TT} > 0$, then there exists a unique solution;
- if $-\mu k_{NT} + k_{TT} \leq 0$, then, depending on the direction of change of external loading, there does not exist any solution or there are multiple solutions.

Note that $\mu > 0$ and $k_{TT} > 0$ while k_{NT} may have any sign. This later constant represent a stiffness coupling between normal and tangential degrees of freedom, so, clearly, for large such couplings we may experience non-uniqueness or non-existence. The non-existence of right hand time derivatives means that the quasistatic problem cannot have any solutions that are continuous in time.

1.3 CLASSICAL FORMULATION OF THE QUASISTATIC FRICTIONAL CONTACT PROBLEM

We will give a classical formulation of the quasistatic contact problem with Coulomb friction. A linearly elastic body which may come into contact with a fixed rigid obstacle is considered. Two mathematical formulations of the physical conditions of impenetrability and non-adhesion are given. The first one is the Signorini-Fichera complementarity condition. The second one is the so-called normal compliance law.

The classical quasistatic formulation is obtained from the dynamic formulation by neglecting inertial terms, which means that we are looking at a sequence of equilibrium states. The linearly elastic body occupies a bounded Lipschitz domain Ω in R^3. The body is subjected to body forces $\boldsymbol{f} = (f_1, f_2, f_3)$ and to prescribed tractions $\boldsymbol{t} = (t_1, t_2, t_3)$ and displacements $\bar{\boldsymbol{u}}$ on the parts S_t and S_u of the boundary $\partial\Omega$, respectively. The potential contact boundary is S_c. Furthermore, S_c, S_t and S_u will be mutually disjoint, relatively open subsets of $\partial\Omega$. The following classical equations of linear elasticity are valid:

$$\frac{\partial \sigma_{ij}}{\partial x_j} + f_i = 0 \quad \text{in} \quad \Omega, \tag{1.5}$$

$$\sigma_{ij} = E_{ijkl}\frac{\partial u_k}{\partial x_l} \quad \text{in} \quad \Omega, \tag{1.6}$$

$$\sigma_{ij}\nu_j = t_i \quad \text{on} \quad S_t. \tag{1.7}$$

$$\boldsymbol{u} = \bar{\boldsymbol{u}} \quad \text{on} \quad S_u. \tag{1.8}$$

Here $\boldsymbol{u} = (u_1, u_2, u_3)$ is the displacement vector, $\boldsymbol{\sigma} = \{\sigma_{ij}\}$ is the stress tensor and ν_j are the components of the outward unit normal vector. E_{ijkl} are elasticity constants which satisfy the usual symmetry and ellipticity conditions. Furthermore, $i, j, k, l = 1, 2, 3$, the summation convention is used and $(0, x_1, x_2, x_3)$ is the cartesian reference frame.

To state the laws of contact and friction we decompose the displacement and traction vectors on S_c into normal and tangential components:

$$\sigma_N = \sigma_{ij}\nu_i\nu_j, \quad \sigma_{Ti} = \sigma_{ij}\nu_j - \sigma_N\nu_i,$$

$$u_N = u_i\nu_i, \quad u_{Ti} = u_i - u_N\nu_i.$$

The classical Signorini-Fichera contact law may now be stated as

$$\sigma_N \leq 0, \quad u_N - g \leq 0, \quad \sigma_N(u_N - g) = 0 \quad \text{on } S_c, \tag{1.9}$$

where g is the initial gap between the body and the rigid support. Note that there is no sign restriction for g.

The friction law is that of Coulomb, which can be written as

$$|\sigma_T| \leq -\mu\sigma_N, \quad \text{and} \tag{1.10}$$
$$|\sigma| < -\mu\sigma_N \Rightarrow \dot{u}_T = 0, \quad 0 < |\sigma_T| = -\mu\sigma_N \Rightarrow \dot{u}_T = -\lambda\sigma_T, \quad \lambda \geq 0$$

where μ is the friction coefficient, a superposed dot denotes time derivative, $\sigma_T = (\sigma_{T1}, \sigma_{T2}, \sigma_{T3})$ and $u_T = (u_{T1}, u_{T2}, u_{T3})$.

Relations, (1.5)-(1.10) constitute the quasi-static frictional contact problem. The problem is time dependent via the time derivative in Coulomb's friction law. Therefore the external forces $\boldsymbol{f}$ and $\boldsymbol{t}$ should be taken as time dependent.

The Signorini-Fichera conditions (1.9) are approximations to the behavior of contacting surfaces. Although these conditions are very useful in many situations an alternative which somewhat reflects the physical nature of contacting surfaces has been suggested and analysed in the literature. This is the so called normal compliance law

$$-\sigma_N = c_N(u_N - g)_+^{m_N}. \tag{1.11}$$

where c_N and m_N are positive parameters representing the physical characteristics of the interface and $(z)_+ = \max(0, z)$. Since (1.11) is a local relation between σ_N and u_N we can eliminate the former from the friction law (1.10). However, one can also generalize the friction law somewhat in the following way

$$\begin{gathered} |\boldsymbol{\sigma}_T| \leq c_T(u_N - g)_+^{m_T}, \quad \text{and} \\ |\boldsymbol{\sigma}_T| < c_T(u_N - g)_+^{m_T} \Rightarrow \dot{\boldsymbol{u}}_T = 0, \\ 0 < |\boldsymbol{\sigma}_T| = c_T(u_N - g)_+^{m_T} \Rightarrow \dot{\boldsymbol{u}}_T = -\lambda\boldsymbol{\sigma}_T, \quad \lambda \geq 0 \end{gathered} \tag{1.12}$$

where c_T and m_T are new physical parameters. One retrieves (1.10) from (1.12) by putting $c_T = \mu c_N$ and $m_T = m_N$.

Relations, (1.5)-(1.8), (1.11) and (1.12) constitute the quasistatic frictional contact problem with normal compliance.

1.4 THE STATIC PROBLEM

We now formulate a static version of the time dependent problem with the Signorini-Fichera conditions (1.9) of the previous section. Therefore we consider time independent force- and traction fields satisfying equations (1.5)–(1.8). The time dependent friction condition (1.10) is then replaced by

$$\begin{gathered} |\boldsymbol{\sigma}_T| \leq -\mu\sigma_N, \quad \text{and} \\ |\boldsymbol{\sigma}_T| < -\mu\sigma_N \Rightarrow \boldsymbol{u}_T = 0, \ 0 < |\boldsymbol{\sigma}_T| = -\mu\sigma_N \Rightarrow \boldsymbol{u}_T = -\lambda\boldsymbol{\sigma}_T, \quad \lambda \geq 0. \end{gathered} \tag{1.13}$$

Here (1.13) may be considered as a backward difference version of the time derivative condition (1.10).

Using Green's formula, it follows in a straightforward way that the equations (1.5)–(1.8), and (1.13) are equivalent to the following variational problem, provided that the displacement field $\boldsymbol{u}$ and the fields $\boldsymbol{f}$ and $\boldsymbol{t}$ are sufficiently regular.

VPstat: Find $\boldsymbol{u}$ with $\boldsymbol{u}|_{S_u} = \bar{\boldsymbol{u}}$ and $u_N \leq g$ such that

$$\begin{aligned} a(\boldsymbol{u}, \boldsymbol{v} - \boldsymbol{u}) &- \int_{S_c} \mu\sigma_N(|\boldsymbol{v}_T| - |\boldsymbol{u}_T|)\, dS \\ &\geq \int_\Omega \boldsymbol{f}\cdot(\boldsymbol{v} - \boldsymbol{u})\, dx + \int_{S_t} \boldsymbol{t}\cdot(\boldsymbol{v} - \boldsymbol{u})\, dS \end{aligned} \tag{1.14}$$

for all $\boldsymbol{v}$ with $\boldsymbol{v}|_{S_u} = \bar{\boldsymbol{u}}$ and $v_N \leq g$. Here $a(u,v) = \int_\Omega E_{ijkl} \frac{\partial u_i}{\partial x_j} \frac{\partial v_k}{\partial x_l} \, dx$ denotes the bilinear energy functional.

For the problem **VPstat** to have sense we must specify appropriate function spaces. Let $V = \{\boldsymbol{u} \in (H^1(\Omega))^3 : \boldsymbol{u}|_{S_u} = \bar{\boldsymbol{u}}\}$ and $K = \{\boldsymbol{u} \in V : u_N|_{S_c} \leq g\}$, where restriction of a function in the Sobolev space $H^1(\Omega)$ should be interpreted in the sense of traces and where, for the gap function g we assume that $g \in H^{1/2}(\partial\Omega)$. Then V will be a closed linear submanifold of $(H^1(\Omega))^3$ and K a closed convex subset of V.

For the force-field and the traction-field we will require that $\boldsymbol{f} \in (L^2(\Omega))^3$ and that $\boldsymbol{t} \in (H^{-1/2}(\Omega))^3$.

Under the assumptions that $\boldsymbol{u}$ and $\boldsymbol{v} \in K$, the second and fourth terms in (1.14), are not properly defined, since σ_N is in $H^{-1/2}(\partial\Omega)$ and not necessarily in $L^2(\partial\Omega)$. Therefore we should take, *e.g.*,

$$\int_{S_c} \mu\sigma_N |v_T| \, dS := \langle \mu\sigma_N, |v_T|\psi_c \rangle_{H^{-1/2}(\partial\Omega), H^{1/2}(\partial\Omega)}$$

and

$$\int_{S_t} \boldsymbol{t} \cdot \boldsymbol{v} \, dS := \langle \boldsymbol{t}, \boldsymbol{v} \rangle_{(H^{-1/2}(\partial\Omega))^3, (H^{1/2}(\partial\Omega))^3}$$

where $\psi_c \in C_0^\infty(R^3)$ is some cut-off function such that $\psi_c = 1$ in a neighbourhood of $\bar{S}_c$ and $\psi_c = 0$ in a neigbourhood of $\bar{S}_t \cup \bar{S}_u$. Since $\sigma_N = 0$ outside $\bar{S}_t \cup \bar{S}_u \cup \bar{S}_c$ the value of $\langle \mu\sigma_N, |v_T|\psi_c \rangle$ will be independent of the particular choice of ψ. We will keep the notation of (1.14) also for the similar variational problems formulated in the sequel. This is to some extent justified by the fact that many existence theorems will give, a posteriori, enough regularity on σ_N for the integrals to be well defined.

The exact formulation of the problem **VPstat** is then the following.

VPstat: Find $\boldsymbol{u} \in K$ such that the inequality (1.14) is valid for all $\boldsymbol{v} \in K$.

The contact problem without friction, *i.e.*, **VPstat** with $\mu = 0$, can be formulated as a problem of minimizing a convex (quadratic) functional over a convex domain of definition, and was treated by Fichera in [8, 9].

Contact problems with friction were first considered by Duvaut and Lions [10] and somewhat later formulated in the book [11] (english translation [12]). They gave however a simplified version of the variational problem **VPstat**, where the normal pressure σ_N was prescribed on the contact surface and where no non-penetration condition of the form $u_N \leq g$ was present. Under these simpler conditions a proof of existence and uniqueness was given.

An early existence and uniqueness result for a modified **VPstat** problem was given by Duvaut, [7]. He introduced a nonlocal frictional law by replacing σ_N in (1.14) by a mollified function $\sigma_N^* \in L^2(\partial\Omega)$, where the mapping

$$H^{-1/2}(\partial\Omega) \ni \sigma_N \mapsto \sigma_N^* \in L^2(\partial\Omega)$$

was assumed to be bounded and linear. Using a fixed point argument, Duvaut showed existence of a solution and, if the coefficient of friction was taken small enough, its uniqueness.

Similar results on existence and uniqueness were given by Cocu, [6], and Demkowitz and Oden, [8], using different methods. Cocu made use of an abstract theory of inequalities in Banach spaces, due to Ky Fan, [14], and Brézis, Nirenberg and Stampacchia, [5].

The first result for the the problem **VPstat**, without any regularization of the friction term, was given by Nečas, Jarušek and Haslinger in [35], where the existence of a solution was proved, under the condition that the coefficient was *small enough.* The proof in [35] was rather complicated and based on a shifting technique for estimating Sobolev norms, introduced by Fichera, see *e.g.* [16, 9]. Also for technical reasons the treatment in [35] was restricted to the situation when the domain Ω was an infinite strip in the plane. In a later article, [21], Jarušek, considered the more general situation with a domain $\Omega \in R^3$, where the part S_c was locally $C^{2,1}$ (meaning that for some local coordinate-system the boundary was represented as the graph of a function with bounded second derivatives). Moreover, it was assumed that $0 \leq \mu$, $\text{supp}\mu \in S_c$ and that $\mu \in C^1(\Omega)$. For the elasticity tensor it was assumed that the functions E_{ijkl} are Lipschitz continuous and satisfy the usual positivity and symmetry conditions. Then existence was proved for small enough values of the sup-norm $||\mu||_\infty$. The proof which supplied explicit bounds for the sup-norm of μ was quite technical. In the articles [35] and [21] it was assumed that a part S_u of the boundary had a prescribed motion (as in the formulation of **VPstat**). This means that, by Korn's inequality, the energy functional $a(u)$ is coercive, in contrast to the case when the set S_c is empty where one has to take into account that rigid body motions might be possible. This non-coercive case was treated by Jarušek in [22] where a similar existence result as in [21] was given.

A slight generalization of the results in [35, 21, 22] was given by Kato, [23], who proved existence for **VPstat**. In particular one should note that Kato did not impose the restriction that the coefficient of friction, μ, should have compact support in the open set S_c. Kato assumed that the gap function g is zero, that the boundary $\partial\Omega$ is smooth and that $\mu \in C^1(\bar{S}_c)$, with sufficiently small sup-norm $||\mu||_\infty$. He also gave some results of regularity for the solution.

Recently Eck, [13], gave an alternative and somewhat simpler proof of the results in [35, 21], by using a penalty formulation of the normal compliance type and taking limits as the penalty parameter tends to infinity. Eck also considered the case when μ was dependent of the tangential displacement.

One may note that essentially the same shifting technique is used in all of the works [35, 21, 22, 23, 13] in order to obtain necessary compactness arguments. One may also note that no uniqueness results seem to be known for **VPstat** in its original form without any form of regularization of the friction term.

1.5 STEADY SLIDING PROBLEM

A substantial simplification of the quasistatic problem appears if it can be argued that the direction of tangential sliding on S_c is known. This occurs if the rigid obstacle has prescribed motion or if the obstacle is fixed and a superimposed rigid body motion with zero acceleration is assumed for the body. This latter situation may be viewed as a *steady sliding* problem.

In such a case Coulomb's law of friction may be replaced by

$$\sigma_T = \mu \sigma_N \boldsymbol{e} \tag{1.15}$$

where $\boldsymbol{e}$ is a unit vector defining the direction of sliding. The steady sliding problem is now defined by (1.5)-(1.9) and (1.15).

The variational formulation of this problem becomes much similar to that of **VPstat**, but the nondifferentiable second term on the left hand side is replaced by a bilinear form:

VPsteady-sliding: Find $\boldsymbol{u} \in K$ such that

$$a(\boldsymbol{u}, \boldsymbol{v}-\boldsymbol{u}) - \int_{S_c} \mu \sigma_N(\boldsymbol{u}-\boldsymbol{v}) \cdot \boldsymbol{e} \, dS \geq \int_\Omega \boldsymbol{f} \cdot (\boldsymbol{v}-\boldsymbol{u}) \, dx + \int_{S_t} \boldsymbol{t} \cdot (\boldsymbol{v}-\boldsymbol{u}) \, dS$$

for all $\boldsymbol{v} \in K$.

For the coercive case, i.e. when S_c is large enough to prevent any rigid body motions, modifications of this problem was treated by Pires and Trabucho [36] and Rabier, Martins, Oden and Campos [37]. In the first of these papers the nonlocal friction law of Duvaut was used and the existence of a unique solution for sufficiently small friction coefficient or "sufficiently nonlocal" friction law was showed. In [37] the normal compliance regularization was used which transforms **VPsteady-sliding** into a variational equality. The existence of a locally unique solution for small friction coefficient or small load was shown. The semi-coercive version of the problem, i.e. when S_c is empty, has been considered only for special geometries. Gastaldi and Martins [18] considered the sliding on a flat surface. Andersson and Klarbring [4] considered unidirectional sliding occurring, for instance, when uncorking a wine bottle. It may be noted that no regularization of the friction term was used in [18] and [4]. The discrete version of the steady sliding problem has been studied in Klarbring [27] and Klarbring and Pang [32].

1.6 EXISTENCE RESULTS FOR QUASISTATIC FRICTION PROBLEMS

We will now formulate a variational problem corresponding to the time dependent problem with the Signorini-Fichera conditions (1.9) of section 1.3. Therefore we consider time dependent force- and traction fields satisfying equations (1.5)–(1.8).

Using Green's formula, as in section 1.4, we are led to formulate the following time dependent variational problem.

VPQ: Find a time dependent displacement field u such that $\sigma_N|_{S_c} \in L^2(S_c)$ and so that for all (or almost all) t we have

$$a(u, v-\dot{u}) - \int_{S_c} \mu\sigma_N(|v_T| - |\dot{u}_T|)\, dS \geq \int_\Omega f \cdot (v - \dot{u})\, dx + \int_{S_t} t \cdot (v - \dot{u})\, dS \quad (1.16)$$

for all $v \in K$. Further we require that

$$\sigma_N|_{S_c} \leq 0, \qquad \text{and } \sigma_N(u_N - g) \geq 0, \text{ almost everywhere on } S_c. \quad (1.17)$$

In this generality no existence or uniqueness results are known, as far as we know. If however, as in section 1.4, we introduce a bounded linear mapping

$$H^{-1/2}(\partial\Omega) \ni \sigma_N \mapsto \sigma_N^* \in L^2(\Omega)$$

and replace σ_N by σ_N^* in the integral over S_c in (1.16), then an existence result has been given by Cocu, Pratt and Raous, [7], for the case that the coefficient of friction is chosen sufficiently small. We note that, in their work, μ is allowed to be time dependent. Their proof is based on an incremental formulation of the problem **VPQ** and employs a technique for passing to the limit which is similar to that in [1].

Martins *et al.* have studied the quasistatic (as well as the dynamic) Signorini problem for an elastic system with only two degrees of freedom in a series of articles, see for instance [33] and [34]. They consider the case with a large coefficient of friction, when, due to the examples of Klarbring, [24, 26], described above in section 1.2, one may expect discontinuities. Existence results are given for the quasistatic case, where the solution $u(t)$ may exhibit jump discontinuous as a function of time. They also show that these solutions may be obtained as limits of problems with inertia terms and viscosity terms, as the masses and the viscosity tend to zero. The results by Martins *et al.* are related to a similar result for continuous systems and a normal compliance model, [2].

For the case with a normal compliance model, we obtain by the aid of Green's formula the following time dependent quasistatic problem.

VPQNC: Find a time dependent displacement field u such that for all (or almost all) t we have

$$a(u, v - \dot{u}) + \varphi(u, v - \dot{u}) + j(u, v) - j(u, \dot{u}) \geq \langle F, v - \dot{u} \rangle \quad (1.18)$$

for all $v \in V$.

Here we have introduced the notation $\varphi(u, v) = \int_{S_c} (u_N - g)_+^{m_N} v_N\, dS$ and $j(u, v) = \int_{S_c} (u_N - g)_+^{m_T} |v_T|\, dS$, and $\langle F, u \rangle = \int_\Omega f \cdot v\, dx + \int_{S_t} t \cdot v\, dS$.

The problem **VPQNC** was formulated by Klarbring, Mikelić and Shillor in [28]. They gave a proof of existence for the so called rate problem, *i.e.*, the problem to calculate the derivative $\dot{u}$ at a fixed time with a given field u, by the aid of (1.18). They also formulated a related incremental version of **VPQNC**, *i.e.*, a time discretized problem. They observed that the incremental problem was essentially a sequence of static problems, similar to **VPstat**, but with a normal compliance frictional term, and proved a theorem of existence. In this proof they made the assumptions $1 \leq m_N$, $m_T < 3$ (in three space dimensions). There were however no restrictions on the size of the coefficients c_N and c_T, nor on the coefficient of friction μ, for the case that $m_N = m_T$ and $c_T = \mu c_N$.

In [29] the same authors gave a local proof of uniqueness, under certain additional restrictions on the size of c_N and c_T (with bounds depending on $\boldsymbol{f}$, $\boldsymbol{t}$ and g). In [30] they investigated a modified version of the time dependent problem **VPQNC**, where the terms $c_N(u_N - g)_+^{m_N}$ and $c_T(u_T - g)_+^{m_T}$ were replaced by mollifications backwards in time.

In [1], Andersson treated the problem **VPQNC** without any additional regularization, and gave a proof of existence under similar restrictions on m_N, m_T, on the size of the loads and the size of coefficients c_N and c_T as above. An estimate of the form $|\dot{\boldsymbol{u}}(t)| \leq C|\dot{\boldsymbol{f}}(t)|$ for the solution, valid a.e. in t, was proved.

In a later paper, [2], a similar existence result was obtained without any restrictions on the size of the loads and the size of coefficients c_N and c_T. For this solution $\boldsymbol{u}(t)$ the conclusion regarding time regularity was that $\boldsymbol{u} \in BV(0, t_0; V)$, *i.e.*, that it was of bounded variation in time. It is therefore not necessarily differentiable everywhere and might even have countably many jump discontinuities.

In [3] the problem **VPQNC** was treated with $m_T = m_N$ and

$$\varphi(\boldsymbol{u}, \boldsymbol{v}) = \lambda \int_{S_c} (u_N - g)_+^{m_N} v_N \, dS \quad \text{and} \quad j(\boldsymbol{u}, \boldsymbol{v}) = \lambda \int_{S_c} \mu (u_N - g)_+^{m_N} |\boldsymbol{v}_T| \, dS$$

where μ is the coefficient of friction and λ is a, possibly large, penalization parameter. For small enough values on the Lipschitz norm $||\mu||_{Lip}$ of μ, the existence of a solution was proved. The bound on $||\mu||_{Lip}$ was however independent of the penalty parameter λ and the load as well as on the distances between the sets S_c, S_t and S_u.

1.7 CONCLUSION

For frictional contact problems there are surprisingly many open problems, some of which will be commented upon here.

First we note that for quasistatic evolution problems, as described in section 1.6, without any viscosity terms, there does not seem to be any uniqueness results at all available. This is true also if we introduce various regularizations of the friction terms, such as normal compliance or mollification of the normal pressure.

If we consider quasistatic evolution problems with Signorini-Fichera contact conditions and without regularization, no results of existence are known for continuous elastic systems. For systems with finitely many degrees of freedom it seems likely that existence results might be obtained by the weak solution techniques used in [1, 2, 7], for small coefficients of frictions. For the case with two degrees of freedom the results by Martins *et al.*, [33, 34], provide results of existence, where the solutions for large coefficients or friction may have discontinuities. To obtain uniqueness one probably has to impose additional assumptions on the solutions. This is still an open problem.

For static problems, treated in section 1.4, there are, as mentioned above, uniqueness results given for the case of a small coefficient of friction, μ, and with some kind of mollification, [7, 6, 8]. For the more general problem with Signorini-Fichera conditions and without regularization, treated by Nečas, Jarušek and Haslinger, [35], no uniqueness result is known. Here one might add that for μ large, even existence has not been established.

As appropriate areas of future research in contact problems with friction, we would like to identify the following.

The work by Martins *et al.* for quasistatic problems with two degrees of freedom should be extended to systems with arbitrarily many degrees of freedom. If possible, questions of uniqueness for their discontinuous solutions for large coefficients of frictions should be investigated. For small coefficients of friction one should establish a theory of existence and uniqueness.

For continuous elastic systems with Signorini-Fichera contact conditions (and without regularization of the frictional terms and without viscosity) questions of existence in the static case should be settled.

More importantly, one should establish existence results for quasistatic problems for continuous elastic systems with Signorini-Fichera contact conditions (without regularization of the frictional terms and without viscosity).

Ultimately, questions of uniqueness for elastic quasistatic evolution problems are of great importance, as well as the question whether the (unique?) quasistatic solutions appear as limits of dynamic solutions or of solutions with viscosity present, as the mass or the viscosity tends to zero, respectively. For the simplest case with two degrees of freedom Martins *et al.*, [33, 34] have some results of this kind.

Most of the problems treated here have been for linearly elastic systems. It is natural to try to extend them to cases with large deformations.

Acknowledgments

This work was supported by the Swedish Research Council for Engineering Sciences under contract no. 223-96-716.

References

[1] L.-E. Andersson, "A quasistatic frictional problem with normal compliance", *Nonlinear Analysis, Theory Methods & Applications* 16 (1991) 347-369.

[2] L.-E. Andersson, "A global existence result for a quasistatic problem with friction", *Advances in Mathematical Sciences and Applications* 1 (1995) 249-286.

[3] L.-E. Andersson, "A quasistatic frictional problem with a normal compliance penalization term", to appear in *Nonlinear Analysis, Theory Methods & Applications.*

[4] L.-E. Andersson and A. Klarbring, "On a Class of Limit States of Frictional Joints: Formulation and Existence Theorem", *Quarterly of Applied Mathematics* 55(1) (1997) 69-87.

[5] H. Brézis, L. Nirenberg, G. Stampacchia, *Bolletino U. M. I.*, 6 (1972) 293.

[6] M. Cocu, "Existence of solutions of Signorini problems with friction", *Int. J. Engng. Sci.* 22 (1984) 567-575.

[7] M. Cocu, E. Pratt, M. Raous, "Formulation and approximation of quasistatic frictional contact", *Int. J. Engng. Sci.* 34 (1996) 783-798.

[8] I. Demkowitz, J.T. Oden, "On some existence and uniqueness results in contact problems with nonlocal friction", *Nonlinear Analysis, Theory Methods & Applications* 6 (1982) 1075-1093.

[9] G. Duvaut, "Équilibre d'un solide élastique avec contact unilatéral et frottement de Coulomb", *C. r. hebd. Séanc. Acad. Sci. Paris* 290 A (1980) 263-265.

[10] G. Duvaut, J.L. Lions, "Elasticité avec frottement", *J. Mécanique*, 10, (1971) 409-420.

[11] G. Duvaut, J.L. Lions, *Les inéquations en mécanique et en physique*, Paris: Dunod (1972).

[12] G. Duvaut, J.L. Lions, *Inequalities in Mechanics and Physics*, Springer, Berlin (1976).

[13] C. Eck, "Existenz und Regularität der Lösungen für Kontaktprobleme mit Reibung", *Dissertation, Matematisches Institut A der Universität Stuttgart* 1996.

[14] K. Fan, *Inequalities III*, Ed. O. Shisha, Academic Press, New York (1972) 103-113.

[15] G. Fichera, "Problemi elastici con vincoli unilaterali il problema die Signorini con ambigue condizioni al contorno", *Mem. Accad. Naz. Lincei* 8 (1964) 91-140.

[16] G. Fichera, "Existence theorems in elasticity", *Handbuch der Physik* VI a/2, Springer, Berlin-Heidelberg-New York (1972) 347-389 .

[17] G. Fichera, "Boundary value problems of elasticity with unilateral constraints", *Handbuch der Physik* VI a/2, Springer, Berlin-Heidelberg-New York (1972) 391-424.

[18] F. Gastaldi and J.A.C. Martins, "A noncoercive steady-sliding problem with friction", *Istituto di Analisi Numerica, Pubblicazioni N. 650*, Pavia 1988.

[19] V. Janovský, "Catastrophic features of Coulomb friction model", *Technical Report KNM-0105044/80, Charles University, Prague* (1980).

[20] V. Janovský, "Catastrophic features of Coulomb friction model", in *The Mathematics of Finite Elements and Applications*, ed. J.R. Whiteman, Academic Press, London (1981) 259-264.

[21] J. Jarušek, "Contact problems with bounded friction. Coercive case", *Czechoslovak Math. J.* 33 (1983) 237-261.

[22] J. Jarušek, "Contact problems with bounded friction. Semicoercive case", *Czechoslovak Math. J.* 34 (1984) 619-629.

[23] Y. Kato, "Signorini's problem with friction in linear elasticity", *Japan J. Appl. Math.* 4 (1987) 237-268.

[24] A. Klarbring, "Contact problems with friction by linear complementarity", in *Unilateral Problems in Structural Analysis, Vol. 2 (CISM Courses and Lectures, No. 304)*, eds. G. Del Piero, F. Maceri, Springer, Wien (1987) 197-219.

[25] A. Klarbring, "Derivation and analysis of rate boundary-value problems with friction", *European Journal of Mechanics/A* 1 (1990) 211-226.

[26] A. Klarbring, "Examples of non-uniqueness and non-existence of solutions to quasistatic contact problems with friction", *Ingenieur-Archiv* 60 (1990) 529-541.

[27] A. Klarbring, "Steady sliding and linear complementarity", in M.C. Ferris and J.S. Pang, eds., *Complementarity and Variational Problems: State of the Art*, SIAM Publications, Philadelphia (1997) 132–147.

[28] A. Klarbring, A. Mikelić, M. Shillor, "Frictional contact problems with normal compliance", *Int. J. Engng. Sci* 26 (1988) 811-832.

[29] A. Klarbring, A. Mikelić, M. Shillor, "On friction problems with normal compliance", *Nonlinear Analysis, Theory, Methods & Applications* 13 (1989) 935-955.

[30] A. Klarbring, A. Mikelić, M. Shillor, "A global existence result for the quasistatic frictional contact problem with normal compliance", *International Series of Numerical Mathematics* 101 (1991) 85-111.

[31] A. Klarbring, J.S. Pang, "Existence of solutions to discrete semicoercive frictional contact problems", to appear in *SIAM Journal on Optimization.*

[32] A. Klarbring, J.-S. Pang,"The discrete steady sliding problem", submitted.

[33] J.A.C. Martins, M.D.P.M. Marques, F. Gastaldi, "On an example of non-existence of solution to a quasistatic frictional contact problem", *Eur. J. Mech., A/Solids* 13 (1994) 113-133.

[34] J.A.C. Martins, F.M.F. Simões, F. Gastaldi, M.D.P.M. Marques, "Dissipative graph solutions for a 2 degree-of-freedom quasistatic frictional contact problem", *Int. J. Engng. Sci* 33 (1995) 1959-1986.

[35] J. Nečas, J. Jarušek, J. Haslinger, "On the solution of the variational inequality to the Signorini problem with small friction", *Bolletino U. M. I.* 5, (1980) 796-811.

[36] E.B. Pires and L. Trabucho, "The steady sliding problem with nonlocal friction", *Int. J. Engng Sci.*, 27(7) (1990) 631-641.

[37] P. Rabier, J.A.C. Martins, J.T. Oden and L. Campos, "Existence and local uniqueness of solutions to contact problems in elasticity with nonlinear friction laws", *Int. J. Engng Sci.*, 24(11) (1986) 1755-1768.

[38] D.E. Stewart, "Existence of solutions to rigid body dynamics and the paradoxes of Painlevé", *Comptes Rendus Acad. Paris* 325 (1997) 689-693.

2 A SURVEY OF RECENT RESULTS ON THE ENCLOSURE AND EXTREMALITY OF SOLUTIONS FOR QUASILINEAR HEMIVARIATIONAL INEQUALITIES

Siegfried Carl

Martin-Luther-Universität Halle-Wittenberg
Fachbereich Mathematik und Informatik, Institut für Analysis
06099 Halle, Germany

Dedicated to the memory of Professor Gaetano Fichera

2.1 INTRODUCTION

Hemivariational inequalities arise e.g. in the variational formulation of boundary value problems in Mechanics and Engineering governed by nonconvex, possibly nonsmooth energy functionals (so-called superpotentials). This kind of energy functionals appear if nonmonotone, possibly multivalued constitutive laws are taken into account, cf. e.g. [17, 20, 22]. An abstract formulation of a hemivariational inequality reads as follows:

Let V be a reflexive Banach space and V^* its dual, let $A : V \to V^*$ be some pseudomonotone and coercive operator and let $h \in V^*$ be some given element. Find $u \in V$ such that

$$\langle h - Au, v - u\rangle + J^o(u, v - u) \geq 0\,, \quad \text{for all } v \in V \tag{2.1}$$

where $J^o(u, v)$ denotes the directional differential in the sense of Clarke of a locally Lipschitz functional $J : V \to \mathbb{R}$, cf. e.g. [20]. An equivalent multivalued formulation of (2.1) is given by

$$Au \in \partial_c J(u) + h \tag{2.2}$$

R.P. Gilbert et al. (eds.), From Convexity to Nonconvexity, 15–28.

where $\partial_c J(u)$ denotes Clarke's generalized gradient, cf. [16, Chapter 2]. Abstract existence results for (2.1) (resp. (2.2)) can be found e.g. in [20].

In this note we consider concrete realizations of (2.2) and its corresponding dynamic counterpart in the form

$$\frac{\partial u}{\partial t} + Au \in \partial_c J(u) + h \tag{2.3}$$

where A is assumed to be a quasilinear (in case of (2.3) also time-dependent) elliptic differential operator of Leray-Lions type. In case that the functional J on the right-hand side of (2.2) and (2.3) is generated by some locally bounded function $f : \mathbb{R} \to \mathbb{R}$ and assuming appropriately defined upper and lower solutions general existence and enclosure results have been obtained recently for both the static and dynamic hemivariational inequalities (2.2) and (2.3), respectively, in [7, 12]. However, in order to get more informations about the solution set enclosed by the assumed upper and lower solutions some more structure of the energy functional involved is needed.

The main goal of this paper is to prove the existence of extremal solutions within a sector formed by upper and lower solutions for a wide class of hemivariational inequalities whose nonconvex and nonsmooth superpotential can be expressed as the difference of two convex functions, so-called d.c. functions, cf. [17], i.e., functions Φ of the form

$$\Phi(u) = \Phi_1(u) - \Phi_2(u)\,, \tag{2.4}$$

where Φ_1 and Φ_2 are convex, possibly nonsmooth functions. By means of (2.4) indeed most practical engineering problems can be characterized such as nonmonotone zig-zag friction, delamination problems, plasticity with softening etc., cf. [22].

In this paper we concentrate on dynamic hemivariational inequalities involving d.c. functions and formulate the corresponding results for static problems.

In section 2 we give some notations and formulate our main result. Section 3 provides necessary auxiliary results, and in section 4 we prove our main result, discuss further generalizations on the basis of recently obtained results and give an application.

Our approach is based on results for elliptic and parabolic differential inclusions involving discontinuous nonlinearities obtained by the author and S. Heikkilä in [3, 4, 9, 11, 13]. The main tools used in the proofs are comparison techniques, new results on evolution equations governed by Leray-Lions operators, and fixed point results in partially ordered sets.

2.2 NOTATIONS, HYPOTHESES AND THE MAIN RESULT

Let $\Omega \subset \mathbb{R}^N$ be a bounded domain with Lipschitz boundary $\partial\Omega$, $Q = \Omega \times (0,\tau)$ and $\Gamma = \partial\Omega \times (0,\tau)$, and $\tau > 0$. We consider dynamic hemivariational inequalities whose energy function is given as the difference of two convex, possibly nonsmooth functions. The corresponding multivalued version that we are dealing with is the following:

$$\frac{\partial u}{\partial t} + Au + \Theta\Phi(u) \ni h \quad \text{in } Q\,, \quad u = \psi \quad \text{on } \Omega\,, \quad u = 0 \quad \text{on } \Gamma\,, \tag{2.5}$$

where $\Phi(s) = \Phi_1(s) - \Phi_2(s)$, $\Phi_i : \mathbb{R} \to \mathbb{R}$, $i = 1, 2$ are convex functions, and where the multivalued function $\Theta\Phi : \mathbb{R} \to 2^{\mathbb{R}} \setminus \emptyset$ is the difference of the subdifferentials $\partial\Phi_1$ and $\partial\Phi_2$ of Φ_1 and Φ_2, respectively, i.e.,

$$\Theta\Phi(s) := \partial\Phi_1 - \partial\Phi_2 = \{w = w_1 - w_2 \mid w_1 \in \partial\Phi_1(s),\ w_2 \in \partial\Phi_2(s)\}\,. \tag{2.6}$$

It is well known that the subdifferentials $\partial\Phi_i : \mathbb{R} \to 2^{\mathbb{R}} \setminus \emptyset$, $i = 1, 2$, are generated by nondecreasing functions $f_i : \mathbb{R} \to \mathbb{R}$ in the following way:

$$\partial\Phi_i(s) = [f_i(s-), f_i(s+)]\,. \tag{2.7}$$

Here $f_i(s\pm)$ denote the one-sided limits given by $f_i(s\pm) := \lim_{\varepsilon\downarrow 0} f_i(s \pm \varepsilon)$.

We shall assume that for each $s \in \mathbb{R}$ one of the subdifferentials $\partial\Phi_i(s)$ is a singleton. In this case we obtain $\Theta\Phi(s) = \partial_c\Phi(s)$ for all $s \in \mathbb{R}$ so that

$$\Theta\Phi(s) = \partial_c\Phi(s) = \partial\Phi_1(s) - \partial\Phi_2(s) \tag{2.8}$$

holds for all $s \in \mathbb{R}$. This is because convexity of $\Phi_i : \mathbb{R} \to \mathbb{R}$ implies that Φ_i are locally Lipschitzian, and thus from [16, Proposition 2.2.4] it follows that by our assumption for each $s \in \mathbb{R}$ one of the functions Φ_i must be strictly differentiable in s. Using [16, Chapter 2.3, Corollary 2] the latter implies

$$\partial_c\Phi(s) = \partial_c\Phi_1(s) - \partial_c\Phi_2(s)$$

which yields in view of $\partial_c\Phi_i(s) = \partial\Phi_i(s)$ the relation (2.8).

Further, we assume that A is a second order quasilinear differential operator in divergence form of Leray-Lions type given by

$$Au(x,t) = -\sum_{i=1}^{N} \frac{\partial}{\partial x_i} a_i(x, t, \nabla u(x,t))\,. \tag{2.9}$$

Let $W^{1,p}(\Omega)$ denote the usual Sobolev space and $(W^{1,p}(\Omega))^*$ its dual space. For the sake of simplicity we shall assume $p \geq 2$, and $q \in \mathbb{R}$ being the dual real satisfying $1/p + 1/q = 1$. Then $W^{1,p}(\Omega) \subset L^2(\Omega) \subset (W^{1,p}(\Omega))^*$ forms an evolution triple with all the embeddings being continuous, dense and compact, cf. [23].

We set $\mathcal{V} = L^p(0,\tau; W^{1,p}(\Omega))$, denote its dual space by $\mathcal{V}^* = L^q(0,\tau; (W^{1,p}(\Omega))^*)$, and define the function space $\mathcal{W}$ by

$$\mathcal{W} = \{w \in \mathcal{V} \mid \frac{\partial w}{\partial t} \in \mathcal{V}^*\}\,,$$

where the derivative $\partial/\partial t$ is understood in the sense of vector-valued distributions, cf. [23]. The space $\mathcal{W}$ endowed with the norm

$$\|w\|_{\mathcal{W}} = \|w\|_{\mathcal{V}} + \|\partial w/\partial t\|_{\mathcal{V}^*}$$

is a Banach space which is separable and reflexive due to the separability and reflexivity of $\mathcal{V}$ and $\mathcal{V}^*$, respectively, and the embedding $\mathcal{W} \subset C([0,\tau], L^2(\Omega))$ is continuous, cf. [23]. Finally, because $W^{1,p}(\Omega) \subset L^p(\Omega)$ is compactly embedded, we have a compact embedding of $\mathcal{W} \subset L^p(Q)$, cf. [23].

By $W_0^{1,p}(\Omega)$ we denote the subspace of $W^{1,p}(\Omega)$ whose elements have generalized homogeneous boundary values. Let $W^{-1,q}(\Omega)$ denote the dual space of $W_0^{1,p}(\Omega)$. Then obviously $W_0^{1,p}(\Omega) \subset L^2(\Omega) \subset W^{-1,q}(\Omega)$ forms an evolution triple and all statements made above remain true also in this situation when setting
$\mathcal{V}_0 = L^p(0,\tau; W_0^{1,p}(\Omega))$, $\mathcal{V}_0^* = L^q(0,\tau; W^{-1,q}(\Omega))$ and $\mathcal{W}_0 = \{w \in \mathcal{V}_0 \mid \frac{\partial w}{\partial t} \in \mathcal{V}_0^*\}$.

We impose the following standard conditions of Leray-Lions type on the coefficient functions $a_i : Q \times \mathbb{R}^N \to \mathbb{R}$, $i = 1, \dots, N$, e.g. cf. [18].

(A1) Each $a_i(x,t,\xi)$ is a Carathéodory function, i.e. measurable in $(x,t) \in Q$ for all $\xi \in \mathbb{R}^N$ and continuous in $\xi \in \mathbb{R}^N$ for a.a. $(x,t) \in Q$.

(A2) There exists a constant $c_0 \geq 0$ and a function $k_0 \in L_+^q(Q)$ such that $|a_i(x,t,\xi)| \leq k_0(x,t) + c_0 |\xi|^{p-1}$ for a.a. $(x,t) \in Q$ and for all $\xi \in \mathbb{R}^N$.

(A3) $\sum_{i=1}^N (a_i(x,t,\xi) - a_i(x,t,\xi'))(\xi_i - \xi_i') > 0$ for a.a. $(x,t) \in Q$ and for all ξ, $\xi' \in \mathbb{R}^N$ with $\xi \neq \xi'$.

(A4) There exists a positive constant c_1 and a function $k_1 \in L^1(Q)$ such that

$$\sum_{i=1}^N a_i(x,t,\xi)\,\xi_i \geq c_1|\xi|^p + k_1(x,t), \quad \text{for a.a. } (x,t) \in Q \text{ and for all } \xi \in \mathbb{R}^N.$$

Denote by $\langle \cdot, \cdot \rangle$ the duality pairing between the elements of $\mathcal{V}^*$ and $\mathcal{V}$ (respectively $\mathcal{V}_0^*$ and $\mathcal{V}_0$). Then as a consequence of (A1) and (A2) the semilinear form a associated with the operator A by

$$\langle Au, \varphi \rangle := a(u,\varphi) = \sum_{i=1}^N \int_Q a_i(x,t,\nabla u) \frac{\partial \varphi}{\partial x_i}\, dxdt$$

is well-defined on $\mathcal{V} \times \mathcal{V}$, and the operator $A : \mathcal{V} \to \mathcal{V}^* \subset \mathcal{V}_0^*$ is continuous and bounded. The norm (strong) convergence is denoted by $\to$, and the weak convergence by $\rightharpoonup$.

A partial ordering in $L^p(Q)$ is introduced by $u \leq w$ if and only if $w - u$ belongs to the set $L_+^p(Q)$ of all nonnegative elements of $L^p(Q)$. This induces a corresponding partial ordering also in the subset $\mathcal{W}$ of $L^p(Q)$, and if $\underline{u}$, $\bar{u} \in \mathcal{W}$ with $\underline{u} \leq \bar{u}$ then

$$[\underline{u}, \bar{u}] = \{u \in \mathcal{W} \mid \underline{u} \leq u \leq \bar{u}\}$$

denotes the order interval formed by $\underline{u}$ and $\bar{u}$. We assume $\psi \in L^2(\Omega)$ and $h \in \mathcal{V}_0^*$.

Let us introduce the notion of a (weak) solution of the initial-boundary value problem (IBVP) (2.5).

Definition 2.1. A function $u \in \mathcal{W}_0$ is called a *solution* of problem (2.5) if there is a function triple $(u, w_1, w_2) \in \mathcal{W}_0 \times L^q(Q) \times L^q(Q)$ such that

(i) $u(\cdot, 0) = \psi(\cdot)$ in Ω,

(ii) $w_i(x,t) \in \partial\Phi_i(u(x,t))$ for a.e. $(x,t) \in Q$,

(iii) $\langle \frac{\partial u}{\partial t}, \varphi\rangle + a(u,\varphi) + \int_Q (w_1 - w_2)\,\varphi\, dx\, dt = \langle h, \varphi\rangle$ for all $\varphi \in \mathcal{V}_0$.

Let $\bar{f}_i, \underline{f}_i : \mathbb{R} \to \mathbb{R}$ be defined by

$$\bar{f}_i(s) := f_i(s+), \quad \underline{f}_i(s) := f_i(s-)$$

and denote by $\bar{F}_i$ and $\underline{F}_i$ the Nemytskij operators related with $\bar{f}_i$ and $\underline{f}_i$, respectively. Then we introduce the notion of upper and lower solution for the IBVP (2.5) as follows.

Definition 2.2. A function $\bar{u} \in \mathcal{W}$ is called an *upper solution* for the IBVP (2.5) if there is a pair $(\bar{u}, \bar{w}_1) \in \mathcal{W} \times L^q(Q)$ such that

(i) $\bar{u}(\cdot, 0) \geq \psi(\cdot)$ in Ω, $\bar{u} \geq 0$ on Γ,

(ii) $\bar{w}_1(x,t) \in \partial\Phi_1(\bar{u}(x,t))$ for a.e. $(x,t) \in Q$,

(iii) $\langle \frac{\partial \bar{u}}{\partial t}, \varphi\rangle + a(\bar{u},\varphi) + \int_Q w_1\,\varphi\, dx\, dt \geq \int_Q \bar{F}_2(\bar{u})\,\varphi\, dx\, dt + \langle h, \varphi\rangle$
for all $\varphi \in \mathcal{V}_0 \cap L^p_+(Q)$.

Similarly a function $\underline{u} \in \mathcal{W}$ is a *lower solution* for the IBVP (2.5) if there is a pair $(\underline{u}, \underline{w}_1) \in \mathcal{W} \times L^q(Q)$ such that the reversed inequalities hold in (i) and (iii) of Definition 2.2 with $\underline{w}_1 \in \partial\Phi_1(\underline{u})$ and $\bar{F}_2$ replaced by $\underline{F}_2$. Further we assume the following hypotheses.

(H1) There is a constant $\alpha > 0$ such that $\bar{F}_1(\bar{u}+\alpha)$, $\underline{F}_1(\underline{u}-\alpha)$, $\bar{F}_2(\bar{u})$, $\underline{F}_2(\underline{u}) \in L^q(Q)$ for given upper and lower solutions $\bar{u}$ and $\underline{u}$, respectively, satisfying $\underline{u} \leq \bar{u}$.

(H2) The functions $\Phi_i : \mathbb{R} \to \mathbb{R}$, $i = 1, 2$, are convex, and for each $s \in \mathbb{R}$ one of the subdifferentials $\partial\Phi_i(s)$ is a singleton.

Remark 2.1. As shown above hypothesis (H2) implies $\Theta\Phi(s) = \partial_c\Phi(s)$ for all $s \in \mathbb{R}$, i.e., $w \in \partial_c\Phi(s)$ if and only if there are $w_i \in \partial\Phi_i(s)$, $i = 1, 2$, such that $w = w_1 - w_2$. Thus any solution of the IBVP (2.5) according to Definition 2.1 must be a solution of the hemivariational inequality

$$\frac{\partial u}{\partial t} + Au + \partial_c\Phi(u) \ni h \quad \text{in } Q, \quad u = \psi \quad \text{on } \Omega, \quad u = 0 \quad \text{on } \Gamma, \tag{2.10}$$

and vice versa.

Definition 2.3. A solution u^* is called the *greatest solution* within the order interval $[\underline{u}, \bar{u}]$ if for any solution $u \in [\underline{u}, \bar{u}]$ we have $u \leq u^*$. Similarly, u_* is the *least solution* in $[\underline{u}, \bar{u}]$ if for any solution $u \in [\underline{u}, \bar{u}]$ it holds $u_* \leq u$. The least and greatest solutions are called the *extremal* ones within the sector $[\underline{u}, \bar{u}]$.

The main result of this paper reads as follows.

Theorem 2.1. *Let $\bar{u}$ and $\underline{u}$ be upper and lower solutions of the IBVP (2.5) satisfying $\underline{u} \leq \bar{u}$, and let hypotheses (A1)-(A4) and (H1), (H2) be satisfied. Then problem (2.5) (respectively (2.10)) possesses extremal solutions within the sector $[\underline{u}, \bar{u}]$. Moreover, these extremal solutions can be characterized as specific solutions of some evolution variational inequalities involving discontinuous nonlinearities.*

2.3 AUXILIARY RESULTS

Throughout this section we shall assume that the hypotheses of Theorem 2.1 are satisfied.

Let $\tilde{f}_2 : \mathbb{R} \to \mathbb{R}$ be any singlevalued selection of $\partial\Phi_2$, i.e., $\tilde{f}_2(s) \in \partial\Phi_2(s)$ for all $s \in \mathbb{R}$. For given $v \in L^p(Q)$ consider the following IBVP

$$\frac{\partial u}{\partial t} + Au + \partial\Phi_1(u) \ni \tilde{F}_2(v) + h \quad \text{in } Q\,, \quad u = \psi \quad \text{on } \Omega\,, \quad u = 0 \quad \text{on } \Gamma\,, \tag{2.11}$$

where $\tilde{F}_2$ denotes the Nemytskij operator related with $\tilde{f}_2$. Then we prove the following result.

Lemma 2.1. *Let $\bar{U}$ and $\underline{U}$ be any upper and lower solution of the IBVP (2.11), respectively, satisfying $\underline{U} \leq \bar{U}$ and which belong to the interval $[\underline{u}, \bar{u}]$. Then for any $v \in [\underline{U}, \bar{U}]$ the evolution variational inequality (2.11) has a unique solution $U \in [\underline{U}, \bar{U}]$.*

Proof:

a) Existence

From Hypothesis (H1) and the inequality $\underline{f}_2(s) \leq \tilde{f}_2(s) \leq \bar{f}_2(s)$ for all $s \in \mathbb{R}$ as well as due to the monotonicity of $\tilde{f}_2$ it follows that $\tilde{F}_2(v) \in L^q(Q)$ for any $v \in [\underline{U}, \bar{U}]$. Hence the right-hand side of (2.11) is an element of $\mathcal{V}_0^*$ and the existence of a solution $U \in [\underline{U}, \bar{U}]$ can easily be deduced from a recent result of the author proved in [13].

b) Uniqueness

Let U and $\hat{U}$ be two solutions of (2.11), i.e., we have

$$\frac{\partial U}{\partial t} + AU + W = \tilde{F}_2(v) + h \quad \text{in } Q\,,$$

$$\frac{\partial \hat{U}}{\partial t} + A\hat{U} + \hat{W} = \tilde{F}_2(v) + h \quad \text{in } Q\,,$$

where $W \in \partial\Phi_1(U)$, $\hat{W} \in \partial\Phi_1(\hat{U})$ and both solutions satisfy the same initial and boundary conditions. Subtracting the equations for U and $\hat{U}$ from each other and taking in the resulting weak formulation for the difference $U - \hat{U}$ as special test function this difference itself we get

$$\langle \frac{\partial(U-\hat{U})}{\partial t}, U-\hat{U}\rangle + \langle AU - A\hat{U}, U-\hat{U}\rangle + \int_Q (W-\hat{W})\,(U-\hat{U})\,dx\,dt = 0\,.$$

Using integration by parts for the first term on the left-hand side and the monotonicity of the operators A and $\partial\Phi_1$ we obtain for any $t \in [0,\tau]$ the inequality

$$0 \le \frac{1}{2}\,\|U(t)-\hat{U}(t)\|^2_{L^2(\Omega)} = \int_0^t < \frac{\partial(U(s)-\hat{U}(s))}{\partial s}, U(s)-\hat{U}(s) > ds \le 0\,,$$

where $< \cdot,\cdot >$ under the integral sign denotes the duality pairing between $W^{-1,q}(\Omega)$ and $W_0^{1,p}(\Omega)$. Thus it follows $U(x,t) = \hat{U}(x,t)$ a.e. in Q, i.e., uniqueness. ∎

By means of Lemma 2.1 we are going to prove an extremality result for a related evolution variational inequality involving a discontinuous nonlinear term, namely the following one:

$$\frac{\partial u}{\partial t} + Au + \partial\Phi_1(u) \ni \tilde{F}_2(u) + h \quad \text{in } Q\,, \quad u = \psi \quad \text{on } \Omega\,, \quad u = 0 \quad \text{on } \Gamma\,, \tag{2.12}$$

where the Nemytskij operator $\tilde{F}_2$ is generated by any singlevalued selection $\tilde{f}_2$ of $\partial\Phi_2$. Hence $\tilde{f}_2 : \mathbb{R} \to \mathbb{R}$ is a monotone nondecreasing possibly discontinuous function. In the proof of the extremality result for (2.12) which will be given by Lemma 2.3 we apply the following fixed point result for increasing mappings in partially ordered sets due to S. Heikkilä, cf. [19, Proposition 1.2.1]

Lemma 2.2. *Given a nondecreasing mapping $P : \mathcal{Z} \mapsto \mathcal{Z}$ of a partially ordered set $\mathcal{Z}$ to itself and let $\bar{u} \in \mathcal{Z}$. Then there exists a unique inversely well-ordered chain C in $\mathcal{Z}$, called an i.w.o. chain of P-iterations of $\bar{u}$, satisfying*

$$\bar{u} = \max C \quad \text{and if } u < \bar{u} \text{ then } \; u \in C \;\text{ iff }\; u = \inf P\{z \in C \mid z > u\}.$$

If $u^* = \inf P[C]$ exists and $P\bar{u} \le \bar{u}$, then u^* is the greatest fixed point of P in

$$(\bar{u}] := \{z \in \mathcal{Z} \mid z \le \bar{u}\}.$$

Lemma 2.3. *Let $\bar{v}, \underline{v} \in \mathcal{W}$ be any upper and lower solutions of the IBVP (2.12) satisfying $\underline{u} \le \underline{v} \le \bar{v} \le \bar{u}$ where $\bar{u}$ and $\underline{u}$ are the given upper and lower solutions of the original problem (2.5). Then for any fixed singlevalued selection $\tilde{f}_2$ of $\partial\Phi_2$ the IBVP (2.12) possesses extremal solutions within the interval $[\underline{v}, \bar{v}]$.*

Note that, in particular, $\bar{u}$ and $\underline{u}$ are also upper and lower solutions, respectively, of problem (2.12).

Proof of Lemma 2.3: In the proof we focus on the existence of the greatest solution of (2.12) within $[\underline{v}, \bar{v}]$, since the existence of the least solution can be shown in a similar way. Lemma 2.1 and Lemma 2.2 will be the main tools used in the proof. We introduce the following partially ordered set $\mathcal{Z}$:

$$\mathcal{Z} := \{z \in \mathcal{W} \mid z \in [\underline{v}, \bar{v}] \text{ and } z \text{ is an upper solution of the IBVP (2.12)}\}\,,$$

and define an operator P that assigns to each $z \in \mathcal{Z}$ the solution of the IBVP

$$\frac{\partial u}{\partial t} + Au + \partial\Phi_1(u) \ni \tilde{F}_2(z) + h \quad \text{in } Q\,, \quad u = \psi \quad \text{on } \Omega\,, \quad u = 0 \quad \text{on } \Gamma\,. \tag{2.13}$$

Due to Lemma 2.1 there is a unique solution of (2.13) so that the operator P is well defined, i.e., $u := Pz$. First we show that the inequality

$$\underline{v} \le Pz \le z\,. \tag{2.14}$$

is satisfied for any $z \in \mathcal{Z}$. By definition z satisfies

$$\frac{\partial z}{\partial t} + Az + w_1 \ge \tilde{F}_2(z) + h \quad \text{in } Q\,, \quad z \ge \psi \quad \text{on } \Omega\,, \quad z \ge 0 \quad \text{on } \Gamma\,, \tag{2.15}$$

where $w_1 \in \partial\Phi_1(z)$, and $u := Pz$ fulfils

$$\frac{\partial u}{\partial t} + Au + \hat{w}_1 = \tilde{F}_2(z) + h \quad \text{in } Q\,, \quad u = \psi \quad \text{on } \Omega\,, \quad u = 0 \quad \text{on } \Gamma\,, \tag{2.16}$$

with $\hat{w}_1 \in \partial\Phi_1(u)$. Subtracting (2.15) from (2.16) we get

$$\begin{aligned} \frac{\partial(u-z)}{\partial t} + Au - Az + (\hat{w}_1 - w_1) \le 0 \quad \text{in } Q\,, u - z &\le 0 \quad \text{on } \Omega\,, \\ u - z &\le 0 \quad \text{on } \Gamma\,. \end{aligned} \tag{2.17}$$

Passing to the corresponding weak formulation of (2.17) and taking as special nonnegative test function $\varphi = (u-z)^+ \in \mathcal{V}_0 \cap L^p_+(Q)$ where $(u-z)^+ := \max\{(u-z), 0\}$, we obtain

$$\langle \frac{\partial(u-z)}{\partial t}, (u-z)^+ \rangle + \langle Au - Az, (u-z)^+ \rangle + \int_Q (\hat{w}_1 - w_1)(u-z)^+ \, dx\, dt \le 0\,.$$

The monotonicity of A and $\partial\Phi_1$ implies that the second and third term on the left-hand side of the last inequality are nonnegative which yields

$$0 \ge \langle \frac{\partial(u-z)}{\partial t}, (u-z)^+ \rangle = \frac{1}{2} \int_\Omega [(u-z)^+(x,t)]^2 \, dx \ge 0, \quad \text{for any } t \in [0,\tau]\,.$$

Hence it follows that $(u-z)^+(x,t) = 0$ for a.e. $(x,t) \in Q$, and thus $u(x,t) \le z(x,t)$ for a.e. $(x,t) \in Q$. The inequality $\underline{v} \le Pz$ can be proved in just the same way. Moreover, the monotonicity of $\tilde{F}_2$ along with (2.14) imply that the solution $u = Pz$ is again an upper solution of (2.12) such that P is a mapping from $\mathcal{Z}$ into itself, i.e., $P : \mathcal{Z} \to \mathcal{Z}$.

Next we show that the operator P is nondecreasing. To this end let $z, \hat{z} \in \mathcal{Z}$ satisfy $z \le \hat{z}$. We have to show that $Pz \le P\hat{z}$ holds. This, however, is again an immediate consequence of the monotonicity of $\tilde{F}_2$ and can be done analogously to the proof of (2.14). In order to apply the fixed point result given by Lemma 2.2 we still have to show the following:

Whenever $\mathcal{C}$ is the inversely well ordered (i.w.o.) chain of P-iterations generated by $\bar{v}$ as defined in Lemma 2.2 then $u^* = \inf P[\mathcal{C}] \in \mathcal{Z}$ exists.

Let $\mathcal{C}$ be the i.w.o. chain of P-iterations of $\bar{v}$. Because P is nondecreasing, then $P[\mathcal{C}]$ is also an inversely well-ordered chain in $[\underline{v}, \bar{v}]$. Since $P[\mathcal{C}]$ is uniformly $L^p(Q)$-bounded, we get from (A2) - (A4) and (H1) that $P[\mathcal{C}]$ is uniformly bounded also in $\mathcal{W}_0$ which yields by [6, Lemma 3.1] the existence of a nonincreasing sequence $(u_n)_{n=0}^{\infty}$ in $P[\mathcal{C}]$ converging to $u^* = \inf P[\mathcal{C}] \in \mathcal{W}_0$ weakly in $\mathcal{W}_0$ and strongly in $L^p(Q)$ as $n \to \infty$. By definition, $u_n = Pz_n$ satisfies $\underline{v} \le u_n \le z_n$ and

$$\frac{\partial u_n}{\partial t} + Au_n + w_1^n = \tilde{F}_2(z_n) + h \quad \text{in } Q\,, \quad u_n = \psi \quad \text{on } \Omega\,, \quad u_n = 0 \quad \text{on } \Gamma\,, \tag{2.18}$$

where $w_1^n \in \partial\Phi_1(u_n)$. Since $u_n \in [\underline{v}, \bar{v}]$, it follows from (H1) that (w_1^n) is bounded in $L^q(Q)$ and thus there is a subsequence of (w_1^n) (again denoted by (w_1^n)) which is weakly convergent in $L^q(Q)$ to w_1^*. Since the convex function $\Phi_1 : \mathbb{R} \to \mathbb{R}$ is continuous (even locally Lipschitzian), by [1, Prop. 2.7, Chapter 2] it follows that $w_1^n(x,t) \in \partial\Phi_1(u_n(x,t))$ if and only if $w_1^n \in \partial J(u_n)$ where the functional $J : L^p(Q) \to (-\infty, \infty]$ defined by

$$J(u) = \begin{cases} \int_Q \Phi_1(u(x,t))dx\,dt & \text{if } \Phi_1(u) \in L^1(Q) \\ +\infty & \text{otherweise} \end{cases}$$

is convex and lower-semicontinuous on $L^p(Q)$. Thus for any $\varphi \in L^p(Q)$ and all n we get

$$J(\varphi) \ge J(u_n) + \int_Q w_1^n(\varphi - u_n)\,dx\,dt$$

which yields as $n \to \infty$

$$J(\varphi) \ge J(u^*) + \int_Q w_1^*(\varphi - u^*)\,dx\,dt$$

or $w_1^*(x,t) \in \partial\Phi(u^*(x,t))$, for a.e. $(x,t) \in Q$. From (2.18) we obtain by the monotonicity of $\tilde{F}_2$ along with the inequality $\underline{v} \le u^* \le u_n \le z_n$ the following inequality

$$\frac{\partial u_n}{\partial t} + Au_n + w_1^n \ge \tilde{F}_2(u^*) + h \quad \text{in } Q\,, \quad u_n = \psi \quad \text{on } \Omega\,, \quad u_n = 0 \quad \text{on } \Gamma\,, \tag{2.19}$$

for all n. Since $u_n \in M := \{z \in \mathcal{W}_0 \mid z(x,0) = \psi(x)\}$ and $M \subset \mathcal{W}_0$ is closed and convex, it follows that M is weakly closed in $\mathcal{W}_0$ and hence $u^* \in M$. Furthermore, from (2.18) we obtain with the special test function $\varphi = u_n - u^*$

$$\begin{aligned} a(u_n, u_n - u^*) &= -\langle\frac{\partial(u_n - u^*)}{\partial t}, u_n - u^*\rangle - \langle\frac{\partial u^*}{\partial t}, u_n - u^*\rangle + \langle h, u_n - u^*\rangle \\ &+ \int_Q (\tilde{F}_2(z_n) - w_1^n)\,(u_n - u^*)\,dx\,dt\,. \end{aligned} \tag{2.20}$$

Because of $\langle\frac{\partial(u_n - u^*)}{\partial t}, u_n - u^*\rangle \ge 0$ from (2.20) we get

$$\limsup_{n\to\infty} a(u_n, u_n - u^*) \le 0$$

which yields the strong convergence $u_n \to u^*$ in $\mathcal{V}_0$ due to the (S_+) property of the operator $A : \mathcal{V}_0 \to \mathcal{V}_0^*$, cf. e.g. [2, 18]. Finally, the convergence properties of (u_n) and (w_1^n) allow to pass to the limit in (2.19) as $n \to \infty$ which yields

$$\frac{\partial u^*}{\partial t} + Au^* + w_1^* \geq \bar{F}_2(u^*) + h \quad \text{in } Q\,, \quad u^* = \psi \quad \text{on } \Omega\,, \quad u^* = 0 \quad \text{on } \Gamma\,. \tag{2.21}$$

From (2.21) we see that $u^* = \inf P[\mathcal{C}]$ is an upper solution of (2.12), i.e., $u^* \in \mathcal{Z}$. Now Lemma 2.2 can be applied which proves that u^* is greatest fixed point of P in $(\bar{v}] = \{z \in \mathcal{Z} \mid z \leq \bar{v}\}$. Hence it follows that u^* must be greatest solution of (2.12) within the interval $[\underline{v}, \bar{v}]$. ■

Now we are able to prove our main result

2.4 PROOF OF THE MAIN RESULT

Let $\bar{u}$ and $\underline{u}$ be the given upper and lower solutions of the original IBVP (2.5), respectively, according to Definition 2.2. Consider the following IBVPs:

$$\frac{\partial u}{\partial t} + Au + \partial\Phi_1(u) \ni \bar{F}_2(u) + h \quad \text{in } Q\,, \quad u = \psi \quad \text{on } \Omega\,, \quad u = 0 \quad \text{on } \Gamma\,, \tag{2.22}$$

and

$$\frac{\partial u}{\partial t} + Au + \partial\Phi_1(u) \ni \underline{F}_2(u) + h \quad \text{in } Q\,, \quad u = \psi \quad \text{on } \Omega\,, \quad u = 0 \quad \text{on } \Gamma\,, \tag{2.23}$$

where the Nemytskij operators $\bar{F}_2$ and $\underline{F}_2$ are related with the special single-valued selection $\bar{f}_2$ and $\underline{f}_2$ of $\partial\Phi_2$, respectively. It can easily be seen that $\bar{u}$ and $\underline{u}$ are upper and lower solutions, respectively, for both problems (2.22) and (2.23). According to Lemma 2.3 there exist the least and greatest solution $\bar{u}_*$ and $\bar{u}^*$ of (2.22) and $\underline{u}_*$ and $\underline{u}^*$ of (2.23), respectively, within the order interval $[\underline{u}, \bar{u}]$. (Notice that any solution of (2.22) or (2.23) is also a solution of the original problem (2.5).)

We shall show that the greatest solution $\bar{u}^*$ of (2.22) and the least solution $\underline{u}_*$ of (2.23) are the greatest and least solutions of our original problem (2.5), respectively, within the interval $[\underline{u}, \bar{u}]$. In this way we are able to provide a characterization of the extremal solutions. Let $u \in [\underline{u}, \bar{u}]$ be any solution of (2.5), i.e., there is a triple $(u, w_1, w_2) \in \mathcal{W}_0 \times L^q(Q) \times L^q(Q)$ satisfying

$$\frac{\partial u}{\partial t} + Au + w_1 = w_2 + h \quad \text{in } Q\,, \quad u = \psi \quad \text{on } \Omega\,, \quad u = 0 \quad \text{on } \Gamma\,, \tag{2.24}$$

where $w_i \in \partial\Phi_i(u)$, $i = 1, 2$. Let us show $u \leq \bar{u}^*$. Since $w_2 \in \partial\Phi_2(u)$, it follows $w_2 \leq \bar{F}_2(u)$, and hence u is a lower solution for problem (2.22). Then by Lemma 2.3 there exist extremal solutions of (2.22) with respect to the interval $[u, \bar{u}]$. However, $\bar{u}^*$ is the greatest solution of (2.22) with respect to the interval $[\underline{u}, \bar{u}] \supset [u, \bar{u}]$ which implies that $\bar{u}^*$ exceeds also any solution of (2.22) out of $[u, \bar{u}]$ and thus $u \leq \bar{u}^*$.

In a similar way one can prove $\underline{u}_* \leq u$. This completes the proof. ■

Hint: Notice that the comparison of u and $\bar{u}^*$ as given in the proof above requires the extremality result of Lemma 2.3 and cannot be done by using comparison based on differential inequality techniques or maximum principles.

2.4.1 *Static problem*

Consider the corresponding static problem to the IBVP (2.5) in the form

$$Au + \Theta\Phi(u) \ni h \quad \text{in } \Omega\,, \qquad u = 0 \quad \text{on } \partial\Omega\,, \tag{2.25}$$

where $\Theta\Phi$ is as in Section 2 and $A : V_0 \to V_0^*$ is an operator of Leray-Lions type acting from the Sobolev space $V_0 := W_0^{1,p}(\Omega)$ into its dual and given by (2.9) where the coefficients a_i do not depend on t. According to the Definition 2.1 we call $u \in V_0$ a *solution* of (2.25) if there is a triple $(u, w_1, w_2) \in V_0 \times L^q(\Omega) \times L^q(\Omega)$ satisfying

(i) $w_i \in \partial\Phi_i(u)$

(ii) $a(u, \varphi) + \int_\Omega w_1\, \varphi\, dx \;=\; \int_\Omega w_2\, \varphi\, dx + < h, \varphi >$ for all $\varphi \in V_0$,

and analogously according to Definition 2.2 we call $\bar{u}$ an upper solution of (2.25) if there is a pair $(\bar{u}, \bar{w}_1) \in V \times L^q(\Omega)$ such that

(i) $\bar{u} \geq 0$ on $\partial\Omega$ and $\bar{w}_1 \in \partial\Phi_1(\bar{u})$

(ii) $a(\bar{u}, \varphi) + \int_\Omega w_1\, \varphi\, dx \;\geq\; \int_\Omega \bar{F}_2(\bar{u})\, \varphi\, dx + < h, \varphi >$ for all $\varphi \in V_0 \cap L_+^p(\Omega)$,

where $V := W^{1,p}(\Omega)$. Replacing Q by Ω then under hypothses (A1)-(A4) and (H1), (H2) our main result still holds true for the boundary value problem (2.25), i.e., there exist extremal solutions of (2.25) within the sector $[\underline{u}, \bar{u}]$. The proof, however, is more envolved and will be given elsewhere.

2.4.2 *Example*

Nonconvex semipermeability problems were investigated e.g. in [17, 20, 21] and arise for instance in heat conduction, in electrostatics, and in flow problems through porous media. As a typical model of such kind of problems we consider the following so-called dynamic interior semipermeability problem

$$\frac{\partial u}{\partial t} - k\Delta u + \partial_c\Phi(u) \ni g \quad \text{in } Q\,, \qquad u = \psi \quad \text{on } \Omega\,, \qquad u = 0 \quad \text{on } \Gamma\,, \tag{2.26}$$

where k is some positive constant and $\partial_c\Phi(u)$ is given by the following graph, cf. e.g. [17, 20]

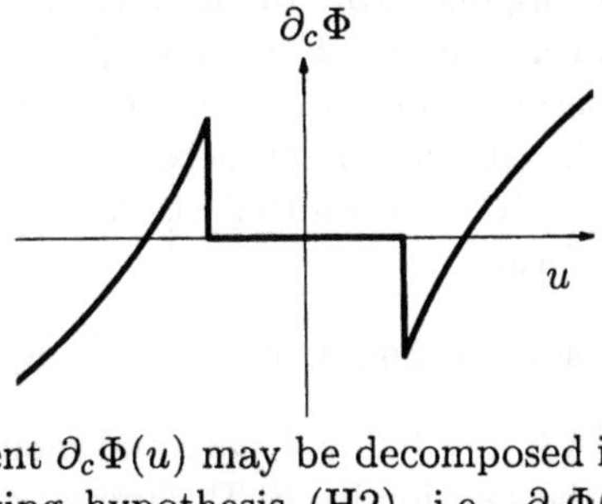

The generalized gradient $\partial_c\Phi(u)$ may be decomposed into the difference of two subdifferentials satisfying hypothesis (H2), i.e., $\partial_c\Phi(u) = \partial\Phi_1(u) - \partial\Phi_2(u)$ whose graphs are given in the following figures.

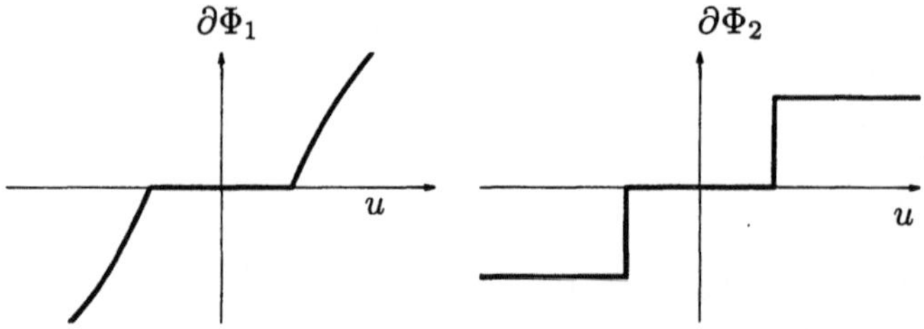

Let $p = q = 2$ and assume $\psi \in L^\infty(\Omega)$, $g \in L^\infty(Q)$ then there exist always upper and lower solutions for (2.26), namely $\bar{u}(x,t) \equiv \alpha$ with $\alpha \geq \max\{\|\psi\|_{L^\infty(\Omega)}, \|g\|_{L^\infty(Q)}\}$ sufficiently large, and $\underline{u}(x,t) \equiv \beta$ with $\beta \leq -\max\{\|\psi\|_{L^\infty(\Omega)}, \|g\|_{L^\infty(Q)}\}$ and sufficiently small. Whence Theorem 2.1 can be applied. The corresponding problems (2.22) and (2.23) that characterize the extremal solution can be calculated in this case by a numerical method developed in [15]. If more informations on the data are available then better bounds for the solutions may be obtained, e.g., in case that $\psi \geq 0$ and $g \geq 0$ then $\underline{u} = 0$ is a lower solution.

2.4.3 Concluding remarks

The proof of our main result is essentially based on Lemma 2.3, and the definition of the fixed point operator P in the proof of Lemma 2.3 relies on the unique solvability of the related IBVP (2.11) which is a consequence of the monotonicity of the operators A and $\partial\Phi_1$. However, in case A is replaced by a general divergence operator of Leray-Lions type in the form

$$Au = -\sum_{i=1}^{N} \frac{\partial}{\partial x_i} a_i(x,t,u,\nabla u) \tag{2.27}$$

whose coefficients a_i depend also on u then the corresponding IBVP (2.11) is, in general, not uniquely solvable. Nevertheless one can extend the main result also to this case by proving that instead of a unique solution the problem (2.11) possesses extremal solutions. The fixed point operator P is then defined by assigning to each $z \in \mathcal{Z}$ the greatest (respectively smallest) solution of the auxiliary problem (2.11). The existence of extremal solutions for quasilinear parabolic problems involving operators of the form (2.27) has been proved in a recent paper by the author in [8]. Even though the result in [8] has been proved under Dirichlet-periodic boundary conditions it holds likewise also for the initial-boundary conditions considered here.

A further generalization (to be submitted elsewhere) is concerned with multivalued functions of the form

$$\Theta\Phi(u) := \partial\Phi_1(u,u) - \partial\Phi_2(u,u)$$

where $\partial\Phi_i(u,\cdot) : \mathbb{R} \to 2^{\mathbb{R}} \setminus \emptyset$, $i = 1, 2$, are maximal monotone graphs in $\mathbb{R}^2$ depending, in addition, on the solution itself. Preliminary results in this direction have been obtained in the static case in [5, 14]. However, the method we have used in the static case cannot be applied in the dynamic case. This is basically because the underlying Sobolev spaces in the static case have lattice structure while the space $\mathcal{W}$ used in the dynamic case does not possess lattice structure.

Furthermore, within the context of extremality of solutions one can consider also multivalued boundary conditions of the form

$$\frac{\partial u}{\partial \nu} \in \Theta\Phi(u) \quad \text{on } \Gamma$$

where the multivalued function Θ is of the form (2.8), and $\partial/\partial\nu$ denotes the exterior conormal derivative related with the operator A. Relevant examples of this type from nonsmooth Mechanics can be found in [17, Chapter 3], in particular, Example 3.12 of [17, Chapter 3]. Also mixed boundary conditions can be taken into account. A first extremality result with nonlinear discontinuous boundary conditions was given in [10].

References

[1] V. Barbu and Th. Precupanu, "Convexity and Optimization in Banach Spaces", Sijthoff & Noordhoff, International Publishers, 1978.

[2] J. Berkovits and V. Mustonen, Topological degree for perturbations of linear maximal monotone mappings and applications to a class of parabolic problems, Rend. Mat. Appl. (7) 12 (1992), 597-621.

[3] S. Carl, A combined variational-monotone iterative method for elliptic boundary value problems with discontinuous nonlinearity, Appl. Anal. 43 (1992), 21-45.

[4] S. Carl and S. Heikkilä, On extremal solutions of an elliptic boundary value problem involving discontinuous nonlinearities, Differential and Integral Equations 5 (1992), 581-589.

[5] S. Carl, S. Heikkilä and V. Lakshmikantham, Nonlinear elliptic differential inclusions governed by state-dependent subdifferentials, Nonlinear Anal. 25 (1995), 729-745.

[6] S. Carl, On the existence of extremal weak solutions for a class of quasilinear parabolic problems, Differential Integral Equations, 6 (1993), 1493-1505.

[7] S. Carl, Enclosure of solutions for quasilinear dynamic hemivariational inequalities, Nonlinear World 3 (1996), 281-298.

[8] S. Carl, Existence of extremal periodic solutions for quasilinear parabolic equations, Abstract Applied Analysis, 2, Nos.3-4 (1997), 257-270.

[9] S. Carl and S. Heikkilä, An existence result for elliptic differential inclusions with discontinuous nonlinearity, Nonlinear Anal. 18 (1992), 471-479.

[10] S. Carl and S. Heikkilä, On the existence of extremal solutions for discontinuous elliptic equations under discontinuous flux conditions, Nonlinear Anal. 23 (1994), 1499-1506.

[11] S. Carl and S. Heikkilä, Extremal solutions of quasilinear parabolic boundary value problems with discontinuous nonlinearities, Dynamic Systems and Applications 3 (1994), 251-258.

[12] S. Carl and H. Dietrich, The weak upper and lower solution method for quasilinear elliptic equations with generalized subdifferentiable perturbations, Appl. Anal. 56 (1995), 263-278.

[13] S. Carl, An existence result for a class of parabolic hemivariational inequalities, Proceedings of Dynamic Systems and Applications, Vol. 2 (1996), 91-98.

[14] S. Carl, Leray-Lions operators perturbed by state-dependent subdifferentials, Nonlinear World 3 (1996), 505-518.

[15] S. Carl, Ch. Grossmann and C.V. Pao, Existence and monotone iterations for parabolic differential inclusions, Commun. Appl. Nonlinear Anal. 3 (1996), 1-24.

[16] F.H. Clarke, "Optimization and Nonsmooth Analysis", Wiley, New York, 1983.

[17] F. Dem'yanov, G.E. Stavroulakis, L.N. Polyakova and P.D. Panagiotopoulos, "Quasidifferentiability and Nonsmooth Modelling in Mechanics, Engineering and Economics, Kluwer Academic Publishers, Dordrecht, 1996.

[18] J. Deuel and P. Hess, Nonlinear parabolic boundary value problems with upper and lower solutions, Israel J. Math. 29 (1978), 92-104.

[19] S. Heikkilä and V. Lakshmikantham, "Monotone Iterative Techniques for Discontinuous Nonlinear Differential Equations", Marcel Dekker, New York (1994).

[20] Z. Naniewicz and P.D. Panagiotopoulos, "Mathematical Theory of Hemivariational Inequalities and Applications", Marcel Dekker, New York, 1995.

[21] M. Miettinen and P.D. Panagiotopoulos, On parabolic hemivariational inequalities and applications, Nonlinear Anal. 35 (1998).

[22] P.D. Panagiotopoulos, "Hemivariational Inequalities: Applications in Mechanics and Engineering", Springer-Verlag, New York, 1993.

[23] E. Zeidler, "Nonlinear Functional Analysis and its Applications II/B: Nonlinear Monotone Operators", Springer-Verlag, New York, 1990.

3 A SURVEY ON NONSMOOTH CRITICAL POINT THEORY AND APPLICATIONS

Marco Degiovanni

Dipartimento di Matematica
Università Cattolica del Sacro Cuore
Via Trieste, 17
I 25121 Brescia, Italy*

m.degiovanni@dmf.bs.unicatt.it

Abstract: In the recent years, new advances have been obtained in critical point theory for nonsmooth functionals and in applications to nonlinear differential equations. Here we provide a survey on some of such progresses.

3.1 INTRODUCTION

Let X be a Banach manifold of class C^2 and $f : X \to \mathbf{R}$ a function of class C^1. Classical critical point theory deals with the existence and multiplicity of points u's in X with $f'(u) = 0$. Since the pioneering papers of R. S. PALAIS [42, 43, 44], this abstract framework has provided plenty of applications, in particular to ordinary and partial differential equations. We refer the reader to A. AMBROSETTI [1], K. C. CHANG [13], K. DEIMLING [29], J. MAWHIN and M. WILLEM [41], P. H. RABINOWITZ [46], M. STRUWE [51] and E. ZEIDLER [53] for detailed expositions of the abstract and applied aspects of the subject.

During the 80's, also the case where f is not smooth has received more and more attention. The first development in this direction was due to K. C. CHANG [12], who considered the situation where f is locally Lipschitzian. In such a case, u is said a critical point if $0 \in \partial f(u)$, where ∂f is the subdifferential of f in the sense of F. H. CLARKE [14]. Particular classes of lower semicontinuous functions were also considered by E. DE GIORGI, M. DEGIOVANNI, A.

*Partially supported by Ministero dell'Università e della Ricerca Scientifica e Tecnologica (40% - 1999).

R.P. Gilbert et al. (eds.), From Convexity to Nonconvexity, 29–42.

MARINO, D. SCOLOZZI and M. TOSQUES [20, 22, 25, 38] and by A. SZULKIN [52]. To introduce the notion of critical point, it was there simpler and more natural to consider the Fréchet subdifferential (see e. g. E. DE GIORGI, A. MARINO and M. TOSQUES [21]).

More recently, the case where X is a metric space and f is continuous has been independently treated by J.-N. CORVELLEC, M. DEGIOVANNI and M. MARZOCCHI [19, 26] and by A. IOFFE, G. KATRIEL and E. SCHWARTZMAN [31, 33]. Some classes of lower semicontinuous functions, including those of [20, 22, 25, 52], have been also considered in [19, 26]. In this setting, the notion of critical point is introduced through the equation $|df|(u) = 0$, where $|df|$ is a suitable generalization of the norm of the derivative. Moreover, a related notion of subdifferential has been introduced by I. CAMPA and M. DEGIOVANNI [7]. In the first two sections, we will expose the main points of this abstract framework.

To give a first model example, consider a bounded open subset Ω of $\mathbf{R}^n$ and a functional $f : W_0^{1,2}(\Omega) \to \mathbf{R}$ of the form

$$f(u) = \frac{1}{2} \int_\Omega \sum_{i,j=1}^{n} a_{ij}(x,u) D_{x_i} u D_{x_j} u \, dx - \int_\Omega G(x,u)\, dx \, .$$

If the a_{ij}'s are independent of u, it is well known that the solvability of the associated Euler equation can be studied by techniques of critical point theory. We may expect that some results will be true also when the a_{ij}'s depend on u. However, a technical difficulty arises: under reasonable assumptions on a_{ij} and G, f turns out to be well defined and continuous and the associated Euler equation still makes sense in $\mathcal{D}'(\Omega)$, but f is not of class C^1, unless $n = 1$. More precisely, if f were locally Lipschitzian, one would have

$$\forall u \in W_0^{1,2}(\Omega) : \quad \sum_{i,j=1}^{n} D_s a_{ij}(x,u) D_{x_i} u D_{x_j} u \in W^{-1,2}(\Omega) \, .$$

We may expect such a property only if $n = 1$ or the a_{ij}'s are independent of u.

As an alternative to this direct approach, one can define f only on $W_0^{1,2}(\Omega) \cap L^\infty(\Omega)$ endowed with the family of norms

$$\|u\|_\varepsilon = \|u\|_{W_0^{1,2}} + \varepsilon \|u\|_{L^\infty} \qquad 0 < \varepsilon \le 1 \, .$$

In such a case, f can be of class C^1, but cannot satisfy the usual Palais-Smale condition. This is the approach of M. STRUWE [50] and, with some variants, of D. ARCOYA and L. BOCCARDO [4]. The direct approach, based on the critical point theory for continuous functionals, has been started by A. CANINO [8, 9, 10] and developed by G. ARIOLI, M. CONTI, J.-N. CORVELLEC, M. DEGIOVANNI, F. GAZZOLA and B. PELLACCI [5, 6, 15, 18, 45]. In the third and fourth sections, we will provide a somewhat general framework and expose a jumping result taken from [9, 10].

Another model example is given by functionals $f : BV(\Omega) \to \mathbf{R}$ which come from the relaxation of

$$f(u) = \int_\Omega \sqrt{1+|\nabla u|^2}\,dx - \int_\Omega G(x,u)\,dx \qquad \text{if } u \in W_0^{1,1}(\Omega)\,.$$

Under suitable assumptions on G, the functional is continuous and has the form already considered in [52]. However, the norm convergence of $BV(\Omega)$ turns out to be too strong, to allow the Palais-Smale condition to hold. In such a case, it is more convenient to extend the functional to a suitable L^p space with the value $+\infty$ outside BV. In this way, the structure required in [52] is still maintained and the Palais-Smale condition holds. This is however an interesting case where a modified lower semicontinuous functional is better behaved than the original continuous functional. This kind of approach has been developed by M. MARZOCCHI [39, 40] and will be addressed in the last section.

Finally, we want to mention that the nonsmooth critical point theory of [19, 26] has been applied also to perturbation problems by M. DEGIOVANNI and S. LANCELOTTI [23, 24, 35, 36], to geodesic theory by M. DEGIOVANNI and L. MORBINI [27] and to nonlinear continuum mechanics by M. DEGIOVANNI and F. SCHURICHT [28].

3.2 CRITICAL POINT THEORY IN METRIC SPACES

In the following, $\overline{\mathbf{R}}$ will denote the extended real line $\mathbf{R} \cup \{-\infty, +\infty\}$. Let X be a metric space endowed with the metric d and let $f : X \to \overline{\mathbf{R}}$ be a function. If $Y \subseteq X$, we denote by $\overline{Y}$, $\mathrm{int}\,(Y)$ and ∂Y the closure, the interior and the boundary of Y in X, respectively. We also denote by $\mathrm{B}_r\,(u)$ the open ball of center u and radius r and we put

$$\mathrm{epi}\,(f) = \{(u,\lambda) \in X \times \mathbf{R} :\, f(u) \le \lambda\}\,.$$

The set $X \times \mathbf{R}$ will be endowed with the metric

$$d\,((u,\lambda)\,,(v,\mu)) = \left(d(u,v)^2 + (\lambda-\mu)^2\right)^{\frac{1}{2}}$$

and $\mathrm{epi}\,(f)$ with the induced metric.

Definition 3.1. *For every $u \in X$ with $f(u) \in \mathbf{R}$, we denote by $|df|\,(u)$ the supremum of the σ's in $[0,+\infty[$ such that there exist $\delta > 0$ and a continuous map*

$$\mathcal{H} : (\mathrm{B}_\delta\,(u, f(u)) \cap \mathrm{epi}\,(f)) \times [0,\delta] \to X$$

satisfying

$$d(\mathcal{H}((w,\mu),t),w) \le t\,, \qquad f(\mathcal{H}((w,\mu),t)) \le \mu - \sigma t\,,$$

whenever $(w,\mu) \in \mathrm{B}_\delta\,(u,f(u)) \cap \mathrm{epi}\,(f)$ and $t \in [0,\delta]$.

The extended real number $|df|\,(u)$ is called the weak slope *of f at u.*

As the notation suggests, the weak slope is a generalization of the norm of the derivative. Actually, if X is a Finsler manifold of class C^1 and $f : X \to \mathbf{R}$ a function of class C^1, then for every $u \in X$ we have $|df|(u) = \|df(u)\|$.

Definition 3.1 can be simplified, when f is real valued and continuous.

Proposition 3.2. *Let $f : X \to \mathbf{R}$ be continuous. Then $|df|(u)$ is the supremum of the σ's in $[0, +\infty[$ such that there exist $\delta > 0$ and a continuous map*

$$\mathcal{H} : \mathrm{B}_\delta(u) \times [0, \delta] \to X$$

satisfying

$$d(\mathcal{H}(w,t), w) \le t, \qquad f(\mathcal{H}(w,t)) \le f(w) - \sigma t,$$

whenever $w \in \mathrm{B}_\delta(u)$ and $t \in [0, \delta]$.

By means of the weak slope, we can now introduce the two main notions of critical point theory.

Definition 3.3. *We say that $u \in X$ is* a (lower) critical point *of f, if $f(u) \in \mathbf{R}$ and $|df|(u) = 0$. We say that $c \in \mathbf{R}$ is* a (lower) critical value *of f, if there exists a (lower) critical point $u \in X$ of f with $f(u) = c$.*

Definition 3.4. *Let $c \in \mathbf{R}$. We say that f satisfies* the Palais-Smale condition at level c *($(PS)_c$ for short), if every sequence (u_h) in X with $f(u_h) \to c$ and $|df|(u_h) \to 0$ admits a convergent subsequence (u_{h_k}).*

The next theorems are adaptations to the continuous or lower semicontinuous case of well known minimax results (see e.g. [1, 3, 46, 51]).

Theorem 3.5. *Suppose that X is complete and $f : X \to \mathbf{R}$ is continuous. Let $v_0, v_1 \in X$ and $r > 0$ be such that $d(v_1, v_0) > r$,*

$$\inf\{f(u) : u \in X,\, d(u, v_0) = r\} > \max\{f(v_0), f(v_1)\},$$

$$\Gamma := \{\gamma \in C([0,1]; X) : \gamma(0) = v_0,\, \gamma(1) = v_1\} \ne \emptyset, \quad c_1 := \inf_{\mathrm{B}_r(v_0)} f > -\infty,$$

and assume that f satisfies the Palais-Smale condition at the two levels c_1 and

$$c_2 := \inf_{\gamma \in \Gamma} \max_{[0,1]} (f \circ \gamma).$$

Then $c_1 < c_2$ and there exist a critical point (actually, a local minimum) u_1 of f with $d(u_1, v_0) < r$ and $f(u_1) = c_1$, and another critical point u_2 of f with $f(u_2) = c_2$.

Proof. See [26, Theorem (3.12)] and [9, Theorem 1.3]. ∎

The next result is an improvement of [52, Theorem 4.4].

Theorem 3.6. *Suppose that X is a Banach space and $f : X \to \mathbf{R} \cup \{+\infty\}$ is lower semicontinuous and even. Assume also that* *(a)* $f = f_0 + f_1$ *with* $f_0 : X \to \mathbf{R} \cup \{+\infty\}$ *convex and* $f_1 : X \to \mathbf{R}$ *of class* C^1;

(b) there exist a strictly increasing sequence (V_h) *of finite-dimensional subspaces of* X, *a closed subspace* Z *of* X, $\rho > 0$, $\alpha > f(0)$ *and a sequence* (R_h) *in* $]\rho, +\infty[$ *with*

$$X = V_0 \oplus Z\,, \qquad \forall u \in Z : \|u\| = \rho \Longrightarrow f(u) \geq \alpha\,;$$

$$\forall u \in V_h : \|u\| \geq R_h \Longrightarrow f(u) \leq f(0)\,;$$

(c) f *satisfies* $(PS)_c$ *for any* $c \geq \alpha$.

Then there exists a sequence (u_h) *of critical point of* f *with* $\lim_h f(u_h) = +\infty$.

Proof. By [26, Theorem (2.11)] assumption (c) agrees with the Palais-Smale condition considered in [39]. Then the assertion follows from [39, Theorem (1.2)]. ■

Many other extensions of well known minimax results and deformation techniques can be considered. We refer the reader to [11, 16, 17, 19, 31, 32, 33, 39]. Developments in the direction of multifunctions have also been treated in [30].

3.3 SUBDIFFERENTIAL CALCULUS

Throughout this section, X will denote a real Banach space and $f : X \to \overline{\mathbf{R}}$ a function. Being a generalization of the norm of the derivative, the weak slope cannot have a rich calculus. For this reason we introduce, following [7], a subdifferential which will be conveniently related to the weak slope and will provide sufficient calculus rules for many purposes.

Definition 3.7. *Let* $u \in X$ *with* $f(u) \in \mathbf{R}$. *For every* $v \in X$ *and* $\varepsilon > 0$, *let* $f^0_\varepsilon(u; v)$ *be the infimum of the* r*'s in* $\mathbf{R}$ *such that there exist* $\delta > 0$ *and a continuous map*

$$\mathcal{V} : (\mathrm{B}_\delta(u, f(u)) \cap \mathrm{epi}(f)) \times]0, \delta] \to \mathrm{B}_\varepsilon(v)$$

satisfying

$$f(w + t\mathcal{V}((w, \mu), t)) \leq \mu + rt$$

whenever $(w, \mu) \in \mathrm{B}_\delta(u, f(u)) \cap \mathrm{epi}(f)$ *and* $t \in]0, \delta]$ *(we agree that* $\inf \emptyset = +\infty$*). Let also*

$$f^0(u; v) := \sup_{\varepsilon > 0} f^0_\varepsilon(u; v) = \lim_{\varepsilon \to 0+} f^0_\varepsilon(u; v)\,.$$

We say that $f^0(u; v)$ *is the* generalized directional derivative *of* f *at* u *with respect to* v.

If we drop the continuity condition on $\mathcal{V}$, we get exactly the generalized directional derivative in the sense of Rockafellar (see [14, 49]). Therefore $f^0(u; v)$ is greater than or equal to the corresponding notion of Rockafellar.

Proposition 3.8. *For any $u \in X$ with $f(u) \in \mathbf{R}$, the function $\{v \longmapsto f^0(u;v)\}$ is convex, lower semicontinuous and positively homogeneous of degree* 1. *Moreover, we have $f^0(u;0) \in \{0, -\infty\}$.*

Definition 3.9. *For every $u \in X$ with $f(u) \in \mathbf{R}$, we set*

$$\partial f(u) := \{\alpha \in X^* : \langle \alpha, v\rangle \le f^0(u;v) \quad \forall v \in X\} .$$

We say that $\partial f(u)$ is the subdifferential *of f at u, while the elements of $\partial f(u)$ are called* subgradients *of f at u.*

It turns out that $\partial f(u)$ is convex, weak* closed and contains the subdifferential in the sense of Clarke-Rockafellar (see [14, 49]). Moreover, one has $\partial f(u) = \emptyset$ if and only if $f^0(u;0) = -\infty$.

Theorem 3.10. *The following assertions hold:*

(a) *if f is convex, then, for every $u \in X$ with $f(u) \in \mathbf{R}$, $\partial f(u)$ agrees with the subdifferential of convex analysis, i.e. we have*

$$\partial f(u) = \{\alpha \in X^* : f(v) \ge f(u) + \langle \alpha, v - u\rangle \quad \forall v \in X\} ;$$

(b) *if $f : X \to \mathbf{R}$ is locally Lipschitzian, then $\partial f(u)$ and $f^0(u;\cdot)$ agree with the corresponding notions of Clarke (see [14]).*

The next result is the main motivation for the introduction of such a new subdifferential.

Theorem 3.11. *For every $u \in X$ with $f(u) \in \mathbf{R}$, we have $|df|(u) < +\infty$ if and only if $\partial f(u) \neq \emptyset$. Moreover, if $|df|(u) < +\infty$, we have*

$$|df|(u) \ge \min\{\|\alpha\| : \alpha \in \partial f(u)\} .$$

Remark 3.12. *The above inequality may be strict, even if f is Lipschitzian (so that the usual Clarke subdifferential is involved). For a counterexample, see [47].*

Let us recall that Theorem 3.11 is not true for the Clarke-Rockafellar subdifferential, even if f is real valued and continuous. We refer the reader to [7] for a discussion of this point.

As an example, let us provide a simple calculus rule.

Theorem 3.13. *Let $g : X \to \mathbf{R}$ be locally Lipschitzian and let $u \in X$ with $f(u) \in \mathbf{R}$.*

Then the following facts hold:

(a) *for every $v \in X$, we have $(f+g)^0(u;v) \le f^0(u;v) + g^0(u;v)$;*

(b) *$\partial f(u) + \partial g(u)$ is weak* closed in X^* and we have*

$$\partial(f+g)(u) \subseteq \partial f(u) + \partial g(u) ;$$

(c) *if g is of class C^1, we have*

$$\partial(f+g)(u) = \{\alpha + g'(u) : \alpha \in \partial f(u)\} .$$

We refer the reader to [7] for more refined calculus rules.

3.4 FUNCTIONALS OF THE CALCULUS OF VARIATIONS

Let Ω be an open subset of $\mathbf{R}^n$, let $1 < p < n$ and let $L : \Omega \times \mathbf{R} \times \mathbf{R}^n \to \mathbf{R}$ be a function such that

$(L.1)$ for a.e. $x \in \Omega$, $\{(s,\xi) \longmapsto L(x,s,\xi)\}$ is of class C^1 on $\mathbf{R} \times \mathbf{R}^n$ and for every $(s,\xi) \in \mathbf{R} \times \mathbf{R}^n$, $\{x \longmapsto L(x,s,\xi)\}$ is measurable on Ω;

$(L.2)$ there exist $a \in L^1_{loc}(\Omega)$ and $b \in L^\infty_{loc}(\Omega)$ such that

$$|D_{\xi_j} L(x,s,\xi)| \leq a(x) + b(x)\left(|s|^{\frac{np}{n-p}} + |\xi|^p\right),$$

$$|D_s L(x,s,\xi)| \leq a(x) + b(x)\left(|s|^{\frac{np}{n-p}} + |\xi|^p\right).$$

Let $f : W_0^{1,p}(\Omega) \to \overline{\mathbf{R}}$ be the functional defined by

$$f(u) = \int_\Omega L(x,u,\nabla u)\, dx,$$

where, as in [48], we agree that $f(u) = +\infty$ whenever

$$\int_\Omega (L(x,u,\nabla u))^+\, dx = \int_\Omega (L(x,u,\nabla u))^-\, dx = +\infty.$$

From the Sobolev Theorem, it is easy to deduce that

$$D_{\xi_j} L(x,u,\nabla u) \in L^1_{loc}(\Omega), \qquad D_s L(x,u,\nabla u) \in L^1_{loc}(\Omega)$$

for every $u \in W_0^{1,p}(\Omega)$, so that

$$-\sum_{j=1}^n D_{x_j}\left[D_{\xi_j} L(x,u,\nabla u)\right] + D_s L(x,u,\nabla u)$$

defines a distribution on Ω.

Theorem 3.14. *Let $u \in W_0^{1,p}(\Omega)$ with $f(u) \in \mathbf{R}$. Then the following assertions hold:*

(a) *for every $v \in C_c^\infty(\Omega)$ we have*

$$f^0(u;v) \leq \int_\Omega \left(\sum_{j=1}^n D_{\xi_j} L(x,u,\nabla u)\, D_{x_j} v + D_s L(x,u,\nabla u)\, v\right) dx;$$

(b) *if $\partial f(u) \neq \emptyset$, we have*

$$-\sum_{j=1}^n D_{x_j}\left[D_{\xi_j} L(x,u,\nabla u)\right] + D_s L(x,u,\nabla u) \in W^{-1,p'}(\Omega),$$

$$\partial f(u) = \left\{ -\sum_{j=1}^{n} D_{x_j}\left[D_{\xi_j} L(x,u,\nabla u)\right] + D_s L(x,u,\nabla u) \right\}.$$

Proof. See [7, Theorem 6.1]. ■

The previous result provides a useful link between the abstract notion of subdifferential and usual notions of *PDE*'s. For instance, if $0 \in \partial f(u)$, it follows that u is a weak solution of a partial differential equation.

3.5 FUNCTIONALS WITH QUADRATIC DEPENDENCE ON THE GRADIENT

Let Ω be a connected bounded open subset of $\mathbf{R}^n$ with $n \geq 3$. We consider functionals $f : W_0^{1,2}(\Omega) \to \mathbf{R}$ of the form

$$f(u) = \frac{1}{2}\int_\Omega \sum_{i,j=1}^{n} a_{ij}(x,u) D_{x_i} u D_{x_j} u \, dx - \int_\Omega G(x,u)\, dx - \langle \omega, u \rangle\,, \tag{3.15}$$

where $\omega \in W^{-1,2}(\Omega)$ and $a_{ij} : \Omega \times \mathbf{R} \to \mathbf{R}$ $(1 \leq i, j \leq n)$, $G : \Omega \times \mathbf{R} \to \mathbf{R}$ satisfy the following conditions:

(a.1) for every $s \in \mathbf{R}$, $a_{ij}(\cdot, s)$ is measurable and for a.e. $x \in \Omega$, $a_{ij}(x,\cdot)$ is of class C^1; moreover, we have $a_{ij}(x,s) = a_{ji}(x,s)$;

(a.2) there exists $C > 0$ such that $|a_{ij}(x,s)| \leq C$, $|D_s a_{ij}(x,s)| \leq C$;

(a.3) there exists $\nu > 0$ such that

$$\sum_{i,j=1}^{n} a_{ij}(x,s)\xi_i\xi_j \geq \nu|\xi|^2\,;$$

(a.4) there exists $R > 0$ such that

$$|s| \geq R \Longrightarrow \sum_{i,j=1}^{n} s D_s a_{ij}(x,s)\xi_i\xi_j \geq 0\,;$$

(a.5) there exists a uniformly Lipschitzian bounded function $\vartheta : \mathbf{R} \to [0, +\infty[$ such that

$$\sum_{i,j=1}^{n} s D_s a_{ij}(x,s)\xi_i\xi_j \leq s\vartheta'(s) \sum_{i,j=1}^{n} a_{ij}(x,s)\xi_i\xi_j\,;$$

(a.6) for a.e. $x \in \Omega$ we have

$$\lim_{s\to-\infty} a_{ij}(x,s) = \lim_{s\to+\infty} a_{ij}(x,s) \tag{3.16}$$

(these limits exist by $(a.4)$);

$(g.1)$ $G(x,s) = \int_0^s g(x,t)\,dt$, where $g : \Omega \times \mathbf{R} \to \mathbf{R}$ is a Carathéodory function;

$(g.2)$ there exist $a \in L^{\frac{2n}{n+2}}(\Omega)$ and $b \in L^{\frac{n}{2}}(\Omega)$ such that

$$|g(x,s)| \le a(x) + b(x)|s|\,;$$

$(g.3)$ there exist α, $\beta \in \mathbf{R}$ such that for a.e. $x \in \Omega$, we have

$$\lim_{s\to-\infty} \frac{g(x,s)}{s} = \alpha\,, \qquad \lim_{s\to+\infty} \frac{g(x,s)}{s} = \beta.$$

Denote by $A_{ij}(x)$ the limit in (3.16), by λ_1 the first eigenvalue of the linear operator $-\sum D_{x_j}(A_{ij} D_{x_i} u)$ with homogeneous Dirichlet condition and by φ_1 a nonnegative eigenfunction corresponding to λ_1.

We are interested in a jumping problem of Ambrosetti-Prodi type [2]. For further results in the semilinear case, see [37] and references therein.

Theorem 3.17. *Assume that $\beta < \lambda_1 < \alpha$. Then for every $\omega_0 \in W^{-1,2}(\Omega)$ there exist $\underline{t}, \overline{t} \in \mathbf{R}$ such that the quasilinear equation*

$$-\sum_{i,j=1}^{n} D_{x_j}(a_{ij}(x,u)D_{x_i}u) + \frac{1}{2}\sum_{i,j=1}^{n} D_s a_{ij}(x,u) D_{x_i}u D_{x_j}u = g(x,u) + t\varphi_1 + \omega_0$$

has at least two weak solutions in $W_0^{1,2}(\Omega)$ for every $t > \overline{t}$ and no weak solutions in $W_0^{1,2}(\Omega)$ for every $t < \underline{t}$. Moreover, if $\omega_0 \in W^{-1,p}(\Omega)$ with $p > n$ and $(g.2)$ is satisfied with $a, b \in L^q(\Omega)$, $q > n/2$, then each weak solution u belongs to $W_0^{1,2}(\Omega) \cap L^\infty(\Omega)$.

Proof. The above result was first proved in [9, Corollary 2.3] under a stronger version of $(a.4)$ and then, in the present form, in [10, Corollary 6.3]. Here we only sketch the proof of the assertion concerning $t > \overline{t}$, just to show the application of the abstract notions we have introduced.

Consider the functional f in (3.15) with $\omega = t\varphi_1 + \omega_0$. Then f is continuous and its first term has the form considered in Theorem 3.14, while the last two terms are of class C^1. A quite technical argument shows that f satisfies $(PS)_c$ for any $c \in \mathbf{R}$. Moreover, if t is sufficiently large, also the geometric assumptions of Theorem 3.5 are fulfilled. Therefore we get the existence of two distinct critical points u_1, u_2 for f. From Theorem 3.11 we deduce that $0 \in \partial f(u_k)$, $k = 1, 2$, and from Theorem 3.13 we infer that $g(x, u_k) + t\varphi_1 + \omega_0$ is a subgradient of the first term of f at u_k. From Theorem 3.14 we conclude that u_k is a weak solution of the quasilinear equation. ■

Let us recall that, for weak solutions $u \in W_0^{1,2}(\Omega) \cap L^\infty(\Omega)$, further regularity informations can be obtained by [34].

3.6 AREA-TYPE FUNCTIONALS

Let Ω be a bounded open subset of $\mathbf{R}^n$ ($n \geq 2$) with Lipschitz boundary. We consider here functionals $f : BV(\Omega) \to \mathbf{R}$ of the form

$$f(u) = \mathcal{A}(u) - \int_\Omega G(x,u)\,dx\,,$$

$$\mathcal{A}(u) = \int_\Omega \Psi\left(\nabla u^a\right)dx + \int_\Omega \Psi^\infty\left(\frac{\nabla u^s}{|\nabla u^s|}\right)d|\nabla u^s| + \int_{\partial\Omega} \Psi^\infty(u\nu)\,d\mathcal{H}^{n-1}\,,$$

where $\nabla u = \nabla u^a\,dx + \nabla u^s$ is the Lebesgue decomposition of ∇u, Ψ^∞ is the recession functional associated with Ψ and ν is the outer normal to Ω. A model example is given by the functional

$$f(u) = \int_\Omega \sqrt{1+|\nabla u^a|^2}\,dx + |\nabla u^s|(\Omega) + \int_{\partial\Omega} |u|\,d\mathcal{H}^{n-1} - \int_\Omega G(x,u)\,dx\,,$$

which comes, under conditions (g.1) and (g.4) below, from the relaxation of

$$f(u) = \int_\Omega \sqrt{1+|\nabla u|^2}\,dx - \int_\Omega G(x,u)\,dx \qquad \text{if } u \in W_0^{1,1}(\Omega)\,.$$

We assume (g.1) as before and the following conditions:

(Ψ) the function $\Psi : \mathbf{R}^n \to \mathbf{R}$ is convex, even and there exist $c, d > 0$ such that

$$d|\xi| - c \leq \Psi(\xi) \leq c(1+|\xi|)\,;$$

(g.4) there exist $a \in L^n(\Omega)$, $b \in \mathbf{R}$ and $p \in]1, n/(n-1)[$ such that

$$|g(x,s)| \leq a(x) + b|s|^{p-1}\,;$$

(g.5) $G(x,\cdot)$ is even and there exist $q > 1$ and $R > 0$ such that

$$|s| \geq R \Longrightarrow 0 < q\,G(x,s) \leq s\,g(x,s)\,.$$

From Theorems 3.11, 3.13 and 3.10 it easily follows that every critical point $u \in BV(\Omega)$ of f satisfies the inequality

$$\forall v \in BV(\Omega) : \mathcal{A}(v) \geq \mathcal{A}(u) + \int_\Omega g(x,u)(v-u)\,dx\,.$$

Actually, the converse also holds.

The next theorem is related to a well known result of Ambrosetti-Rabinowitz [3].

Theorem 3.18. *There exists a sequence (u_h) in $BV(\Omega)$ of critical points of f with $\lim_h f(u_h) = +\infty$.*

Proof. The result was first proved in [39, Theorem (1.1)] under the stronger assumption that $a \in L^{p'}(\Omega)$. The adaptations, to obtain the present version, are contained in [40]. The idea of the proof is to extend f to $L^{\frac{n}{n-1}}(\Omega)$ with value $+\infty$ outside of $BV(\Omega)$ and to apply Theorem 3.6. ■

References

[1] AMBROSETTI A., Critical points and nonlinear variational problems, *Mém. Soc. Math. France (N.S.) No. 49* (1992), 139 pp.

[2] AMBROSETTI A., AND G. PRODI, On the inversion of some differentiable mappings with singularities between Banach spaces, *Ann. Mat. Pura. Appl.* **93** (1972), 231-247.

[3] AMBROSETTI A., AND P. H. RABINOWITZ, Dual variational methods in critical point theory and applications, *J. Funct. Anal.* **14** (1973), 349-381.

[4] ARCOYA D., AND L. BOCCARDO, Critical points for multiple integrals of the Calculus of Variations, *Arch. Rational Mech. Anal.* **134** (1996), 249-274.

[5] ARIOLI G., AND F. GAZZOLA, Weak solutions of quasilinear elliptic *PDE*'s at resonance, *Ann. Fac. Sci. Toulouse*, **6** 6 (1997), 573-589

[6] ARIOLI G., AND F. GAZZOLA, Quasilinear elliptic equations at critical growth, *NoDEA Nonlinear Differential Equations Appl.*, **5** (1998), 83-97

[7] CAMPA I., AND M. DEGIOVANNI, Subdifferential calculus and nonsmooth critical point theory, *SIAM J. Optim*, **10** (2000), 1020-1048

[8] CANINO A., Multiplicity of solutions for quasilinear elliptic equations, *Topol. Methods Nonlinear Anal.* **6** (1995), 357–370.

[9] CANINO A., On a jumping problem for quasilinear elliptic equations, *Math. Z.* **226** (1997), 193-210

[10] CANINO A., On a variational approach to some quasilinear problems, in *Well-posed Problems and Stability in Optimization* (Marseille, 1995), *Serdica Math. J.* **22** (1996), 399-426.

[11] CANINO A., AND M. DEGIOVANNI, Nonsmooth critical point theory and quasilinear elliptic equations, in *Topological Methods in Differential Equations and Inclusions* (Montreal, PQ, 1994), 1–50, *NATO Adv. Sci. Inst. Ser. C Math. Phys. Sci.*, **472**, Kluwer Acad. Publ., Dordrecht, 1995.

[12] CHANG K. C., Variational methods for non-differentiable functionals and their applications to partial differential equations, *J. Math. Anal. Appl.* **80** (1981), 102-129.

[13] CHANG K. C., *Infinite Dimensional Morse Theory and Multiple Solution Problems*, Progress in Nonlinear Differential Equations and their Applications, **6**, Birkhäuser, Boston, 1993.

[14] CLARKE F. H., *Optimization and Nonsmooth Analysis*, Canadian Mathematical Society Series of Monographs and Advanced Texts. A Wiley-Interscience Publication. John Wiley & Sons, Inc., New York, 1983.

[15] CONTI M., AND F. GAZZOLA, Positive entire solutions of quasilinear elliptic problems via nonsmooth critical point theory, *Topol. Methods Nonlinear Anal.* **8** (1996), 275–294.

[16] CORVELLEC J.-N., Morse theory for continuous functionals, *J. Math. Anal. Appl.* **196** (1995), 1050-1072.

[17] CORVELLEC J.-N., Quantitative deformation theorems and critical point theory, *Pacific J. Math.*, **187** (1999), 263-279

[18] CORVELLEC J.-N., AND M. DEGIOVANNI, Nontrivial solutions of quasilinear equations via nonsmooth Morse theory, *J. Differential Equations* **136** (1997), 268–293.

[19] CORVELLEC J.-N., M. DEGIOVANNI, AND M. MARZOCCHI, Deformation properties for continuous functionals and critical point theory, *Topol. Methods Nonlinear Anal.* **1** (1993), 151-171.

[20] DE GIORGI E., M. DEGIOVANNI, A. MARINO AND M. TOSQUES, Evolution equations for a class of non-linear operators, *Atti Accad. Naz. Lincei Rend. Cl. Sci. Fis. Mat. Natur.* (8) **75** (1983), 1-8 (1984).

[21] DE GIORGI E., A. MARINO, AND M. TOSQUES, Problemi di evoluzione in spazi metrici e curve di massima pendenza, *Atti Accad. Naz. Lincei Rend. Cl. Sci. Fis. Mat. Natur.* (8) **68** (1980), 180–187.

[22] DEGIOVANNI M., Homotopical properties of a class of nonsmooth functions, *Ann. Mat. Pura Appl.* (4) **156** (1990), 37–71.

[23] DEGIOVANNI M., AND S. LANCELOTTI, Perturbations of even nonsmooth functionals, *Differential Integral Equations* **8** (1995), 981–992.

[24] DEGIOVANNI M., AND S. LANCELOTTI, Perturbations of critical values in nonsmooth critical point theory, in: *Well-Posed Problems and Stability in Optimization* (Marseille, 1995), *Serdica Math. J.* **22** (1996), 427–450.

[25] DEGIOVANNI M., A. MARINO, AND M. TOSQUES, Evolution equations with lack of convexity, *Nonlinear Anal.* **9** (1985), 1401–1443.

[26] DEGIOVANNI M., AND M. MARZOCCHI, A critical point theory for nonsmooth functionals, *Ann. Mat. Pura Appl.* (4) **167** (1994), 73–100.

[27] DEGIOVANNI M., AND L. MORBINI, Closed geodesics with Lipschitz obstacle, *J. Math. Anal. Appl.* **233** (1999), 767-789.

[28] DEGIOVANNI M., AND F. SCHURICHT, Buckling of nonlinearly elastic rods in the presence of obstacles treated by nonsmooth critical point theory, *Math. Ann.* **311** (1998), 675-728.

[29] DEIMLING K., *Nonlinear functional analysis*, Springer-Verlag, Berlin-New York, 1985.

[30] FRIGON M., On a critical point theory for multivalued functionals and application to partial differential inclusions, *Nonlinear Anal.* **31** (1998), 735-753.

[31] IOFFE A., AND E. SCHWARTZMAN, Metric critical point theory 1. Morse regularity and homotopic stability of a minimum, *J. Math. Pures Appl.* **75** (1996), 125-153.

[32] IOFFE A., AND E. SCHWARTZMAN, Metric critical point theory. II. Deformation techniques, in New Results in Operator Theory and its Applications, 131-144, *Oper. Theory Adv. Appl.* **98** Birkhäuser, Basel, 1997

[33] KATRIEL G., Mountain pass theorems and global homeomorphism theorems, *Ann. Inst. H. Poincaré Anal. Non Linéaire* **11** (1994), 189-209.

[34] LADYZHENSKAYA O. A., AND N. N. URAL'TSEVA, *Linear and Quasilinear Elliptic Equations*, Nauka Press, Moscow, 1964; Academic Press, New York, 1968.

[35] LANCELOTTI S., Perturbations of symmetric constraints in eigenvalue problems for variational inequalities, *Nonlinear Anal.* **27** (1996), 633–644.

[36] LANCELOTTI S., Infinitely many solutions for one-dimensional eigenvalue problems for variational inequalities, *Ann. Univ. Ferrara Sez. VII (N. S.).* **41** (1995), 33–44.

[37] MARINO A., A. M. MICHELETTI, AND A. PISTOIA, Some variationals results on semilinear elliptic problems with asymptotically nonsymmetric behaviour, *Nonlinear Analysis. A tribute in honour of Giovanni Prodi*, 243-256, Scuola Normale Superiore, Pisa, 1991.

[38] MARINO A., AND D. SCOLOZZI, Geodetiche con ostacolo, *Boll. Un. Mat. Ital. B* (6) **2** (1983), 1-31.

[39] MARZOCCHI M., Multiple solutions of quasilinear equations involving an area-type term, *J. Math. Anal. Appl.* **196** (1995), 1093-1104.

[40] MARZOCCHI M., Nontrivial solutions of quasilinear equations in BV, *Well-posed Problems and Stability in Optimization* (Marseille, 1995), *Serdica Math. J.* **22** (1996), 451-470.

[41] MAWHIN J., AND M. WILLEM, *Critical Point Theory and Hamiltonian Systems*, Applied Mathematical Sciences, **74**, Springer-Verlag, New York-Berlin, 1989.

[42] PALAIS R. S., Morse theory on Hilbert manifolds, *Topology* **2** (1963), 299–340.

[43] PALAIS R. S., Lusternik-Schnirelman theory on Banach manifolds, *Topology* **5** (1966), 115–132.

[44] PALAIS R. S., Critical point theory and the minimax principle, *Global Analysis* (Proc. Sympos. Pure Math., Vol. **XV**, Berkeley, Calif., 1968), 185–212. Amer. Math. Soc., Providence, R.I., 1970.

[45] PELLACCI B., Critical points for non differentiable functionals, *Boll. Un. Mat. Ital. B* (7) **11** (1997), 733-749.

[46] RABINOWITZ P. H., *Minimax Methods in Critical Point Theory with Applications to Differential Equations*, CBMS Regional Conference Series in Mathematics, **65**. Published for the Conference Board of the Mathematical Sciences, Washington, D.C.; by the American Mathematical Society, Providence, R.I., 1986.

[47] RIBARSKA N. K., TS. Y. TSACHEV, AND M. I. KRASTANOV, Speculating about mountains, in *Well-posed Problems and Stability in Optimization* (Marseille, 1995), *Serdica Math. J.* **22** (1996), 341-358.

[48] ROCKAFELLAR R. T., Integral functionals, normal integrands and measurable selections, in *Nonlinear Operators and the Calculus of Variations* (Bruxelles, 1975), 157-207, *Lecture Notes in Math.*, **543**, Springer, Berlin-New York, 1976.

[49] ROCKAFELLAR R. T., Generalized directional derivatives and subgradients of nonconvex functions, *Canad. J. Math.* **32** (1980), 257-280.

[50] STRUWE M., Quasilinear elliptic eigenvalue problems, *Comment. Math. Helv.* **58** (1983), 509-527.

[51] STRUWE M., *Variational Methods*, Springer-Verlag, Berlin, 1990.

[52] SZULKIN A., Minimax principles for lower semicontinuous functions and applications to nonlinear boundary value problems, *Ann. Inst. H. Poincaré Anal. Non Linéaire* **3** (1986), 77-109.

[53] ZEIDLER E., *Nonlinear functional analysis and its applications. III. Variational methods and optimization*, Springer-Verlag, New York-Berlin, 1985.

4 EXHAUSTIVE FAMILIES OF APPROXIMATIONS REVISITED

V.F.Demyanov

Applied Mathematics Dept., St.Petersburg State University
Staryi Peterhof St Petersburg, 198904 RUSSIA

A.M.Rubinov

School of Information Technology, and Mathematical Sciences
University of Ballarat, Victoria 3353, Australia

Abstract: The notions of exhaustive families of upper convex and lower concave approximations (in the sense of B.N.Pschenichnyi) were introduced in 1982. For some classes of nonsmooth functions, these tools appeared to be very productive and constructive (e.g., in the case of quasidifferentiable functions). In the present paper we introduce notions of upper exhanster and lower exhanster. It is demonstrated how to employ these notions to describe necessary optimality conditions and to find directions of steepest ascent and descent. If a proper exhauster is known, the above problems are reduced to the problems of finding the nearest points to convex sets.

4.1 DIRECTIONAL DERIVATIVES AND GENERALIZATIONS

The notion of directional derivative (or its generalization) plays an essential role in Nonsmooth Analysis and Nondifferentiable Optimization.

[1]The research was supported by the Russian Foundation for Fundamental Studies (grant RFFI No. 97-01-00499)

R.P. Gilbert et al. (eds.), From Convexity to Nonconvexity, 43–50.

Let $f : X \to \mathbb{R}, X \subset \mathbb{R}^n$ be an open set. The function$f$ is called Dini directionally differentiable (D.-d.d.) at $x \in X$ if for every $g \in \mathbb{R}^n$ there exists the finite limit

$$f'_{\mathcal{D}}(x,g) = \lim_{\alpha\downarrow 0} \frac{f(x+\alpha g) - f(x)}{\alpha}. \tag{1}$$

The function f is called Hadamard directionally differentiable (H.-d.d.) at $x \in X$ if for every $g \in \mathbb{R}^n$ there exists the finite limit

$$f'_H(x,g) = \lim_{[\alpha,g']\to[+0,g]} \frac{1}{\alpha}\left[f(x+\alpha g') - f(x)\right]. \tag{2}$$

Of course, if f is H-d.d. at x, then it is $\mathcal{D}$-d.d. at x and $f'_{\mathcal{D}}(x,g) = f'_H(x,g)$. The inverse statement is not true. The quantity $f'_{\mathcal{D}}(x,g)(f'_H(x,g))$ is called the Dini (Hadamard) derivative of f at x in the direction g.

Let $x \in X, g \in \mathbb{R}^n$. The quantity

$$f^{\uparrow}_{\mathcal{D}}(x,g) = \limsup_{\alpha\downarrow 0} \frac{1}{\alpha}\left[f(x+\alpha g) - f(x)\right] \tag{3}$$

is called the Dini upper derivative of f at x in the direction g.

The quantity

$$f^{\downarrow}_{\mathcal{D}}(x,g) = \liminf_{\alpha\downarrow 0} \frac{1}{\alpha}\left[f(x+\alpha g) - f(x)\right] \tag{4}$$

is called the Dini lower derivative of f at x in the direction g.

The quantity

$$f^{\uparrow}_H(x,g) = \limsup_{[\alpha,g']\to[+0,g]} \frac{1}{\alpha}\left[f(x+\alpha g') - f(x)\right] \tag{5}$$

is called the Hadamard upper derivative of f at x in the direction g. The quantity

$$f^{\downarrow}_H(x,g) = \liminf_{[\alpha,g']\to[+0,g]} \frac{1}{\alpha}\left[f(x+\alpha g') - f(x)\right] \tag{6}$$

is called the Hadamard lower derivative of f at x in the direction g. The limits in (3)-(6) always exist (unlike the ones in (1) and (2)) but may be not finite.

The role played by the above derivatives in Nonsmooth Analysis and Nondifferentiable Optimization can hardly be overestimated.

Let us recall, e.g., the following well-known properties (see, e.g., [3]).

Lemma 1. Let a function $f : X \to \mathbb{R}$ be given, $X \subset \mathbb{R}^n$ be an open set. For a point $x^* \in X$ to be a local minimizer of f on X it is necessary that

$$f^{\downarrow}_{\mathcal{D}}(x^*,g) \geq 0 \quad \forall g \in \mathbb{R}^n \tag{7}$$

and

$$f^{\downarrow}_H(x^*,g) \geq 0 \quad \forall g \in \mathbb{R}^n. \tag{8}$$

The condition

$$f^{\downarrow}_H(x^*,g) > 0 \quad \forall g \neq 0_n \tag{9}$$

is sufficient for $x^* \in X$ to be a strict local minimizer of f on X.

Lemma 2. Let a function $f : X \to \mathbb{R}$ be given, $X \subset \mathbb{R}^n$ be an open set. For a point $x^{**} \in X$ to be a local maximizer of f on X it is necessary that

$$f_D^{\uparrow}(x^{**}, g) \leq 0 \quad \forall g \in \mathbb{R}^n \tag{10}$$

and

$$f_H^{\uparrow}(x^{**}, g) \leq 0 \quad \forall g \in \mathbb{R}^n . \tag{11}$$

The condition

$$f_H^{\uparrow}(x^{**}, g) < 0 \quad \forall g \neq 0_n \tag{12}$$

is sufficient for x^{**} to be a strict local maximizer of f on X.

Note, that (8) implies (7) and (11) implies (10) but the inverses are not true.

A point $x^* \in X$ satisfying (7) ((8)) is called a Dini (Hadamard) inf - stationary point of f. A point $x^{**} \in X$ satisfying (10) ((11)) is called a Dini (Hadamard) sup - stationary point of f. If a point $x_0 \in X$ is not Dini inf-stationary then a direction $g_0 \in \mathbb{R}^n$, $||g_0|| = 1$, such that

$$f_D^{\downarrow}(x_0, g_0) = \inf_{||g||=1} f_D^{\downarrow}(x_0, g)$$

is called a Dini steepest descent direction of f at x_0.

If a point $x_0 \in X$ is not Hadamard inf-stationary then a direction $g_0 \in \mathbb{R}^n$, $||g_0|| = 1$, such that

$$f_H^{\downarrow}(x_0, g_0) = \inf_{||g||=1} f_H^{\downarrow}(x_0, g)$$

is called an Hadamard steepest descent direction of f at x_0.

It may happen that a steepest descent (ascent) direction is not unique (this is not possible in the smooth case).

Analogously one can define Dini and Hadamard steepest ascent directions.

Note that all the functons defined by (1)-(6) are positively homogeneous (p.h.) of the first degree as functions of g (a function $h(g)$ is p.h., if

$$h(\lambda g) = \lambda h(g) \quad \forall \lambda \geq 0, \quad \forall g \in \mathbb{R}^n .)$$

4.2 EXHAUSTIVE FAMILIES OF UPPER AND LOWER APPROXIMATIONS

B.N.Pshenichnyi in [5] introduced the notions of upper convex and lower concave approximations. Let a function $h : \mathbb{R}^n \to \mathbb{R}$ be positively homogeneous (p.h.) of the first degree. A convex p.h. function $\overline{h} : \mathbb{R}^n \to \mathbb{R}$ is called an upper convex approximation (u.c.a) of the function h if

$$h(g) \leq \overline{h}(g) \quad \forall g \in \mathbb{R}^n . \tag{13}$$

A concave p.h. function $\underline{h} : \mathbb{R}^n \to \mathbb{R}$ is called an lower concave approximation (l.c.a) of the function h if

$$\underline{h}(g) \leq h(g) \quad \forall g \in \mathbb{R}^n . \tag{14}$$

Exhaustive families of approximations were introduced in [1] (see also [2] and [3]):

A set Λ^* of u.c.a.'s of the function h is called an exhaustive family of u.c.a's (U.E.F) of h if

$$h(g) = \inf_{\overline{h}\in\Lambda^*} \overline{h}(g) \quad \forall g \in \mathbb{R}^n . \tag{15}$$

A set Λ_* of l.c.a.'s of the function h is called an exhaustive family of l.c.a's (L.E.F) of h if

$$h(g) = \sup_{\underline{h}\in\Lambda_*} \underline{h}(g) \quad \forall g \in \mathbb{R}^n . \tag{16}$$

The existence of exhaustive families is proved, e.g., in [3], Theorem III.5.1. In particular, they exist if h is continuous in g on $\mathbb{R}^n$. For example, if f is a Lipschitz function then its Hadamard derivatives coincide with the respective Dini ones and are continuous as functions of direction.

Since (see [6]) every convex p.h. function $\overline{h}(g)$ can be represented in the form

$$\overline{h}(g) = \max_{v\in\overline{C}(\overline{h})} (v, g) \quad \forall g \in \mathbb{R}^n$$

where $\overline{C}(\overline{h})$ is the subdifferential of $\overline{h}$ at 0_n (it is a compact convex set of $\mathbb{R}^n$), then (15) may be rewritten as

$$h(g) = \inf_{C\in E^*} \max_{v\in C}(v, g) \quad \forall g \in \mathbb{R}^n \tag{17}$$

where

$$E^* = \{C \mid C = \overline{C}(\overline{h}),\ \overline{h} \in \Lambda^*\}.$$

The set $E^* = E^*(h)$ is called an upper exhauster of h.

Analogously, since every concave p.h.function $\underline{h}$ can represented in the form

$$\underline{h}(g) = \min_{v\in\underline{C}(\underline{h})} (v, g) \quad \forall g \in \mathbb{R}^n$$

where $\underline{C}(\underline{h})$ is the superdifferential of $\underline{h}$ at 0_n (it is also a compact convex set of $\mathbb{R}^n$), then (16) takes the form

$$h(g) = \sup_{C\in E_*} \inf_{w\in C)} (w, g) \quad \forall g \in \mathbb{R}^n \tag{18}$$

where

$$E_* = \{C \mid C = \underline{C}(\underline{h}),\ \underline{h} \in \Lambda_*\}.$$

The set $E_* = E_*(h)$ is called a lower exhauster of h.

Lemma 3:

Let $h : \mathbb{R}^n \to \mathbb{R}$ be a p.h.function and there exist an upper exhauster $E^*(h)$ of h. Then the following statements are equivalent:

$$i) \qquad h(g) \geq 0 \quad \forall g \in \mathbb{R}^n, \tag{19}$$

$$ii) \qquad 0_n \in C \quad \forall C \in E^*(h), \tag{20}$$

$$iii) \quad 0_n \in L^*(h) = \bigcap\{C \mid C \in E^*(h)\}. \tag{21}$$

Lemma 4:
Let $h : \mathbb{R}^n \to \mathbb{R}$ be a p.h.function and there exist an upper exhauster $E^*(h)$ of h. Then the following statements are equivalent:

$$i) \qquad h(g) > 0 \quad \forall g \in \mathbb{R}^n, \quad g \neq 0_n, \tag{22}$$

$$ii) \qquad 0_n \in \text{ int } C \quad \forall C \in E^*(h), \tag{23}$$

$$iii) \quad 0_n \in \text{ int } L^*(h) = \bigcap\{C \mid C \in E^*(h)\}. \tag{24}$$

Lemma 5:

Let $h : \mathbb{R}^n \to \mathbb{R}$ be a p.h.function and there exist a lower exhauster $E_*(h)$ of h. Then the following statements are equivalent:

$$i) \qquad h(g) \leq 0 \quad \forall g \in \mathbb{R}^n, \tag{25}$$

$$ii) \qquad 0_n \in C \quad \forall C \in E_*(h), \tag{26}$$

$$iii) \quad 0_n \in L_*(h) = \bigcap\{C \mid C \in E_*(h)\}. \tag{27}$$

Lemma 6:

Let $h : \mathbb{R}^n \to \mathbb{R}$ be a p.h.function and there exist an lower exhauster $E_*(h)$ of h. Then the following statements are equivalent:

$$i) \qquad h(g) < 0 \quad \forall g \in \mathbb{R}^n, \quad g \neq 0_n, \tag{28}$$

$$ii) \qquad 0_n \in \text{int } C \quad \forall C \in E_*(h), \tag{29}$$

$$iii) \quad 0_n \in \text{int } L_*(h) = \bigcap\{C \mid C \in E_*(h)\}. \tag{30}$$

Remark 1. If follows from Lemma 1 and Lemmas 3 and 4 that the problem of verifying conditions (8) and (9) is reduced to checking conditions (20) and (23) (or (21) and (24)) where $E^*(h)$ is an upper exhauster of the function $h(g) = f_H^{\downarrow}(x^*, g)$. To check condition (7) it is necessary to check conditions (20) or (21) where $E^*(h)$ is an upper exhauster of the function $h(g) = f_{\mathcal{D}}^{\downarrow}(x^*, g)$.

Analogously the verification of conditions (10),(11) or (12) is replaced by the problem of checking conditions (26) and (29) (or (29) and (30)) for the set $E_*(h)$ where $E_*(h)$ is a lower exhauster of the function $h(g) = f_H^{\uparrow}(x^{**}, g)$ (or $f_{\mathcal{D}}^{\uparrow}(x^{**}, g)$ for condition (10)).

Remark 2. Let $h : \mathbb{R}^n \to \mathbb{R}$ be a p.h.function and there exist an upper exhauster $E^*(h)$ of h. Assume that condition (19) (or, equivalently (20) or (21)) doesn't hold. Then there exists $C \in E^*(h)$ such that $0_n \neq C$. Find

$$d(C) = \min_{v \in C} ||v|| = ||v_C|| > 0$$

and take $g_C = -v_C \Big/ ||v_C||$. It is clear that

$$h(g_C) \leq -d(C) < 0. \tag{31}$$

If $C^* \in E^*(h)$ is such that

$$d(C^*) = \sup_{C \in E^*(h)} d(C) \tag{32}$$

then the direction $g^* = g_{C^*}$ is a steepest descent direction of h at 0_n. (Since C^* satisfying (32) is not uniquely defined then a steepest descent direction is not unique.) If h is the Dini or Hadamard upper directional derivative of some function f at some point x_0 then the direction g^* thus constructed is a steepest descent direction of f at x_0 while the direction g_C satisfying (31) is a descent direction.

Analogously if $h : \mathbb{R}^n \to \mathbb{R}$ be a p.h.function and a lower exhauster $E_*(h)$ of h exists and if condition (25) doesn't hold then for any $C \in E_*(h)$ such that $0_n \neq C$ let us find

$$d(C) = \min_{v \in C} ||v|| = ||v_C|| > 0$$

and take $g_C = v_C \Big/ ||v_C||$. It is clear that

$$h(g_C) \geq d(C) > 0. \tag{33}$$

If $C_* \in E_*(h)$ is such that

$$d(C_*) = \sup_{C \in E_*(h)} d(C) \tag{34}$$

then the direction $g_* = g_{C_*}$ is a steepest ascent direction of h at 0_n (such a direction is also not unique). If h is the Dini or Hadamard lower directional derivative of some function f at some point x_0 then the direction g_* thus constructed is a steepest ascent direction of f at x_0 while the direction g_C satisfying (33) is an ascent direction.

Remark 3. It follows from the aforesaid that the problems of verifying necessary and/or sufficient optimality conditions and computing descent, steepest descent, ascent and steepest ascent directions of some (generally speaking nondifferentiable) function are reduced to solving some geometric problems and finding nearest points to some convex sets. To employ the above results it is required to be able to construct the corresponding exhausters. For some classes of functions such tools are evailable. For example, if f is a quasidifferentiable function then [3] its directional derivative at a point x is represented as

$$f'(x,g) = \max_{v \in \underline{\partial} f(x)} (v,g) + \min_{w \in \overline{\partial} f(x)} (w,g) \tag{35}$$

where $\underline{\partial} f(x),)\overline{\partial} f(x)) \subset \mathbb{R}^n$ are compact convex sets. It is clear from (35) that

$$f'(x,g) = \min_{w \in \overline{\partial} f(x)} \max_{v \in w + \underline{\partial} f(x)} (v,g) = \max_{v \in \underline{\partial} f(x)} \min_{w \in v + \overline{\partial} f(x)} (w,g).$$

Hence, for the function $h(g) = f'(x, g)$ we get

$$E^*(h) = \{C = w + \underline{\partial} f(x) \mid w \in \overline{\partial} f(x)\}, \tag{36}$$

$$E_*(h) = \{C = v + \overline{\partial} f(x) \mid v \in \underline{\partial} f(x)\}. \tag{37}$$

Condition (20) and relation (36) (applied to the function $h(g) = f'(x^*, g)$) provide the following necessary condition for a minimum of a quasidifferentiable function

$$0_n \in w + \underline{\partial} f(x^*) \quad \forall w \in \overline{\partial} f(x^*)$$

which is equivalent to (see [3]) $-\overline{\partial} f(x^*) \subset \underline{\partial} f(x^*)$.

Condition (26) and relation (37) (applied to the function $h(g) = f'(x^{**}, g)$) provide the following necessary condition for a maximum of a quasidifferentiable function

$$0_n \in v + \overline{\partial} f(x^{**}) \quad \forall w \in \overline{\partial} f(x^{**})$$

which is equivalent to (see [3]) $-\underline{\partial} f(x^{**}) \subset \underline{\partial} f(x^{**})$.

Remark 4. An example of a p.h. function which is not Lipschits (and, hence, is not quasidifferentiable) was given by B.M.Glover et al [4].

Remark 5. Observe, in passing, that conditions (21), (24), (27) and (30) are not "constructive" since if (19), (22), (25) or (28) are not yet satisfied, then the sets $L^*(h)$ and/or $L_*(h)$ may happen to be empty, and we get no "useful" information related to the behaviour of the function under consideration and descent or ascent directions. At the same time conditions (20), (23), (26) or (29) are constructive (we were able to find directions of descent or ascent).

Remark 6. The aim of our paper is to show that practically every function may be treated by means of existing tools developed, e.g., for quasidifferentiable functions, by employing the notions of upper and lower exhausters. The problem now is to be able to construct the mention exhausters. Some elements of a Calculus of exhausters were indicated in [2].

References

[1] V.F. Demyanov and A.M. Rubinov. Some approaches to a nonsmooth optimization problem. *Econom. i Mat. Metody*, 17(6):1153–1174, 1981. In Russian.

[2] V.F. Demyanov and A.M. Rubinov. Elements of quasidifferential calculus. In V.F. Demyanov, editor, *Nonsmooth Problems of Optimization Theory and Control*, chapter 1, pages 5–127. Leningrad University Press, Leningrad, 1982.

[3] V.F. Demyanov and A.M. Rubinov. *Constructive Nonsmooth Analysis.* Verlag Peter Lang, Frankfurt a/M, 1995.

[4] B.M. Glover, Y. Ishizuka, V. Jeyakummar, and H.D Tuan. Complete characterizations of global optimality for problems involving the pointwise minimum of sublinear functions. *SIAM J. Optimization*, 6(2):362–372, 1996.

[5] B.N. Pshenichnyi. *Convex analysis and Extremal Problems.* Nauka Publishers, Moscow, 1980. In Russian.

[6] R.T. Rockafellar. *Convex Analysis.* Princeton University Press, Princeton N.J., 1970.

5 A SURVEY ON OPTIMAL SHAPE DESIGN PROBLEMS FOR SYSTEMS DESCRIBED BY PARTIAL DIFFERENTIAL EQUATIONS AND HEMIVARIATIONAL INEQUALITIES

Zdzisław Denkowski

Jagellonian University, Institute of Computer Science,
ul. Nawojki 11, 30072 Cracow, Poland

denkowsk@softlab.ii.uj.edu.pl

Abstract: OSD problems are formulated for physical systems described by PDEs, VIs and HVIs. By inspection of the direct method we get an abstract theorem (Thm 4.2) assuring the existence of optimal shapes. Basing on this theorem we give a survey of different approaches to the existence results in OSD problems (e.g. the mapping method applied to various situations). In the case where there are no "classical" optimal shapes we present the Buttazzo, Dal Maso, Mosco approach to the relaxation for systems described by Dirichlet problems. A theorem on lsc property for cost functionals in OSD problems is provided.

5.1 INTRODUCTION

Roughly speaking by optimal shape design (OSD) problems we mean a class of control problems for physical systems where controls are geometric domains which are the admissible shapes of systems.

Motivation for considering such kind of problems comes from engineering: electrostatics (control in coefficients, homogenization e.g. the best arrangements of given two materials in a conductor, ...), fluid dynamics (the minimal drag problem, optimal shape of airplane wings, ...), structure mechanics (minimization of stresses, weight,...of elements in a construction), electrochemical machining problems and so on.

R.P. Gilbert et al. (eds.), From Convexity to Nonconvexity, 51–65.

Starting in 70-ties with papers of Cea, Chenais, Miele, Murat and Simon, Pironneau,... the theory developped quickly and by now there is an extensive literature on the subject, see monographs by Murat and Simon [25], Pironneau [29], Sokołowski and Zolesio [32], and the literature therein, proceedings of conferences (e.g. edited by Da Prato and Zolesio [2]), and many original papers. We mention only that the optimal shape design problems for PDEs were considered also by Allaire, Ambrosio, Bucur, Buttazzo, Dal Maso, Fujii, Tartar, Sverak, and many others, while for VIs by Butt, Haslinger, Liu and Rubio, Neittaanmäki, Tiihonen, and others. For computational aspects of shape design problems, one can consult Haslinger, Masmoudi, Mäkelä, Miettinen, Neittaanmäki, Salmenjoki, ... We are far from giving complete references at the end of the paper.

The main research subjects of OSD are: existence of optimal solutions, their characterization (neccesary and sufficient conditions of optimality), numerical approximations, and sensitivity analysis. In the case of lack of optimal solutions there is also a need for relaxation (i.e. looking for a new kind of generalized solutions).

In this paper we present (in a very limited manner for obvious reasons) only some approaches (based on the direct method) to the existence problems and relaxation (putting stress on systems described by HVIs). For results concerning characterization of optimal solutions and sensitivity analysis we refer to [25], [32], [5], [29].

Because of really rich existing material on the subject our review is far from being complete (and it does not pretend to be such in any aspect).

The paper is organized as follows: after preliminaries of Section 2 we recall in Section 3 some relations describing physical systems based on PDEs, VIs and HVIs. In Section 4 we inspect the direct method in application to the abstract OSD problem and we formulate the main existence theorem. In Section 5 we briefly describe the mapping method due to Murat-Simon and we quote some its applications to OSD for systems desribed by PDEs, VIs, and HVIs. In Section 6 we mention some other methods and other classes of admissible shapes. In Section 7 we describe an approach to the relaxation due to Buttazzo, Dal Maso, Mosco and their collaborators. Finally in Section 8 we give some information concerning the lower semicontinuity of integral functionals with respect also to the domain of integration.

5.2 PRELIMINARIES

We introduce here some notions and notations which are indispensable for the sequel. Now we fix an open and bounded set $\Omega \subset \mathbb{R}^N$ but in the next Sections it will be changing in a family $\mathcal{B}$ of all admissible shapes $\mathcal{O} = \mathcal{O}_{ad}$. Let $V = V(\Omega), H = H(\Omega), V' = V'(\Omega)$ be a Gelfand-Lions triplet of Sobolev type spaces of functions defined on Ω; i.e. $V \subset H \subset V'$ with dense and compact embeddings. Assume $a : V \times V \longrightarrow \mathbb{R}$ be a continuous coercive bilinear form and $A : V \longrightarrow V'$ be associated linear operator; i.e. $a(u, v) =< Au, v >$ for every $u, v \in V$ ($<\cdot,\cdot>$ being the duality pairing). Let $K = K(\Omega)$ be a convex closed subset of V and f be a fixed element of V'.

We recall (see e.g. [26]) that for a locally Lipschitz function $j : \mathbb{R} \longrightarrow \mathbb{R}$ we can define its generalized Clarke directional derivative and subdifferential, respectively, by the formulae:

$$j^0(\xi;\gamma) = \limsup_{h\to 0,\ t\downarrow 0} \frac{j(\xi + h + t\gamma) - j(\xi + h)}{t} \quad \text{for all } \xi, \gamma \in \mathbb{R}. \tag{2.1}$$

$$\partial j(\xi) = \{\eta \in \mathbb{R} \mid j^0(\xi;\gamma) \geq \eta\gamma, \ \ \forall\, \gamma \in \mathbb{R}\} \quad \text{for all } \xi \in \mathbb{R}. \tag{2.2}$$

It is well known (see e.g. [29]) that in the family of all closed subsets of $\mathbb{R}^N$ the function

$$d_H(A, B) = \max\{\sup_{x\in A} \inf_{y\in B} |x - y|, \sup_{y\in B} \inf_{x\in A} |x - y|\} \tag{2.3}$$

defines the metric called the Hausdorff distance. It can be used also to define the "complementary Hausdorff distance" d_{H^c} in the family

$$\mathcal{A}(\Omega_0) = \{\Omega \subset \Omega_0 |\ \Omega \text{ is open}\}, \tag{2.4}$$

where Ω_0 is a given open and bounded subset of $\mathbb{R}^N$; i.e.

$$d_{H^c}(\Omega_1, \Omega_2) = d_H(\bar{\Omega}_0 \setminus \Omega_1, \bar{\Omega}_0 \setminus \Omega_2) \ \ (\Omega_i \in \mathcal{A}(\Omega_0), i = 1, 2). \tag{2.5}$$

Given a sequence $\{K_n\}$ of sets in a Banach space X we define the upper and lower limits (in the sense of Kuratowski) in the topology τ (τ stands for s = strong or w = weak topology) and the Mosco convergence, resp., by the formulae:

$$\tau - \overline{\lim}_{n\to\infty} K_n = \{v \in X \mid \exists\ n_1 < n_2 < ..., v_k \in K_{n_k},\ v_k \xrightarrow{\tau} v\} \tag{2.6}$$

$$\tau - \underline{\lim}_{n\to\infty} K_n = \{v \in X \mid \exists v_n \in K_n,\ v_n \xrightarrow{\tau} v\} \tag{2.7}$$

$$K_n \xrightarrow{M} K \iff w - \overline{\lim}_{n\to\infty} K_n \subset K \subset s - \underline{\lim}_{n\to\infty} K_n \tag{2.8}$$

Setting $K_\infty = K$ we say that the sequence $\{K_n\}$ is $\tau - usc$ (in Kuratowski sense)(at ∞) iff $\tau - \overline{\lim}_{n\to\infty} K_n \subset K_\infty$. For a set K in X we denote by 1_K its characteristic function; i.e. $1_K(x) = 1$, if $x \in K$ and $K(x) = 0$, if $x \notin K$.

5.3 STATE RELATIONS FOR PHYSICAL SYSTEMS

In the paper the states of physical systems will be described as the solutions to some boundary value problems (bvp) for partial differential equations (PDE) or variational inequalities (VI) or hemivariational inequalities (HVI), mainly of elliptic type (but generalizations to parabolic and hyperbolic type are also possible - see the last relation below).

Namely, under the notations of Section 2, we shall consider the following simple model problems (the boundary conditions and constraints being incorporated in the definitions of V and K, resp.,):

$$(PDE) \qquad Au = f, \quad u \in V(\Omega),$$

$$(VI) \qquad a(u, v-u) + \Phi(v) - \Phi(u) \ \geq \ < f, v-u > \quad \forall v \in K(\Omega),$$

$$(HVI) \qquad a(u, v-u) + \int_\Omega j^0(u, v-u)\, dx \ \geq \ < f, v-u > \quad \forall v \in K(\Omega),$$

$$(HVI_p) \qquad < u'(t), v-u(t) > + a(u(t), v-u(t)) + \int_\Omega j^0(u(t), v-u(t))\, dx \geq$$
$$\geq \ < f(t), v-u(t) >, \qquad \forall v \in K(\Omega), \text{and } a.e. \ t \in (0, I)$$

The theory of (bvp) for PDEs is well founded by now with numerous monographs (e.g. by Lions-Magenes, Dautray-Lions, Zeidler,...) and original papers (e.g. excelent paper by Fichera [16]) so we pass here to some remarks and comments only on (VI) and (HVI).

The theory of VIs was started in the 60-ties with papers of Baiocchi, Brezis, Duvaut, Lions, Stampacchia,... It provides good mathematical models for problems concerning the elasticity theory, the fluid flow by porous media, semipermeable media, etc., ... Based on the variational methods for PDEs it permits to characterize solutions of minimization problems for convex and differentiable functionals on closed and convex sets. In this context the VIs can be regarded as more general description of systems than PDEs (e.g. VIs characterizing minimizers of quadratic functionals become PDEs - their Euler equations provided the sets on which minimization is performed are vector subspaces).

In the next decades the theory of VIs has been considerably enlarged in many directions. For instance, in mechanical formulations of unilateral constraints J.J. Moreau [24] has introduced the notion of superpotential for convex but generally nondifferentiable and nonfinite energy functionals. It permitted the derivation of VIs interpreted physically as expressions of the principle of virtual work or of d'Alembert's principle in dynamical problems.

In 1981 P. Panagiotopoulos [27] has introduced HVIs which generalize VIs in the sense that they permit to cover (bvp) for systems with nonmonotone, nonconvex, possibly multivalued laws (relations: reactions-displacemnts, forces-velocities,...) The involved superpotentials need not be convex but only locally Lipschitz so nonsmooth analysis can be used. For the problems where the Clarke subdifferential is pseudomonotone operator we refer to [26].

Example 3.1 (see [26] for more details). Consider the contact problem in linear elasticity ($\sigma_{ij} = C_{ijkl}\varepsilon_{kl}$, where σ, ε, C denote, resp., the stress, strain tensors and elasticity matrix). Assume $f = \bar{f} + \hat{f}$, $\bar{f}$ being the external volume forces (e.g. gravity) and $\hat{f}$ introduces the so called "skin effects" (e.g. friction, adhesive contact) on $\Omega' \subset \Omega$ or at least on some part of the boundary $\Gamma = \partial\Omega$. Suppose we have one-dimensional nonconvex, nonmonotone law:

$$(3.1) \qquad -\hat{f}_N(x) \in \hat{\beta}(u_N(x)), \quad x \in \Gamma,$$

where by $\hat{\beta}$ we denoted the multifunction obtained from a function $\beta \in L^{\infty}_{loc}(\mathbb{R})$ by " filling in the gaps" operation, while $\hat{f}_N(x)$ and $u_N(x)$ denote, resp., the exterior normal component of $\hat{f}(x)$ and of the displacement $u(x)$ at point $x \in \Gamma$.

Due to Chang theorem we get the existence of the unique (up to an additive constant) superpotential $j : \mathbb{R} \mapsto \mathbb{R}$ such that

$$(3.2) \qquad \partial j(\xi) = \hat{\beta}(\xi).$$

Then after the standard calculations on the equilibrium equation $\sigma_{ij} + f_i = 0$ and using the definition of the Clarke subdifferential we arrive at (HVI) above (with integral taken over Γ).

In the sequel let E stands for PDE or VI or HVI describing the state relation on the given set Ω. By a solution of the state relation (E) we mean the standard "weak solution" in the case of E = PDE or E = VI, while for E = HVI we admit the first of the two definitions below.

Definition 3.2 An element $u \in K = K(\Omega)$ is said to be a solution of (HVI) if there exists $\chi \in L^2(\Omega)$ such that

$$(i) \qquad a(u, v - u) + \int_{\Omega} \chi \cdot (v - u)\, dx \ \geq \ < f, v - u > \quad \forall v \in K$$

$$(ii) \qquad \chi(x) \in \partial j(u(x)) \ \text{ a.e. in } \Omega.$$

Definition 3.3 The set $S_E(\Omega) = \{u \in V(\Omega) |\ u \text{ is a solution to (E)}\}$ will be called the solution set to the state relation (E).

Remark 3.4 Note that in the case where E = PDE or E = VI for the so called "well posed problems" the set $S_E(\Omega)$ consits of the unique element, while in general the set $S_{HVI}(\Omega)$ contains more than one element (if any) as the assumption of convexity of superpotentials is rejected. However, due to some result of [28], we have $S_{HVI}(\Omega) \neq \emptyset$.

Remark 3.5 Let us quote the observation (see Prop. 3.8 in [26]) that if superpotential j (linked with the function β by the relation (3.2)) is locally Lipschitz and ∂-regular and the growth condition below holds:

$$(3.3) \qquad |\beta(\xi)| \leq C(1 + |\xi|) \ \ \forall \xi \in \mathbb{R} \ \ (\text{for some } C > 0),$$

then the solutions to HVI are substationarity points of the energy functional; i.e. we have:

$$u \in S_{HVI}(\Omega) \iff 0 \in \partial\mathcal{E}(u),$$

where $\mathcal{E}(v) = \frac{1}{2}a(v,v) + \Phi(v) - < f, v >$ and $\Phi(v) = \int_\Omega j(v(x))\, dx$. Analogous relations are valid for VIs and PDEs but then the Clarke subdifferential reduces, resp., to the subdifferential in the sense of convex analysis and to the gradient (the right hand side of the last relation becomes then the Euler equation for the energy $\mathcal{E}$).

5.4 ABSTRACT OSD AND DIRECT METHOD

An abstract OSD problem for system E (E stands for PDE or VI or HVI) can be formally written down as follows:

$$(OSD)_E \qquad \min\{J(\Omega, u_\Omega) \mid \Omega \in \mathcal{B}, u_\Omega \in S_E(\Omega)\},$$

which often is written as

$$(OSD)_E^{'} \qquad \min_{\Omega \in \mathcal{B}} \min_{u \in S_E(\Omega)} J(\Omega, u),$$

and in the case of well posed problems ($S_E(\Omega) = \{u_\Omega\}$) it simply reduces to the standard formulation

$$(OSD)_E^{''} \qquad \min_{\Omega \in \mathcal{B}} J(\Omega, u_\Omega), \quad u_\Omega \text{ is the solution to (E).}$$

Above $\mathcal{B}$ denotes a subclass of the family of all admissible shapes (subsets of $\mathbb{R}^N$), $S_E(\Omega)$ is the solution set for system E (see Def. 3.3), and J is a cost functional.

The general cost functional J (for elliptic type OSD problems) takes the form

$$(4.1) \quad J(\Omega, u) = \int_\Omega L(x, u(x), \nabla u(x))\, dx + \int_\Gamma L_0(x, u(x), \nabla u(x))\, d\Gamma + \alpha \mathcal{P}(\Omega)$$

for some given functions L, L_0 and a positive constant α ($\mathcal{P}$ being a regularizing term assuring the existence of optimal solutions). In the particular cases of minimization of weight or deviation of states (or stresses, velocities) from the desired functions it can be specified, resp., to the following functionals: $J_1(\Omega, u) = \int_\Omega \rho(x)\, dx$, $J_2(\Omega, u) = \int_\Omega |u - u_d|^2\, dx$, $J_3(\Omega, u) = \int_\Omega |\nabla(u - u_d)|^2\, dx$.

We admit the definition

Definition 4.1 By the optimal solution to $(OSD)_E$ we mean any pair $(\Omega^*, u^*) \in \bigcup_{\Omega \in \mathcal{B}} (\Omega \times S_E(\Omega))$ such that

$$J(\Omega^*, u^*) \le J(\Omega, u), \quad \forall\, (\Omega, u) \in \bigcup_{\Omega \in \mathcal{B}} (\Omega \times S_E(\Omega)).$$

A natural way of getting an existence result in any optimization problem is to apply the direct method of the calculus of variations which assures the existence of optimal solutions if we have compactness of minimizing sequences and lower semicontinuity (lsc) of functionals. In the case of OSD the problem is that we have no natural topology in the set $\mathcal{O}$ of admissible shapes, while in a universal state space V the topology τ_V is imposed by the state relation (E).

Thus, in the set $\mathcal{O}$ we have to introduce a topology $\tau_{\mathcal{O}}$ so that we are able to prove the following properties.
1° The family $\mathcal{B}$ is $\tau_{\mathcal{O}}$-closed subset of $\mathcal{O}$.
2° The minimizing sequence $(\Omega_n, u_n) \in \bigcup_{\Omega\in\mathcal{B}}(\Omega \times S_E(\Omega))$, $n = 1, 2, ...$ is compact in the sense that it possesses a convergent subsequence in the product topology $\tau_{\mathcal{O}} \times \tau_V$.
3° The multifunction $\mathcal{O} \ni \Omega \mapsto S_E(\Omega) \subset V$ (with nonempty values !) is $\tau_{\mathcal{O}}$-usc (see Section 2); i.e. we have

$$(4.2) \qquad \tau_V - \overline{\lim}_{n\to\infty} S_E(\Omega_n) \subset S_E(\Omega), \;\; \forall \, \{\Omega_n\} \subset \mathcal{B}, \Omega_n \overset{\tau_{\mathcal{O}}}{\to} \Omega.$$

4° The functional

$$J : \bigcup_{\Omega\in\mathcal{B}} (\Omega \times S_E(\Omega)) \ni (\Omega, u) \mapsto J(\Omega, u) \in \bar{\mathbb{R}}$$

is sequentially lsc in the product topology $\tau_{\mathcal{O}} \times \tau_V$; i.e.

$$(4.3) \qquad J(\Omega, u) \leq \underline{\lim}_{n\to\infty} J(\Omega_n, u_n), \; \forall (\Omega_n, u_n) \longrightarrow (\Omega, u).$$

The direct method for OSD problems can be summarized as follows:
Theorem 4.2 If for some topologies $\tau_{\mathcal{O}}$ and τ_V the properties $1^o - 4^o$ above hold, then the problem $(OSD)_E$ has at least one optimal solution.
Remark 4.3 Let us observe that there is a "conflict" of topologies in the existence result above. The stronger are the topologies $\tau_{\mathcal{O}}$ and τ_V, the easier we get the lsc property for functional J in 4°, while the weaker they are, the easier we obtain the compactness properties for minimizing sequences in 2°. Thus, in order to obtain an existence result we have to find a compromise between these two tendencies.

One can make a survey of papers dealing with existence results accordingly to various approaches to the properties $1^o - 4^o$ above. In next sections we describe shortly some of them.

As concerns the properties in 1° and 2° we shall mention only the mapping method and the Dellacherie theorem (see Section 5 and Section 6).
Remark 4.4 Let us observe that the Kuratowski usc property of 3° in the case of well posed problems (for $(OSD)_{PDE}$ or $(OSD)_{VI}$) reduces to some asymptotic problems concerning the dependence of solutions to (bvp) for (E) on domains. For systems described by PDEs of elliptic type there are nice papers [11] [12] on the subject with applications to OSD (see Section 7).

There are also papers (see [31], [17], [22], [6]) concerning lsc-property of 4°, we come back to this subject in the last Section 8.

5.5 MAPPING METHOD AND ITS APPLICATIONS

The mapping method introduced by Murat-Simon ([25]) provides us with both: the set of admissible shapes, gained as images of a fixed set, and a topology on it. Let C be a bounded open domain in $\mathbb{R}^N$ with a boundary ∂C of class $W^{i,\infty}$, $i \geq 1$ and such that $int\,\bar{C} = C$. For $k \geq 1$ we put

$$W^{k,\infty}(\mathbb{R}^N, \mathbb{R}^N) = \{\varphi \,|\, D^\alpha\varphi \in L^\infty(\mathbb{R}^N, \mathbb{R}^N) \text{ for all } \alpha, 0 \leq |\alpha| \leq k\},$$

where derivatives $D^\alpha\varphi$ are understood in the distributional sense, and we supply this set with the norm

$$||\varphi||_{k,\infty} = \operatorname*{ess\,sup}_{x\in\mathbb{R}^N} \left(\sum_{0\leq|\alpha|\leq k} |D^\alpha\varphi|^2_{\mathbb{R}^N} \right)^{1/2}.$$

Now, as admissible shapes we admit the class of open and bounded subsets of $\mathbb{R}^N$ which are isomorphic with C; i.e. we set $\mathcal{O}_{ad} = \mathcal{O}^{k,\infty}$, where

$$\mathcal{O}^{k,\infty} = \{\Omega \mid \Omega = T(C),\ T \in \mathcal{F}^{k,\infty}\},$$

$$\mathcal{F}^{k,\infty} = \{T\colon \mathbb{R}^N \to \mathbb{R}^N \mid T \text{ is bijective and } T-I, T^{-1}-I \in W^{k,\infty}(\mathbb{R}^N, \mathbb{R}^N)\}.$$

As Murat and Simon observed in [25] the function

$$\delta_{k,\infty}(\Omega_1,\Omega_2) = \inf_{T\in\mathcal{F}^{k,\infty}, T(\Omega_1)=\Omega_2} \left(||T-I||_{k,\infty} + ||T^{-1}-I||_{k,\infty}\right).$$

defines a pseudodistance in $\mathcal{O}^{k,\infty}$ (as it does not satisfy the triangle identity - see section 2.4 in [25] for a justification) but the following propositions hold.

Proposition 5.1 For $k \geq 1$ we have:
(*i*) There exists a positive constant μ_k such that $d_{k,\infty} = (\delta_{k,\infty} \wedge \mu_k)^{1/2}$ defines a metric on $\mathcal{O}^{k,\infty}$.
(*ii*) The space $(\mathcal{O}^{k,\infty}, d_{k,\infty})$ is a complete metric space.
(*iii*) If $k \geq 2$, then the injection from $\mathcal{O}^{k,\infty}$ into $\mathcal{O}^{k-1,\infty}$ is compact. More precisely, if $k \geq 2$ and $\mathcal{B}$ is a bounded (in $\delta_{k,\infty}$), closed subset of $\mathcal{O}^{k,\infty}$, then for any sequence $\{\Omega_m\} \subset \mathcal{B}$, there exist a subsequence $\{\Omega_{m_\nu}\}$ of $\{\Omega_m\}$ and a set $\Omega \in \mathcal{B}$ such that $\Omega_{m_\nu} \to \Omega$ in $\mathcal{O}^{k-1,\infty}$.

Proposition 5.2 Let be $k \geq 1$. Then the implications below hold.

(*i*)
$$\Omega_n \xrightarrow{d_{k,\infty}} \Omega \implies 1_{\Omega_n} \xrightarrow{s-L^2(\mathbb{R}^N)} 1_\Omega.$$

(*ii*)
$$\Omega_n \xrightarrow{d_{k,\infty}} \Omega \implies \Omega_n \xrightarrow{d_{H^c}} \Omega.$$

(*iii*)
$$\Omega_n \xrightarrow{d_{H^c}} \Omega \implies \forall\, G \subset\subset \Omega,\ \exists\, n_G \in \mathbb{N} \text{ such that } G \subset \Omega_n\ \forall n \geq n_G.$$

Observe that Proposition 5.1 enables us to implement immediately the point 1° and partially 2° of Theorem 4.2, while the remaining points are much more involved. We formulate also remarks:

Remark 5.3 The condition (ii) in Prop. 5.2 means that the Hausdorff complementary topology is weaker than that given by the distance $d_{k,\infty}$.

Remark 5.4 The condition (iii) of Prop. 5.2 assures the important (for lsc of J in 4°) property of covering of "compacts" (see Section 8).

The mapping method was successfuly applied by Murat-Simon to $(OSD)_{PDE}$ with $A = -\Delta$ for Dirichlet and Neumann problems and for quadratic cost functionals of type J_2 and J_3 from Section 4. In order to pass from changing with Ω state spaces to the universal one they used a prolongation operator (see also Chenais [8]) for Dirichlet problems and the change of variable ($x = T(X), T : C \mapsto \Omega, T \in \mathcal{F}^{k,\infty}$) for Neumann problems (see Ch.5 and Ch.6 in [25]).

Next, this method was also used for $(OSD)_{VI}$ (with $\Phi = 0$) by Liu-Rubio (see [21]) who then applied the existence results to some electrochemical machining problems. Roughly speaking they looked for an optimal shape of the cell forming the cathode in order to obtain the desired shape of the metallic anode in the quasi-steady or evolutionary electrochemical proccesses.

Some existence results for $(OSD)_{HVI}$ were obtained by S. Migórski and the present author (see [14], [15]) with applications to some linear and nonlinear elasticity problems. The results were generalized then by L. Gasiński ([18]) for $(OSD)_{HVI_p}$. For the lack of space we quote here as an example only a "draft" version of an existence result (for the precise formulation see Thms.4.1 and 4.2. in [15]).

Theorem 5.5 Under the standard continuity and coercivity assumptions concerning the bilinear form a which are uniform with respect to $\Omega \in \mathcal{B}$ we assume:
(i) $\mathcal{B}$ is a closed and bounded (in $\delta_{k,\infty}$) subset of $\mathcal{O}_{k,\infty}$ with $k \geq 2$, and $f \in L^2(\mathbb{R}^N)$,
(ii) The superpotential j comes from a function β (as in Remark 3.5) which satisfies (3.1) and "ultimately increases" (see [26])
(iii) The cost functional J given by (4.1) with $L_0 = \mathcal{P} = 0$ is lsc with respect to the "local convergence" of Serrin.
Then there exists at least one solution to $(OSD)_{HVI}$ problem provided the set K is independent of T; i.e. $K = T(K(\Omega)) \subset H^1(C)$ for all $T \in \mathcal{F}^{k,\infty}$, or we have the Mosco convergence:

$$T_n(K(\Omega_n)) \xrightarrow{M} T(K(\Omega)) \text{ in } H^1(C),$$

for all $T_n \to T, T_n^{-1} \to T^{-1}$ in $[W^{1,\infty}(\mathbb{R}^N)]^N$ such that $\Omega_n = T_n(C)$ and $\Omega = T(C)$.

As concerns the proof of this theorem we confine ourselves to the remark that the most difficult part of the proof, based on Theorem 4.2, concerns the point 3°, that is the usc (or "closed graph") property. In our case it can be stated as follows (see Proposition 3.1 in [15]):

Proposition 5.6 Under the assumptions of Theorem 5.5 the map $\mathcal{B} \ni \Omega \mapsto S(\Omega) \subset V(\Omega)$ has a closed graph in the following sense: if $\Omega_m, \Omega_0 \in \mathcal{B}$, $\Omega_m \to \Omega_0$ in $\mathcal{O}^{k,\infty}$, $u_m \in S(\Omega_m)$, $\widehat{u}_m = u_m \circ T_m$, $\widehat{u}_m \to u^*$ weakly in $H^1(C)$, then $u^* = u_0 \circ T_0$ for some $u_0 \in S(\Omega_0)$, where $\Omega_m = T_m(C)$ and $\Omega_0 = T_0(C)$.

5.6 SOME REMARKS ON OTHER METHODS FOR OSD PROBLEMS

An obvious drawback of the mapping method is that it deals with the limited classes of admissible shapes, for instance, no "extra holes" are admitted. So this method is not applicable to the case of "highly perforated" domains whose boundaries oscillate strongly. This is important case of the homogenization ("control in coefficients") related e.g. to the problem of finding the "best arrangement" in a conductor of the two (or more) materials (given in a fixed proportion) with different conductivities (see e.g. [34]).

Moreover, in the case of periodically perforated domains there are examples (see [9] quoted also in [29], [11],...) showing that for a sequence of Dirichlet problems in the limit problem appears an "extra term". Thus, the usc property of the mapping $\Omega \mapsto S_E(\Omega)$ in point 3° of Theorem 4.1 is violated (as the limit of solutions to the Dirichlet problems is a solution to the different kind of equation) and in consequence there are no solutions to some corresponding OSD problems (see also relaxation below).

In order to avoid such oscillations of the boundaries, as in the case of perforated domains, Chenais [8], Pironneau [29] required the so called ε - cone property accordingly to the definition:

Definition 6.1 We say that Ω has ε - cone property if for every $x \in \Gamma = \partial\Omega$ there exists a direction $\xi(x)$ such that $C(\varepsilon, \xi(x), y) \subset \Omega$ for all $y \in B(x,\varepsilon) \cap \Omega$, where $C(\varepsilon, \xi(x), y)$ denotes the half-cone of angle ε, direction $\xi(x)$ and vertex y, intersected with $B(x,\varepsilon)$ (the ball centered at x and with radius ε).

As admissible shapes they considered the following class of sets.

$$(6.1) \quad \mathcal{O}_\varepsilon = \{\Omega \subset \mathbb{R}^N |\ D \subset \Omega \subset C,\ \Omega \text{ is open and has } \varepsilon - \text{cone property}\},$$

where D and C are given sets, C being bounded.

Using results of Chenais (Th.IV.1 in [8]) and Pironneau (Theorems 2 and 5 in [29]) we may formulate theorem:

Theorem 6.2 For a given function $u_d \in L^2(D)$ the two OSD problems below (for Dirichlet and Neumann problems):

$$(OSD)_{DP} \qquad \min_{\Omega \in \mathcal{O}_\varepsilon} \{\int_D |\nabla u - u_d|^2\, dx\ ;\ -\Delta u = f \text{ in } \Omega,\ u|_\Gamma = 0\},$$

$$(OSD)_{NP} \qquad \min_{\Omega \in \mathcal{O}_\varepsilon} \{\int_D |\nabla u - u_d|^2\, dx\ ;\ u - \Delta u = f \text{ in } \Omega,\ \frac{\partial u}{\partial n}|_\Gamma = 0\},$$

admit at least one solution.

The proof can be obtained by the direct method (Thm 4.2) using the Hausdorff complementary topology given by metric d_{H^c}, the Dellacherie (see Thm 1 in Ch.3 in [29]) compactness result for $\mathcal{O}_\varepsilon$, and some results (see Prop. 1, Ch. 3 in [29]) concerning the continuous dependence (in $s - L^2(D)$ topology) of solutions to the Dirichlet, respectively the Neumann problems on domains.

If one would like to escape from any constraints on the boundaries, then in order to get an existence result it is necessary to impose other type constraints, for instance: (i) on the volume (e.g. Thm 3 in [29]), (ii) on the perimeter (e.g. [32] - transmission problem), (iii) on capacity (e.g. [4] in the case of "quasi open" sets).

Other classes of admissible shapes are given by "hypographs" [32]. One can parametrize them by functions and then deal with them by methods of calculus of variations or of optimal control theory ([19],[30],...).

5.7 RELAXATION IN OSD PROBLEMS

We follow an approach of Buttazzo, Dal Maso, Mosco ([5],[6], [13],[7], see also [3]), although the relaxation in OSD was considered earlier by Murat-Tartar, Kohn-Strang ([20]), Vogelius, and others...

Assume that linear operator $A, f \in H^{-1}(\Omega)$ and the family of admissible shapes $\mathcal{A}(\Omega_0)$ are defined as in Preliminaries, the function $L : \Omega_0 \times \mathbb{R} \mapsto \mathbb{R}$ is of Caratheodory type and satisfies some growth condition (making the integral functional below continuous in $L^2(\Omega_0)$). We consider the OSD problem:

$$(OSD)_{DP} \qquad \min_{\Omega \in \mathcal{A}(\Omega_0)} \int_\Omega L(x, u_\Omega(x))\, dx,$$

where u_Ω is the solution to Dirichlet problem

$$(DP) \qquad Au = f \text{ in } \Omega, \quad u \in H_0^1(\Omega)$$

extended by zero on $\Omega_0 \setminus \Omega$. Sometimes we shall write it down also as $u_\Omega \in S_{DP}(\Omega)$ although in this case the solution set consists of one element.

It is well known (see e.g. Ex. 4.3 in [5]) that $(OSD)_{DP}$ problem in general does not possess a solution. The reason is that even the sequence $\{u_{\Omega_n}\}$ corresponding to a minimizing sequence $\{\Omega_n\}$ has a limit point $u \in H_0^1(\Omega_0)$ in the $w - H_0^1(\Omega_0)$ - topology, the u need not be the solution to (DP) for any open set $\Omega \in \mathcal{A}(\Omega_0)$ (i.e. there is no open set Ω such that $u \in S_{DP}(\Omega)$).

Moreover, (as we have already mentioned) some counterexamples ([9], [10],...) show that the limit function can be the solution to the different kind of equation or, in other words, in the limit equation may appear an "extra term" (e.g. for $A = -\Delta$ the limit function u is the unique solution to equation $-\Delta u + cu = f$ with some $c > 0$, provided the sequence $\{\Omega_n\}$ is obtained by appropriate regular periodic perforations).

Hence much wider class of problems arises; e.g.

- What is the asymptotic behaviour of solutions to (DP) with respect to Ω changing in $\mathcal{A}(\Omega_0)$?
- Under what conditions the limit points of sequences of solutions $\{u_{\Omega_n}\}$ are still solutions to (DP) ? (this means exactly the "usc property" in point 3° of Thm 4.2 for $\{S_{DP}(\Omega_n)\}$).
- Is it possible to identify and characterize a new class of differential equations which are satisfied by the all limit points of sequences $\{u_{\Omega_n}\}$?

There is an extensive literature on the subject. Specially Buttazzo, Dal Maso, Mosco, Murat ([5], [6], [7], [13],... and their collaborators have obtained many interesting results in this direction. We mention some of them underlining the applications to OSD problems.

Namely, denoting by $\mathcal{M}_0(\Omega_0)$ the family of all Borel measures on Ω_0 which are nonnegative, possibly infinte and vanishing on all subsets of the harmonic capacity zero, we may describe roughly the relaxed Dirichlet problems in the following steps.

1. From a given minimizing sequence $\{\Omega_n\}$ to $(OSD)_{DP}$ problem we can choose by a diagonal procedure some subsequence (we keep the same notation) in such a way that corresponding solutions $\{u_{\Omega_n}\}$ to (DP) (with arbitrary $f \in H^{-1}(\Omega_0)$) converge (in the weak topology of $H_0^1(\Omega_0)$) to a limit point $u \in H_0^1(\Omega_0)$.

2. In the case there is no $\Omega \in \mathcal{A}(\Omega_0)$ such that $u \in S_{DP}(\Omega)$, it is possible (see [6]) to find a "relaxed solution" to $(OSD)_{DP}$. This is a measure $\mu \in \mathcal{M}_0(\Omega_0)$ which depends only on $\{\Omega_n\}$, on operator A and on functional J, and is such that the function u is the solution of the new equation:

   $$Au + \mu u = f, \quad u \in X_\mu =: \; H_0^1(\Omega_0) \cap L_\mu^2(\Omega_0). \tag{7.1}$$

   More precisely we have

   $$< Au, v > + \int_{\Omega_0} uv \; d\mu \; = < f, v > \quad \forall \, v \in X_\mu, \tag{7.2}$$

   where $< \cdot, \cdot >$ denotes the duality pairing between $H_0^1(\Omega_0)$ and $H^{-1}(\Omega_0)$.

3. "Classical" solution $\Omega \in \mathcal{A}(\Omega_0)$ corresponds to measure μ_Ω defined by the formula:

   $$\mu_\Omega(E) = \begin{cases} 0, & \text{if } \operatorname{cap}(E \setminus \Omega) = 0, \\ +\infty, & \text{if } \operatorname{cap}(E \setminus \Omega) > 0. \end{cases}$$

4. We have a "density" result: for every $\mu \in \mathcal{M}_0(\Omega_0)$ there exists a sequence $\{\Omega_n\}$ in $\mathcal{A}(\Omega_0)$ such that the sequence of corresponding solutions $\{u_{\Omega_n}\}$ to (DP) converges (in the weak topology of $H_0^1(\Omega_0)$) to a function $u \in H_0^1(\Omega_0)$ and $u_\mu := u$ is the solution to (7.1) with the given μ.

5. Accordingly to the general theory of relaxation ([10]) we can formulate the relaxed (OSD) problem as:

   $$(ROSD)_{DP}, \qquad \min_{\mu \in \mathcal{M}_0(\Omega_0)} \bar{J}(\mu),$$

   where the functional $\bar{J}$, the sequential lsc regularization of J in the so-called γ topology of $\mathcal{M}_0(\Omega_0)$ ($J(\Omega, u_\Omega) = \int_\Omega L(x, u_\Omega(x))dx$), is given by

   $$\bar{J}(\mu) = \inf\{\liminf_{n\to\infty} J(\Omega_n, u_{\Omega_n}) \mid \Omega_n \in \mathcal{A}(\Omega_0), \; \mu_{\Omega_n} \xrightarrow{\gamma} \mu\}.$$

6. Moreover, the problems $(OSD)_{DP}$ and $(ROSD)_{DP}$ have the same values; i.e. we have

   $$\min_{\mu \in \mathcal{M}_0(\Omega_0)} \bar{J}(\mu) = \inf_{\Omega \in \mathcal{A}(\Omega_0)} J(\Omega, u_\Omega).$$

Remark 7.4 Another approach to the relaxation is given in [3]. It consists in completion of the metric space obtained from the set of admissible shapes by the transport of the metric corresponding to the Γ convergence in the set of all lower semicontinuous and equicoercive functionals on the underlying vector space of states.

5.8 LOWER SEMICONTINUITY OF FUNCTIONALS IN OSD PROBLEMS

As far as concerns the lsc property (see 4° in Thm 4.2) for the cost functionals which depend explicitely on the domain of integration we refer to already classic paper of Serrin [31], where the notion of local convergence was introduced and then to papers by Fujii [17], Buttazzo, Dal Maso, Garroni, Malusa [6] and Liu-Rubio [22], who adapted and extended the results to OSD problems.

Here we confine ourselves to quote from Liu-Rubio [22] a general abstract theorem on the subject. To this end we need first some notions:

Definition 8.1 Given a sequence $\{\Omega_n\}_{n=1}^{\infty}$ of open sets in $\mathbb{R}^N$ we say that it has the property G iff for every open set $G \subset\subset \Omega_\infty$ there exists $n_G > 0$ such that $G \subset \Omega_n$ for every $n > n_G$.

For a sequence of open sets contained in a bounded set C we have the implication (see [8], [17],[22]):

$$\Omega_n \xrightarrow{d_{H^c}} \Omega_\infty \implies \{\Omega_n\}_{n=1}^{\infty} \text{ has the property G.}$$

Definition 8.2 Assume $\{\Omega_n\}_{n=1}^{\infty}$ is a sequence of bounded sets in $\mathbb{R}^N$ with the property G and such that $u_n \in H^1(\Omega_n), n = 1, 2, ..., \infty$. We say the sequence $\{u_n\}$ *locally weakly (strongly)* converges to u_∞ in $H^1(\Omega_\infty)$ and we write it down as

$$u_n \xrightarrow{L} u_\infty \text{ weakly (strongly) in } H^1(\Omega_\infty)$$

iff the sequence $\{||u_n||_{H^1(\Omega_n)}\}$ is bounded and for any open set $G \subset\subset \Omega_\infty$ there is $n_G > 0$ such that $u_n \in H^1(G)$ for $n > n_G$ and $||u_n - u_\infty||_{L^2(G)} \longrightarrow 0$ ($||u_n - u_\infty||_{H^1(G)} \longrightarrow 0$).

Theorem 8.3 Let $\{\Omega_n\}_{n=1}^{\infty}$ be a sequence of bounded open sets in $\mathbb{R}^N$ with the property G and $u_n \in H^1(\Omega_n), n = 1, 2, ..., \infty$. If the nonnegative measurable function $L : \mathbb{R}^N \times \mathbb{R} \times \mathbb{R}^N \mapsto \mathbb{R}$ is such that for every open set $G \subset\subset \Omega_\infty$ the functional $u \mapsto \int_G L(x, u, \nabla u)\, dx$ is weakly lsc in $H^1(G)$, then

$$\int_{\Omega_\infty} L(x, u_\infty, \nabla u_\infty)\, dx \le \underline{\lim}_{n\to\infty} \int_{\Omega_n} L(x, u_n, \nabla u_n)\, dx,$$

provided $u_n \xrightarrow{L} u_\infty$ weakly in $H^1(\Omega_\infty)$.

For the conditions which assure the above assumption that the functional $u \mapsto \int_G L(x, u, \nabla u)\, dx$ is weakly lsc in $H^1(G)$ one can consult for instance the monograph [10].

References

[1] M. Belloni, G. Buttazzo and L. Freddi, *Completion by Gamma-convergence for optimal control problems,* Ann. Fac. Sci. Toulouse, 2 (1993), 149-162.

[2] *Partial Differential Equation Methods in Control and Shape Analysis,* G. Da Prato and J.-P. Zolesio Eds.,Marcel Dekker Inc., New York, 1997.

[3] R. Butt, *Optimal shape design of systems governed by variational inequalities,* Math. Comput. Modelling, 25 (10) (1997), 47-55.

[4] G. Buttazzo and G. Dal Maso, *An existence result for a class of shape optimization problems,* Arch. Rat. Mech. Anal., 122 (1993), 183-195.

[5] G. Buttazzo and G. Dal Maso, *Shape optimization for Dirichlet problems: relaxed formulation and optimality conditions,* Appl. Math. Optim., 23 (1991), 17-49.

[6] G. Buttazzo, G. Dal Maso, A. Garroni and A. Malusa, *On the relaxed formulation of some shape optimization problems,* Adv. Math. Sci. Appl., to appear.

[7] G. Buttazzo, G. Dal Maso and U. Mosco, *Asymptotic behaviour for Dirichlet problems in domains bounded by thin layers,* in: Partial Differential equations and Calculus of Variations, Essays in Honor of Ennio De Giorgi, 193-249, Birkhäuser, Boston, 1989.

[8] D. Chenais, *On the existence of a solution in a domain identification problem,* J. Math. Anal. Appl., 52 (1975), 189-219.

[9] D. Cioranescu and F. Murat, *Un terme étrange venu d'ailleurs, I and II,* Nonlinear Partial Differential Equations and Their Applications. Collége de France Seminar. Vol. II, 98-138, and Vol. III, 154-178, Res. Notes in Math., 60 and 70, Pitman, London, 1982 and 1983.

[10] G. Dal Maso, *An Introduction to Γ-convergence,* Birkhäuser, Boston, 1993.

[11] G. Dal Maso, *Comportamento asintotico delle soluzioni di problemi di Dirichlet,* Preprint 10/97/M, SISSA.

[12] G. Dal Maso and F. Murat, *Dirichlet problems in perforated domains for homogeneous monotone operators on* H_0^1, in: Proc. Conference *Calculus of variations, homogenization and continuum mechanics,* June 21-25, 1993, Marseille, Bouchitte et al. Eds., World Scientific, 1994, 177-202.

[13] G. Dal Maso and U. Mosco, *Wiener's criterion and Γ-convergence,* Appl. Math. Optim., 15 (1987), 15-63.

[14] Z. Denkowski and S. Migórski, *Optimal Shape Design for Elliptic Hemivariational Inequalities in Nonlinear Elasticity,* Proc. *12th Conference on Variational Calculus, Optimal Control and Applications,* Trassenheide, Germany, September 23-27, ISNM, Vol. 124, 1998, Birkhäuser-Verlag, 1999, 31-40

[15] Z. Denkowski and S. Migórski, *Optimal shape design problems for a class of systems described by hemivariational inequalities,* Journal of Global Optimization, Vol.12, No. 1 (1998), 37-59.

[16] G. Fichera, *Existence theorems in elasticity,* in: Handbuch der Physik, Band VIa/2, C. Truesdell Ed., Springer-Verlag, Berlin, 347-389, 1972.

[17] N. Fujii, *Lower semicontinuity in domain optimization problems,* J. Optim. Theory Appl., Vol. 59, No. 3 (1988), 407-422.

[18] L. Gasiński, *Optimal shape design problems for a class of systems described by parabolic hemivariational inequality,* J. Global Optim., Vol. 12 (1998), 299-317

[19] J. Haslinger and P. Neittaanmäki, *On the existence of optimal shapes in contact problems,* Numer. Funct. Anal. Optim., 7, (1984-85), 107-124.

[20] R. V. Kohn and G. Strang, *Optimal design and relaxation of variational problems, I, II, III,* Comm. Pure Appl. Math. 39 (1986), 113-137, 139-182, 353-377.

[21] W. B. Liu and J. E. Rubio, *Optimal Shape Design for Systems Governed by Variational Inequalities, Part 1: Existence Theory for Elliptic Case, Part 2: Existence Theory for Evolution Case,* J. Optim. Th. Appl., 69 (1991), 351-371, 373-396.

[22] W. B. Liu and J. E. Rubio, *Local convergences and optimal shape design,* SIAM J. Control Optim., Vol. 30, No. 1 (1992), 49-62.

[23] M. Miettinen, M. M. Mäkelä and J. Haslinger, *On numerical solution of hemivariational inequalities by nonsmooth optimization methods,* J. Global Optim., 6 (1995), 401-425.

[24] J. J. Moreau, *La notion de superpotentiel et les liaisons unilatérales en élastostatique,* C. R. Acad. Sci. Paris, 167A (1968), 954-957.

[25] F. Murat and J. Simon, *Sur le Controle par un Domaine Geometrique,* Preprint no. 76015, University of Paris 6, 1976.

[26] Z. Naniewicz and P. D. Panagiotopoulos, *Mathematical Theory of Hemivariational Inequalities,* Dekker, New York, 1995.

[27] P. D. Panagiotopoulos, *Nonconvex superpotentials in the sense of F. H. Clarke and applications,* Mech. Res. Commun., 8 (1981), 335-340.

[28] P. D. Panagiotopoulos, *Coercive and semicoercive hemivariational inequalities,* Nonlinear Anal. Theory Methods Appl., 16 (1991), 209–231.

[29] O. Pironneau, *Optimal Shape Design for Elliptic Systems,* Springer-Verlag, New York, 1984.

[30] K. Salmenjoki, *On Numerical Methods for Shape Design Problems,* Thesis, University of Jyväskylä, Finland, 1991.

[31] J. Serrin, *On the definition and properties of certain variational integrals,* Trans. Amer. Math. Soc., 101 (1961), 139-167.

[32] J. Sokołowski and J.-P. Zolesio, *Introduction to Shape Optimization,* Springer-Verlag, Berlin, Heidelberg, 1992.

[33] V. Sverak, *On optimal shape design,* J. Math. Pures Appl., 72 (1993), 537-551.

[34] L. Tartar, *Remarks on optimal design problems,* in: Proc. Conference *Calculus of variations, homogenization and continuum mechanics,* June 21-25, 1993, Marseille, Bouchitte et al. Eds., World Scientific, 1994, 279-296.

6 DUALITY IN NONCONVEX FINITE DEFORMATION THEORY: A SURVEY AND UNIFIED APPROACH

David Yang Gao

Department of Mathematics
Virginia Polytechnic Institute & State University
Blacksburg, VA 24061, USA

gao@math.vt.edu

Abstract: This paper presents a brief survey and a unified approach to the complementary-dual variational extremum principles in nonconvex finite deformation theory. By using an abstract deformation operator and its tangent decomposition, the well-known classical complementary energy principles can be written in a general form, and their extremum properties are clarified. Based on a new pure complementary energy principle (involving the stress only), it is shown that the nonlinear equilibrium equation in finite deformation problems can be transformed into a tensor equation. Hence a general analytic solution for 3-dimensional large deformation problems can be constructed. The properties of this general solution are characterized by a triality extremum principle.

6.1 INTRODUCTION

Nonconvex optimization and nonsmooth mechanics have experienced significant development in the last decade. Many problems arising in finite deformation theory, hysteresis and phase transitions, composite materials and smart structures, contact mechanics and inequality variational problems, post-bifurcation and stability analysis, require the considerations of nonconvexity and nondifferentiablity for their total potentials and mechanical modelling. Several monographs have documented the basic theory, methods, algorithms and applications of nonsmooth mechanics (cf. e.g. Panagiotopoulos, 1985, 1993; Dem'yanov-Stavroulakis-Polyakova-Panagiotopoulos, 1996). Traditional analytic methods for solving nonsmooth, nonconvex large deformation problems are usually very difficult. Direct variational approaches can only provide

R.P. Gilbert et al. (eds.), From Convexity to Nonconvexity, 67–83.

one-side approximations to the problems. However, the complementary-dual variational methods have beautiful theoretical properties, powerful alternative performance and pleasing relationships to classical energy principles in nonlinear boundary value problems.

Duality in nonlinear optimization falls principally into two categories: (i) the Fenchel-Rockafellar duality; (ii) the Lagrangian duality. The first duality was originally developed in convex analysis (cf. e.g. Rockafellar, 1967). Most of complementary energy principles in linear elasticity belong to this category. The second duality can be used in general nonconvex variational problems (cf. e.g. Toland, 1979; Auchmuty, 1982) as well as constrained global optimization (see Dür-Horst, 1997). In convex systems, these two dualities are usually equivalent to each other (see Ekeland-Temam, 1976) and the symmetry between primal and dual problems is amazingly beautiful (see, Strang, 1986; Sewell, 1987). However, in geometrically nonlinear systems, such a symmetry is lost because of the nonlinearity of the finite deformation operator (Gao-Strang, 1989). If the system is nonconvex, there exists a dual gap between the primal and dual problems. In finite deformation theory, the well-known Levinson-Zubov principle is actually a Fenchel-Rockafellar dual problem, which is not valid unless the stored energy is convex in deformation gradient.

The study of dual variational methods in finite deformation theory has a long history. The well-known Hellinger-Reissner principle was first proposed in 1914. Since the boundary condition was clarified by Reissner in 1953, the complementary energy principles and mixed variational methods (i.e. Lagrange duality) have been studied extensively during the last fifty years (cf. e.g. Koiter, 1976; Nemat-Nasser, 1977; Stumpf, 1978; Lee & Shield, 1980; Bufler, 1983; Atluri, 1984; Ogden, 1984 and much more). Many monographs have been published in applied mathematics (Arthurs, 1980; Sewell, 1987), continuum mechanics (Washizu, 1968; Oden-Reddy, 1983) and linear dynamics (Tabarrok-Rimrott, 1994). Since the Hellinger-Reissner principle involves both the second Piola-Kirchhoff stress $\mathbf{T}$ and the displacement $\mathbf{u}$, it is not considered as a pure complementary energy principle. In order to apply this principle, one needs to understand the property of the critical points of the Hellinger-Reissner energy. But until recently this property was not adequately understood. In finite deformation theory, the existence of a pure complementary energy principle dual to the potential variational problem has been argued for many years. In 1989, Gao and Strang discovered that in geometrically nonlinear systems, there exists a so-called complementary gap function between the total potential energy and the Fenchel-Rockafellar dual functional. They proved that if this gap function is positive on equilibrium admissible space, the generalized Hellinger-Reissner energy $L(\mathbf{u}, \mathbf{T})$ is a saddle functional. In this case, the total potential is convex and there exists only one dual problem. However, if this gap function is negative, the total potential is nonconvex and the system has two dual problems (Gao, 1997a). A triality extremum theory in nonconvex problems was discovered recently and a pure complementary energy in finite elasticity is proposed (Gao, 1997, 1998a,b, 1999 a, b).

Duality algorithms and finite element approximations of mixed variational problems have received sustained attention again in nonconvex optimization (Auchmuty, 1989), nonsmooth plasticity (Gao, 1988, 1996b; Han-Reddy, 1995; Repin-Xanthis, 1996) and in computational mechanics (cf. e.g. Pian-Tong, 1980; Felippa, 1987; Gallagher, 1993; Wataru-Atluri, 1995). In the past decade the primal-dual algorithms have emerged as the most important and useful algorithms from linear programming (see, Wright, 1996). In nonconvex finite deformation theory, because the convexity condition of the Hellinger-Reissner energy was an open problem, the major difficulties encountered in mixed finite element approaches centre around the algorithm and the construction of finite element bases which are compatible in the sense that they give rise to stable and convergent methods.

The theory of variational inequalities is a relatively young discipline. Apparently, one of the main bases for its development was the paper by Fichera (1964) on the Signorini problem in elasticity. The dual variational inequality was first studied by Mosco (1972). Since then many papers and monographs have been published in geometrically linear problems (cf. e.g. Haslinger-Panagiotopoulos, 1984; Hlavacek-Haslinger-Necas-Lovisek, 1988; Klarbring-Mikelic-Shillor, 1990; Reddy, 1992) and finite deformation structural analysis (cf. e.g. Yau-Gao, 1992; Gao, 1996). Actually, by introducing a so-called super-potential, the variational inequality in conservative systems can usually be written as a nonsmooth variational problem (cf. e.g. Panagiotopoulos, 1985 and 1993). Hence, many problems can be studied within the framework of nonsmooth optimization.

The purpose of this paper is to provide a unified approach to the complementar dual variational principles in nonconvex finite deformation theory. By using an abstract deformation operator and a super-potential, a framework in fully nonlinear boundary value problems is presented, and the well-known classical complementary energy principles can be written in a general form. A new pure complementary energy principle (involving only the stress tensors) is proposed. It is shown that by using this principle, the nonlinear equilibrium equation in finite deformation problems can be transformed into a tensor equation. Hence a general analytic solution for 3-dimensional large deformation problems can be constructed. The properties of this general solution are characterized by a triality extremum principle proposed recently.

6.2 FRAMEWORK AND ABSTRACT BOUNDARY VALUE PROBLEM

Let $\Omega \subset \mathbb{R}^n$ be an open, simply connected, bounded domain with boundary $\partial\Omega = \Gamma_t \cup \Gamma_u$, $\Gamma_t \cap \Gamma_u = \emptyset$. On Γ_t and Γ_u, the surface traction and the displacement are prescribed, respectively. Let $\mathcal{U}$ be a general displacement space, $\mathcal{F}$ a general force space. The bilinear form $\langle *, * \rangle : \mathcal{U} \times \mathcal{F} \to \mathbb{R}$ puts the spaces $\mathcal{U}$ and $\mathcal{F}$ in duality. If $\bar{\mathbf{f}} \in \mathcal{F}$ is specified as the body force $\bar{\mathbf{b}}$ in Ω, and the surface traction $\bar{\mathbf{t}}$ on Γ, then this bilinear form

$$\langle \mathbf{u}, \bar{\mathbf{f}} \rangle = \int_\Omega \mathbf{u} \cdot \bar{\mathbf{b}} \, d\Omega + \int_\Gamma \mathbf{u} \cdot \bar{\mathbf{t}} \, d\Gamma$$

represents the *external work*. We use Lin to denote the linear space of all second-order tensors. For a given admissible deformation $\mathbf{u} \in \mathcal{U}$, the deformation gradient $\mathbf{F} = \mathbf{I} + \nabla\mathbf{u}$ should be a nonsingular, two-point tensor such that $\mathbf{F} = \mathbf{R} \cdot \mathbf{U} = \mathbf{V} \cdot \mathbf{R}$, where $\mathbf{U}$ and $\mathbf{V}$ are the right and left stretch tensors, $\mathbf{R}$ is a rotation tensor, $\mathbf{R}^t = \mathbf{R}^{-1}$.

The generalized strain space $\mathcal{E}$ and its conjugate stress space $\mathcal{S}$ should be symmetrical subspaces of Lin. The second bilinear form

$$\langle \mathbf{e}; \mathbf{s} \rangle = \int_\Omega \mathbf{e} : \mathbf{s} \, d\Omega = \int_\Omega e_{ij} s_{ij} \, d\Omega,$$

puts $\mathcal{E}$ and $\mathcal{S}$ in duality.

For hyperelasticity, we assume that the stored strain energy W is a convex, lower semicontinuous (l.s.c) function of the generalized strain tensor $\mathbf{e}$ such that the *constitutive equation* can be described in subdifferential form:

$$\mathbf{s} \in \partial W(\mathbf{e}) := \{\mathbf{s} \in \mathcal{S} | \;\; W(\boldsymbol{\epsilon}) - W(\mathbf{e}) \geq \langle \boldsymbol{\epsilon} - \mathbf{e}; \mathbf{s} \rangle \;\; \forall \boldsymbol{\epsilon} \in \mathcal{E}\}. \tag{6.1}$$

If W is Gâteaux differentiable, $\partial W(\mathbf{e}) = \{DW(\mathbf{e})\}$, and $DW(\mathbf{e}) = \partial W / \partial \mathbf{e}$ stands for the Gâteaux-derivative of W at $\mathbf{e}$. The complementary energy can be obtained by the Legendre-Fenchel transformation:

$$W^*(\mathbf{s}) := \sup_{\mathbf{e}} \{\mathbf{s} : \mathbf{e} - W(\mathbf{e})\}, \tag{6.2}$$

which is always convex and the following equivalent relations hold (cf. Ekeland & Temam, 1976):

$$\mathbf{s} \in \partial W(\mathbf{e}) \;\Leftrightarrow\; \mathbf{e} \in \partial W^*(\mathbf{s}) \;\Leftrightarrow\; W(\mathbf{e}) + W^*(\mathbf{s}) = \mathbf{s} : \mathbf{e}. \tag{6.3}$$

Let $\Lambda : \mathcal{U} \to \mathcal{E}$ be a finite deformation operator, the generalized strain tensor $\mathbf{e} \in \mathcal{E}$ is then determined by the *geometric equation*

$$\mathbf{e} = \Lambda \mathbf{u}. \tag{6.4}$$

According to (Gao & Strang, 1989a), the following decomposition law is very important in finite deformation theory.

$$\Lambda = \Lambda_t + \Lambda_n, \tag{6.5}$$

where $\Lambda_t = D\Lambda$ is the Gâteaux derivative of the operator Λ at $\bar{\mathbf{u}}$ in the direction $\mathbf{u} \in \mathcal{U}$, defined by

$$\Lambda_t(\bar{\mathbf{u}})\mathbf{u} = \delta \mathbf{e}(\bar{\mathbf{u}}; \mathbf{u}) = \lim_{\theta \to 0^+} \frac{\mathbf{e}(\bar{\mathbf{u}} + \theta \mathbf{u}) - \mathbf{e}(\bar{\mathbf{u}})}{\theta}. \tag{6.6}$$

$\Lambda_n = \Lambda - \Lambda_t$ is the complementary operator of Λ_t.

Let $\mathcal{U}_a$ be a kinetically admissible field. In this paper, we simply let

$$\mathcal{U}_a = \{\mathbf{u} \in \mathcal{U} | \;\; \mathbf{u} = 0 \;\text{ on } \Gamma_u\}. \tag{6.7}$$

Using the Gauss-Green law, the virtual work principle can be written as

$$\langle \delta \mathbf{e}(\mathbf{u};\delta\mathbf{u});\mathbf{s}\rangle = \langle \Lambda_t(\mathbf{u})\delta\mathbf{u};\mathbf{s}\rangle = \langle \delta\mathbf{u}, \Lambda_t^*(\mathbf{u})\mathbf{s}\rangle = \langle \delta\mathbf{u}, \bar{\mathbf{f}}\rangle \ \ \forall \delta\mathbf{u} \in \mathcal{U}_a, \tag{6.8}$$

where $\Lambda_t^* : \mathcal{S} \to \mathcal{F}$ is the adjoint operator of Λ_t. For a given $\bar{\mathbf{f}} \in \mathcal{F}$, the *equilibrium equation* can be obtained as:

$$\Lambda_t^*(\mathbf{u})\mathbf{s} = \bar{\mathbf{f}}. \tag{6.9}$$

Using the operator decomposition, the relation between the pairing spaces $(\mathcal{U}, \mathcal{F}^*)$ and $(\mathcal{E}, \mathcal{S}^*)$ should be

$$\langle \mathbf{e};\mathbf{s}\rangle = \langle \Lambda_t\mathbf{u};\mathbf{s}\rangle + \langle \Lambda_n\mathbf{u};\mathbf{s}\rangle = \langle \mathbf{u}, \bar{\mathbf{f}}\rangle - G(\mathbf{u},\mathbf{s}), \tag{6.10}$$

where the so-called *complementary gap function*, introduced by Gao-Strang in 1989, is defined as

$$G(\mathbf{u},\mathbf{s}) = \langle -\Lambda_n(\mathbf{u})\mathbf{u};\mathbf{s}\rangle = -\int_\Omega (\Lambda_n(\mathbf{u})\mathbf{u}) : \mathbf{s}\, d\Omega \tag{6.11}$$

which plays a central role in dual variational principles of finite deformation theory.

The mixed boundary value problem (BVP for short) in finite deformation theory can be proposed as the following:

Problem 1. *(BVP) For the given external force* $\bar{\mathbf{f}} \in \mathcal{F}$, *to find* $\mathbf{u} \in \mathcal{U}_a$ *and associated* $\mathbf{e}$ *and* $\mathbf{s}$ *such that*

$$\Lambda_t^*(\mathbf{u})\partial_{\mathbf{e}} W(\Lambda\mathbf{u}) \ni \bar{\mathbf{f}}. \tag{6.12}$$

The framework for this general finite deformation problem is shown in Fig. 1.

$$\begin{array}{ccccc}
\mathbf{u} \in \mathcal{U} & \longleftarrow & \langle \mathbf{u}, \mathbf{f}\rangle & \longrightarrow & \mathcal{F} \ni \mathbf{f} \\
\big\downarrow \ \Lambda = \Lambda_t + \Lambda_n & & & & \Lambda_t^* = \Lambda^* - \Lambda_n^* \ \big\uparrow \\
\mathbf{e} \in \mathcal{E} & \longleftarrow & \langle \mathbf{e};\mathbf{s}\rangle & \longrightarrow & \mathcal{S} \ni \mathbf{s}
\end{array}$$

Fig. 1. Framework in finite deformation theory.

6.3 CONJUGATE STRESS-STRAIN TENSORS AND GAP FUNCTIONS

In finite deformation theory, complementary formulations mainly depend on the choice of the operator Λ. A critical survey on conjugate strain and stress measures was given by Curnier-Rakotomanana (1991). For any $m \in \mathbb{R}$, the Seth-Hill strain family $\mathbf{e} = \mathbf{E}^{(m)}$ (Hill, 1978) can be written as:

$$\mathbf{E}^{(m)} = \Lambda \mathbf{u} = \frac{1}{2m}[(\mathbf{F}^t \cdot \mathbf{F})^m - \mathbf{I}]. \tag{6.13}$$

If the stored strain energy W is a strictly convex, Gateaux-differentiable function of $\mathbf{E}^{(m)}$, the conjugate stress tensor can be given uniquely by the constitutive equation

$$\mathbf{T}^{(m)} = DW(\mathbf{E}^{(m)}), \tag{6.14}$$

which is equivalent to the following Legendre transformation

$$W^*(\mathbf{T}^{(m)}) = \mathbf{T}^{(m)} : \mathbf{E}^{(m)} - W(\mathbf{E}^{(m)}). \tag{6.15}$$

For different conjugate pairings $(\mathbf{E}^{(m)}, \mathbf{T}^{(m)})$, the general governing equations, gap functions and complementary formulations were discussed in (Gao, 1992). Here we only consider some important cases.

For $m = 1$, $\mathbf{E}^{(1)}$ is the best-known Green-St. Venant strain tensor, denoted by

$$\mathbf{E} = \mathbf{E}^{(1)} = \frac{1}{2}[(\mathbf{F}^t \cdot \mathbf{F}) - \mathbf{I}\,] = \frac{1}{2}[(\nabla \mathbf{u})^t + \nabla \mathbf{u} + (\nabla \mathbf{u})^t \cdot (\nabla \mathbf{u})]. \tag{6.16}$$

Its conjugate stress should be the second Piola-Kirchhoff stress tensor, denoted by $\mathbf{T} = \mathbf{T}^{(1)}$. In this case, $\Lambda : \mathcal{U} \to \mathcal{E}$ is a quadratic operator, and

$$\begin{aligned} \Lambda_t(\mathbf{u})\delta\mathbf{u} &= \frac{1}{2}[\nabla\delta\mathbf{u} + (\nabla\delta\mathbf{u})^t + (\nabla\mathbf{u})^t \cdot (\nabla\delta\mathbf{u}) + (\nabla\delta\mathbf{u})^t \cdot (\nabla\mathbf{u})], \\ \Lambda_n(\mathbf{u})\delta\mathbf{u} &= -\frac{1}{4}[(\nabla\mathbf{u})^t \cdot (\nabla\delta\mathbf{u}) + (\nabla\delta\mathbf{u})^t \cdot (\nabla\mathbf{u})]. \end{aligned}$$

Then the equilibrium equation in this case can be written as

$$\Lambda_t^*(\mathbf{u})\mathbf{T} = \begin{cases} -\nabla \cdot [(\mathbf{I} + \nabla\mathbf{u}) \cdot \mathbf{T}] = \bar{\mathbf{b}} & \text{in } \Omega \\ \mathbf{n} \cdot [(\mathbf{I} + \nabla\mathbf{u}) \cdot \mathbf{T}] = \bar{\mathbf{t}} & \text{on } \Gamma_t, \end{cases} \tag{6.17}$$

where $\mathbf{n}$ is a unit vector normal to the boundary $\Gamma = \partial\Omega$. The gap function for $\Lambda\mathbf{u} = \mathbf{E}$ is then

$$G(\mathbf{u}, \mathbf{T}) = \langle -\Lambda_n(\mathbf{u})\mathbf{u}; \mathbf{T} \rangle = \int_\Omega \frac{1}{2}[(\nabla\mathbf{u})^t \cdot (\nabla\mathbf{u})] : \mathbf{T} \, d\Omega. \tag{6.18}$$

For $m = \frac{1}{2}$, we have the right-stretch strain tensor (or the engineering strain):

$$\mathbf{E}^{(\frac{1}{2})} = (\mathbf{F}^t \cdot \mathbf{F})^{\frac{1}{2}} - \mathbf{I} = \mathbf{U} - \mathbf{I}. \tag{6.19}$$

Its conjugate $\mathbf{T}^{(\frac{1}{2})}$ is the so-called Biot-Jaumann stress tensor, denoted by $\mathbf{S}$. In this case, Λ is a nonlinear operator, and

$$\Lambda_t \mathbf{u} = \mathbf{U} - \frac{1}{2}(\mathbf{R} + \mathbf{R}^t), \quad \Lambda_n \mathbf{u} = \frac{1}{2}(\mathbf{R} + \mathbf{R}^t) - \mathbf{I}.$$

The gap function for $\mathbf{E}^{(\frac{1}{2})}$ should be (see Gao, 1992):

$$G_{\mathbf{S}}(\mathbf{u},\mathbf{S}) = \int_{\Omega} \mathrm{tr}[\mathbf{S}\cdot(\mathbf{I}-\mathbf{R}(\mathbf{u}))]d\Omega. \tag{6.20}$$

Since Λ_t is a nonlinear operator, the equilibrium equation in terms of $\mathbf{S}$ is very complicated.

If $m=0$, $\mathbf{E}^{(0)} = \ln \mathbf{U}$ is the logarithmic strain. Its conjugate stress $\mathbf{T}^{(0)}$ was first discovered by Hoger in 1987. If $\mathbf{T}^{(m)} : \mathbf{E}^{(m)} = \mathbf{E}^{(m)} : \mathbf{T}^{(m)}$, $\mathbf{T}^{(0)} = \mathbf{TU}$ (see Ogden, 1984). For $m=-1$, $\mathbf{E}^{(-1)} = \frac{1}{2}[\mathbf{I}-(\mathbf{F}^t\cdot\mathbf{F})^{-1}]$ is the Almansi strain tensor. Its conjugate stress is the so-called weighted *convected stress.*

In finite deformations, the quadratic operator Λ plays an important role in duality theory:

$$\Lambda\mathbf{u} = a_2(\nabla\mathbf{u})^t\cdot(\nabla\mathbf{u}) + a_1[\nabla\mathbf{u}+(\nabla\mathbf{u})^t] + a_0\mathbf{I},\quad a_2>0. \tag{6.21}$$

If $a_2=a_1=a_0=1$, $\mathbf{e}=\Lambda\mathbf{u}$ is the so-called *right Cauchy-Green strain tensor*, denoted as $\mathbf{C}$ and we have

$$\mathbf{C} = 2\mathbf{E}+\mathbf{I} = \mathbf{F}^t\cdot\mathbf{F}. \tag{6.22}$$

By the definition of the conjugate stress, we know that the conjugate stress of $\mathbf{C}$ is equal to $\frac{1}{2}\mathbf{T}$.

If Λ is an affine operator $\Lambda\mathbf{u} = \nabla\mathbf{u}+\mathbf{I}$, then $\Lambda\mathbf{u}$ is the deformation gradient $\mathbf{F}$. Since $\mathbf{F}$ is not a strain measure, the stored energy $W(\mathbf{e}) = \bar{W}(\mathbf{F})$ is usually not convex in $\mathbf{F}$ (cf. Ogden, 1984). The conjugate variable of $\mathbf{F}$ is the first Piola-Kirchhoff stress tensor:

$$\boldsymbol{\tau} = D\bar{W}(\mathbf{F}), \tag{6.23}$$

which is also a two-point tensor. In this case,

$$\Lambda_t\mathbf{u} = \nabla\mathbf{u},\quad \Lambda_n\mathbf{u} = \mathbf{I},$$

and the abstract equilibrium equation can be simply written as

$$\Lambda_t^*\boldsymbol{\tau} = \begin{cases} -\nabla\cdot\boldsymbol{\tau} = \bar{\mathbf{b}} & \text{in } \Omega, \\ \mathbf{n}\cdot\boldsymbol{\tau} = \bar{\mathbf{t}} & \text{on } \Gamma_t. \end{cases} \tag{6.24}$$

The gap function in this case depends on $\boldsymbol{\tau}$ only:

$$G(\mathbf{u},\boldsymbol{\tau}) = G_{\mathbf{F}}(\boldsymbol{\tau}) = \langle -\Lambda_n\mathbf{u};\boldsymbol{\tau}\rangle = -\int_{\Omega}\mathrm{tr}\boldsymbol{\tau}d\Omega. \tag{6.25}$$

The complementary energy defined by the Fenchel transformation

$$\bar{W}^*(\boldsymbol{\tau}) = \sup_{\mathbf{F}}\{\boldsymbol{\tau}:\mathbf{F}-\bar{W}(\mathbf{F})\} \tag{6.26}$$

is always convex in $\boldsymbol{\tau}$, but the constitutive law (6.23) is usually irreversible. The following result was obtained in (Gao, 1992).

Lemma 6.1. *Suppose that W is strictly convex in* $\mathbf{E}$. *For any given* $\mathbf{u} \in \mathcal{U}$ *such that* $\mathbf{T}(\mathbf{u}) = \partial W(\mathbf{E}(\mathbf{u}))/\partial \mathbf{E}$, *if the density of the gap function*

$$g(\mathbf{v}, \mathbf{T}) := \frac{1}{2}\mathbf{T}(\mathbf{u}) : [(\nabla \mathbf{v}) \cdot (\nabla \mathbf{v})^t] \geq 0 \ \forall \mathbf{v} \in \mathcal{U}, \tag{6.27}$$

the constitutive relation $\boldsymbol{\tau} = D_{\mathbf{F}} \bar{W}(\mathbf{F}(\mathbf{u}))$ *is reversible:*

$$\mathbf{F}(\mathbf{u}) = D\bar{W}^*(\boldsymbol{\tau}), \tag{6.28}$$

and

$$\bar{W}^*(\boldsymbol{\tau}(\mathbf{u})) = W^*(\mathbf{T}(\mathbf{u})) + g(\mathbf{u}, \mathbf{T}). \tag{6.29}$$

The following equation always holds (see, Gao, 1992):

$$W^*(\mathbf{T}) = W^*(\mathbf{T}^{(-1)}) = W^*(\mathbf{S}) + \mathbf{T} : (\mathbf{U} - \mathbf{E} - \mathbf{I}). \tag{6.30}$$

The relations among $\mathbf{S}$, $\boldsymbol{\tau}$ and $\mathbf{T}$ are given by

$$\boldsymbol{\tau} = \mathbf{F} \cdot \mathbf{T}, \tag{6.31}$$

$$\mathbf{S} = \frac{1}{2}[\boldsymbol{\tau}^t \cdot \mathbf{R} + \mathbf{R}^t \cdot \boldsymbol{\tau}] = \frac{1}{2}(\mathbf{T} \cdot \mathbf{U} + \mathbf{U} \cdot \mathbf{T}). \tag{6.32}$$

6.4 POTENTIAL EXTREMUM PRINCIPLE

Let $F : \mathcal{U}_a \to \mathbb{R}$ be the external energy: $F(\mathbf{u}) = \langle \mathbf{u}, \bar{\mathbf{f}} \rangle$. For any geometric operator $\Lambda : \mathcal{U} \to \mathcal{E}$, the total potential energy of the mixed boundary value problem can be written as

$$P(\mathbf{u}) = \int_\Omega W(\Lambda \mathbf{u}) d\Omega - F(\mathbf{u}). \tag{6.33}$$

If P is Gâteaux differentiable, then we have[1]

$$\delta P(\bar{\mathbf{u}}; \delta \mathbf{u}) = 0 \ \forall \delta \mathbf{u} \in \mathcal{U}_a \ \Leftrightarrow \ \Lambda_t^* D_{\mathbf{e}} W(\Lambda \bar{\mathbf{u}}) = \bar{\mathbf{f}}. \tag{6.34}$$

Since the total potential $P : \mathcal{U}_a \to \mathbb{R}$ is usually nonconvex, the solution of (BVP) may not be a global minimizer of P. The following extremum principle was proved by Gao & Strang (1989a).

Theorem 6.1. *Suppose that* $\Lambda : \mathcal{U} \to \mathcal{E}$ *is a quadratic operator such that* W *is a convex in* $\mathbf{e} = \Lambda \mathbf{u}$, *and that* $\bar{\mathbf{u}}$ *is a critical point of* P *and* $\bar{\mathbf{s}} = DW(\Lambda \bar{\mathbf{u}})$ *is the associated general stress tensor. If the gap function* $G(\mathbf{u}, \bar{\mathbf{s}}) \geq 0 \ \forall \mathbf{u} \in \mathcal{U}_a$, *then* $\bar{\mathbf{u}}$ *is a solution of the following global minimization problem:*

$$(\mathcal{P}_{\inf}) : \quad P(\bar{\mathbf{u}}) = \inf P(\mathbf{u}) \quad \forall \mathbf{u} \in \mathcal{U}_a. \tag{6.35}$$

If $\mathcal{U}_a$ *is a bounded, closed convex set of a reflexive Banach space, the solution of the problem* ($\mathcal{P}_{inf}$) *exists. The solution is unique if the gap function is strictly positive.*

However, if the gap function has a negative sign, the total potential energy is nonconvex. Then the solution $\mathbf{u}$ of the (BVP) could be a local minimizer or a local maximizer of the total potential P on a subset of $\mathcal{U}_a$. Let $\mathcal{U}_b \subset \mathcal{U}_a$ be a neighborhood of the critical point of P. The following sup-primal problem plays an important role in phase transitions and post-buckling analysis.

$$(\mathcal{P}_{\text{sup}}): \quad P(\bar{\mathbf{u}}) = \sup P(\mathbf{u}) \quad \forall \mathbf{u} \in \mathcal{U}_b. \tag{6.36}$$

By traditional direct methods, this locally unstable problem is much more difficult than the inf-primal problem. But the duality methods are potentially powerful for solving such nonconvex, unstable problems.

6.5 CLASSICAL COMPLEMENTARY ENERGY PRINCIPLES

For any geometrical operator $\Lambda : \mathcal{U} \to \mathcal{E}$, a unified complementary energy was proposed by Gao and Strang in 1989 as the following:

$$P^c(\mathbf{s}, \mathbf{u}) := F^*(\Lambda_t^*(\mathbf{u})\mathbf{s}) - \int_\Omega W^*(\mathbf{s})\, \mathrm{d}\Omega - G(\mathbf{u}, \mathbf{s}), \tag{6.37}$$

where $F^*(\mathbf{f})$ is the conjugate function of F. If $\mathcal{U}_a$ is given by (6.7), then

$$F^*(\mathbf{f}) = \inf_{\mathbf{u} \in \mathcal{U}_a} \{\langle \mathbf{f}, \mathbf{u} \rangle - \mathbf{F}(\mathbf{u})\} = \begin{cases} 0 & \text{if } \mathbf{f} = \bar{\mathbf{f}} \\ -\infty & \text{if } \mathbf{f} \neq \bar{\mathbf{f}}. \end{cases}$$

If the operator Λ is linear, then $\Lambda_t^* = \Lambda^*$, and the gap function $G = 0$. Eqn (6.37) shows that for nonlinear operator Λ, there exists a gap between P^c and the Fenchel-Rockafellar dual energy $F^* - \int_\Omega W^* \mathrm{d}\Omega$. Let $\mathcal{S}_a \subset \mathcal{S}$ be a range of the constitutive mapping $DW(\mathcal{E})$. We introduce the so-called equilibrium admissible space:

$$\mathcal{S}_{\mathbf{u}} := \{(\mathbf{s}, \mathbf{u}) \in \mathcal{S}_a \times \mathcal{U}_a | \ \Lambda_t^*(\mathbf{u})\mathbf{s} = \bar{\mathbf{f}}\}. \tag{6.38}$$

Then on $\mathcal{S}_{\mathbf{u}}$, the total complementary energy P^c is finite:

$$P^c(\mathbf{s}, \mathbf{u}) = -\int_\Omega W^*(\mathbf{s}) \mathrm{d}\Omega - G(\mathbf{u}, \mathbf{s}). \tag{6.39}$$

The following theorem was given by Gao and Strang (1989a).

Theorem 6.2. (Complementary energy principle) *Let $\Lambda : \mathcal{U} \to \mathcal{E}$ be a finite deformation operator such that W is convex in $\Lambda \mathbf{u}$. Then the solution $(\bar{\mathbf{s}}, \bar{\mathbf{u}})$ of the following constrained variational problem solves the (BVP):*

$$(\mathcal{P}^c): \qquad \delta P^c(\bar{\mathbf{s}}, \bar{\mathbf{u}}; \mathbf{s}, \mathbf{u}) = 0 \ \ \forall (\mathbf{s}, \mathbf{u}) \in \mathcal{S}_{\mathbf{u}}. \tag{6.40}$$

Moreover, if $\Lambda : \mathcal{U} \to \mathcal{E}$ is quadratic and the gap function $G(\mathbf{u}, \mathbf{s}) \geq 0$ $\forall (\mathbf{s}, \mathbf{u}) \in \mathcal{S}_{\mathbf{u}}$, then any critical point $(\bar{\mathbf{s}}, \bar{\mathbf{u}})$ of P^c maximizes P^c over $\mathcal{S}_{\mathbf{u}}$ and

$$\inf_{\mathbf{u} \in \mathcal{U}_a} P(\mathbf{u}) = \sup_{(\mathbf{s}, \mathbf{u}) \in \mathcal{S}_{\mathbf{u}}} P^c(\mathbf{s}, \mathbf{u}). \tag{6.41}$$

If $\Lambda\mathbf{u}$ is the Green-St Venant strain $\mathbf{E}$, $P^c(\mathbf{s},\mathbf{u})$ is the well-known *Hellinger-Reissner complementary energy*:

$$P^c_{H-R}(\mathbf{T},\mathbf{u}) = -\int_\Omega W^*(\mathbf{T})d\Omega - \int_\Omega \frac{1}{2}\mathbf{T} : [(\nabla\mathbf{u})^t \cdot (\nabla\mathbf{u})]d\Omega. \tag{6.42}$$

But its two variables $\mathbf{T}$ and $\mathbf{u}$ are not independent. If the gap function $G(\mathbf{u},\mathbf{T})$ is positive on $\mathcal{S}_\mathbf{u}$, $P^c_{H-R} : \mathcal{S}_\mathbf{u} \to \mathbb{R}$ is concave.

If $\Lambda\mathbf{u} = \mathbf{E}^{(\frac{1}{2})} = \mathbf{U}(\mathbf{u}) - \mathbf{I}$, then P^c can be written as

$$P^c_{\frac{1}{2}}(\mathbf{S},\mathbf{u}) = -\int_\Omega W^*(\mathbf{S})d\Omega - \int_\Omega \mathrm{tr}[\mathbf{S}\cdot(\mathbf{I} - \mathbf{R}(\mathbf{u}))]d\Omega. \tag{6.43}$$

Since the equilibrium equation in terms of $\mathbf{S}$ is very complicated, this complementary energy functional has not been found direct application. If $\mathbf{S}$ is replaced by $\frac{1}{2}[\mathbf{U}\cdot\mathbf{T} + \mathbf{T}\cdot\mathbf{U}]$, $W^*(\mathbf{S})$ is replaced by (6.30), $P^c_{\frac{1}{2}}$ is just the Hellinger-Reissner energy. However, if $\mathbf{S}$ is replaced by $\frac{1}{2}[\boldsymbol{\tau}^t\cdot\mathbf{R} + \mathbf{R}\cdot\boldsymbol{\tau}]$, and instead of $\mathbf{u}$, we consider the rotation tensor $\mathbf{R}$ as the variational argument, $P^c_{\frac{1}{2}}$ is the well-known *Fraeijs de Veubeke energy*:

$$P^c_{FdV}(\boldsymbol{\tau},\mathbf{R}) = -\int_\Omega W^*(\mathbf{S}(\boldsymbol{\tau},\mathbf{R}))d\Omega - \int_\Omega \mathrm{tr}[\boldsymbol{\tau}\cdot(\mathbf{R} - \mathbf{I})]d\Omega. \tag{6.44}$$

In terms of $\boldsymbol{\tau}$, the equilibrium admissible space $\mathcal{S}_\mathbf{u}$ can be written as

$$\mathcal{T}_a := \{\boldsymbol{\tau} \in \mathrm{Lin}| \ -\nabla\cdot\boldsymbol{\tau} = \bar{\mathbf{b}} \ \text{ in } \Omega; \ \mathbf{n}\cdot\boldsymbol{\tau} = \bar{\mathbf{t}} \ \text{ on } \Gamma_t\}. \tag{6.45}$$

So for any $\boldsymbol{\tau} \in \mathcal{T}_a$ and independent rotation tensor $\mathbf{R} \in \mathrm{Lin}$, the critical point of P^c_{FdV} solves the (BVP). Since the Fraeijs de Veubeke energy involves both the first Piola-Kirchhoff stress $\boldsymbol{\tau}$ and the rotation tensor $\mathbf{R}$, it is not considered as a pure complementary energy either.

If $\Lambda\mathbf{u} = \mathbf{F}$, the deformation gradient, then P^c is the *generalized Levinson-Zubov energy,*

$$P^c_{L-Z}(\boldsymbol{\tau}) = -\int_\Omega \bar{W}^*(\boldsymbol{\tau})d\Omega - \int_\Omega \mathrm{tr}\boldsymbol{\tau}d\Omega, \tag{6.46}$$

which depends on the first Piola-Kirchhoff stress only. By Lemma 1, we have

Theorem 6.3. *For any given external load $\bar{\mathbf{f}} \in \mathcal{F}$, if $\mathcal{T}_a$ is a nonempty subspace of a reflexive Banach space, the complementary variational problem*

$$(\mathcal{P}^c_{L-Z}): \qquad \delta P^c_{L-Z}(\bar{\boldsymbol{\tau}};\boldsymbol{\tau}) = 0 \ \ \forall\boldsymbol{\tau} \in \mathcal{T}_a \tag{6.47}$$

has at least one solution, which is a global maximizer of P^c_{L-Z} on $\mathcal{T}_a$ and the following inequality is always hold:

$$\inf P(\mathbf{u}) \geq \sup P^c_{L-Z}(\boldsymbol{\tau}) \ \ \forall\mathbf{u} \in \mathcal{U}_a, \ \ \forall\boldsymbol{\tau} \in \mathcal{T}_a. \tag{6.48}$$

The equality is hold if and only if $\bar{W}(\mathbf{F})$ is convex in $\mathbf{F}$. In this case, the critical point $\bar{\boldsymbol{\tau}}$ of P^c_{L-Z} solves (BVP). The solution is unique if $\bar{W}$ is strictly convex,

The proof of this theorem was given in (Gao, 1992). The inequality (6.48) shows that there exists a dual gap between the primal problem $(\mathcal{P}_{\inf})$ and the generalized Levinson-Zubov variational problem $(\mathcal{P}^c_{L-Z})$ if the complementary gap function is negative. In this case, the critical point of P^c_{P-Z} is only a relaxed solution of (BVP).

6.6 GENERALIZED VARIATIONAL PRINCIPLES AND TRIALITY THEORY

Using the Legendre transform $F^*(\Lambda_t^*(\mathbf{u})\mathbf{s}) = \langle \Lambda_t^*(\mathbf{u})\mathbf{s}, \mathbf{u}\rangle - F(\mathbf{u})$ to replace F^* in P^c, the Lagrangian $L : \mathcal{U}_a \times \mathcal{S}_a \to \mathbb{R}$ can be written as

$$L(\mathbf{u}, \mathbf{s}) = \int_\Omega \mathbf{s} : (\Lambda \mathbf{u}) d\Omega - \int_\Omega W^*(\mathbf{s})\, d\Omega - F(\mathbf{u}). \tag{6.49}$$

This is the so-called generalized complementary energy in finite deformation theory. If L is Gâteaux differentiable, the variational problem $(\mathcal{P}_L)$

$$(\mathcal{P}_L): \quad \delta L(\bar{\mathbf{u}}, \bar{\mathbf{s}}; \mathbf{u}, \mathbf{s}) = 0 \;\; \forall (\mathbf{u}, \mathbf{s}) \in \mathcal{U}_a \times \mathcal{S}_a \tag{6.50}$$

gives the following canonical Euler-Lagrange equations:

$$\Lambda_t^*(\bar{\mathbf{u}})\bar{\mathbf{s}} = DF(\bar{\mathbf{u}}), \quad \Lambda \bar{\mathbf{u}} = DW^*(\bar{\mathbf{s}}). \tag{6.51}$$

Generalized complementary energy principle: *Among all $(\mathbf{u}, \mathbf{s}) \in \mathcal{U}_a \times \mathcal{S}_a$, the critical point $(\bar{\mathbf{u}}, \bar{\mathbf{s}})$ of the Lagrangian L solves the (BVP).*

If $\Lambda : \mathcal{U} \to \mathcal{E}$ is a quadratic operator, the critical property of L depends on the sign of the gap function (Gao, 1997):

Theorem 6.4. *Suppose that $(\bar{\mathbf{u}}, \bar{\mathbf{s}})$ is a critical point of L. Then $(\bar{\mathbf{u}}, \bar{\mathbf{s}})$ is a saddle point of L, i.e.*

$$L(\bar{\mathbf{u}}, \mathbf{s}) \le L(\bar{\mathbf{u}}, \bar{\mathbf{s}}) \le L(\mathbf{u}, \bar{\mathbf{s}}) \quad \forall (\mathbf{u}, \mathbf{s}) \in \mathcal{U}_a \times \mathcal{S}_a \tag{6.52}$$

if and only if the gap function $G(\mathbf{u}, \bar{\mathbf{s}}) \ge 0 \;\; \forall \mathbf{u} \in \mathcal{U}_a$. However, the point $(\bar{\mathbf{u}}, \bar{\mathbf{s}})$ is a super-critical point of L, i.e.

$$L(\bar{\mathbf{u}}, \mathbf{s}) \le L(\bar{\mathbf{u}}, \bar{\mathbf{s}}) \ge L(\mathbf{u}, \bar{\mathbf{s}}) \quad \forall (\mathbf{u}, \mathbf{s}) \in \mathcal{U}_a \times \mathcal{S}_a, \tag{6.53}$$

if and only if the gap function $G(\mathbf{u}, \bar{\mathbf{s}}) \le 0 \;\; \forall \mathbf{u} \in \mathcal{U}_a$.

Moreover, the gap function $G(\mathbf{u}, \mathbf{s})$ is positive on $\mathcal{S}_{\mathbf{u}}$ if and only if all critical points of L are saddle points. If the gap function is strictly positive on $\mathcal{S}_{\mathbf{u}}$, and $W^* : \mathcal{S} \to \mathbb{R}$ is strictly convex, then L has at most one saddle point.

For $\Lambda \mathbf{u} = \mathbf{E}$, the Lagrangian $L(\mathbf{u}, \mathbf{T})$ is the well-known *generalized Hellinger-Reissner functional*:

$$L(\mathbf{u}, \mathbf{T}) = \int_\Omega [\mathbf{T} : \mathbf{E}(\mathbf{u}) - W^*(\mathbf{T})] d\Omega - F(\mathbf{u}). \tag{6.54}$$

Its extremum property was discovered recently (Gao, 1997):

Theorem 6.5. *(Triality Theorem) Suppose that $(\bar{\mathbf{u}}, \bar{\mathbf{T}})$ is a critical point of $L(\mathbf{u}, \mathbf{T})$ and $\mathcal{U}_b \times \mathcal{S}_b$ is a neighborhood of $(\bar{\mathbf{u}}, \bar{\mathbf{T}})$ such that on $\mathcal{U}_b \times \mathcal{S}_b$, $(\bar{\mathbf{u}}, \bar{\mathbf{T}})$ is the only one critical point of L. If the gap function $G(\mathbf{u}, \mathbf{T}) \geq 0 \ \forall (\mathbf{u}, \mathbf{T}) \in \mathcal{S}_{\mathbf{u}}$, then*

$$\inf_{\mathbf{u} \in \mathcal{U}_a} \sup_{\mathbf{T} \in \mathcal{S}_a} L(\mathbf{u}, \mathbf{T}) = \sup_{\mathbf{T} \in \mathcal{S}_a} \inf_{\mathbf{u} \in \mathcal{U}_a} L(\mathbf{u}, \mathbf{T}). \tag{6.55}$$

However, if $G(\bar{\mathbf{u}}, \bar{\mathbf{T}}) \leq 0$ then

$$\inf_{\mathbf{u} \in \mathcal{U}_b} \sup_{\mathbf{T} \in \mathcal{S}_b} L(\mathbf{u}, \mathbf{T}) = \inf_{\mathbf{T} \in \mathcal{S}_b} \sup_{\mathbf{u} \in \mathcal{U}_b} L(\mathbf{u}, \mathbf{T}), \tag{6.56}$$

$$\sup_{\mathbf{u} \in \mathcal{U}_b} \sup_{\mathbf{T} \in \mathcal{S}_b} L(\mathbf{u}, \mathbf{T}) = \sup_{\mathbf{T} \in \mathcal{S}_b} \sup_{\mathbf{u} \in \mathcal{U}_b} L(\mathbf{u}, \mathbf{T}). \tag{6.57}$$

In the Lagrangian $L(\mathbf{u}, \mathbf{s})$, if we replace W^* by the Fenchel transformation (6.2), then we have

$$L(\mathbf{u}, \mathbf{s}) = \inf_{\mathbf{e}} L_p(\mathbf{u}, \mathbf{s}, \mathbf{e}), \tag{6.58}$$

where L_p is the so-called pseudo-Lagrangian (see Gao-Strang, 1989a)

$$L_p(\mathbf{u}, \mathbf{s}, \mathbf{e}) = \langle \Lambda \mathbf{u} - \mathbf{e}; \mathbf{s} \rangle + \int_\Omega W(\mathbf{e}) d\Omega - F(\mathbf{u}). \tag{6.59}$$

For the Green-St Venant finite deformation operator $\Lambda \mathbf{u} = \mathbf{E}$, $L_p(\mathbf{u}, \mathbf{T}, \mathbf{E})$ is the well-known *Hu-Washizu energy*. Its extremum property follows easily from the triality theorem.

6.7 PURE COMPLEMENTARY ENERGY PRINCIPLES AND MINIMAX THEORY

For any quadratic operator $\Lambda : \mathcal{U} \to \mathcal{E}$ such that W is convex in $\mathbf{e} = \Lambda \mathbf{u}$, the pure complementary energy $P^*(\mathbf{s})$ can be constructed by the following theorem (Gao, 1997):

Theorem 6.6. *For a given $\mathbf{s} \in \mathcal{S}_a$, if $\mathcal{S}_{\mathbf{u}}$ is not empty, the pure complementary energy can be obtained by*

$$P^*(\mathbf{s}) = \begin{cases} \inf_{\mathbf{u}} L(\mathbf{u}, \mathbf{s}) & \text{if } G(\mathbf{u}, \mathbf{s}) \geq 0 \ \forall (\mathbf{s}, \mathbf{u}) \in \mathcal{S}_{\mathbf{u}}, \\ \sup_{\mathbf{u}} L(\mathbf{u}, \mathbf{s}) & \text{if } G(\mathbf{u}, \mathbf{s}) \leq 0 \ \forall (\mathbf{s}, \mathbf{u}) \in \mathcal{S}_{\mathbf{u}}. \end{cases} \tag{6.60}$$

The critical point of P^* solves the (BVP).

The following results show that P^* is actually a dual functional of P.

Theorem 6.7. *If $(\bar{\mathbf{u}}, \bar{\mathbf{s}})$ is a critical point of L, then $DP(\bar{\mathbf{u}}) = 0$, $DP^*(\bar{\mathbf{s}}) = 0$, and*

$$P(\bar{\mathbf{u}}) = L(\bar{\mathbf{u}}, \bar{\mathbf{s}}) = P^*(\bar{\mathbf{s}}). \tag{6.61}$$

Theorem 6.8. *(Minimax Theorem) Suppose that* $\bar{\mathbf{u}}$ *and* $\bar{\mathbf{s}}$ *are critical points of* P *and* P^**, respectively.* $\mathcal{U}_b$ *and* $\mathcal{S}_b$ *are their neighborhoods such that* P *and* P^* *have only one critical point on* $\mathcal{U}_b$ *and* $\mathcal{S}_b$*, respectively. If* $G(\bar{\mathbf{u}}, \bar{\mathbf{s}}) \geq 0$*, then*

$$P(\bar{\mathbf{u}}) = \inf_{\mathbf{u}\in\mathcal{U}_b} P(\mathbf{u}) \;\Leftrightarrow\; P^*(\bar{\mathbf{s}}) = \sup_{\mathbf{s}\in\mathcal{S}_b} P^*(\mathbf{s}). \tag{6.62}$$

If $G(\bar{\mathbf{u}}, \bar{\mathbf{s}}) \leq 0$*, then*

$$P(\bar{\mathbf{u}}) = \inf_{\mathbf{u}\in\mathcal{U}_b} P(\mathbf{u}) \quad\Leftrightarrow\quad P^*(\bar{\mathbf{s}}) = \inf_{\mathbf{s}\in\mathcal{S}_b} P^*(\mathbf{s}), \tag{6.63}$$

$$P(\bar{\mathbf{u}}) = \sup_{\mathbf{u}\in\mathcal{U}_b} P(\mathbf{u}) \quad\Leftrightarrow\quad P^*(\bar{\mathbf{s}}) = \sup_{\mathbf{s}\in\mathcal{S}_b} P^*(\mathbf{s}). \tag{6.64}$$

The proofs of these theorems were given in (Gao, 1998a). Let $\mathcal{S}_{\mathbf{T}} \subset \mathcal{S}_a$ be the so-called dual admissible stress space:

$$\mathcal{S}_{\mathbf{T}} = \{\mathbf{T} \in \mathcal{S}_a | \;\det \mathbf{T}(\mathbf{x}) \neq 0 \;\forall \mathbf{x} \in \Omega\}. \tag{6.65}$$

By Theorem 6, for any $\mathbf{T} \in \mathcal{S}_{\mathbf{T}}$ and $\tau \in \mathcal{T}_a$, a pure complementary energy theorem was proposed recently (Gao, 1997b, 1998b):

$$P^*(\mathbf{T}, \tau) = -\int_\Omega W^*(\mathbf{T})\mathrm{d}\Omega - \int_\Omega \frac{1}{2}\mathrm{tr}[\mathbf{T}^{-1} \cdot (\tau^t \cdot \tau) - 2\tau + \mathbf{T}]\mathrm{d}\Omega. \tag{6.66}$$

Theorem 6.9. *For a given external force* $\bar{\mathbf{f}} \in \mathcal{F}$ *such that* $\mathcal{T}_a$ *is not empty, then the solution of the following dual variational problem solves the (BVP).*

$$(\mathcal{P}^*): \qquad \delta P^*(\bar{\mathbf{T}}, \bar{\tau}; \mathbf{T}, \tau) = 0 \;\;\forall \mathbf{T} \in \mathcal{S}_{\mathbf{T}}, \;\; \tau \in \mathcal{T}_a. \tag{6.67}$$

Moreover, $\bar{\mathbf{u}}$ *is a critical point of* $P(\mathbf{u})$ *if and only if the associated* $\bar{\mathbf{T}} = D_{\mathbf{E}}W(\mathbf{E}(\bar{\mathbf{u}})) \in \mathcal{S}_{\mathbf{T}}$ *and* $\bar{\tau} = \mathbf{F}(\bar{\mathbf{u}})\bar{\mathbf{T}} \in \mathcal{T}_a$ *are critical point of* P^**, and*

$$P(\bar{\mathbf{u}}) = P^*(\bar{\mathbf{T}}, \bar{\tau}). \tag{6.68}$$

Proof. By introducing a Lagrange multiplier $\mathbf{u} \in \mathcal{U}_a$ to relax the equilibrium constrain in $\mathcal{T}_a$, the Lagrangian associated with P^* is

$$L_G(\mathbf{u}, \mathbf{T}, \tau) = \langle \nabla \mathbf{u}; \tau \rangle + P^*(\mathbf{T}, \tau) - F(\mathbf{u}). \tag{6.69}$$

The critical condition $DL_G(\bar{\mathbf{u}}, \bar{\mathbf{T}}, \bar{\tau}) = 0$ gives the well-known relations

$$-\nabla \cdot \bar{\tau} = \bar{\mathbf{f}} \;\text{ in } \Omega, \;\; \mathbf{n} \cdot \bar{\tau} = \bar{\mathbf{t}} \;\text{ on } \Gamma_t, \tag{6.70}$$

$$\bar{\tau} = (\nabla \bar{\mathbf{u}} + \mathbf{I}) \cdot \bar{\mathbf{T}}, \tag{6.71}$$

and a *dual Euler-Lagrange equation*

$$\bar{\mathbf{T}} \cdot [2DW^*(\bar{\mathbf{T}}) + \mathbf{I}] \cdot \bar{\mathbf{T}} = \bar{\tau}^t \cdot \bar{\tau}. \tag{6.72}$$

This is an algebraic equation. Replacing $\bar{\tau}$ by $(\nabla\bar{\mathbf{u}} + \mathbf{I}) \cdot \bar{\mathbf{T}}$, it follows the constitutive equation $DW^*(\bar{\mathbf{T}}) = \mathbf{E}(\bar{\mathbf{u}})$, which is equivalent to the $\bar{\mathbf{T}} = D_{\mathbf{E}}W(\mathbf{E}(\bar{\mathbf{u}}))$.

So the critical point of L_G solves the (BVP). ∎

Based on this pure complementary energy principle, a generalized analytic solution for (BVP) can be obtained.

Theorem 6.10. *For any given* $\bar{\mathbf{f}} \in \mathcal{F}$ *and* $\bar{\mathbf{u}}(\mathbf{x}_0)$ $\forall \mathbf{x}_0 \in \Gamma_u$ *such that* $\boldsymbol{\tau} \in \mathcal{T}_a$ *is a critical point of* P^c_{L-Z} *and* $\mathbf{T} \in \mathcal{S}_{\mathbf{T}}$ *is a solution of the dual Euler-Lagrange equation (6.72). If* $\nabla \times (\boldsymbol{\tau} \cdot \mathbf{T}^{-1}) = 0$*, then along any integral path from* $\mathbf{x}_0$ *to* $\mathbf{x}$*, the displacement given by*

$$\mathbf{u}(\mathbf{x}) = \int_{\mathbf{x}_0}^{\mathbf{x}} (\boldsymbol{\tau} \cdot \mathbf{T}^{-1} - \mathbf{I}) \cdot d\mathbf{x} + \bar{\mathbf{u}}(\mathbf{x}_0) \tag{6.73}$$

provides a generalized solution to (BVP) in the sense that $\mathbf{u}$ *satisfies the equilibrium conditions and the geometrical-constitutive equation* $\mathbf{E}(\mathbf{u}) = DW^*(\mathbf{T})$*. If* $\mathbf{T}(\mathbf{x})$ *is positive definite everywhere in* Ω*, then* $\mathbf{u}(\mathbf{x})$ *is a unique global minimizer of* P*.*

Detailed discussion for this generalized analytic solution and applications in post-buckling analysis were given in author's recent papers (see Gao, 1997, 1998a,b, 1999a,b).

Notes

1. In the proof given in (Gao-Strang, 1989a), the stationary condition $\delta P(\bar{\mathbf{u}}; \delta\mathbf{u}) = 0$ was considered to be equivalent to the subdifferential condition $0 \in \partial P(\bar{\mathbf{u}})$. This is true only when the gap function is positive. See the following theorem

References

[1] Arthurs, AM (1980), *Complementary Variational Principles*, Second edition, Clarendon Press.

[2] Atluri, SN (1984), Alternate stress and conjugate strain measures, and mixed variational formulations involving rigid rotations, for computational analysis of finitely deformed solids, with application to plates and shells -I, Theory, *Computers & Structures*, **18**, no. 1, pp. 93-116.

[3] Auchmuty, G (1983), Duality for non-convex variational principles, *J. Diff. Equations*, **vol. 50**, pp 80-145

[4] Auchmuty, G (1989), Duality algorithms for nonconvex variational principles, *Numer. Funct. Anal. and Optim.*, **10**, pp. 211-264.

[5] Bufler, H (1983), On the work theorems for finite and incremental elastic deformations with discontinuous fields: a unified treatment of different versions, *Comput. Meth. Appl. Mech. Eng.*, **36**, 95-124.

[6] Curnier, A and Rakotomanana, L (1991), Generalized strain and stress measures: Critical survey and new results, *Engineering Transactions*, **39**, no. 3-4, pp. 461-538.

[7] Dem'yanov, VF, Stavroulakis, GE, Polyakova, LN and Panagiotopoulos, PD (1996), *Quasidifferentiability and Nonsmooth Modelling in Mechanics, Engineering and Economics*, Kluwer Academic Publishers, Dordrecht / Boston / London, 348pp.

[8] Dür, M and Horst, R (1997), Lagrange duality and partitioning techniques in nonconvex global optimization, *J. Optim. Theory Appl.*, **95**, no. 2, pp. 347-369.

[9] Ekeland, I and Temam, R (1976), *Convex Analysis and Variational Problems*, North-Holland, 1976

[10] Felippa, C (1987), Will the force method come back? *Trans. ASME, J. Appl. Mech.* **54**, pp. 726-728.

[11] Fichera, G (1964), Problemi elastostatici con vincoli unilaterali; il problema di Signorini con ambigue condizioni al controno, *Mem. Accad. Naz. Lincei*, **8**, no. 7, pp. 91-140.

[12] Gallagher, RH (1993), Finite element structural analysis and complementary energy, *Finite Element in Analysis and Design*, **13**, pp. 115-126.

[13] Gao, DY (1988), Panpenalty finite element programming for limit analysis, *Computers & Structures*, **28**, pp. 749-755.

[14] Gao, DY (1992), Global extremum criteria for finite elasticity, *ZAMP*, **43**, 924-937.

[15] Gao, DY (1994), Stability and extremum principles for post yield analysis of finite plasticity, *Acta Mech Sinica*, **10**, 311-325.

[16] Gao, DY, (1995), Duality theory in nonlinear buckling analysis for von Karman equations, *Studies in Applied Mathematics*, **94**, pp. 423-444.

[17] Gao, DY (1996a), Nonlinear elastic beam theory with application in contact problems and variational approaches, *Mech. Research Commun.*, **23**, no. 1, pp. 11-17.

[18] Gao, DY (1996b), Complementary finite-element method for finite deformation nonsmooth mechanics, *J. Eng. Math.*, **30**, pp. 339-353.

[19] Gao, DY (1997), Dual extremum principles in finite deformation theory with applications in post-buckling analysis of extended nonlinear beam model, *Appl. Mech. Reviews, ASME*, **50**, no. 11, part 2, November, S64-S71.

[20] Gao, DY (1998), Complementary extremum principles in nonconvex parametric variational problems with applications, *IMA J. of Applied Math*, 61, 199–235

[21] Gao, DY (1999), Pure complementary energy principle and triality theory in finite elasticity, *Mech. Research Commun.*, 26, 31–37

[22] Gao, DY (1999a), General analytic solutions and complementary variational principle for large deformation nonsmooth mechanics, *Meccanica*, 34, 169-198

[23] Gao, DY (1999b), Duality Principles in Nonconvex systems, Theory, Methods and Applications, Kluwer Academic Publishers, 454 pp.

[24] Gao, DY and Strang, G (1989a), Geometric nonlinearity: Potential energy, complementary energy and the gap function, *Quartl Appl Math*, **47**(3), 487-504.

[25] Gao, DY and Strang, G (1989b), Dual extremum principles in finite deformation elastoplasitc analysis, *Acta Applicandae Mathematicae,* **17**, 357-267

[26] Han, W and Reddy, BD (1995), Computational plasticity: The variational basis and numerical analysis, *Comput. Mechanics Advances*, **2**, pp. 283-400.

[27] Haslinger, J and Panagiotopoulos, PD (1984), The reciprocal variational approach to the Signorini problem with friction, approximation results, *Proc. Roy Soc. Edinburgh Sect A*, **98**, pp. 365-383.

[28] Hoger, A (1987), The stress conjugate to logarithmic strain, *Int. J. Solids and Struct.*, **23**, no. 12, 1645-1656.

[29] Koiter, WT (1976), On the complementary energy theorem in nonlinear elasticity theory, *Trends in Appl of Pure Math to Mech*, Ed by G Fichera, Pitman, London, pp 207-32.

[30] Lee, SJ and Shield, RT (1980), Variational principles in finite elastics, *J Appl Math Physics (ZAMP)*, **31**, pp 437-453.

[31] Mosco, U (1972), Dual variational inequalities, *J. Math. Analy. Appl.*, **40**, 202-206.

[32] Nemat-Nasser, S (1972), General variational principles in nonlinear and linear elasticity with applications, *Mechanics Today*, **1**, 214-61.

[33] Oden, JT and Reddy, JN (1983), *Variational Methods in Theoretical Mechanics*, Springer-Verlag.

[34] Ogden, RW (1984), *Non-linear elastic deformations*, Ellis Horwood Ltd, Chichester, 417pp.

[35] Panagiotopoulos, PD (1985), *Inequality Problems in Mechanics and Applications*, Birkhäuser, Boston.

[36] Panagiotopoulos, PD (1993), *Hemivariational Inequalities: Applications in Mechanics and Engineering*, Springer-Verlag, 451pp.

[37] Pian, THH and Tong, P (1980), Reissner's principle in finite element formulations. In: S. Nemat-Nasser (ed.) *Mechanics Today*, **5**, Pergamon Press, pp. 377-395.

[38] Reddy, BD (1992), Mixed variational inequalities arising in elastoplasticity, *Nonlinear Analysis, Theory, Methods & Applications,* **19**, no. 11, pp. 1071-1089.

[39] Reissner, E. (1996), *Selected works in applied mechanics and mathematics*, Jones and Bartlett Publishers, Boston, MA, 601pp.

[40] Repin, SI and Xanthis, LS (1996), Aposteriori error estimation for elastoplastic problems based on duality theory, *Comput. Methods Appl. Mech. Engrg.* **138**, pp. 317-339.

[41] Rockafellar, RT (1974), *Conjugate Duality and Optimization*, SIAM.

[42] Sewell, MJ (1987), *Maximum and Minimum Principles*, Cambridge Univ. Press, 468pp.

[43] Strang, G (1986), *Introduction to Applied Mathematics*, Wellesley-Cambridge Press, 758pp

[44] Stumpf, H (1978), Dual extremum principles and error bounds in nonlinear elasticity theory, *J. Elasticity*, **8**, no. 4, 425-438.

[45] Toland, JF (1979), A duality principle for non-convex optimization and the calculus of variations, *Arch. Rational Mech. Anal.* **71,** 41-61.

[46] Tonti, E (1972), A mathematical model for physical theories, *Rend. Accad. Lincei*, **Vol. LII**, I, pp. 133-139; II, pp. 350-356.

[47] Wataru, S and Atluri, SN (1995), On newly developed assumed stress finite element formulations for geometrically and materially nonlinear problems, *Finite Elem. Anal. Des*, **21**, no. 1-2, pp. 75-110.

[48] Wright, SJ (1996), *Primal-Dual Interior-Point Methods*, SIAM, Philadelphia, PA, 289pp.

[49] Yau, ST and Gao, DY (1992), Obstacle problem of von Karman equations, *Adv. Appl. Math.*, **13**, pp. 123-141.

7 CONTACT PROBLEMS IN MULTIBODY DYNAMICS A REVIEW

Friedrich Pfeiffer and Christoph Glocker

Lehrstuhl B für Mechanik
Technische Universität München
D-85747 Garching, Germany

pfeiffer@lbm.mw.tu-muenchen.de
glocker@lbm.mw.tu-muenchen.de

Abstract: The interest in problems of contact mechanics came up in the sixties with questions of statics and elastomechanics. Most of the mathematical tools were developed and applied in those fields. With increasing pressure from the practical side of vibration and noise connected with contact processes methods of multibody dynamics with unilateral constraints have been elaborated. The paper gives a review of these activities mainly based on findings from the authors' institute.

7.1 INTRODUCTION

Contact events appear in dynamical systems very often due to the fact that the world of dynamics usually happens to be as much unilateral as it is bilateral. Walking, grasping, climbing are typically unilateral processes, the operation of machines and mechanisms includes a large variety of unilateral aspects. From this there emerges a need to extend multibody theory by contact phenomena.

All contact processes have some characteristic features in common. If a contact is closed, a motion changes from slip to stick, we come out with some additional constraints generating constraint forces. We then call the contact active. Otherwise it is passive. Obviously transitions in such contacts depend on the dynamics of the system under consideration. The beginning of such a contact event is indicated by kinematical magnitudes like relative distances or relative velocities, the end by kinetic magnitudes like normal force or friction

R.P. Gilbert et al. (eds.), From Convexity to Nonconvexity, 85–109.

force surplus. This will deliver a basis for the mathematical formulation to follow.

A large variety of possibilities exists in modeling local contact physics, from Newton's, Poisson's and Coulomb's laws to a discretization of local behavior by FE- or BE-methods. But, simulations of large dynamical systems require compact contact laws. Therefore, we shall concentrate on the first types of laws which inspite of their simple structure still are able to describe realistically a large field of applications.

Literature covers aspects like contact laws, FEM- and BEM-analysis, contact statics, contact dynamics and a large body of various applications. With respect to multibody systems with multiple unilateral contacts most of the mathematical fundamentals, though firstly regarding statical problems only, were laid down by European scientists. First considerations were started by MOREAU and his school [11], which in the meantime continues his efforts in a remarkable way [1]. MOREAU introduced convex analysis into multibody dynamics and reformulated the classical equations of motion in terms of measure differential inclusions in order to cover both, impact free motion and shocks as they appear in frictional contact problems.

The scientific community includes in addition scientists like PANAGIOTOPOULOS and LÖTSTEDT who developed powerful methods for statical and dynamical problems of contact mechanics [12], [9]. Especially PANAGIOTOPOULOS established a general theory on unilateral problems in mechanics. LÖTSTEDT developed an advanced index-two-type integration algorithm for planar contact problems in rigid body systems [9]. The Swedish school in that field is continued with remarkable results by KLARBRING, who focusses his work to problems of FEM- and BEM-modeling [7].

At the autor's insitute research has been performed in that field for more than ten years which is mostly summarized in the book by PFEIFFER, GLOCKER [15]. Some newer results on nonlinear complementarity problems may be found in WÖSLE [20]. As the author's institute is one of engineering mechanics, most of the research work deals with a transfer of the demanding mathematical fundamentals to an engineering and application-friendly level.

7.2 THE EVOLUTION OF A THEORY

Woodpecker Toy

The first example with structure-variant properties, which was analysed at the author's institute was a woodpecker toy which operates by self-excited vibrations (Fig. 7.1 and [13]). The theory applied 1984 was at that time incomplete because no reasonable concept for impacts with friction was available. Therefore friction was included empirically. The results compared very well with measurements. In the meantime an impact-with-friction-theory exists [3], [4].

Its application to the woodpecker example confirms the simpler theory at the same time achieving improvements with respect to friction modeling. Figure 7.2 depicts some results in the form of phase portraits [3]. The data used for the woodpecker may be seen from [15].

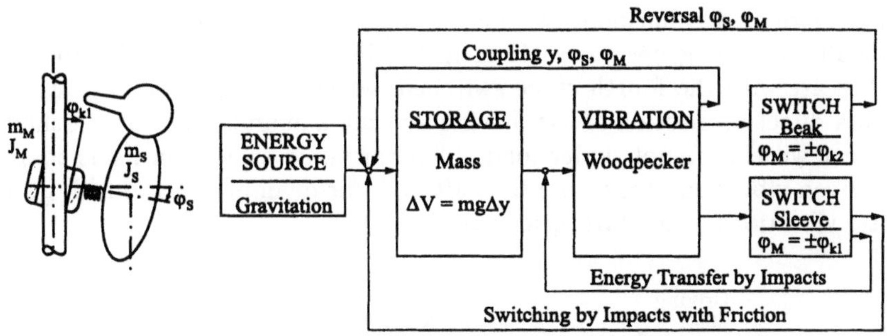

Figure 7.1 Woodpecker's self-excited vibration.

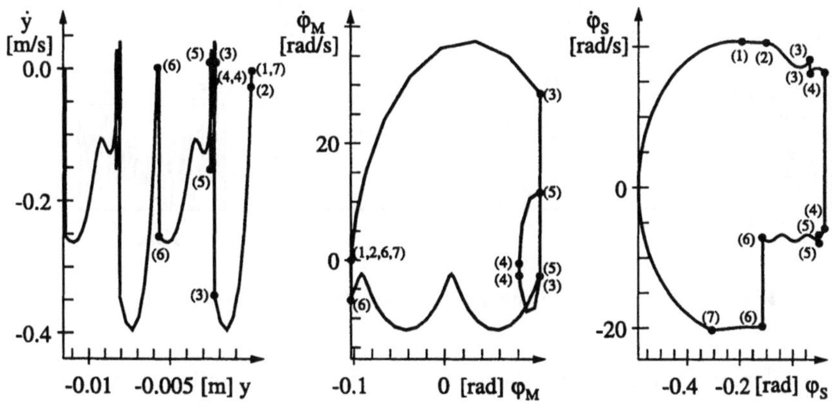

Figure 7.2 Phase Space Portraits [3].

The main phases of motion are (Fig. 7.2): (1 – 2) sliding, (2) loss of contact, (2 – 3) free fall with high frequency oscillation of 73 Hz, (3) inelastic impact of the upper edge of the sleeve, (4) beak impact, (5 – 6) free fall downward, (6 – 7) transition to sleeve sticking and self-locking, (7 – 1) woodpecker angular motion with a low frequency oscillation of around 9 Hz, which has also been approved by measurements.

Gear Rattling

Gear rattling is a noise problem appearing in gear stages not under load. Examples are change-over gears or gear-driven balancers. Figure 7.3 illustrates a five-stage gear configuration where the driving shaft moves the countershaft, and according to the switched gear, the countershaft drives the main shaft. All gear wheels mesh. Switching is performed by synchronizers connecting the appropriate wheel with the main shaft. The wheel is loaded and all other wheels are not under load and can rattle due to backlash in the mesh of the gears. The driving shaft is excited by the torsional vibrations of the drive system, which are more or less harmonical. These excitations are transported to the countershaft where it generates rattling in the loose gear wheels. Figure 7.3 shows the case with the fourth stage switched, which for this type of gearbox results in a direct connection of the driving and main shafts. The countershaft and all gear wheels are not under load. The harmonic excitation leads to an impact driven vibration of the countershaft and, for example, in the fifth gear to chaotic vibrations of the fifth gear wheel.

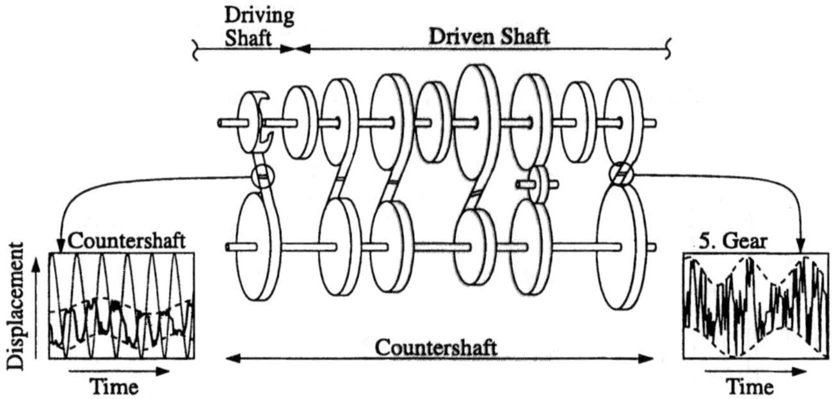

Figure 7.3 Five-stage change-over-gear, fourth gear switched.

The problem is analyzed with the same theory as being applied to the woodpecker example in an early stage [13]. For rattling a multibody approach with multiple impacts turns out to be sufficient. Following [15] we describe a multibody configuration with constraints in normal direction by

$$\boldsymbol{M}\ddot{\boldsymbol{q}} - \boldsymbol{h} - \boldsymbol{W}_N\boldsymbol{\lambda}_N = 0; \quad \dot{\boldsymbol{g}}_N = \boldsymbol{W}_N^T\dot{\boldsymbol{q}} + \hat{\boldsymbol{w}}_N \tag{7.1}$$

where $\boldsymbol{q} \in \Re^f$ are generalized coordinates, $\boldsymbol{M} \in \Re^{f,f}$ is the mass matrix, $\boldsymbol{h}$ includes all forces, $\boldsymbol{\lambda}_N \in \Re^{n_s}$ are constraint forces in normal contact direction, $\boldsymbol{W}_N \in \Re^{f,n_s}$ is the constraint matrix, $\dot{\boldsymbol{g}}_N \in \Re^{n_s}$ the relative velocities in the contacts and $\hat{\boldsymbol{w}}_N$ are usually excitation terms. Assuming very short impacts we reduce the force-acceleration level to an impulse-velocity level:

$$\boldsymbol{M}(\dot{\boldsymbol{q}}_E - \dot{\boldsymbol{q}}_A) - \boldsymbol{W}_N\boldsymbol{\Lambda}_N = 0 \quad \text{with } \boldsymbol{\Lambda}_N = \lim_{t_E \to t_A} \int_{t_A}^{t_E} \boldsymbol{\lambda}_N dt \tag{7.2}$$

(indices A and E for the beginning and the end of the impact, respectively). Introducing Newton's law of impact

$$\dot{\boldsymbol{g}}_{NE} = -\boldsymbol{\epsilon}\dot{\boldsymbol{g}}_{NA} \tag{7.3}$$

(ϵ matrix of coefficients of restitution), we may combine the equations (7.1), (7.2), (7.3) to give

$$\begin{aligned} \dot{g}_{NE} &= \dot{g}_{NA} + G_N \Lambda_N \\ \Lambda_N &= -G_N^{-1}(E + \epsilon_N)\dot{g}_{NA} \\ \dot{q}_E &= \dot{q}_A - M^{-1} W_N G_N^{-1}(E + \epsilon_N)\dot{g}_{NA} \\ G_N &= W_N^T M^{-1} W_N. \end{aligned} \tag{7.4}$$

It turns out that an excellent measure for the noise intensity of gear rattling can be achieved by a sum over all impulses Λ_N

$$\text{Noise Intensity} \sim \sum |\Lambda_N|, \tag{7.5}$$

which represents a relative measure, not an absolute one. Many applications to real gears confirm however the fact, that the measure (7.5) describes all parameter dependencies correctly.

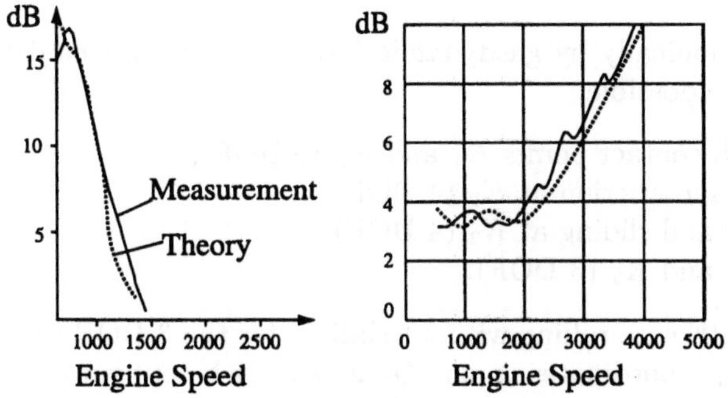

Figure 7.4 Rattling theory versus measurements.

Figure 7.4 depicts two examples for the change-over-gears of a pick-up-car (left) and an upper middle-class car (right). Only one measurement point has been considered for adapting a proportionality factor for eq. (7.5).

Turbine Blade Damper

The two examples of the woodpecker toy and gear rattling afford only impact theory with certain extensions to friction. The operation of a turbine blade damper is completely based on stick-slip processes, which was at that time a step into new research efforts [5], [14].

Turbine blade dampers are parabolic sheet steel devices often used in gas turbines with the requirement to damp the blade vibrations by relative motion through dry friction. Figure 7.5 shows a typical configuration. The dampers are located between the blade platforms of two neighboring blades and pressed against the oblique plane by centrifugal forces.

The system was approximated in this case by a two-dimensional model with seven degrees of freedom and two contact zones (see Fig. 7.5). The blades are

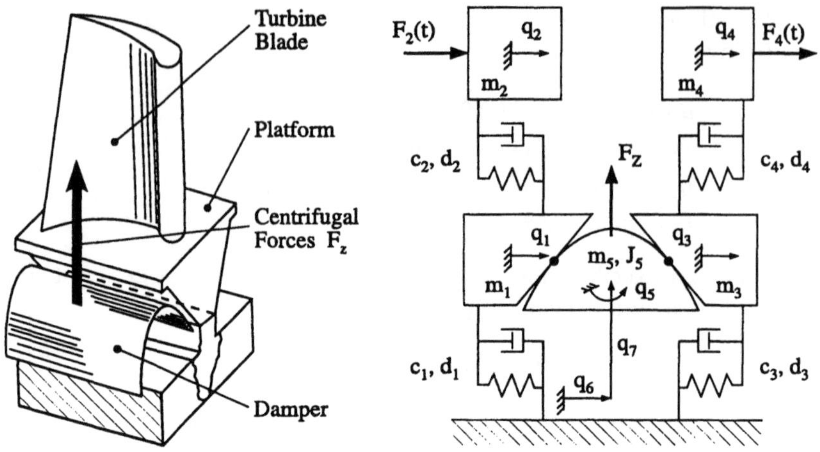

Figure 7.5 Turbine blade damper and its mechanical model.

excited approximately harmonically by gasdynamic forces. For such a model the following movements are possible:

Sliding at both contact points K_1 and K_2 (5 DOF),
Sliding at K_1 and stiction at K_2 (4 DOF),
Stiction at K_1 and sliding at K_2 (4 DOF),
Stiction at K_1 and K_2 (3 DOF).

Stiction means for the 4 DOF case rolling without sliding, for the 3 DOF case a reduction of the (q_1, q_3, q_5) coordinates to one DOF. For each configuration a set of equations of motion can be established, which describe the motion as long as no transition occurs. In the case of a blade damper this transitions follow typical stick-slip behavior:

- *Transition from sliding to stiction:*
  Relative tangential velocity in one of the contact becomes zero, and at the same time the static friction force must be larger than the tangential constraint force.

- *Transition from stiction to sliding:*
  The tangential constraint force becomes equal to the static friction force, and at the same time the tangential acceleration starts to be non-zero.

After such a transition event the corresponding set of equations of motion has been selected for further evaluation up to the next transition [5]. For such evaluation processes computing time can be reduced significantly by first determining the frictionless equilibria, which may be achieved by solving a set of nonlinear algebraic equations resulting from the equations of motion with velocities and accelerations equal to zero [15].

Theory has been compared with measurements. For these experiments the centrifugal force is replaced by a spring force, the excitation is realized elec-

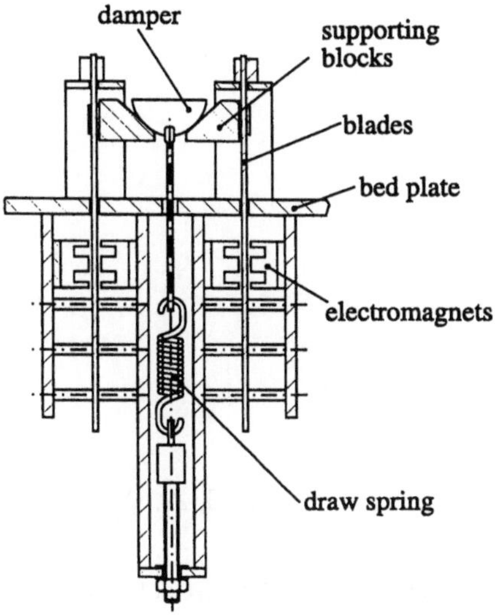

Figure 7.6 Test-set-up for a turbine blade damper.

tromagnetically, and the blades are represented by two bars. The geometry of the damper and the blade platforms is the same as for real dampers, but on a larger scale. The amplitudes of the damper and the blades are measured by six inductive displacement transducers and the spring force by a strain gauge arrangement. Figure 7.6 illustrates the setup and the measurement scheme to experimentally determine the stick-slip motion.

One must keep in mind that the centrifugal forces in gas turbines are so large that the dampers move only occasionally. With respect to some time unit they move only about 10% of that time. Therefore, a slight improvement is worth an effort. One important magnitude of a damper is the excitation force amplitude F_E necessary to break away from a stiction situation in both contact points K_1, K_2 in Fig. 7.5. These results give a measure for the damping capability of the devices. Figure 7.7 shows three curves for three different values of the spring force (centrifugal force) and for a contact angle of $\gamma = 50^o$. The excitation force amplitude $F_E[N]$ must be augmented for larger centrifugal (spring) forces F_S. The very small F_E values for an excitation frequency at 61 Hz can be explained in the following way. The value of $f_E = 61$ Hz corresponds to an eigenfrequency of the blocked blade-damper system. This means that an excitation with these frequencies leads to large amplitudes of the blades and as a consequence to a considerable reduction of stiction. Therefore, the damper devices move from stick to slip at very small excitation force amplitudes.

Landing Impact of an Airplane

The landing impact of an elastic airplane is governed by the airplane configuration and its elasticity and by the main and nose carriages. The practical

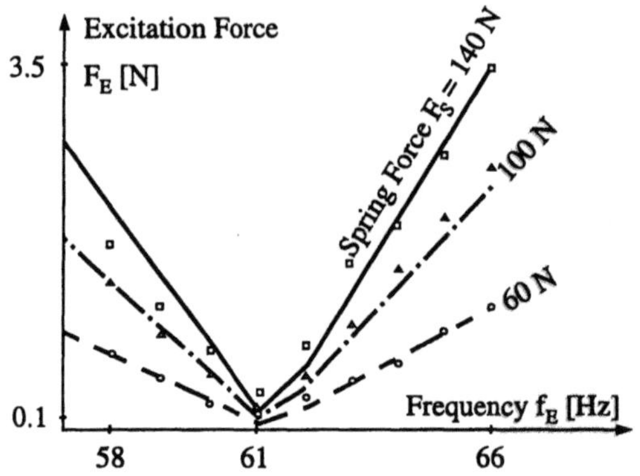

Figure 7.7 Comparison theory (—,- - -, -··-) and measurement (□, △, ○).

example of a landing airplane was the first one where the different sets of equations of motion could not be writen down explicitly, at least not with reasonable effort, but where some theoretical algorithm had to be established selecting the appropriate equations automatically [2], [17]. The internal structure of the carriage systems leads to impacts and to stick-slip phenomena, which influence the landing dynamics significantly.

A plane elastic model was established with the following bodies [15]:

$i = 1 \rightarrow$ elastic fuselage,
$i = 2 \rightarrow$ elastic main carriage,
$i = 3 \rightarrow$ elastic nose carriage,
$i = 4 \rightarrow$ rigid main carriage wheels,
$i = 5 \rightarrow$ rigid nose carriage wheels.

The principal configuration is depicted in Fig. 7.8. For the shock absorbers it will be assumed that their bending deformation is homogeneous over the variable length, which means that the two components of the absorber do not bend in a different way. It turns out that this will be a sufficient approximation.

During operation we have to consider various continuous and discontinuous nonlinearities. The tires possess a stiffness characteristic which increases progressively with the tire deformation. Moreover, the spin-up process of the wheels is governed by a highly nonlinear friction-slippage relationship. The shock absorber itself reveals some nonlinear features. The two gas chambers follow a polytropic compression and expansion law, the fluid flow losses induce a quadratic damping behavior, and the friction forces from ground reaction forces are nonlinear also.

Discontinuous nonlinearities enter through the upper and lower stops within the shock absorber and through stick-slip processes between the two shock absorber cylinders (Fig. 7.9). Mechanically and mathematically they follow the theory as presented in the next section (see also [15]).

The degrees of freedom resulting from this model include rigid degrees of freedom (number f_r) and elastic degrees of freedom (number f_e). With $\boldsymbol{q}_r \in$

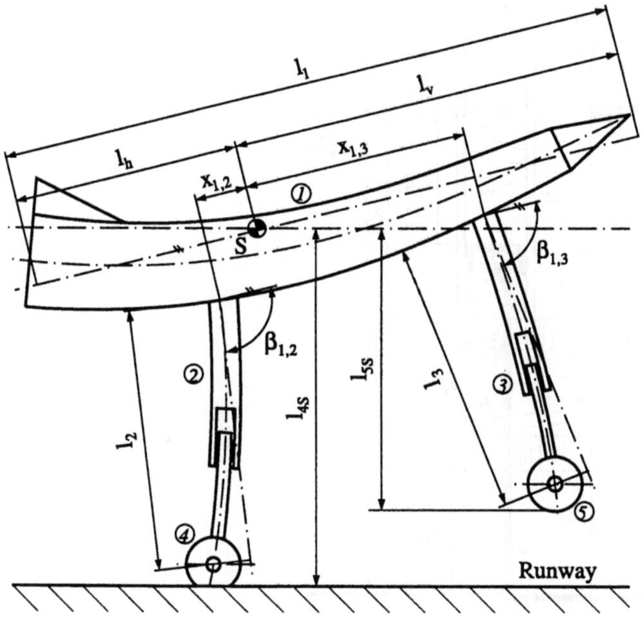

Figure 7.8 Planar mechanical model of a landing elastic airplane.

$\Re^{f_r}$ and $\boldsymbol{q}_e \in \Re^{f_e}$ we obtain

$$\begin{aligned} \boldsymbol{q}^T &= [\boldsymbol{q}_r^T, \boldsymbol{q}_e^T] \\ &= [x, y, \alpha, y_{22}, y_{23}, \varphi_2, \varphi_3, \boldsymbol{q}_1^T, \boldsymbol{q}_2^T, \boldsymbol{q}_3^T] \in \Re^{f=f_r+f_e} \\ \boldsymbol{q}_i^T &= [q_{i1}, q_{i2}, \ldots, q_{in_{ai}}], \quad i = 1,2,3 \\ n_{ai} &: \text{ number of shape functions of body } i \\ f &: \text{ number of minimal coordinates.} \end{aligned} \tag{7.6}$$

The $\boldsymbol{q}_i \in \Re^{n_{ai}}$ are the elastic coordinates, (x, y, α) the three rigid degrees of freedom of the fuselage, (y_{22}, y_{23}) the translational and (φ_2, φ_3) the rotational carriage deflections. The derivation of the equations of motion follows the theory presented below and in [15]. Details may also be found in [2], [17].

Simulations consider a landing impact on a plane surface with a medium general aviation propeller-driven airplane consisting of a rigid fuselage and rigid carriages. The touchdown velocity is 4.3 m/s, and the horizontal velocity is 60 m/s. The landing angle of attack is $\alpha = 13^o$, and the angular velocity is $\dot{\alpha} = 0$. The initial values of the geometric magnitudes y, y_{22}, y_{23} are chosen in such a way that touchdown of the main carriage wheels starts with zero forces. The nose wheel has no ground contact for these initial values ($\alpha = 13^o$) [2], [15].

Elasticity has been considered by a Ritz approach and by taking into account the first five eigenfrequencies for the fuselage and the first two eigenfrequencies for the main and nose carriages. All other data were the same as for the rigid airplane.

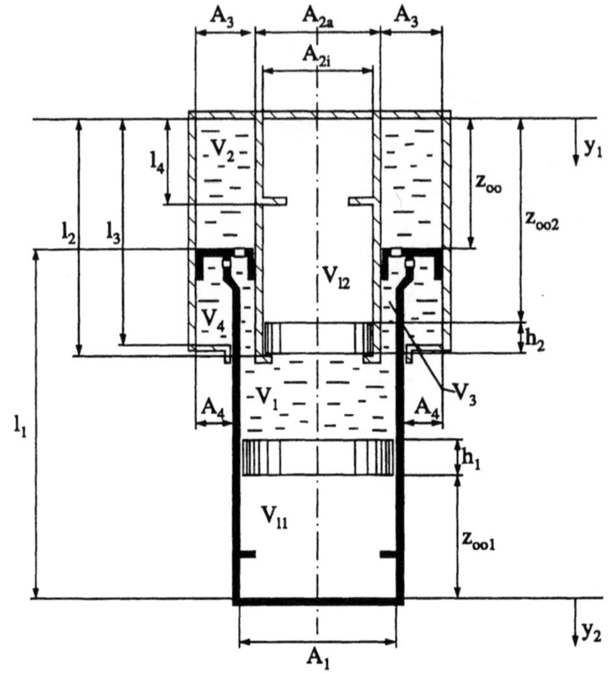

Figure 7.9 Two-chamber shock absorber model.

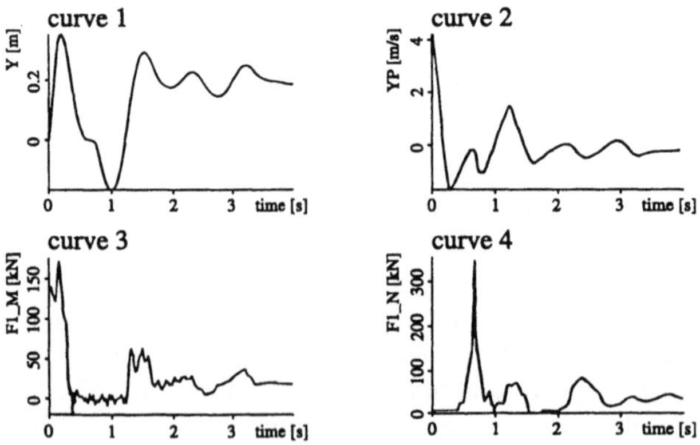

Figure 7.10 Elastic airplane landing impact — airplane results [2].

Figure 7.10 illustrates the situation of the airplane for the above landing impact. The center of mass moves about 300 mm (curve 1), and its velocity has a maximum at the beginning (curve 2). The forces acting on the fuselage from the main and nose carriages reach a maximum of about 170 kN for the main absorbers (curve 3) and a peak value of about 450 kN for the nose, where the nose load builds up later (curve 4).

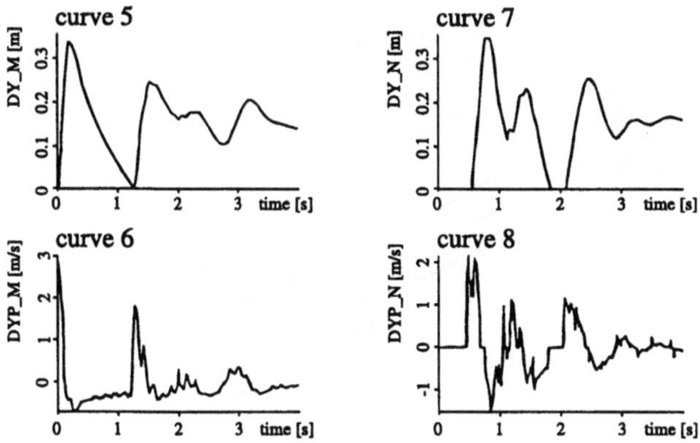

Figure 7.11 Elastic airplane landing impact — carriage results [2].

The displacements are more or less the same as in the rigid case, the diagrams of which are not given here (see [15]). The nose wheel performs an internal impact at the upper stop (curve 7), and in the force and stroke velocity diagrams the lowest elastic fuselage eigenfrequency (~7 Hz) can be detected. The elasticity neither increases nor decreases the tendencies for impacts and stick-slip (curves 8 in Fig. 7.11), but it reduces significantly the maximum load on the nose carriage.

Assembly Processes

Mating parts together involves various and sometimes quickly changing combinations of impacts with and without friction, of stick-slip processes and even of jamming. As these combinations and the transitions between them cannot be predicted beforehand because they depend on the current state, we need a theory which is able to deal with such processes. At the autor's institute this exactly has been the stage to enter into the methods of complementarity, sub-differentials and convex analysis [3], [15], [16].

Within the practical application of these to a certain extend abstract theories we usually work on different levels: The event of a transition between contact configurations is either indicated by magnitudes of relative kinematics (at the beginning) or by those of constraint forces (at the end). The relative kinematical magnitudes undergo a metamorphosis from indicators to constraint equations as components of the constraint matrices valid for the contact configuration to come. The transition itself is governed by the complementarity of relative kinematics and constraint forces: If one group is zero the other is not. Considerations of that kind in combination with the accompanying mathematics [12], [15] allow an efficient modeling of assembly processes as described above. As an example we regard the peg-in-hole insertion performed by a manipulator [10], [18].

A robotic manipulator is usually considered as a tree-like multibody system with rigid and elastic components. Links very often are modelled as rigid

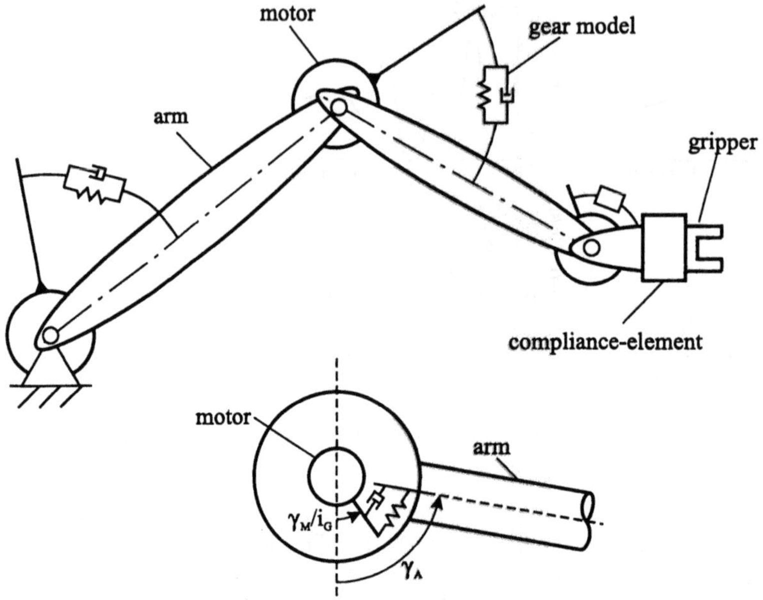

Figure 7.12 Mechanical models of links and joints. $\tau_A = c(\gamma_M/i_G - \gamma_A) + d(\dot{\gamma_M}/i_G - \dot{\gamma_A})$, where τ_A is the torque at robot arm. γ_M, γ_A are the coordinates of motor and arm, respectively, c the joint stiffness, d the joint damping coefficient, and i_G the gear ratio.

bodies, only for special tasks as elastic bodies. Joints may ideally follow a given torque law, or they may be elastic with own degrees of freedom (see Fig. 7.12). Therefore we may subdivide our total number n of degrees of freedom into motor or internal degrees of freedom (n_M), external ones (n_A) and elastic ones (n_C), which results in

$$\gamma = \begin{pmatrix} \gamma_M \\ \gamma_A \\ \gamma_C \end{pmatrix} \in \Re^n, \qquad n = n_M + n_A + n_C. \tag{7.7}$$

The equations of motion are then derived by applying Jourdain's principle which says that constraint forces in the joints are perpendicular to the direction of motion and do not affect power. With given kinematics of the manipulator under consideration, the mobility is specified by the Jacobians $\boldsymbol{J}_T = \partial \boldsymbol{v}/\partial\dot{\gamma}$, $\boldsymbol{J}_R = \partial\boldsymbol{\omega}/\partial\dot{\gamma}$, where $\boldsymbol{v}, \boldsymbol{\omega}$ are the translational and rotational velocities, respectively. When we premultiply Newton/Euler's equations for each individual body with respective Jacobians the equations of motion can recursively be generated in the final form:

$$\boldsymbol{M}(\gamma)\,\ddot{\gamma} + \hat{\boldsymbol{h}}(\gamma, \dot{\gamma}) = \boldsymbol{B}\hat{\boldsymbol{u}} \tag{7.8}$$

with the inertia matrix $\boldsymbol{M} \in \Re^{n,n}$, centrifugal, coriolis and gravitational forces $\hat{\boldsymbol{h}} \in \Re^n$ and the control input $\boldsymbol{B}\hat{\boldsymbol{u}} \in \Re^n$.

Because most mating tasks in assembly cells are limited to a small area of workpiece interaction the robot motion will be slow and centrifugal and coriolis forces may in the following be neglected compared to gravitational and inertia forces. Hence, the robot dynamics can be linearized around a given system state $(\gamma_0, \dot{\gamma}_0) = 0$ and deviations are given by

$$q = \begin{pmatrix} q_M \\ q_A \\ q_C \end{pmatrix} = \gamma - \gamma_0, \qquad \gamma_0 = \begin{pmatrix} I_G \gamma_{A,0} \\ \gamma_{A,0} \\ 0 \end{pmatrix}, \tag{7.9}$$

where the matrix $I_G \in \Re^{n_M, n_A}$ represents the gear ratio that transforms the arm coordinates of the reference position onto the respective motor coordinates. The linearized equations of motion for the open loop dynamics then are

$$M\ddot{q} + P\dot{q} + Qq = h + Bu, \tag{7.10}$$

where $P \in \Re^{n,n}$ is the velocity proportional matrix containing damping terms of joints and compliance and $Q \in \Re^{n,n}$ is the stiffness matrix. The vector $h \in \Re^n$ contains the entire nominal dynamics and feedforward control with τ_{M_0}, where the motor drive torque results from $\tau_M = \tau_{M_0} + u$, u being the feedback torque. $Bu \in \Re^n$ is feedback control with the controller matrix B. For PD position control the feedback term typically has the form

$$u = -K_P(q_M - q_{M_0}) - K_D(\dot{q}_M - \dot{q}_{M_0}). \tag{7.11}$$

During assembly the workpiece held by the robot comes into various contact situations with the complementary mating part. This is illustrated in Fig. 7.13 for a planar peg-in-hole insertion task. In this process one or more contacts may arise which constrains the gripper motion and thus the robot's mobility by closing the kinematical loop of the manipulator.

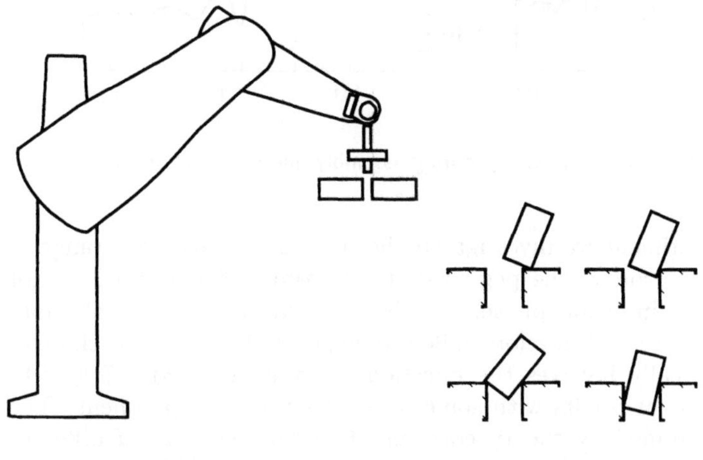

Figure 7.13 Various contact configurations for the planar peg-in-hole problem.

The experimental investigations are focussed on a verification of the theory for various configurations of assembly processes and on a consideration of disturbances during the peg-in-hole insertion. We first consider the perturbed motion during a planar peg-in-hole insertion task where forces only develop in the presence of uncertainties. For the verification of our theoretical approach experiments with a five degrees-of-freedom laboratory robot with a force-torque sensor were carried out. The fixture housing the complementary part is equipped with six distance sensors that measure the gripper's position and orientation and in addition disposes of two translational and three rotational adjustments which allow to produce definite positioning errors with respect to the parts that are to be mated.

In a first experiment we investigated the insertion process when there was a lateral error between the peg and the hole. The respective numerical and experimental results are depicted in Fig. 7.14. The peaks in the force history at $t = 0.9$ s result from the impact when the peg hits the chamfer. When two-point-contact is reached at $t = 1.2$ s the manipulator's motion is slowed down very fast until the jamming condition is reached. Only when the controller torques become large enough the motion is continued with sliding in two and finally even one contact point. This cycle represented by the oscillations in the force history repeats until the end of the trajectory and demonstrates the phenomenon of changing constraints during an assembly operation.

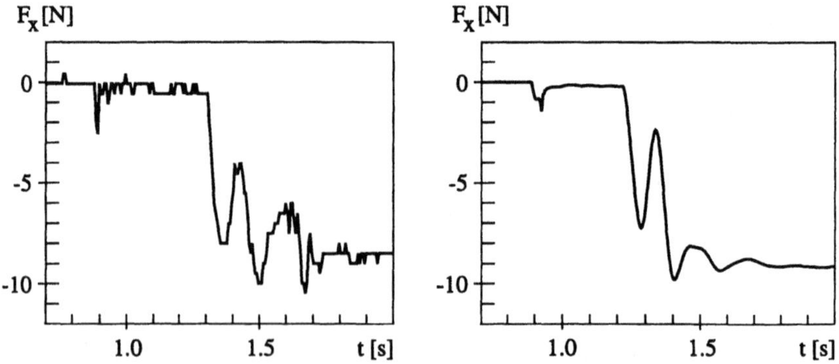

Figure 7.14 Insertion force F_x during assembly due to lateral offset.

In a second experiment we investigated the permissible tolerance range of lateral and angular errors of the peg so that a maximum force of 5 N is not exceeded during the insertion phase. In this context we did not take into consideration the forces that develop while the peg is guided along the chamfer since this automatically involves the question of chamfer design. Fig. 7.15 contrasts the numerical results with some data obtained by experiment. The respective lines identified by theory correspond to the existence of different contact configurations where the maximum insertion force of 5 N is reached and limit the range of permissible positioning tolerances. The difference between numerical and experimental data mainly results from backlash in the last joint of the robot which was not modeled. Note that in both experiments the sensor

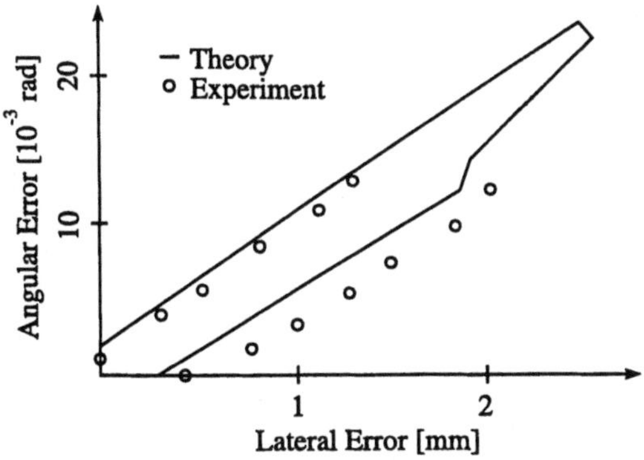

Figure 7.15 Permissible positioning errors for a maximum insertion force of $5N$.

served only as a force measuring device for data-processing, since a simple PD joint controller was used.

7.3 PRESENT MATHEMATICAL FORMULATION

Extensive experience with practical engineering problems has brought up the theory to a level which allows a very straightfoward formulation of contact problems in multibody systems. The formulation is kept as close as possible to the classical theory used in applied dynamics. This enables engineers to capture the main ideas of multivalued force laws quickly, to understand the theory behind them intuitively and to apply them in a correct manner. Even teaching this subject in a graduate course has turned out to be successful. In the following we will give a brief overview of the structure of the equations in the case of spatial Coulomb friction.

Geometry of Surfaces

We consider two bodies moving with $\boldsymbol{v}_{Pi}, \boldsymbol{\Omega}_i$ $(i = 1,2)$ where $\boldsymbol{v}_{Pi}$ denotes the velocity of the body-fixed point P of body i and $\boldsymbol{\Omega}_i$ its angular velocity. Both bodies are assumed to be convex, at least in a neighborhood of a potential contact point, and may be represented by a parametrization of their surfaces in the form $\boldsymbol{r}_{P\Sigma i} = \boldsymbol{r}_{P\Sigma i}(\boldsymbol{q}, \xi_i, \eta_i)$, see Figure 7.16 [10].

Note that $\boldsymbol{r}_{P\Sigma}$ only depends on the surface parameters ξ and η when stated in the corresponding body-fixed frame whereas an additional dependence on the generalized coordinates $\boldsymbol{q}$ is observed when using arbitrary frames. For each of the bodies one defines an orthonormal basis $(\boldsymbol{u}, \boldsymbol{v}, \boldsymbol{n})$ via the tangents

$$\boldsymbol{s} := \partial_\xi \boldsymbol{r}_{P\Sigma}, \quad \boldsymbol{t} := \partial_\eta \boldsymbol{r}_{P\Sigma} \tag{7.12}$$

in order to obtain

$$\boldsymbol{n} := \frac{\boldsymbol{s} \times \boldsymbol{t}}{|\boldsymbol{s} \times \boldsymbol{t}|}, \quad \boldsymbol{u} := \frac{\boldsymbol{s}}{|\boldsymbol{s}|}, \quad \boldsymbol{v} := \boldsymbol{n} \times \boldsymbol{u} \tag{7.13}$$

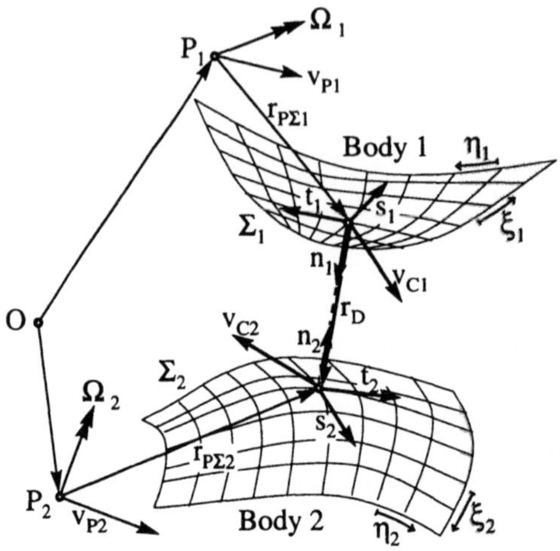

Figure 7.16 The spatial contact problem.

where we agree that (ξ, η) are ordered such that $\boldsymbol{n}$ becomes the outward normal. The (absolute) changes in time of the contour vector $\dot{\boldsymbol{r}}_{P\Sigma}$ and the trihedral $(\dot{\boldsymbol{u}}, \dot{\boldsymbol{v}}, \dot{\boldsymbol{n}})$ are

$$\begin{aligned} \dot{\boldsymbol{r}}_{P\Sigma} &= \boldsymbol{\Omega} \times \boldsymbol{r}_{P\Sigma} + \boldsymbol{R}\dot{\boldsymbol{\zeta}} \quad \text{with } \boldsymbol{R} = \partial_\zeta \boldsymbol{r}_{P\Sigma} = (\boldsymbol{s}, \boldsymbol{t}) \\ \dot{\boldsymbol{k}} &= \boldsymbol{\Omega} \times \boldsymbol{k} \quad + \boldsymbol{K}\dot{\boldsymbol{\zeta}} \quad \text{with } \boldsymbol{K} = \partial_\zeta \boldsymbol{k} \end{aligned} \tag{7.14}$$

where $\boldsymbol{\zeta} = (\xi, \eta)^T$ and $(\boldsymbol{k}, \boldsymbol{K})$ is any of $(\boldsymbol{u}, \boldsymbol{U})$, $(\boldsymbol{v}, \boldsymbol{V})$, $(\boldsymbol{n}, \boldsymbol{N})$. The velocity $\boldsymbol{v}_\Sigma$ of a point moving on the surface in terms of the rigid body motion $(\boldsymbol{v}_P, \boldsymbol{\Omega})$ is thus given by $\boldsymbol{v}_\Sigma = \boldsymbol{v}_P + \dot{\boldsymbol{r}}_{P\Sigma}$ which yields

$$\boldsymbol{v}_\Sigma = \boldsymbol{v}_C + \boldsymbol{R}\dot{\boldsymbol{\zeta}} \quad \text{with } \boldsymbol{v}_C = \boldsymbol{v}_P + \boldsymbol{\Omega} \times \boldsymbol{r}_{P\Sigma} \tag{7.15}$$

Here, $\boldsymbol{v}_C$ denotes the rigid body velocity portion of $\boldsymbol{v}_\Sigma$. Analogously we may find the acceleration $\boldsymbol{a}_C = \dot{\boldsymbol{v}}_C$ by differentiation of the second equation in (7.15),

$$\boldsymbol{a}_C = \boldsymbol{a}_Q + \boldsymbol{\Omega} \times \boldsymbol{R}\dot{\boldsymbol{\zeta}} \quad \text{with } \boldsymbol{a}_Q = \boldsymbol{a}_P + \dot{\boldsymbol{\Omega}} \times \boldsymbol{r}_{P\Sigma} + \boldsymbol{\Omega} \times (\boldsymbol{\Omega} \times \boldsymbol{r}_{P\Sigma}). \tag{7.16}$$

Finally we express $\boldsymbol{v}_C$ and $\boldsymbol{a}_C$ by the generalized magnitudes $\dot{\boldsymbol{q}}$ and $\ddot{\boldsymbol{q}}$,

$$\begin{aligned} \boldsymbol{v}_C &= \boldsymbol{J}_C \dot{\boldsymbol{q}} + \hat{\boldsymbol{\iota}}_C \\ \boldsymbol{a}_C &= \boldsymbol{J}_C \ddot{\boldsymbol{q}} + \bar{\boldsymbol{\iota}}_C, \qquad \bar{\boldsymbol{\iota}}_C = \dot{\boldsymbol{J}}_C \dot{\boldsymbol{q}} + \dot{\hat{\boldsymbol{\iota}}}_C, \end{aligned} \tag{7.17}$$

which involves the Jacobian $\boldsymbol{J}_C$ that will later turn out to be the Jacobian of the contact point.

Contact Kinematics

Within the framework as presented in the foregoing section one is able to derive the main equations of contact kinematics in dynamics which are: The definition of the contact points, their distance, their relative velocities with respect to the normal and the tangential directions, and the corresponding relative accelerations. In order to do the first step one introduces a distance vector

$$\boldsymbol{r}_D := \boldsymbol{r}_{O\Sigma 2} - \boldsymbol{r}_{O\Sigma 1}, \tag{7.18}$$

cp. [3], where $\boldsymbol{r}_{O\Sigma i}$ are vectors pointing from a common inertially fixed point O to the surfaces under consideration.

Next, the two tangent planes spanned by $(\boldsymbol{u}_i, \boldsymbol{v}_i)$ are adjusted parallel to each other such that the distance vector becomes parallel to each of the normals. The surface parameters $(\boldsymbol{\zeta}_1^\star, \boldsymbol{\zeta}_2^\star)$ corresponding with that situation solve the four nonlinear equations

$$\boldsymbol{n}_1^T\boldsymbol{u}_2 = 0, \quad \boldsymbol{n}_1^T\boldsymbol{v}_2 = 0, \quad \boldsymbol{r}_D^T\boldsymbol{u}_1 = 0, \quad \boldsymbol{r}_D^T\boldsymbol{v}_1 = 0 \tag{7.19}$$

and are called the contact parameters; the corresponding points at the surfaces given by $\boldsymbol{r}_{P\Sigma i}(\boldsymbol{\zeta}_i^\star)$ are called the contact points. When the bodies are moving the conditions (7.19) for the contact points should not change [3], [10], thus

$$\begin{aligned} \dot{\boldsymbol{n}}_1^T\boldsymbol{u}_2 + \boldsymbol{n}_1^T\dot{\boldsymbol{u}}_2 &= 0, & \dot{\boldsymbol{r}}_D^T\boldsymbol{u}_1 + \boldsymbol{r}_D^T\dot{\boldsymbol{u}}_1 &= 0 \\ \dot{\boldsymbol{n}}_1^T\boldsymbol{v}_2 + \boldsymbol{n}_1^T\dot{\boldsymbol{v}}_2 &= 0, & \dot{\boldsymbol{r}}_D^T\boldsymbol{v}_1 + \boldsymbol{r}_D^T\dot{\boldsymbol{v}}_1 &= 0 \end{aligned} \tag{7.20}$$

must hold which constitutes a system of four linear equations in $\dot{\boldsymbol{\zeta}}_i^\star$ when substituting $(\dot{\boldsymbol{n}}_i, \dot{\boldsymbol{u}}_i, \dot{\boldsymbol{v}}_i)$ from (7.14) and $\boldsymbol{v}_{\Sigma i}$ from (7.15) with $\dot{\boldsymbol{r}}_D = \boldsymbol{v}_{\Sigma 2} - \boldsymbol{v}_{\Sigma 1}$.

After the solution $(\boldsymbol{\zeta}_1^\star, \boldsymbol{\zeta}_2^\star)$ of (7.19) has been found the distance g_N of the contact points can be calculated as

$$g_N = \boldsymbol{n}_1^T\boldsymbol{r}_D \tag{7.21}$$

with values greater than zero for separation and values less than zero for overlapping. The relative velocities of the contact points with respect to the three directions $(\boldsymbol{u}_1, \boldsymbol{v}_1, \boldsymbol{n}_1)$ are then given by

$$\dot{g}_K = (\boldsymbol{v}_{C2} - \boldsymbol{v}_{C1})^T\boldsymbol{k}_1, \qquad (K, \boldsymbol{k}_1) = (U, \boldsymbol{u}_1), (V, \boldsymbol{v}_1), (N, \boldsymbol{n}_1) \tag{7.22}$$

where it can be easily verified that differentiation of (7.21) with respect to time results in (7.22) for $(K, \boldsymbol{k}_1) = (N, \boldsymbol{n}_1)$. The temporal changes of the relative velocities $\dot{g}_K$ in (7.22) are given by

$$\ddot{g}_K = (\boldsymbol{a}_{C2} - \boldsymbol{a}_{C1})^T\boldsymbol{k}_1 + (\boldsymbol{v}_{C2} - \boldsymbol{v}_{C1})^T\dot{\boldsymbol{k}}_1, \tag{7.23}$$

where $\boldsymbol{a}_{Ci}$ is known from (7.16), $\dot{\boldsymbol{k}}_1$ may be taken from (7.14), and $\dot{\boldsymbol{\zeta}}_i^\star$ involved in $\boldsymbol{a}_{Ci}$ and $\dot{\boldsymbol{k}}_1$ is the solution of (7.20).

Substituting $\boldsymbol{v}_{Ci}$ and $\boldsymbol{a}_{Ci}$ from (7.17) into (7.22), (7.23) yields a representation of the relative velocities and accelerations in the form

$$\begin{aligned} \dot{g}_K &= \boldsymbol{w}_K^T\dot{\boldsymbol{q}} + \hat{w}_K && \text{with } \boldsymbol{w}_K = (\boldsymbol{J}_{C2} - \boldsymbol{J}_{C1})^T\boldsymbol{k}_1; \quad \hat{w}_K = (\hat{\boldsymbol{\iota}}_{C2} - \hat{\boldsymbol{\iota}}_{C1})^T\boldsymbol{k}_1 \\ \ddot{g}_K &= \boldsymbol{w}_K^T\ddot{\boldsymbol{q}} + \bar{w}_K && \text{with } \bar{w}_K = \dot{\boldsymbol{w}}_K^T\dot{\boldsymbol{q}} + \dot{\hat{w}}_K. \end{aligned} \tag{7.24}$$

Finally we arrange the two tangential relative velocities in the vector $\dot{\boldsymbol{g}}_T$, i.e. $\dot{\boldsymbol{g}}_T^T := (\dot{g}_U, \dot{g}_V)$ which gives us the sliding direction of the contact points in the common tangent plane. With the abbreviations $\boldsymbol{W}_T := (\boldsymbol{w}_U, \boldsymbol{w}_V)$, $\hat{\boldsymbol{w}}_T^T = (\hat{w}_U, \hat{w}_V)$, $\bar{\boldsymbol{w}}_T^T = (\bar{w}_U, \bar{w}_V)$ we rewrite (7.24) in the form

$$\begin{aligned} \dot{g}_N &= \boldsymbol{w}_N^T \dot{\boldsymbol{q}} + \hat{w}_N, & \dot{\boldsymbol{g}}_T &= \boldsymbol{W}_T^T \dot{\boldsymbol{q}} + \hat{\boldsymbol{w}}_T \\ \ddot{g}_N &= \boldsymbol{w}_N^T \ddot{\boldsymbol{q}} + \bar{w}_N, & \ddot{\boldsymbol{g}}_T &= \boldsymbol{W}_T^T \ddot{\boldsymbol{q}} + \bar{\boldsymbol{w}}_T \end{aligned} \tag{7.25}$$

which will serve as the contact kinematic equations in standard form in the sequel.

Kinetics

The dynamics of a multibody system with n bodies and one contact is described by the Newton-Euler equations in the form

$$\sum_{i=1}^{n} \left[\boldsymbol{J}_S^T(\dot{\boldsymbol{p}} - \boldsymbol{F}) + \boldsymbol{J}_R^T(\dot{\boldsymbol{L}} - \boldsymbol{M}) \right]_i - \boldsymbol{J}_{C1}^T \boldsymbol{F}_{C1} - \boldsymbol{J}_{C2}^T \boldsymbol{F}_{C2} = 0, \tag{7.26}$$

see e.g. [3], [15]. The terms $\boldsymbol{p}_i$ and $\boldsymbol{L}_i$ denote the momentum and the moment of momentum of body i, and $\boldsymbol{F}_i$ and $\boldsymbol{M}_i$ are the applied forces and moments, respectively. The contact forces acting at the contact points C_1 and C_2 are denoted by $\boldsymbol{F}_{C1}$ and $\boldsymbol{F}_{C2}$. In order to perform a projection into the configuration space all terms have to be premultiplied by the corresponding Jacobians. In particular, $\boldsymbol{J}_{C1}$ and $\boldsymbol{J}_{C2}$ are the Jacobians of the contact points which have already been introduced in equation (7.17).

Decomposing the contact forces into components with respect to trihedral $(\boldsymbol{u}_1, \boldsymbol{v}_1, \boldsymbol{n}_1)$ admits, together with the cutting principle, a unique representation in the form

$$\boldsymbol{F}_{C2} = -\boldsymbol{F}_{C1} =: \boldsymbol{n}_1 \lambda_N + \boldsymbol{u}_1 \lambda_U + \boldsymbol{v}_1 \lambda_V. \tag{7.27}$$

Putting (7.27) into (7.26) and using the abbreviations $\boldsymbol{w}_K = (\boldsymbol{J}_{C2} - \boldsymbol{J}_{C1})^T \boldsymbol{k}_1$ and $\boldsymbol{W}_T = (\boldsymbol{w}_U, \boldsymbol{w}_V)$ from equation (7.24) and below we arrive with

$$\boldsymbol{M}\ddot{\boldsymbol{q}} - \boldsymbol{h} - \boldsymbol{w}_N \lambda_N - \boldsymbol{W}_T \boldsymbol{\lambda}_T = 0, \tag{7.28}$$

where the vector $\boldsymbol{\lambda}_T$ of the tangential contact forces is defined to be $\boldsymbol{\lambda}_T^T = (\lambda_U, \lambda_V)$. The sum in (7.26) is taken into account by the term $(\boldsymbol{M}\ddot{\boldsymbol{q}} - \boldsymbol{h})$ with the symmetric and positive definite mass matrix $\boldsymbol{M}$ of the system. In the case of n_A contacts eq. (7.28) clearly transforms to

$$\boldsymbol{M}\ddot{\boldsymbol{q}} - \boldsymbol{h} - \sum_{i=1}^{n_A} \boldsymbol{w}_{Ni} \lambda_{Ni} - \boldsymbol{W}_{Ti} \boldsymbol{\lambda}_{Ti} = 0 \tag{7.29}$$

which has still to be completed by the contact laws.

Contact Laws

Generally, every force occuring in a rigid multibody system may be expressed in terms of some relative kinematic magnitudes such as relative displacements and velocities. Classical applied forces are represented by continuous functions whereas classical bilateral constraints already require a representation via set-valued mappings. Contact laws as used in the sequel belong to the latter class and constitute mixed-type force charateristics, acting in some areas as classical applied forces, in other areas as constraints.

With g_{Ni} being the distance of the contact points and λ_{Ni} the corresponding normal force one might express impenetrability of rigid bodies by the Signorini-Fichera condition [1], [3], [7], [9], [12]

$$g_{Ni} \geq 0, \quad \lambda_{Ni} \geq 0, \quad g_{Ni}\lambda_{Ni} = 0 \tag{7.30}$$

which might be combined, for example, with Coulomb-friction in the tangential directions [1], [7], [11], [12], i.e.

$$\begin{cases} \dot{\boldsymbol{g}}_{Ti} = 0 \Rightarrow |\boldsymbol{\lambda}_{Ti}| \leq \mu_{0i}\lambda_{Ni} \\ \dot{\boldsymbol{g}}_{Ti} \neq 0 \Rightarrow |\boldsymbol{\lambda}_{Ti}| = \mu_i\lambda_{Ni}, \\ \qquad\qquad \boldsymbol{\lambda}_{Ti} = -\alpha_i \dot{\boldsymbol{g}}_{Ti}, \ \alpha_i > 0, \end{cases} \tag{7.31}$$

where $\mu_i(\dot{\boldsymbol{g}}_{Ti})$ denotes the coefficient of friction for contact i and $\mu_{0i} = \mu_i(0)$.

By introducing the index sets [3], [15], [20]

$$\begin{aligned} I_A &= \{1, 2, \ldots, n_A\} \\ I_C &= \{i \in I_A : g_{Ni} = 0, \quad \dot{g}_{Ni} \geq 0\} \\ I_N &= \{i \in I_C : \dot{g}_{Ni} = 0\} \\ I_T &= \{i \in I_N : \dot{g}_{Ti} = 0\} \end{aligned} \tag{7.32}$$

and by the use of certain continuity assumptions on the trajectories $\boldsymbol{q}(t)$ one is also able to express the contact laws (7.30), (7.31) on the acceleration level in order to finally solve equation (7.29) for the unknowns $\ddot{\boldsymbol{q}}$. This yields, for the contact law in the normal direction (7.30)

$$\begin{aligned} &\text{for } i \in I_A \setminus I_N : \lambda_{Ni} = 0 \\ &\text{for } i \in I_N : \qquad \ddot{g}_{Ni} \geq 0, \quad \lambda_{Ni} \geq 0, \quad \ddot{g}_{Ni}\lambda_{Ni} = 0 \end{aligned} \tag{7.33}$$

where the complementarity condition in the second line of (7.33) might also be expressed by one of the variational inequalities, cp. [3], [7], [12], [20],

$$\begin{aligned} &\sum_{i \in I_N} \ddot{g}_{Ni}(\lambda_{Ni}^\star - \lambda_{Ni}) \geq 0, \quad \lambda_{Ni} \geq 0, \quad \forall \lambda_{Ni}^\star \geq 0 \\ &\sum_{i \in I_N} (\ddot{g}_{Ni}^\star - \ddot{g}_{Ni})\, \lambda_{Ni} \geq 0, \quad \ddot{g}_{Ni} \geq 0, \quad \forall \ddot{g}_{Ni}^\star \geq 0 \end{aligned} \tag{7.34}$$

for $i \in I_N$. Similarly we obtain from (7.31) the friction law on the acceleration level in the form [20]

$$\begin{array}{ll} \text{for } i \in I_A \setminus I_N : \boldsymbol{\lambda}_{Ti} = 0 & \\ \text{for } i \in I_N \setminus I_T : \boldsymbol{\lambda}_{Ti} = -\boldsymbol{e}_i \mu_i \lambda_{Ni} & \\ \text{for } i \in I_T : & \left\{ \begin{array}{l} \ddot{\boldsymbol{g}}_{Ti} = 0 \;\Rightarrow\; |\boldsymbol{\lambda}_{Ti}| \le \mu_{0i} \lambda_{Ni} \\ \ddot{\boldsymbol{g}}_{Ti} \ne 0 \;\Rightarrow\; |\boldsymbol{\lambda}_{Ti}| = \mu_{0i} \lambda_{Ni}, \\ \qquad\qquad\quad \boldsymbol{\lambda}_{Ti} \;= -\alpha_i \ddot{\boldsymbol{g}}_{Ti},\; \alpha_i > 0, \end{array} \right. \end{array} \tag{7.35}$$

with $\boldsymbol{e}_i$ being the unit vector of the sliding direction, i.e. $\boldsymbol{e}_i = \dot{\boldsymbol{g}}_{Ti}/|\dot{\boldsymbol{g}}_{Ti}|$. We recall that, due to the dependence on the normal forces, only quasivariational inequalities are available for the third portion of the contact law (7.35) which read [20]

$$\begin{array}{l} \sum\limits_{i \in I_T} \ddot{\boldsymbol{g}}_{Ti}^T (\boldsymbol{\lambda}_{Ti}^\star - \boldsymbol{\lambda}_{Ti}) \ge 0, \quad |\boldsymbol{\lambda}_{Ti}| \le \mu_{0i} \lambda_{Ni}, \quad \forall \boldsymbol{\lambda}_{Ti}^\star : |\boldsymbol{\lambda}_{Ti}^\star| \le \mu_{0i} \lambda_{Ni} \\ \sum\limits_{i \in I_T} (\ddot{\boldsymbol{g}}_{Ti}^\star - \ddot{\boldsymbol{g}}_{Ti})^T \boldsymbol{\lambda}_{Ti} \ge \sum_{i \in I_T} \mu_{0i} \lambda_{Ni} (|\ddot{\boldsymbol{g}}_{Ti}| - |\ddot{\boldsymbol{g}}_{Ti}^\star|), \quad \forall \ddot{\boldsymbol{g}}_{Ti}^\star . \end{array} \tag{7.36}$$

Equations (7.21), (7.25), (7.29), (7.33) and (7.35) provide a complete description of impact free non-smooth rigid body motion under the influence of Coulomb friction in the framework of non-smooth analysis. The equations considered above thus allow any further discussion and evaluation by applying available theoretical results and algorithms from this field [1], [7], [11], [12]. Impacts have been excluded. They can be treated in an analogous manner by rewriting (7.26) as an equality of measures [11] and solving them for the impact times. This yields, together with appropriate impact laws, a set of relations similar to those considered here, with force magnitudes playing then the role of impulsions, and accelerations which have to be expressed by jumps in the velocities [3], [4], [15].

Application: Vibratory Feeders

Vibratory feeders are just one example in machine dynamics on which the theory of spatial contact problems comes to a fruitful application [19]. They are the most common devices used to feed small parts in automatic assembly lines. Compared to other machines vibratory feeders seem to be quite simple. The high number of different types of devices based on the vibratory feeding principle, and the large amount of applications suggests that we are looking at a well developed and reliable tool. This impression is wrong. Problems in automated assembly are mostly caused by malfunctions of part feeders. These errors result from purely experimental tuning of the feeders, which is often done without a theoretical background, especially as far as the mechanics of the transportation process is concerned.

The transportation process of a vibratory feeder (linear or bowl feeder, Fig. 7.17) is based on a micro ballistic principle that is driven by an oscillating track. The mechanical model can be split in the dynamics of the base device, mostly represented by an electro magnetic excited oscillator, and the dynamics of the

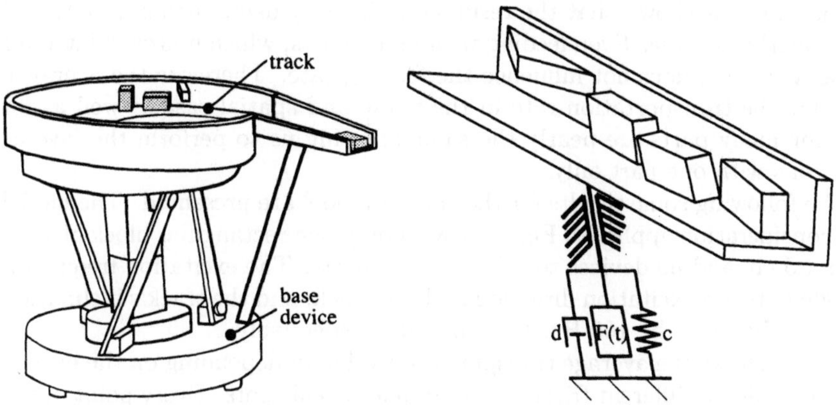

Figure 7.17 Vibratory bowl feeder and mechanical model.

parts transportation process. This yields a coupled system in the sense that the parts to be transported are affected by the vibrations of the track and vice versa.

The base device including the drives and the drive mechanisms can be modeled as a bilaterally coupled system with well-known standard techniques. Nevertheless, this may result in an extensive job. Due to the highly sensitive contact mechanics, all oscillating frequencies and flexible structures have to be analysed very well. Furthermore, interactions with an oscillating environment must be taken into account.

After having computed the vibrations of the base device which now serve as an excitation source for the parts one may concentrate on the modeling of the transportation process. The changes in the contact configurations between each of the parts on one side, and between the parts and the track on the other side are characteristic properties of the feeding process. Closed contacts either occur for entire time intervals or only for discrete points in time, the latter event being the usual outcome of an impact. Friction ist fundamental for the transportation of the parts: Indeed, it is the very feature that makes the feeder work. Consequently, the modeling of the process must be done with respect to friction effects. This leads to a structure variant multibody system with spatial contact laws consisting of unilateral constraints and planar Coulomb friction, the theory of which has been presented in the preceeding section.

The parts to be transported are modeled by rigid bodies with surfaces piecewise approximated by planes. Therefore, unilateral constraints result in point contacts, i.e. contacts between corners and planes or between lines and lines. Contact areas occuring at parallel lines and planes are composed of single point contacts. As one example of intense design parameter studies we present the transportation velocity, because of its outstanding importance for the basic function of the feeder. Not only fast transport but also robustness with respect to unsafe parameters are design criteria.

Simulation has shown that the results obtained by using planar and spatial models nearly coincide. Even feeding with many parts, which was calculated for the planar model, does not influence the feeding rate. Therefore the averaged results for the transportation rate in the plane and spatial case as well as for one or for many parts are nearly the same allowing us to perform the layout-simulations with one part only.

In the following some results for the spatial model are presented. The model under consideration applies to Fig. 7.17 with only one rectangular block of 1 cm x 2 cm x 5 cm and no devices to orientate the parts. The excitation frequency, the angle between excitation direction and the track, and the track's inclination have been chosen to be 100 Hz, 10°, and 20°, respectively.

Fig. 7.18 shows the average transportation velocity depending on the excitation amplitude for different friction- and impact-coefficients. Every point of the curves results from one simulation which was performed until a stable average value was found. Positive velocities mean transportation upwards; negative downwards.

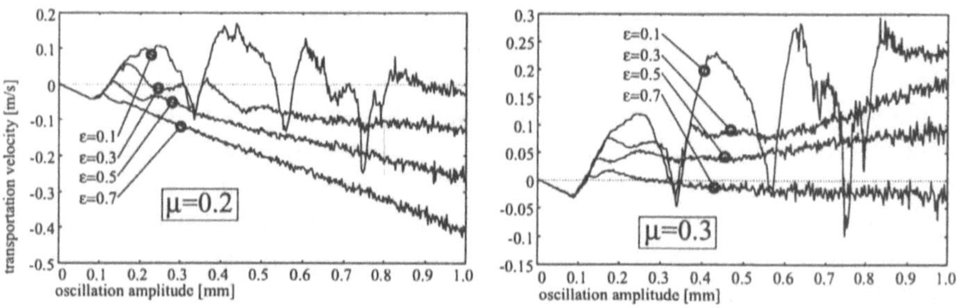

Figure 7.18 Transportation rates of a vibratory feeder.

In the left diagram of Fig. 7.18 we see the results for a friction coefficient $\mu = 0.2$ and four different impact coefficients ε. Small amplitudes cause the conveyor not to work, resulting in pure sliding downwards of the parts. By increasing the excitation amplitude the upwards feeding process starts working, but only in certain regions for small impact coefficients, and based on a ballistic (that means: impact-controlled) mechanism.

The right diagram shows the same investigations but with a bigger coefficient of friction. In this case the transportation rate rises, and the conveyor is working even with bigger impact coefficients.

These results can serve as a basis for the feeder design which allows to choose appropriate materials for the track, to find the best excitation parameters, and to estimate the expected feeding rate. The experimental verification of the presented results is the subject of current activities.

7.4 CONCLUSIONS

In the area of multibody theory regarding unilateral contacts the methodology has reached a state, which allows an application in nearly all fields of mechanical engineering. As unilateral contacts are realized in machines and mechanisms to a very large extend, the use of the corresponding theories is just at the beginning. The paper gives some examples, which demonstrate the typical features of unilateral constraints in combination with machines: They might be applied to achieve some functional performance like the transportation rate in vibratory feeders, or they might generate unwanted vibrations, noise and wear like rattling in gears or wear in chains.

The main problem limiting today's applications is of numerical nature. To describe a machine like the vibration conveyor includes tedious numerical evaluations solving the complementarity problem at each time step. Even with modern high-speed computers we come out with a severe computing time problem. At the time being there are three groups of algorithms available to treat numerics.

The enumerative methods try to find a solution by a kind of intelligent trial and error procedures. They do not satisfy. The pivot-algorithms are related to the well-known simplex-algorithm. The Lemke-method is an example. We apply it with good success, but with large computing times. Within the framework of iterative methods a modified Newton-approach seems to be very promising, even for the nonlinear complementarity problem. At the time being we are working on it.

A second problem, also closely connected with numerics, consists in the solution of nonlinear complementarities as emerging from spatial unilateral contacts. Up to now we have used the following methods: linearization of the friction cones, the augmented Lagrangian methods and NCP-functions as they appear in Mangasarian's theorem. All methods work satisfactorily, all have to be improved, or a better one must be developed. We are also working on it.

As a conclusion, theories and methods available to-day help the engineers to understand much better problems including unilateral features, which in former times were treated only in a very simple way. We are able to analyse machines with unilateral contacts in a way, which five years ago we could not have thought of. Scientists like Professor Fichera have contributed to this development very significantly.

References

[1] Alart, P. and Curnier, A. (1991). "A Mixed Formulation for Frictional Contact Problems Prone to Newton Like Solution Methods," *Comp. Meth. Appl. Mech. Eng.* **92** (3), pp. 353–357.

[2] Braun, J. (1989). *Dynamik und Regelung elastischer Flugzeugfahrwerke.* Thesis at the Dept. of Mech. Eng., Inst. B f. Mech., Tech. Univ. of Munich.

[3] Glocker, Ch. (1995). *Dynamik von Starrkörpersystemen mit Reibung und Stößen.* Fortschrittberichte VDI, Reihe 18, Nr. 182, VDI-Verlag,

Düsseldorf.

[4] Glocker, Ch. and Pfeiffer, F. (1995). "Multiple Impacts with Friction in Rigid Multibody Systems," *Nonlinear Dynam.*, Vol. 7, pp. 471–497.

[5] Hajek, M. (1990). *Reibungsdämpfer für Turbinenschaufeln.* Fortschrittberichte VDI, Reihe 11, Nr. 128, VDI-Verlag, Düsseldorf.

[6] Karagiannis, K. and Pfeiffer, F. (1991). "Theoretical and experimental investigation of gear-rattling," *Nonlinear Dynam.*, Vol. 2, pp. 367–387.

[7] Klarbring, A. and Björkman, G. (1988). "A mathematical programming approach to contact problems with friction and varying contact surface," *Computers & Structures,* Vol. 30, No. 5, pp. 1185–1198.

[8] Küçükay, F. and Pfeiffer, F. (1986). "Über Rasselschwingungen in Kfz-Schaltgetrieben," *Ing.-Arch.*, Vol. 56, pp. 25–37.

[9] Lötstedt, P. (1982). *Numerical simulation of time-dependent contact and friction problems in rigid body mechanics.* Technical Report TRITA-NA-8214, Dept. of Numerical Analysis and Computing Science, The Royal Institute of Technologie, Sweden.

[10] Meitinger, Th. (1997). *Dynamik automatisierter Montageprozesse.* Docotoral Thesis at the Dept. of Mech. Eng., Inst. B f. Mech. Tech. Univ. of Munich.

[11] Moreau, J.J. (1988). "Unilateral contact and dry friction in finite freedom dynamics," CISM courses and lectures, *Nonsmooth mechanics and applications,* Springer-Verlag, Wien, New York.

[12] Panagiotopoulos, P.D. (1993). *Hemivariational Inequalities - Applications in Mechanics and Engineering.* Springer-Verlag, Berlin, Heidelberg, New York.

[13] Pfeiffer, F. (1984). "Mechanische Systeme mit unstetigen Übergängen," *Ing.-Arch.*, Vol. 54, pp. 232–240.

[14] Pfeifer, F. and Hajek, M. (1992). "Stick-slip motion of turbine blade dampers," *Phil. Trans. R. Soc. London,* A 338, pp. 503–517.

[15] Pfeiffer, F. and Glocker, Ch. (1996). *Multibody Danamics with Unilateral Contacts.* John Wiley, New York.

[16] Seyfferth, W. (1993). *Modellierung unstetiger Montageprozesse mit Robotern.* Fortschrittberichte VDI, Reihe 11, Nr. 199, VDI-Verlag, Düsseldorf.

[17] Wapenhans, H. (1989). *Dynamik und Regelung von Flugzeugfahrwerken.* Thesis at the Dept. of Mech. Eng., Inst. B f. Mech., Tech. Univ. of Munich.

[18] Wapenhans, H. (1994). *Optimierung von Roboterbewegungen bei Manipulationsvorgängen.* Fortschrittberichte VDI, Reihe 2, Nr. 304, VDI-Verlag, Düsseldorf.

[19] Wolfsteiner, P. and Pfeiffer, F. (1997). "Dynamics of a Vibratory Feeder," *Proc. of the 1996 ASME 16th Biennal Conference on Vibration and Noise,* Sacramento, California.

[20] Wösle, M. (1997). *Dynamik von räumlichen strukturvarianten Mehrkörpersystemen.* Doctoral Thesis at the Dept. of Mech. Eng. Inst. B f. Mech., Tech. Univ. of Munich.

8 HYPERBOLIC HEMIVARIATIONAL INEQUALITY AND NONLINEAR WAVE EQUATION WITH DISCONTINUITIES

D. Goeleven[1] and D. Motreanu[2]

[1]Department of Mathematics, University of La Réunion
Saint-Dénis, Ile de La Réunion, France
[2]Department of Mathematics, University of Perpignan
Perpignan, France

Abstract: The paper presents existence results for solutions to a nonsmooth hyperbolic problem in the form of a hemivariational inequality separately in the nonresonant and resonant cases.

8.1 INTRODUCTION AND FORMULATION OF NONSMOOTH HYPERBOLIC PROBLEM

Throughout the paper $j : [0,\pi] \times \mathbb{R}^2 \to \mathbb{R}$ denotes a real-valued function $j(x,t,z)$ satisfying the hypothesis

(H_1) $j(x,t,z)$ is 2π-periodic in t, measurable in (x,t), $j(\cdot,\cdot,0) \in L^1((0,\pi) \times (0,2\pi))$, Lipschitz in z, with a bounded generalized gradient

$$|w| \leq M, \quad \forall (x,t,z) \in [0,\pi] \times [0,2\pi] \times \mathbb{R}, \; \forall w \in \partial j(x,t,z).$$

R.P. Gilbert et al. (eds.), From Convexity to Nonconvexity, 111–122.

The notation $\partial j(x,t,z)$ stands for the generalized gradient of $j(x,t,\cdot)$ with respect to $z \in \mathbb{R}$ (see Clarke [6]). The generalized directional derivative of $j(x,t,z)$ with respect to $z \in \mathbb{R}$ (see Clarke [6]) will be denoted by $j^0(x,t,z;\cdot)$.

Let us fix a number $c \in \mathbb{R} \setminus \{0\}$. We formulate the following hyperbolic hemivariational inequality: Find $u \in H^1$ such that

$$(P) \quad \int_0^{\pi}\int_0^{2\pi}(u_x v_x - u_t v_t - cuv + j^0(x,t,u(x,t),v(x,t))dxdt \geq 0, \quad \forall v \in H^1,$$

where H^1 denotes the Hilbert space equal to the closure with respect to the norm

$$||u|| := (||u||^2_{L^2} + ||u_x||^2_{L^2} + ||u_t||^2_{L^2})^{1/2}$$

of the subset of $C^\infty([0,\pi] \times \mathbb{R}, \mathbb{R})$ verifying the boundary conditions

$$u(0,t) = u(\pi,t) = 0, \quad t \in \mathbb{R}, \tag{1}$$

and the periodicity conditions

$$u(x,t+2\pi) = u(x,t), \quad (x,t) \in [0,\pi] \times \mathbb{R}. \tag{2}$$

It is worth to note that a stronger form of Problem (P) is the nonlinear wave equation with discontinuities

$$(P') \qquad u_{xx} - u_{tt} + cu \in \partial j(x,t,u(x,t)), \quad \text{a.e. } (x,t) \in [0,\pi] \times \mathbb{R},$$

under the boundary and periodicity conditions (1), (2).

The paper provides two existence results for the solutions of problem (P'), so of problem (P), which treat separately the cases of nonresonance and resonance in (P') or (P).

In the following we shall need the spectrum

$$\sigma(\Box) = \{k^2 - j^2 : \; k,j \text{ integers}, \; j \neq 0\} \tag{3}$$

of the D'Alembertian $\Box u = u_{tt} - u_{xx}$, with the corresponding eigenfunctions

$$\sin jx(\alpha \cos kt + \beta \sin kt), \quad \alpha, \beta \in \mathbb{R}. \tag{4}$$

We introduce the unbounded linear operator on H^1 by

$$Lu := \Box u - cu. \tag{5}$$

We have the decomposition

$$H^1 = N(L) \oplus N^-(L) \oplus N^+(L), \tag{6}$$

where $N(L)$, $N^-(L)$ and $N^+(L)$ denote the null-space, the space on which L is negative definite and positive definite, respectively.

8.2 FINITE DIMENSIONAL APPROXIMATION

In treating Problem (P') (so (P)) we make use of the idea in Rabinowitz [10] based on a Galerkin approximation argument.

LEMMA 1. For any integer $m \geq 1$ let X_m denote the finite dimensional subspace of H^1 defined by

$$X_m := \text{span}\ \{\sin jx \cos kt,\ \sin jx \sin kt\}_{0 \leq j,k \leq m}. \tag{7}$$

Assuming (H_1) together with

$$c \notin \sigma(\Box), \tag{8}$$

there exist $u_m \in X_m$ and $\xi_m \in X_m$ such that

$$\xi_m(x,t) \in \partial j(x,t,u_m(x,t)) \ \text{ for a.e. } (x,t) \in (0,\pi) \times \mathbb{R} \tag{9}$$

and

$$\int_0^{\pi} \int_0^{2\pi} (u_{mx}\phi_x - u_{mt}\phi_t - cu_m\phi + \xi_m\phi)dxdt = 0,\ \forall \phi \in X_m. \tag{10}$$

PROOF. Assumption (8) implies

$$N(L) = \{0\}. \tag{11}$$

By (6) and (11) the space X_m in (7) is splitted as follows

$$X_m = X_m^- \oplus X_m^+, \tag{12}$$

where $X_m^- := N^-(L) \cap X_m$ and $X_m^+ := N^+(L) \cap X_m$. We apply to the locally Lipschitz functional $I|_{X_m} : X_m \to \mathbb{R}$ the nonsmooth version of Chang ([5], Theorem 3.3) for the Saddle-Point Theorem (see Rabinowitz [10], [11]), with respect to the direct sum decomposition (12), where $I : H^1 \to \mathbb{R}$ is given by

$$I(v) = \int_0^{\pi} \int_0^{2\pi} [\frac{1}{2}(v_x^2 - v_t^2 - cv^2) + j(x,t,v)]dxdt = 0,\ \forall v \in H^1. \tag{13}$$

Notice that by (H_1) and Lebourg's mean value theorem (see Clarke [6]) one has

$$|j(x,t,z)| \leq M|z| + |j(x,t,0)|,\ \ \forall (x,t,z) \in [0,\pi] \times \mathbb{R}^2. \tag{14}$$

From (14) we see that I in (13) is well-defined and locally Lipschitz.

Denote by λ_+ the first positive eigenvalue of L given by (5). Then (7), (12), (13) and (14) show that

$$I(v) \geq \lambda_+ \|v\|_{L^2}^2 - M\|v\|_{L^1} - \int_0^{\pi} \int_0^{2\pi} |j(x,t,0)|dxdt \tag{15}$$

$$\geq c_0\|v\|^2 - c_1\|v\| - c_2,\ \ \forall v \in X_m^+,$$

with positive constants c_0, c_1, c_2, where we have also used that X_m^+ is finite dimensional. Then (15) yields

$$\inf_{X_m^+} I > -\infty. \tag{16}$$

Denote now by λ_- the first negative eigenvalue of the linear operator L in (5). Then (7), (12), (13) and (14) imply the estimate

$$I(v) \leq \lambda_- \|v\|_{L^2}^2 + M\|v\|_{L^1} + \int_0^\pi \int_0^{2\pi} |j(x,t,0)| dx dt \tag{17}$$

$$\leq -d_0\|v\|^2 + d_1\|v\| + d_2, \quad \forall v \in X_m^-,$$

with positive constants d_0, d_1, d_2, where the fact that X_m^- is finite dimensional was used, too. From (17) we obtain

$$I(v) \to -\infty \text{ as } \|v\| \to \infty, \; v \in X_m^-. \tag{18}$$

From (17) and (18) we deduce the existence of $\rho > 0$ such that

$$\sup\{I(v) : \; v \in X_m^-, \; \|v\| = \rho\} < \inf_{X_m^+} I. \tag{19}$$

Let us check the Palais-Smale condition in the sense of Chang [5] for the locally Lipschitz functional $I|_{X_m}$. For this let $\{v_n\} \subset X_m$ be a sequence such that $I(v_m)$ is bounded and there exists a sequence $w_n \in \partial(I|_{X_m})(v_n)$ with $w_n \to 0$ in X_m^*. In view of (5) and (13) we can express

$$w_n = Lv_n + g_n \quad \text{on } X_m,$$

with $g_n \in L^2((0,\pi) \times (0,2\pi))$ such that

$$\langle w_n, v \rangle = \int_0^\pi \int_0^{2\pi} ((Lv_n)v + g_n v) dx dt, \quad \forall v \in X_m,$$

$$g_n(x,t) \in \partial j(x,t,v_n(x,t)) \text{ for a.e. } (x,t) \in (0,\pi) \times (0,2\pi).$$

Since $w_n \to 0$ in X_m^* we can estimate the projections $P_m^- v_n$ and $P_m^+ v_n$ of v_n on X_m^- and X_m^+, respectively, as follows

$$\|P_m^+ v_n\| \geq \int_0^\pi \int_0^{2\pi} ((Lv_n)P_m^+ v_n + g_n P_m^+ v_n) dx dt \geq$$

$$\lambda_+ \|P_m^+ v_n\|_{L^2}^2 - M\|P_m^+ v_n\|_{L^1} \geq c_0\|P_m^+ v_n\|^2 - c_3\|P_m^+ v_n\|$$

and

$$\|P_m^- v_n\| \geq -\int_0^\pi \int_0^{2\pi} ((Lv_n)P_m^- v_n - g_n P_m^- v_n) dx dt \geq$$

$$-\lambda_- \|P_m^- v_n\|_{L^2}^2 - M\|P_m^- v_n\|_{L^1} \geq d_0\|P_m^- v_n\|^2 - d_3\|P_m^- v_n\|,$$

with constants c_3, d_3, and n sufficiently large. The estimates above show that $v_n = P_m^- v_n + P_m^+ v_n \in X_m$ is bounded in X_m. Taking into account that X_m is finite dimensional, we may conclude that $I|_{X_m}$ satisfies the Palais-Smale condition in the sense of Chang [5].

The properties (12), (16), (19) and the Palais-Smale condition allow to apply Saddle-Point Theorem in the nonsmooth version of Chang [5] to the locally Lipschitz functional $I|_{X_m} : X_m \to \mathbb{R}$. It ensures the existence of a critical point $u_m \in X_m$ of $I|_{X_m}$, i.e.,

$$0 \in \partial(I|_{X_m})(u_m). \tag{20}$$

Then (20), in conjuction with Theorem 2.7.5 and Proposition 2.1.2 in Clarke [6], implies (9), (10). This completes the proof of Lemma 1. □

We focus now on the resonant case, i.e.,

$$c \in \sigma(\Box). \tag{21}$$

Taking into account that $c \neq 0$, from (21) we infer that

$$\dim N(L) = p > 0,$$

say

$$N(L) = \text{span }\{\phi_1(x,t), ..., \phi_p(x,t)\}. \tag{22}$$

LEMMA 2. Assume that (H_1) and (21) hold together with either

(H_2^-) $j(x,t,z) \to -\infty$ as $|z| \to \infty$

or

(H_2^+) $j(x,t,z) \to +\infty$ as $|z| \to \infty$,

uniformly with respect to $(x,t) \in (0,\pi) \times (0,2\pi)$.

Then there exist $u_m \in X_m$ and $\xi_m \in X_m$ such that (9) and (10) are verified.

PROOF. To make a choice we suppose that (H_2^-) is valid. Then (H_2^-) yields a number $a_0 > 0$ such that

$$j(x,t,z) \leq 0, \quad \forall (x,t) \in (0,\pi) \times (0,2\pi), \ |z| \geq a_0, \tag{23}$$

and corresponding to each $K > 0$, a number $a_1 = a_1(K) > a_0$ such that

$$j(x,t,z) \leq -K, \quad \forall (x,t) \in (0,\pi) \times (0,2\pi), \ |z| \geq a_1. \tag{24}$$

Let $v \in N(L) \setminus \{0\}$. It can be written as

$$v = \alpha\phi, \ \ ||\phi|| = 1, \ \alpha > 0. \tag{25}$$

Then $I(v)$ can be expressed as follows

$$I(v) = \int_0^\pi \int_0^{2\pi} j(x,t,v(x,t))dxdt = \sum_{i=1}^3 \int_{D_i} j(x,t,v(x,t))dxdt,$$

where, for a fixed $K > 0$ with the corresponding a_1 in (24),

$$D_1 = \{(x,t) \in (0,\pi) \times (0,2\pi) : \ |v(x,t)| \leq a_0\},$$

$$D_2 = \{(x,t) \in (0,\pi) \times (0,2\pi) : \ a_0 < |v(x,t)| < a_1\},$$

$$D_3 = \{(x,t) \in (0,\pi) \times (0,2\pi) : \ |v(x,t)| \geq a_1\}.$$

In view of (25), and since by (22) the function ϕ is a linear combination of generators in (4), we may find some $\alpha = \alpha(K, \phi)$ large enough to have

$$meas(D_3) \geq \pi^2. \tag{26}$$

The compactness of the unit sphere in $N(L)$ (cf. (22)) ensures that in fact (26) holds for all v in (25) with $\alpha \geq \alpha_0 > 0$, where $\alpha_0 = \alpha_0(K)$ denotes a constant which is independent of $\phi \in N(L)$ with $||\phi|| = 1$. Then from (23)-(26) we derive

$$I(v) \leq b_0 - K\pi^2, \ \ \forall v = \alpha\phi \in N(L), \ ||\phi|| = 1, \ \alpha \geq \alpha_0, \tag{27}$$

for a constant $b_0 > 0$. Letting $K \to +\infty$ in (27) we conclude that

$$I(v) \to -\infty \ \text{ as } ||v|| \to \infty, \ v \in N(L). \tag{28}$$

Denoting

$$X_m^0 = X_m \cap N(L),$$

consider the splitting

$$X_m = X_m^- \oplus X_m^0 \oplus X_m^+, \tag{29}$$

where X_m^- and X_m^+ have the meaning in (12). Let $P_m^0 : X_m \to X_m^0$ denote the orthogonal projection of X_m onto X_m^0.

Let $\{v_n\} \subset X_m$ be a sequence such that $I(v_m)$ is bounded and for which there exists a corresponding sequence

$$w_n \in \partial(I|_{X_m})(v_n)$$

with $w_n \to 0$ in X_m^*. By (29) we can write

$$v_n = P_m^- v_n + P_m^0 v_n + P_m^+ v_n.$$

The same argument as in the proof of Lemma 1 shows that the components $\{P_m^- v_n\}$ and $\{P_m^+ v_n\}$ are bounded in X_m. The boundedness of $I(v_n)$ implies that $I(P_m^0 v_n)$ is bounded (here (H_1) and Lebourg's mean value theorem are used). Then (28) ensures that $\{P_m^0 v_n\}$ is bounded. Since X_m is finite dimensional, we obtain that the Palais-Smale condition in the sense of Chang [5] is satisfied in the case of the locally Lipschitz functional $I|_{X_m}$.

With the notations above and using (17) we see that

$$I(v_- + v_0) \leq -d_0||v_-||^2 + \int_0^\pi \int_0^{2\pi} (j(x,t,v_- + v_0) - j(x,t,v_0))dxdt +$$

$$\int_0^\pi \int_0^{2\pi} j(x,t,v_0)dxdt \leq -d_0\|v_-\|^2 + d\|v_-\| + \int_0^\pi \int_0^{2\pi} j(x,t,v_0)dxdt$$

for all $v_- \in X_m^-$ and $v_0 \in X_m^0$, where d_0, d are positive constants. By virtue of (28) it follows

$$I(v) \to -\infty \text{ as } \|v\| \to \infty,\ v \in X_m^- \oplus X_m^0. \tag{30}$$

Taking into account the Palais-Smale condition and relations (29), (15), (30), we can apply to $I|_{X_m}$ the nonsmooth version of Saddle-Point Theorem (see Chang [5], Theorem 3.3). Then there exists a critical point $u_m \in X_m$ of $I|_{X_m}$ in the sense of (20). We complete the proof as in Lemma 1.□

8.3 MAIN RESULTS

Our main results treat separately the nonresonant and resonant case.

THEOREM 1. Let $c \in R\backslash\{0\}$ satisfy the nonresonant condition (8). Assume that the function $j : [0,\pi] \times \mathbb{R}^2 \to \mathbb{R}$ verifies (H_1) and

(H_3) for every $r > 0$ there exist constants $c_1 \in \mathbb{R}$, $c_2 \in \mathbb{R}$, with $c_1 < c$, such that for all $v \in N(\Box) \subset H^1$ with $\|v\|_{L^2} \leq r$ and $\xi \in L^2((0,\pi) \times (0,2\pi))$ satisfying

$$\xi(x,t) \in \partial j(x,t,v(x,t)) \text{ for a.e. } (x,t) \in (0,\pi) \times (0,2\pi)$$

one has

$$|\int_0^\pi \int_0^{2\pi} \xi(x,t)v_{tt}dxdt| \leq c_1\|v_t\|_{L^2}^2 + c_2.$$

Then Problem (P') admits at least a weak solution. In particular, Problem (P) has a solution.

PROOF. Applying Lemma 1 one finds a sequence $\{u_m\}$, with $u_m \in X_m$, for which there exists a corresponding sequence $\{\xi_m\} \in X_m$ satisfying (10), (11). From (10) it is clear that, in $L^2((0,\pi) \times (0,2\pi))$, the equality below holds

$$Lu_m = -\xi_m,\quad m \geq 1. \tag{31}$$

In view of (7) we can write the expansions

$$u_m = \sum_{0 \leq j,|k| \leq m} f_{jk}^m \sin jx e^{ikt},\quad f_{jk} = \overline{f}_{j,-k},$$

and

$$\xi_m = \sum_{0 \leq j,|k| \leq m} h_{jk}^m \sin jx e^{ikt},\quad h_{jk} = \overline{h}_{j,-k},$$

where the coefficients f_{jk}^m and h_{jk}^m are complex numbers. By (8) and (31) we arrive at

$$f_{jk}^m = -(j^2 - k^2 - c)^{-1} h_{jk}^m.$$

Since, by virtue of (8),

$$|j^2 - k^2 - c| \geq a > 0, \quad \forall j, k,$$

with a constant $a > 0$ which is independent of j and k, it follows that u_m and ξ_m entering (31) satisfy

$$||u_m||_{L^2} \leq C_0 ||\xi_m||_{L^2}, \quad \forall m \geq 1, \tag{32}$$

for a constant $C_0 > 0$. On the other hand from (9) and (H_1) we derive the boundedness of $\{\xi_m\}$ in $L^2((0,\pi) \times (0,2\pi))$. Consequently, (32) yields the L^2-boundedness of $\{u_m\}$, say,

$$||u_m||_{L^2} \leq C, \quad \forall m \geq 1. \tag{33}$$

Consider the direct sum decomposition

$$H^1 = N(\Box) \oplus N^- \oplus N^+, \tag{34}$$

where N^- and N^+ represent the subspaces of H^1 on which the D'Alembertian $\Box$ is negative and positive definite, respectively. Thus each $v \in H^1$ can be expressed uniquely as

$$v = Pv + P^-v + P^+v, \tag{35}$$

$$Pv \in N(\Box), \ P^-v \in N^-, \ P^+v \in N^+.$$

Relation (31) can be expressed in the form

$$\Box(P^+u_m + P^-u_m) = c(P^+u_m + P^-u_m) - \xi_m, \quad \forall m \geq 1.$$

Then (33), (H_1) and a standard estimate for the linear wave equation (see, e.g., Brézis [4]) imply

$$||P^+u_m + P^-u_m|| \leq C, \quad \forall m \geq 1, \tag{36}$$

for a new constant $C > 0$.

It remains to estimate the component $Pu_m \in N(\Box)$ (see (35)). Towards this we deduce from (10) that

$$\int_0^\pi \int_0^{2\pi} (cu_m\phi - \xi_m\phi) dx dt = 0, \ \forall \phi \in X_m \cap N(\Box).$$

If we put in the foregoing equality $\phi = -(Pu_m)_{tt}$ we see that

$$c\int_0^\pi \int_0^{2\pi} (Pu_m)_t^2 dx dt = -\int_0^\pi \int_0^{2\pi} \xi_m (Pu_m)_{tt} dx dt. \tag{37}$$

Assumption (H_3) ensures that constants $c_1 > 0$ and $c_2 >$ can be found such that

$$|\int_0^\pi \int_0^{2\pi} \xi_m (Pu_m)_{tt} dx dt| \leq c_1 ||(Pu_m)_t||_{L^2}^2 + c_2,$$

with $c_1 < c$ (here we have used also (33) as well as the fact that $Pu_m \in N(\Box)$). Then (37) yields

$$c\|(Pu_m)_t\|_{L^2}^2 \leq c_1\|(Pu_m)\|_{L^2}^2 + c_2.$$

Since $c_1 < c$, this ensures the boundedness

$$\|(Pu_m)_x\|_{L^2} = \|(Pu_m)_t\|_{L^2} \leq C, \tag{38}$$

for a constant $C > 0$, where the equality in (38) is due to the fact that $Pu_m \in N(\Box)$. Combining (33), (36) and (38) we get the boundedness of $\{u_m\}$ in H^1, so there is a constant $C > 0$ such that

$$\|u_m\| \leq C, \quad \forall m \geq 1. \tag{39}$$

The reflexivity of H^1 and (39) allow to assume, after relabeling, that there exists $u \in H^1$ such that

$$u_m \to u \text{ weakly in } H^1 \text{ as } m \to \infty. \tag{40}$$

Assumption (H_1) implies that, along a subsequence, we have

$$\xi_m \to \xi \text{ weakly in } L^2((0,\pi) \times (0,2\pi)) \text{ as } m \to \infty, \tag{41}$$

for some $\xi \in L^2((0,\pi) \times (0,2\pi))$. We know that $|\xi_m(x,t)| \leq M$, $\forall (x,t) \in [0,\pi] \times [0,2\pi]$ and thus along a subsequence, we may assume that

$$u_m(x,t) \to u(x,t) \text{ a.e. in } (0,\pi) \times (0,2\pi)$$

and

$$\xi_m(x,t) \to \xi(x,t) \text{ a.e. in } (0,\pi) \times (0,2\pi).$$

The weak*-closedness of the generalized gradient (see Clarke [6]) ensures that

$$\xi(x,t) \in \partial j(x,t,u(x,t)) \text{ for a.e. } (x,t) \in (0,\pi) \times (0,2\pi). \tag{42}$$

Let us fix $k \geq 1$ and take $\phi \in X_k$. Then $\phi \in X_m$, $\forall m \geq k$, and we can make use of (10). Since $k \geq 1$ and $\phi \in X_k$ are arbitrary, we deduce from (10), (40), (41) the fromula

$$\int_0^\pi \int_0^{2\pi} (u_x\phi_x - u_t\phi_t - cu\phi + \xi\phi)dxdt = 0, \ \forall \phi \in \cup_{m\geq 1} X_m.$$

By density we derive that $u \in H^1$ is a weak solution of the equation

$$\int_0^\pi \int_0^{2\pi} (u_x\phi_x - u_t\phi_t - cu\phi + \xi\phi)dxdt = 0, \ \forall \phi \in H^1. \tag{43}$$

Therefore u is a solution of Problem (P'). From (43), (42) and the maximality property of $j^0(x,t,\cdot)$ (see Clarke [6]) one obtains that $u \in H^1$ is a solution of Problem (P). $\Box$

We deal now with the resonant case of Problem (P).

THEOREM 2. Assume that the conditions (H_1), (H_3), (21) with $c \neq 0$, and either (H_2^-) or (H_2^+) hold. Then Problem (P') admits at least one solution $u \in H^1$, so Problem (P) has at least a solution.

PROOF. To make a choice let us suppose that $c > 0$. Lemma 2 provides sequences $\{u_m\}$, $\{\xi_m\}$ with $u_m \in X_m$ and $\xi_m \in X_m$ satisfying (9), (10). Then (31) is true. Arguing as in the proof of Theorem 1 we get estimate (36). We claim that

$$\|Pu_m\|_{L^2} \leq C, \quad \forall m \geq 1, \tag{44}$$

for a constant $C > 0$, where P denotes the projection of H^1 onto $N(\Box)$.

The argument relies on the minimax description of the critical value $I(u_m)$ of $I|_{X_m}$. The minimax characterization of the critical point $u_m \in X_m$ of $I|_{X_m}$ in the nonsmooth version of Saddle-Point Theorem of Chang [5] (see Rabinowitz [10], [11] for the smooth case) implies

$$b_m := I(u_m) \geq \inf_{X_m^+} I \geq \inf_{N^+} I > -\infty, \quad \forall m \geq 1, \tag{45}$$

which is valid because $c > 0$. On the other hand from (10) we derive

$$b_m = \frac{1}{2}\int_0^\pi \int_0^{2\pi} (Lu_m)(P^+u_m + P^-u_m)dxdt -$$

$$\int_0^\pi \int_0^{2\pi} (\frac{c}{2}(Pu_m)^2 - j(x,t,u_m))dxdt =$$

$$-\frac{1}{2}\int_0^\pi \int_0^{2\pi} \xi_m(x,t)(P^+u_m + P^-u_m)dxdt -$$

$$\int_0^\pi \int_0^{2\pi} (\frac{c}{2}(Pu_m)^2 - j(x,t,u_m))dxdt.$$

The equality above, (9) and (H_1) show

$$b_m \leq M\|P^+u_m + P^-u_m\|_{L^2} - \int_0^\pi \int_0^{2\pi} (\frac{c}{2}(Pu_m)^2 - j(x,t,u_m))dxdt.$$

In view of (14), (36) and (45), we conclude that the claim in (44) holds true. From now on the proof proceeds as for Theorem 1.□

REMARK 1. This is the first paper studying the nonlinear wave equation with discontinuities without any monotonicity or convexity assumptions. In comparison with the smooth case, i.e., $\xi(x,t)$ in (P) is $g(x,t,u(x,t))$ for $g \in C([0,\pi] \times R^2)$, no additional hypotheses are imposed (see Rabinowitz [10]). Assumption (g_2) in Rabinowitz [10] is slightly stronger that (H_3) even in the smooth case. For continuous, monotone (in u) nonlinearities in (P), as well as for the case of free vibrations, we refer to Brézis [4]. Theorem 1 in Brézis [4] deals with the nonlinear wave equation $\Box u - g(u) = f(x,t)$, where $g : I\!R \to I\!R$ is continuous, monotone, $g(0) = 0$, $|g(u)| \leq \gamma|u| + C$, $\gamma \in [0,3]$, $C \in I\!R$ and $f \in L^\infty$. In our Theorems 1 and 2 we drop the restriction on γ, assuming $c \in I\!R\backslash 0$, but other conditions are added. Necessary conditions of optimality in an optimal control problem subject to the semilinear wave equation above on a bounded domain Ω in $I\!R^N$ for $g \in C^1(I\!R)$, $g' \in L^\infty(I\!R)$, $\|g'\|_{L^\infty} < \lambda_1(-\Delta)$, with Δ on $H_0^1(\Omega)$, have been obtained in Barbu and Pavel [3]. The use of generalized gradients to derive necessary conditions of optimality can be found in Barbu [2] and Clarke [6].

REMARK 2.There is a strong distinction between the nonresonant and resonant cases in Problem (P), i.e., $c \notin \sigma(\Box)$ and $c \in \sigma(\Box)$, respectively. Assumptions (H_2^-), (H_2^+) are specific to the resonant case. In the smooth nonlinear elliptic boundary value problems they go back to Ahmad, Lazer and Paul [1] (see also Rabinowitz [11]). The first nonsmooth version of Landesman-Lazer result [8] for elliptic problems has been obtained by Chang [5]. An extension of Chang's result allowing unbounded nonlinearities has been obtained by Goeleven, Motreanu and Panagiotopoulos [7] in the setting of hemivariational inequalities. A systematic exposition of the general theory of hemivariational inequalities can be found in Panagiotopoulos [9].

References

[1] S. Ahmad, A. C. Lazer and J. L. Paul, Elementary critical point theory and perturbations of elliptic boundary value problems at resonance, Indiana Univ. Math. J. 25 (1976), 933-944.

[2] V. Barbu, Necessary conditions for multiple integral problems in the calculus of variations, Math. Ann. 260 (1982), 175-189.

[3] V. Barbu and N. H. Pavel, Periodic optimal control in Hilbert spaces, Appl. Math. Optimiz. 33 (1996), 169-188.

[4] H. Brézis, Periodic solutions of nonlinear vibrating strings and duality principles, Bull (New Series) Amer. Math. Soc. 8 (1983), 409-426.

[5] K. C. Chang, Variational methods for non-differentiable functionals and their applications to partial differential equations, J. Math. Anal. Appl. 80 (1981), 102-129.

[6] F. H. Clarke, Nonsmooth analysis and optimization, Wiley & Sons, New York, 1984.

[7] D. Goeleven, D. Motreanu and P. D. Panagiotopoulos, Eigenvalue problems for variational-hemivariational inequalities at resonance, Nonlin. Anal. TMA, 33 (1998), 161-185

[8] E. M. Landesman and A. C. Lazer, Nonlinear perturbations of linear elliptic boundary value problems at resonance, J. Math. Mech. 19 (1970), 609-623.

[9] P. D. Panagiotopoulos, Hemivariational Inequalities. Applications in Mechanics and Engineering, Springer-Verlag, Berlin, 1993,

[10] P. H. Rabinowitz, Some minimax theorems and applications to nonlinear partial differential equations, in: Nonlinear Analysis: A collection of papers in honor of E. Röthe, Academic Press, New York, 1978, pp. 161-177.

[11] P. H. Rabinowitz, Minimax methods in critical point theory with applications to differential equations, CBMS Reg. Conf. Ser. Math. 65, Amer. Math. Soc., Providence, R. I., 1986.

9 A SURVEY ON TIME–INTEGRATION ALGORITHMS FOR CONVEX AND NONCONVEX ELASTOPLASTICITY.

Klaus Hackl

Lehrstuhk für Allgemeine Mechanik
Ruhr-Universität Bochum
D-44780 Bochum, Germany

Abstract: Based on an augmented variational principle and on operator–splitting techniques different time–integration schemes for elastoplastic materials are investigated. Existing procedures are put into a consistent framework and a novel approach is suggested.

9.1 INTRODUCTION

A crucial part of any numerical treatment of a flow theory of elastoplasticity is the time–integration of the resulting rate–equations. Traditionally this is done by a so–called "radial–return" algorithm, see [14], which has been established to work well for the classical von Mises yield–criterion. (In the following we will give an operator–theoretic justification for this.) For more general materials there are however some serious problems associated with this approach:

- It is only first–order accurate, the implementation of higher–order methods being a tedious and problematic task.

R.P. Gilbert et al. (eds.), From Convexity to Nonconvexity, 111–122.

- For large strains one has to solve implicit equations for the strain–increments and the consistency parameter which is time–consuming and leads severely complicated expressions.

- For nonconvex yield–surfaces a closest–point projection mapping involves singularities which make the algorithm ill–conditioned.

In order to overcome this problems we suggest an augmented variational principle which allows to approximate the original differential–algebraic problem by a pure differential equation thus reducing the severe stiffness of the system and allowing to use explicit algorithms. Moreover we will introduce operator–splitting techniques in order to enhance the stability of the algorithms.

Additionally both methods introduced above will provide a consistent framework for the investigation of old and newly proposed time–integration schemes, their accuracy, stability and performance.

9.2 AN AUGMENTED PRINCIPLE OF MAXIMUM PLASTIC DISSIPATION

Constitutive Elastoplastic Theory

In [5] a general scheme has been developed which allows to treat arbitrary elastoplastic problems within a consistent variational framework. In the following we will repeat some of the concepts introduced there.

¿From a thermodynamic point of view the constitutive behavior of an elastoplastic material is governed by two scalar quantities: the Helmholtz free energy W and the yield function $\hat{\Phi}$. We assume the free energy to have the form $W = W(\mathbf{E}, \mathbf{P})$, where $\mathbf{E}$ denotes a suitable (total) strain tensor (e.g. Green–Lagrangian–strain) and $\mathbf{P}$ is a general set of internal variables which include plastic (irreversible) strains but may also comprise hardening parameters, quantities associated with material damage, or other variables describing material behavior.

Thermodynamically conjugate forces to $\mathbf{E}$ and $\mathbf{P}$ are defined by

$$\mathbf{S} := \frac{\partial W}{\partial \mathbf{E}}, \qquad \mathbf{Q} := -\frac{\partial W}{\partial \mathbf{P}}, \tag{9.1}$$

where $\mathbf{S}$ is the corresponding stress–tensor.

Let us now assume the yield–function to have the form $\hat{\Phi} = \hat{\Phi}(\mathbf{Q}, \mathbf{P})$. Here $\hat{\Phi}$ is allowed to be a generally nonconvex, locally Lipschitz–continuous functional. Furthermore the evolution in time of the internal variables $\mathbf{P}$ shall take place according to

$$\dot{\mathbf{P}} \in \bar{\partial}_{\mathbf{Q}} \hat{\Phi}, \tag{9.2}$$

where $\bar{\partial}_{\mathbf{Q}}$ denotes the *generalized gradient* in the sense of Clarke, [2], of $\hat{\Phi}$ with respect to $\mathbf{Q}$.

This formulation characterizes the material under investigation as so called *generalized standard medium.* This notion was initially proposed by Halphen and Nguyen, [7]. However 9.2 has been already used in earlier works, see [9],

[11]. In [5] it is shown that essentially any elastoplastic material law can be put into this form. The part of $\mathbf{Q}$ associated with plastic strain coincides with the stress $\mathbf{S}$ in the case of small elastic strains. This is no longer true for large elastic strains, see [5] and example 2 of section 9.5.

In the following we will restrict ourselves to the case where $\hat{\Phi} = I_K$, the indicator of a set given by $K = \{(\mathbf{Q}, \mathbf{P})|\ \Phi(\mathbf{Q}, \mathbf{P}) \leq 0\}$.[1] Then 9.2 assumes the form

$$\dot{\mathbf{P}} = \dot{\lambda}\frac{\partial \Phi}{\partial \mathbf{Q}}, \tag{9.3}$$

subject to the restrictions

$$\Phi \leq 0, \quad \dot{\lambda} \geq 0, \quad \dot{\lambda}\Phi = 0. \tag{9.4}$$

Here $\dot{\lambda}$ are called the ***plastic consistency parameter*** or the ***slip rate*** respectively. Materials laws corresponding to 9.2 or 9.3 and 9.4 are discussed in detail in [10].

Variational Formulation

One task that has to be accomplished by every numerical code for elastoplastic problems is to integrate 9.3, 9.4, provided the evolution of the total strain $\mathbf{E} = \mathbf{E}(t)$ is given. This is what we would like to discuss in the remainder of the present paper. In order to do this in a systematic way, we will have to extend the formalism introduced so far. For this purpose we will reformulate the material laws considered above once again by stating the following ***principle of maximum plastic dissipation:***

The plastic dissipation given by $-\frac{\partial W}{\partial \mathbf{P}} : \dot{\mathbf{P}} = \mathbf{Q} : \dot{\mathbf{P}}$ is maximized by $\mathbf{Q}$ for fixed $\mathbf{P}$ and $\dot{\mathbf{P}}$, given the constraint $(\mathbf{Q}, \mathbf{P}) \in K$.

This means we are looking for stationary values of the functional

$$\dot{\Lambda} := -\mathbf{Q} : \dot{\mathbf{P}} + \dot{\lambda}\, \Phi(\mathbf{Q}, \mathbf{P}), \tag{9.5}$$

with respect to $\mathbf{Q}$ and the Lagrange–parameter $\dot{\lambda}$, subject to the constraints 9.4. Let us refer to this as the *original problem*, which clearly is equivalent to 9.3 and 9.4.

We are now going to extend the above principle by introducing an additional parameter Y and replacing the set K by $K_a = \{(\mathbf{Q}, \mathbf{P}, Y)|\ \Phi(\mathbf{Q}, \mathbf{P}) \leq Y\}$. Moreover, and most important, Y is supposed to satisfy the differential equation

$$\dot{Y} + \beta Y = 0. \tag{9.6}$$

This means we are now looking for stationary values of the augmented functional

$$\dot{\Lambda}_a := -\mathbf{Q} : \dot{\mathbf{P}} + \dot{\lambda}(\Phi(\mathbf{Q}, \mathbf{P}) - Y) + \mu(\dot{Y} + \beta Y), \tag{9.7}$$

with respect to $\mathbf{Q}$, Y and the Lagrange–parameters $\dot{\lambda}$ and μ, subject to the constraints

$$\Phi \leq Y, \quad \dot{\lambda} \geq 0, \quad \dot{\lambda}(\Phi - Y) = 0. \tag{9.8}$$

We are going to refer to this as the *augmented problem*, which in turn is equivalent to 9.3, 9.6 and 9.8.

Remarks:
(i) For the initial condition $Y = 0$ at time t_0 the augmented problem is equivalent to the original problem. The same is true in the limit $\beta \to \infty$. This preservation of the solutions of the original problem is a characteristic feature of any augmented variational principle, see [4].
(ii) As we will see the actual purpose of augmentation is to transform the original problem into a better conditioned one allowing for more efficient and stable numerical algorithms.

9.3 THE EVOLUTION PROBLEM

Let us now assume that $\dot{\lambda} > 0$ during a certain time interval. Using equation 9.1 we express the yield–function in the form $\Phi = \Phi(\mathbf{E}, \mathbf{P})$ and we introduce the abbreviation $\frac{\partial \Phi}{\partial \mathbf{Q}} =: \mathbf{N} = \mathbf{N}(\mathbf{E}, \mathbf{P})$. Then it holds $Y = \Phi(\mathbf{E}, \mathbf{P})$ and $\dot{Y} = \frac{d}{dt}\Phi(\mathbf{E}, \mathbf{P})$. Using 9.3 and 9.6 the augmented problem is governed by the following evolution equations for $\mathbf{P}$ and λ:

$$\begin{aligned} \dot{\mathbf{P}} - \dot{\lambda}\,\mathbf{N} &= 0, && (9.9)\\ \frac{\partial \Phi}{\partial \mathbf{P}} : \dot{\mathbf{P}} &= -\beta\Phi - \frac{\partial \Phi}{\partial \mathbf{E}} : \dot{\mathbf{E}}. && (9.10) \end{aligned}$$

Note that for the original problem, i.e. for $\beta \to \infty$, equation 9.10 is replaced by $\Phi = 0$. So in this case we have to deal with a differential–algebraic system. In contrast to this case, the equations 9.9 and 9.10 can now in principle be integrated using any time–integration–scheme available. So only this formulation provides enough flexibility to be able to compare different algorithms. (This is one essential difference to the original problem which can only be integrated by algorithms especially designed for the differential–algebraic system.)

We are able to eliminate $\dot{\lambda}$ by substituting 9.9 into 9.10 and solving for $\dot{\lambda}$. We obtain

$$\dot{\mathbf{P}} = -\frac{\beta\Phi + \frac{\partial \Phi}{\partial \mathbf{E}} : \dot{\mathbf{E}}}{\frac{\partial \Phi}{\partial \mathbf{P}} : \mathbf{N}}\,\mathbf{N}, \qquad (9.11)$$

which will be the system we are going to refer to in the following.

9.4 TIME–INTEGRATION ALGORITHMS AND THEIR STABILITY PROPERTIES

Stability Functions and Step Size Restrictions

We will discuss now stability and step size restrictions for several integration–algorithms when applied to equation 9.11. Considerations of this kind go back to the paper of Courant, Friedrichs and Lewy, [3]. For linear elastoviscoplastic problems basic studies have been made by Hughes and Taylor [8]. We will present our analysis according to the account in the book by Hairer and Wanner, [6].

Suppose we would like to integrate equation 9.11 over some interval Δt starting at time t_0 by some algorithm. Then according to [6] the stability of this procedure will be closely modeled by the stability of the linearized equation

$$\dot{\mathbf{P}} = \mathbf{J}_0 : \mathbf{P}. \tag{9.12}$$

Here $\mathbf{J}_0 = \mathbf{J}(\mathbf{E}_0, \dot{\mathbf{E}}_0, \mathbf{P}_0, \beta)$ and $\mathbf{E}_0 = \mathbf{E}(t_0)$, $\dot{\mathbf{E}}_0 = \dot{\mathbf{E}}(t_0)$, $\mathbf{P}_0 = \mathbf{P}(t_0)$, where $\mathbf{J}$ denotes the Jacobian of the system 9.11 with respect to $\mathbf{P}$ given by

$$\begin{aligned} \mathbf{J} &= -\frac{1}{\frac{\partial\Phi}{\partial\mathbf{P}} : \mathbf{N}} \mathbf{N} \otimes \left(\beta \frac{\partial\Phi}{\partial\mathbf{P}} + \frac{\partial^2\Phi}{\partial\mathbf{P}\partial\mathbf{E}} : \dot{\mathbf{E}}\right) \\ &+ \frac{\beta\Phi + \frac{\partial\Phi}{\partial\mathbf{E}} : \dot{\mathbf{E}}}{\left(\frac{\partial\Phi}{\partial\mathbf{P}} : \mathbf{N}\right)^2} \mathbf{N} \otimes \left(\frac{\partial^2\Phi}{\partial\mathbf{P}^2} : \mathbf{N} + \frac{\partial\Phi}{\partial\mathbf{P}} : \frac{\partial\mathbf{N}}{\partial\mathbf{P}}\right) - \frac{\beta\Phi + \frac{\partial\Phi}{\partial\mathbf{E}} : \dot{\mathbf{E}}}{\frac{\partial\Phi}{\partial\mathbf{P}} : \mathbf{N}} \frac{\partial\mathbf{N}}{\partial\mathbf{P}}. \end{aligned} \tag{9.13}$$

Now let us assume that $\mathbf{J}_0$ possesses a complete system of eigenvectors $\mathbf{V}_i$ with corresponding (generally complex) eigenvalues ρ_i, $i = 1, \ldots, 6$. Then there exists an expansion $\mathbf{P}_0 = \sum_i \alpha_i \mathbf{V}_i$, and the approximation of equation 9.12 obtained after m steps of stepsize Δt of a specific algorithm is given by

$$\mathbf{P}_m = \sum_i R(\rho_i \Delta t)^m \alpha_i \mathbf{V}_i. \tag{9.14}$$

Here $R(z)$ is the so–called ***stability function*** which can be interpreted as the numerical solution after one step of the algorithm applied to the test problem $\dot{y} = \rho y$, $y_0 = 1$, $z = \rho\Delta t$, see [6]. Stability functions for different algorithms are given in table 9.1. From equation 9.14 we see that $\mathbf{P}_m$ remains bounded if

Table 9.1 stability functions

algorithm	$R(z)$
Euler forward	$1+z$
Runge–Kutta (order p)	$1+z+\frac{z^2}{2!}+\cdots+\frac{z^p}{p!}$
Euler backward	$\frac{1}{1-z}$
implicit midpoint / trapezoidal	$\frac{1+z/2}{1-z/2}$
Hammer–Hollingsworth	$\frac{1+4z/6+z^2/6}{1-z/3}$

$$|R(\rho_i \Delta t)| \leq 1, \qquad \forall \rho_i. \tag{9.15}$$

Inequality 9.15 fundamentally restricts the size of the time–increment Δt.[2] For our purposes, however, we are interested in a slightly different kind of question.

Using the homogeneity of the quantities involved it is easy to show that the eigenvalues have the following functional dependence:

$$\rho_i = \beta f_i(\mathbf{E}_0, \frac{1}{\beta}\dot{\mathbf{E}}_0, \mathbf{P}_0). \tag{9.16}$$

Introducing the *strain–increment* $\Delta\mathbf{E} = \dot{\mathbf{E}}_0\, \Delta t$ and using 9.16 we are able to write 9.15 in the form

$$|R(\kappa f_i(\mathbf{E}_0, \frac{1}{\kappa}\, \Delta\mathbf{E}, \mathbf{P}_0))| \leq 1, \qquad \forall \rho_i, \tag{9.17}$$

where $\kappa = \beta\Delta t$. This means for given initial data $\mathbf{E}_0$ and $\mathbf{P}_0$ the stability of the algorithm depends only on $\Delta\mathbf{E}$ on the one hand and the factor κ on the other hand. Since equation 9.11 is linear in β, however, β is involved in any specific integration algorithm via κ only. (For an example see section 9.1.) Hence, we can consider κ as a *numerical regularization parameter* associated with approximating the original problem by the augmented problem. Thus for a given material law we can study the stability of a specific algorithm by considering the set of all strain increments $\Delta\mathbf{E}$ satisfying inequality 9.17 depending on κ, $\mathbf{E}_0$ and $\mathbf{P}_0$.

The Test Case: Anisotropic Small–Strain von Mises–Plasticity

We would like to test the formalism developed above using an example which is general enough to preserve all essential features while simple enough to stay concise.

Let $\mathbf{E} = \boldsymbol{\epsilon}$, linearized total strain, $\mathbf{P} = \mathbf{p}$, linearized plastic strain, and $\mathbf{S} = \mathbf{Q} = \boldsymbol{\sigma}$, Cauchy–stress. Let the deviatoric parts of $\boldsymbol{\epsilon}$ and $\boldsymbol{\sigma}$ be denoted by $\mathbf{e} = \text{dev}[\boldsymbol{\epsilon}]$ and $\mathbf{s} = \text{dev}[\boldsymbol{\sigma}]$. Then the constitutive laws 9.1 reduce to

$$\mathbf{s} = 2\mu(\mathbf{e} - \mathbf{p}), \tag{9.18}$$

where μ is the shear–modulus. The yield–function is given by

$$\Phi = \sqrt{\mathbf{s} : \mathbf{A} : \mathbf{s}} - r, \tag{9.19}$$

where $r = \sqrt{2/3}\, y_0$ and y_0 is the yield–stress. $\mathbf{A}$ denotes an anisotropy–tensor (a symmetric, positive semidefinite bilinear form on the space of symmetric tensors of rank two). With those definitions equation 9.3 becomes

$$\dot{\mathbf{p}} = \dot{\lambda}\, \frac{\mathbf{A} : \mathbf{s}}{\sqrt{\mathbf{s} : \mathbf{A} : \mathbf{s}}}. \tag{9.20}$$

A straightforward calculation then shows that 9.11 takes the form

$$\dot{\mathbf{p}} = \left[\frac{\mathbf{s} : \mathbf{A} : \dot{\mathbf{e}} + \frac{\beta}{2\mu}\left(\mathbf{s} : \mathbf{A} : \mathbf{s} - r\sqrt{\mathbf{s} : \mathbf{A} : \mathbf{s}}\right)}{\mathbf{s} : \mathbf{A}^2 : \mathbf{s}} \right] \mathbf{A} : \mathbf{s}, \tag{9.21}$$

and the Jacobian becomes

$$\begin{aligned}\mathbf{J} &= -\frac{2\mu}{\mathbf{s}:\mathbf{A}^2:\mathbf{s}}\Bigg[\mathbf{A}:\mathbf{s}\otimes\mathbf{A}:\dot{\mathbf{e}}+\mathbf{s}:\mathbf{A}:\dot{\mathbf{e}}\,\mathbf{A}-2\frac{\mathbf{s}:\mathbf{A}:\dot{\mathbf{e}}}{\mathbf{s}:\mathbf{A}^2:\mathbf{s}}\mathbf{A}:\mathbf{s}\otimes\mathbf{A}^2:\mathbf{s} \\ &+ \frac{\beta}{2\mu}\left(\left(2-\frac{r}{\sqrt{\mathbf{s}:\mathbf{A}:\mathbf{s}}}\right)\mathbf{A}:\mathbf{s}\otimes\mathbf{A}:\mathbf{s}\right. \\ &+ \left.\left(\mathbf{s}:\mathbf{A}:\mathbf{s}-r\sqrt{\mathbf{s}:\mathbf{A}:\mathbf{s}}\right)\left(\mathbf{A}-\frac{2}{\mathbf{s}:\mathbf{A}^2:\mathbf{s}}\mathbf{A}:\mathbf{s}\otimes\mathbf{A}^2:\mathbf{s}\right)\right)\Bigg]. \end{aligned} \tag{9.22}$$

Here "$\otimes$" denotes the usual tensor–product and $\mathbb{I}$ is the fourth–order identity tensor, i.e. $\mathbb{I}:\mathbf{T}=\mathbf{T}$ for and arbitrary second–order tensor $\mathbf{T}$.

$\mathbf{J}_0$ possesses two distinct eigenvalues given by

$$\rho_1=-\beta, \qquad \rho_2=-2\mu\frac{\mathbf{s}_0:\mathbf{A}:\dot{\mathbf{e}}_0}{\mathbf{s}_0:\mathbf{A}:\mathbf{s}_0}-\beta\left(1-\frac{r}{\sqrt{\mathbf{s}_0:\mathbf{A}:\mathbf{s}_0}}\right). \tag{9.23}$$

Now 9.17 yields the conditions

$$0\le\kappa\le\gamma, \tag{9.24}$$

$$0\le 2\mu\frac{\mathbf{s}_0:\mathbf{A}:\Delta\mathbf{e}}{\mathbf{s}_0:\mathbf{A}:\mathbf{s}_0}+\kappa\left(1-\frac{r}{\sqrt{\mathbf{s}_0:\mathbf{A}:\mathbf{s}_0}}\right)\le\gamma, \tag{9.25}$$

where γ is a constant depending on the integration algorithm, defined as the unique solution of $R(-z)=\pm1$, see table 9.2.

Table 9.2 stability bounds

algorithm		γ
Euler forward		2.00000
Runge–Kutta order	2	2.00000
	3	2.52753
	4	2.78529
	5	3.21705
	6	3.55344
all implicit algorithms		∞

A comparison between equations 9.20 and 9.21 shows that the center term of inequality 9.25 is equal to $2\mu\dot{\lambda}\Delta t(\mathbf{s}_0:\mathbf{A}^2:\mathbf{s}_0)/(\mathbf{s}_0:\mathbf{A}:\mathbf{s}_0)^{3/2}$ and therefore always positive. Hence the left hand side inequalities in 9.24 and 9.25 are always identically satisfied.

Thus for explicit algorithms the possible values of the strain increment $\Delta\mathbf{e}$ are restricted by 9.25. Note that for the stresses close to the yield surface, i.e. $\sqrt{\mathbf{s}_0:\mathbf{A}:\mathbf{s}_0}\approx r$, 9.25 is essentially independent of κ. Moreover in this case the size of $\Delta\mathbf{e}$ is approximately restricted by $\gamma r/(2\mu)$, i.e. the maximal admissible strain increment $\Delta\mathbf{e}$ has roughly the size of the yield surface which is in some sense what had to be expected.

So usually inequality 9.25 doesn't present much of a problem because $\Delta\mathbf{e}$ is restricted anyway by accuracy requirements. More serious is the fact that the values of κ are restricted by γ via inequality 9.24. This is what renders explicit algorithms not suitable for the original problem ($\kappa \to \infty$).

Of course, as can be seen from table 9.2, all those restrictions do not apply for implicit algorithms making these (at first sight) better suited for the problems under consideration.

9.5 ALGORITHMS INVOLVING OPERATOR–SPLIT

Operator–Splitting for Evolution Problems

Operator–splitting techniques are widely used in numerical analysis, the idea going back to the famous product–formula for solving the Navier–Stokes–equation by Chorin. The mathematical framework is set by the so-called Lie–Trotter–Kato theorem. For details we refer to [1] and the references therein.

Consider a general evolution problem of the form

$$\dot{y} = f(y), \tag{9.26}$$

where y lies in a suitable Banach–space. Suppose now that the operator f can be additively decomposed into two simpler operators:

$$f(y) = g(y) + h(y). \tag{9.27}$$

Let us now define the *evolution–operator* $\mathcal{E}_f$ via

$$\mathcal{E}_f(t)(y_0) = y(t), \tag{9.28}$$

where $y(t)$ is the solution of equation 9.26 subjected to the initial condition $y(0) = y_0$. In a similar manner we introduce evolution–operators $\mathcal{E}_g$ and $\mathcal{E}_h$. Note the well–known fact that these operators act as a semigroup.

The idea behind operator–splitting is now to approximate $\mathcal{E}_f$ by a sequence of alternating applications of $\mathcal{E}_g$ and $\mathcal{E}_h$. So is a first–order accurate approximation given by

$$\mathcal{E}_f(\Delta t) \approx \mathcal{E}_h(\Delta t) \circ \mathcal{E}_g(\Delta t). \tag{9.29}$$

The simplest decomposition which is second–order accurate is

$$\mathcal{E}_f(\Delta t) \approx \mathcal{E}_g(\Delta t/2) \circ \mathcal{E}_h(\Delta t) \circ \mathcal{E}_g(\Delta t/2). \tag{9.30}$$

(For convenience we will abbreviate 9.29 and 9.30 by $\mathcal{E}_f \approx \mathcal{E}_g \circ \mathcal{E}_h$ and $\mathcal{E}_f \approx 1/2\,\mathcal{E}_g \circ \mathcal{E}_h \circ 1/2\,\mathcal{E}_g$ respectively.)

Elastic–Plastic Split

In the following we are going to assume that $\mathbf{E}$ does not change direction during the time–increment Δt. Note that for many applications this assumption is valid for Δt quite large. This means that $\dot{\mathbf{E}}$ is constant and we can reformulate 9.11 as the autonomous system

$$\dot{\mathbf{P}} = -\frac{\beta\Phi + \frac{\partial\Phi}{\partial\mathbf{E}} : \dot{\mathbf{E}}_0}{\frac{\partial\Phi}{\partial\mathbf{P}} : \mathbf{N}}\mathbf{N}, \qquad \dot{\mathbf{E}} = \dot{\mathbf{E}}_0. \tag{9.31}$$

The system 9.31 can be split in a natural way into two subproblems. The first one of those being trivial:

$$\dot{\mathbf{P}} = 0, \qquad \dot{\mathbf{E}} = \dot{\mathbf{E}}_0, \tag{9.32}$$

the second one being of the form

$$\dot{\mathbf{P}} = -\frac{\beta\Phi + \frac{\partial\Phi}{\partial\mathbf{E}} : \dot{\mathbf{E}}_0}{\frac{\partial\Phi}{\partial\mathbf{P}} : \mathbf{N}}\mathbf{N}, \qquad \dot{\mathbf{E}} = 0. \tag{9.33}$$

Note that while $\mathbf{E}$ is constant in the system 9.33, the parameter $\dot{\mathbf{E}}_0$ still occurs in the equation for $\dot{\mathbf{P}}$.

For obvious reasons we call 9.32 the *elastic step* associated with an evolution operator $\mathcal{E}_e$ and 9.33 the *plastic step* associated with an evolution operator $\mathcal{E}_p$.

Consider now the procedure 9.29 applied in the form $\mathcal{E}_p \circ \mathcal{E}_e$. Here $\mathcal{E}_e$ can be viewed as an elastic predictor while $\mathcal{E}_p$ represents a plastic corrector. A scheme of the form $\mathcal{E}_e \circ \mathcal{E}_p$, although first–order accurate as well, is unfavorable because the second step increases once again the residue to the the yield–condition $\Phi = 0$ which had been decreased before by the first one. For the same reason the second order split 9.30 should be of the form $1/2\,\mathcal{E}_p \circ \mathcal{E}_e \circ 1/2\,\mathcal{E}_p$.

There is a serious drawback to the elastic–plastic split: problem 9.33 has still the same stability bounds than 9.31. Thus explicit algorithms are very limited in performance while the implementation of implicit algorithms is connected with complications which make the task almost unfeasible. However, we would like to mention a frequently occurring and intensively used exception:

We say an elastoplastic material possesses the *radial–projection property* if for any $\mathbf{E}_1$, $\mathbf{P}_0$, fixed in time, and $\alpha \in I\!R$ there exists a $\gamma \in I\!R$ such that

$$\mathbf{N}(\mathbf{E}_1, \mathbf{P}_0 + \alpha\mathbf{N}_1) = \gamma\mathbf{N}_1, \tag{9.34}$$

where $\mathbf{N}_1 = \mathbf{N}(\mathbf{E}_1, \mathbf{P}_0)$, i.e. $\dot{\mathbf{P}}$ does not change direction during the time step.

Substitution into equation 9.33 yields the one–dimensional problem in α:

$$\dot{\alpha} = -\frac{\beta\Phi(\mathbf{E}_1, \mathbf{P}_0 + \alpha\mathbf{N}_1) + \frac{\partial\Phi}{\partial\mathbf{E}}(\mathbf{E}_0, \mathbf{P}_0 + \alpha\mathbf{N}_1) : \dot{\mathbf{E}}_1}{\frac{\partial\Phi}{\partial\mathbf{P}}(\mathbf{E}_1, \mathbf{P}_0 + \alpha\mathbf{N}_1) : \mathbf{N}_1}. \tag{9.35}$$

Now let us be reminded of the fact that we are primarily interested in the original problem, i.e. in the case $\beta \to \infty$. In the limit α in equation 9.35 becomes stationary at the end of the time–step and we can replace problem 9.35 by the algebraic equation in α

$$\Phi(\mathbf{E}_1, \mathbf{P}_0 + \alpha\mathbf{N}_1) = 0, \tag{9.36}$$

and obtain at the end of the time step

$$\mathbf{P}(t_0 + \Delta t) = \mathbf{P}_0 + \alpha \mathbf{N}_1. \tag{9.37}$$

One has to find the smallest positive α which satisfies equation 9.36. For star–shaped yield surfaces this solution depends continuously on $\mathbf{E}_0$ and $\mathbf{P}_0$ and can easily be obtained (for example by a simple newton code).

The split $\mathcal{E}_p \circ \mathcal{E}_e$, in which case $\mathbf{E}_1 = \mathbf{E}_0 + \dot{\mathbf{E}}_0 \Delta t$, in connection with equation 9.36 is in the literature widely known as "Euler–backward–method". This is a slight abuse of name because it differs from the classical Euler–backward–method applied to an ordinary differential equation (although there are of course strong similarities). This approach has been elaborated in a series of papers by Simo, especially in connection with the calculation of consistent tangent operators, see [12]. For large–strain elastoplasticity considerable effort has been made to develop material laws based on logarithmic strains which have (in an approximate sense) the radial–projection property, see [13].

Remarks:
(i) It is tempting to generalize the scheme given by 9.36 and 9.37 to materials which do not have the radial–projection property by establishing an "Euler–backward–method" of the form

$$\mathbf{P} = \mathbf{P}_0 + \alpha \mathbf{N}(\mathbf{E}_1, \mathbf{P}), \qquad \mathbf{\Phi}(\mathbf{E}_1, \mathbf{P}) = 0, \tag{9.38}$$

and solving simultaneously for $\mathbf{P}$ and α. Besides the fact that this approach cannot be motivated within the framework we have established so far, there is another significant problem associated with if. For non–convex yield surfaces the solutions of the system 9.38 in general do not depend continuously on $\mathbf{E}_1$ and $\mathbf{P}_0$ which will render the algorithm poorly conditioned and unstable.
(ii) An interesting observation can be made when applying 9.36 in combination with the split $1/2\,\mathcal{E}_p \circ \mathcal{E}_e \circ 1/2\,\mathcal{E}_p$. Observe that 9.36 has been already solved within the last step of this procedure in the previous time–increment. Therefore the first step of the procedure in the present time–increment becomes trivial and can be omitted. Because $\mathcal{E}_p$ is essentially time–independent now, the remaining two steps can be rescaled to the form $\mathcal{E}_p \circ \mathcal{E}_e$ returning the classical predictor–corrector–scheme. The surprising point is now that the three–step–scheme has been second–order accurate, therefore the two–step–scheme is, too. This is a result that to our knowledge is not known in the literature.

Example 1: the test–case continued.

For the anisotropic material of section 9.1 we have

$$\mathbf{N} = \frac{\mathbf{A} : \mathbf{s}}{\sqrt{\mathbf{s} : \mathbf{A} : \mathbf{s}}}. \tag{9.39}$$

Introducing the rescaled variable $\bar{\alpha} = \alpha / \sqrt{\mathbf{s} : \mathbf{A} : \mathbf{s}}$ we can bring condition 9.34 into the form

$$\bar{\alpha}\, \mathbf{A}^2 : \mathbf{s}_1 = (1 - \gamma)\, \mathbf{A} : \mathbf{s}_1, \tag{9.40}$$

where $\mathbf{s}_1 = \mathbf{e}_1 - \mathbf{p}_0$. This means the material possesses the radial–projection property, if and only if $\mathbf{A}^2 = \mathbf{A}$, i.e. if $\mathbf{A}$ is idempotent in which case all eigenvalues are either zero or one. Equation 9.36 can be solved for $\bar{\alpha}$ explicitly which yields

$$\bar{\alpha} = 1 - \frac{r}{\sqrt{\mathbf{s}_1 : \mathbf{A} : \mathbf{s}_1}}, \tag{9.41}$$

and we obtain at the end of the step

$$\mathbf{P}_1 = \mathbf{P}_0 + \left(1 - \frac{r}{\sqrt{\mathbf{s}_1 : \mathbf{A} : \mathbf{s}_1}}\right) \mathbf{A} : \mathbf{s}_1. \tag{9.42}$$

Example 2: large–strain isotropic von Mises elastoplasticity.

Here $\mathbf{E} = \mathbf{C}$, the total (right) Cauchy–Green strain tensor, and $\mathbf{P} = \mathbf{C}^{p-1}$, where $\mathbf{C}^p$ denotes the plastic Cauchy–Green strain tensor. (For details concerning these quantities, see [5].) A generalization of the von Mises yield function to large strains is given by

$$\Phi = \|\mathrm{dev}[\mathbf{S} \cdot \mathbf{C}]\| - r, \tag{9.43}$$

where $\mathbf{S}$ denotes the second Piola–Kirchhoff stress–tensor. In [5] it is shown that the flow rule 9.3 takes the form

$$\dot{\mathbf{P}} = \dot{\lambda} \frac{\mathrm{dev}[\mathbf{S} \cdot \mathbf{C}]}{\|\mathrm{dev}[\mathbf{S} \cdot \mathbf{C}]\|} \cdot \mathbf{P}, \tag{9.44}$$

and that $\mathbf{S} \cdot \mathbf{C}$ is an isotropic function of $\mathbf{P} \cdot \mathbf{C}$. Hence by the representation theorem for tensor functions we obtain

$$\mathrm{dev}[\mathbf{S} \cdot \mathbf{C}] = B_1 \,\mathrm{dev}[\mathbf{P} \cdot \mathbf{C}] + B_2 \,\mathrm{dev}[(\mathbf{P} \cdot \mathbf{C})^2], \tag{9.45}$$

where B_1 and B_2 depend only on the invariants of $\mathbf{P} \cdot \mathbf{C}$. This yields

$$\mathbf{N} = \frac{B_1 \,\mathrm{dev}[\mathbf{P} \cdot \mathbf{C}] \cdot \mathbf{P} + B_2 \,\mathrm{dev}[\mathbf{P} \cdot \mathbf{C} \cdot \mathbf{P} \cdot \mathbf{C}] \cdot \mathbf{P}}{\|\mathrm{dev}[\mathbf{S} \cdot \mathbf{C}]\|}. \tag{9.46}$$

It is quite obvious that 9.46 does not satisfy the radial–projection property for any nontrivial selection of B_1 and B_2. All integration–methods of "Euler–backward type" for this kind of materials presented e.g. in[12] and [13] rely on further approximations within flow rule and integration scheme.

Radial–Tangential Split

For many materials the radial–projection property does not hold. For those cases we are going to suggest now a novel kind of operator–split which has distinct advantages compared to the elastic–plastic split. Especially it will allow to use explicit integration algorithms while maintaining the stability properties of implicit ones.

For this purpose let us define the projection

$$\boldsymbol{\Pi} = \frac{\hat{\mathbf{N}}}{||\hat{\mathbf{N}}||} \otimes \frac{\hat{\mathbf{N}}}{||\hat{\mathbf{N}}||}, \tag{9.47}$$

where $\hat{\mathbf{N}} = \hat{\mathbf{N}}(t) = \mathbf{N}(\mathbf{E}(t), \mathbf{P}_0)$, and let us split equation 9.31 into the two subproblems defined by

$$\dot{\mathbf{P}} = -\frac{\beta\Phi + \frac{\partial\Phi}{\partial\mathbf{E}} : \dot{\mathbf{E}}_0}{\frac{\partial\Phi}{\partial\mathbf{P}} : \mathbf{N}} \boldsymbol{\Pi} : \mathbf{N}, \qquad \dot{\mathbf{E}} = 0, \tag{9.48}$$

and

$$\dot{\mathbf{P}} = -\frac{\beta\Phi + \frac{\partial\Phi}{\partial\mathbf{E}} : \dot{\mathbf{E}}_0}{\frac{\partial\Phi}{\partial\mathbf{P}} : \mathbf{N}} (\mathbf{I\!I} - \boldsymbol{\Pi}) : \mathbf{N}, \qquad \dot{\mathbf{E}} = \dot{\mathbf{E}}_0. \tag{9.49}$$

The idea is now to treat the parameter β differently in problem 9.49 versus problem 9.50. Note that β plays the role of a numerical regularization parameter which does not influence the exact solution of the original problem and thus can be adjusted at will.

Consider problem 9.49. Here $\dot{\mathbf{P}}$ is orthogonal to $\hat{\mathbf{N}}$. For this reason we will call 9.49 the *tangential step* and denote the corresponding evolution operator by $\mathcal{E}_t$. Because $(\mathbf{I\!I} - \boldsymbol{\Pi}_0) : \mathbf{N}_0 = 0$, where $\mathbf{N}_0 = \mathbf{N}(\mathbf{E}_0, \mathbf{P}_0)$ and $\boldsymbol{\Pi}_0 = \boldsymbol{\Pi}(\mathbf{E}_0, \mathbf{P}_0)$, we have

$$\mathbf{J}_0 = -\frac{\beta\Phi(\mathbf{E}_0, \mathbf{P}_0) + \frac{\partial\Phi}{\partial\mathbf{E}}(\mathbf{E}_0, \mathbf{P}_0) : \dot{\mathbf{E}}_0}{\frac{\partial\Phi}{\partial\mathbf{P}}(\mathbf{E}_0, \mathbf{P}_0) : \mathbf{N}_0} (\mathbf{I\!I} - \boldsymbol{\Pi}_0) : \frac{\partial\mathbf{N}}{\partial\mathbf{P}}(\mathbf{E}_0, \mathbf{P}_0). \tag{9.50}$$

This means 9.49 can be integrated in a stable manner for any algorithm (especially for any explicit one) provided $\kappa = \beta\Delta t$ is small enough (for example zero).

Let us look at problem 9.48 which we will call the ***radial step*** and denote the corresponding evolution operator by $\mathcal{E}_r$. It holds $\boldsymbol{\Pi} : \mathbf{N} = \gamma\mathbf{N}_1$ with $\gamma = \mathbf{N}_1 : \mathbf{N}/||\mathbf{N}_1||^2$, i.e. $\mathcal{E}_r$ possesses the radial–projection property and the method introduced in section 9.5 can be applied as well. Of course this means we have to assume $\beta \to \infty$ here.

The evolution operator $\mathcal{E}_r$ can once again be interpreted as a plastic corrector step, while $\mathcal{E}_t$ plays the role of a, now refined, elastoplastic predictor. By the same argument as in the preceding subsection the split $\mathcal{E}_r \circ \mathcal{E}_t$ is second–order accurate, but this time we have to take care that $\mathcal{E}_t$ itself is integrated by a method which is second–order accurate.

Note that for the evaluation of $\mathcal{E}_t$ equation 9.48 has to be integrated with the value of $\mathbf{E}(t)$ at the end of the previous step as initial value. Consider for example the split $\mathcal{E}_r \circ \mathcal{E}_t$. Here for the first step we have $\hat{\mathbf{N}} = \mathbf{N}_0 = \mathbf{N}(\mathbf{E}_0, \mathbf{P}_0)$ while for the second step it holds $\hat{\mathbf{N}} = \mathbf{N}_1 = \mathbf{N}(\mathbf{E}_0 + \dot{\mathbf{E}}_0\Delta t, \mathbf{P}_0)$. For materials which possess the radial–projection property $\mathcal{E}_t$ becomes trivial and the algorithm described above is equivalent to that one based on the elastic–plastic split.

Example 3: the test–case continued.

For the anisotropic material of section 9.1 equation 9.36 becomes

$$\mathbf{s}_1 : \left(\bar{\alpha}^2 \mathbf{A}^3 - 2\bar{\alpha}\mathbf{A}^2 + \mathbf{A}\right) : \mathbf{s}_1 = r^2. \tag{9.51}$$

If $\mathbf{A}$ is idempotent this clearly reduces to equation 9.41.

9.6 CONCLUSION

We have established a framework which allows to judge and compare all time–integration schemes. We have singled out the radial–tangential split as the most efficient one and shown under which circumstances it is equivalent to other known algorithms. Of course stability and accuracy have been investigated with regard to the linearized system only, not accounting for nonlinear effects. Those will have to be studied via numerical experiments which will be reported in a subsequent paper.

Several generalizations seem to be interesting: to multisurface plasticity, to elasto–viscoplasticity or to dynamical problems. Finally a closer mathematical investigation of the different roles of the parameter β in the treatment of the tangential and the radial problem is desirable.

Notes

1. It is easily possible to extend this formulation to multiple yield–surfaces, see [10].

2. If $\mathbf{J}_0$ does not have a complete set of eigenvectors equation 9.14 is not meaningful. But inequality 9.15 still holds with the right hand side replaced by an arbitrary positive constant $c < 1$.

References

[1] A. J. Chorin, T. J. R. Hughes, M. F. McCracken, and J. E. Marsden. Product formulas and numerical algorithms. *Comm. Pure Appl. Math.*, 31:205–256, 1978.

[2] F. H. Clarke. *Optimization and Nonsmooth Analysis.* Classics in Applied Mathematics. SIAM, Philadelphia, 1990.

[3] R. Courant, K. Friedrichs, and H. Lewy. Über die partiellen Differenzengleichungen der mathematischen Physik. *Math. Ann.*, 100:32–74, 1928.

[4] R. Glowinski and P. Le Tallec. *Augmented Lagrangian and Operator–Splitting Methods in Nonlinear Mechanics.* SIAM Studies in Applied Mathematics. SIAM, Philadelphia, 1989.

[5] K. Hackl. Generalized Standard Media and Variational Principles in Classical and Finite Strain Elastoplasticity. *J. Mech. Phys. Solids*, 45:667–688, 1997.

[6] E. Hairer and G. Wanner. *Solving Ordinary Differential Equations II, Stiff and Differential–Algebraic Problems.* Springer Series in Computational Mathematics 14. Springer, Berlin, Heidelberg, New York, 1991.

[7] B. Halphen and Nguyen Quoc Son. Sur les matériaux standards généralisés. *J. Mécanique*, 14:39–63, 1975.

[8] T. J. R. Hughes and R. L. Taylor. Unconditionally stable algorithms for quasi-static elasto/visco-plastic finite element analysis. *Computers & Structures*, 8:169–173, 1978.

[9] J. J. Moreau. Sur les lois de frottement, de plasticité et de viscosité. *C. R. Acad. Sc. Paris, Série A*, 271:608–611, 1970.

[10] P. D. Panagiotopoulos. *Hemivariational Inequalities, Applications in Mechanics and Engineering*. Springer, Berlin, Heidelberg, New York, 1993.

[11] J. R. Rice. Inelastic Constitutive Relations for Solids: an Internal-Variable Theory and its Application to Metal Plasticity. *J. Mech. Phys. Solids*, 19:433–455, 1971.

[12] J. C. Simo. A Framework for Finite Strain Elastoplasticity Based on Maximum Plastic Dissipation and the Multiplicative Decomposition: Part I. Continuum Formulation. *Comput. Meths. Appl. Mech. Engrg.*, 66:199–219, 1988.

[13] J. C. Simo. Algorithms for static and dynamic multiplicative plasticity that preserve the classical return mapping schemes of the infinitesimal theory. *Comput. Meths. Appl. Mech. Engrg.*, 99:61–112, 1992.

[14] M. L. Wilkins. *Calculation of Elastic-plastic Flow*, volume 3 of *Meth. of Computational Physics*. Academic Press, New York, 1964.

10 CONTACT STRESS OPTIMIZATION

J. Haslinger

KFK MFF UK, Ke Karlovu 5, 121 16,
Praha 2, Czech Republic

Abstract: The contribution deals with a class of contact optimal shape design problems the main goal of which is to drive contact stresses between elastic bodies in mutual contact. The reciprocal variational formulation is used as a tool for their numerical realization. Such a formulation makes possible to approximate directly contact stresses.

10.1 INTRODUCTION

Optimal shape design is an applied branch of the optimal control theory, in which control variables are related to the geometry of a structure, in this case to its shape. When speaking on contact shape optimization we have in mind that the structure consists of several independent substructures being mutual contact. It is known that the mathematical model describing the equilibrium state of such a system is given by variational inequalities (see Fichera 1972). Just this fact makes the whole matter more involved. It is known also that the mapping: *controle variable* $\longrightarrow$ *solution of a state problem* is not continuously differentiable, in general. This phenomena has to be taken into account when realizing optimization problems numerically. It also explains why classical gradient type minimization methods or fail or give unsatisfactory results. Therefore contact shape optimization *is not* a straight-forward extension of known results, when the state relation is given by classical linear elasticity problems.

R.P. Gilbert et al. (eds.), From Convexity to Nonconvexity, 137–145.

One of typical problems arising in practice can be formulated as follows: how to design contact surfaces in order to get properly distributed contact stresses. A natural question arises, namely how to choose a cost functional by means of which one can control the behaviour of contact stresses. This is the goal of the next section.

10.2 FORMULATION OF THE PROBLEM

We start this section by recalling abstract setting of optimal shape design problems.

Let Θ be a family of *admissible domains.* With any $\Omega \in \Theta$ the solution $u(\Omega) \in V(\Omega)$ of a *state problem* $(P(\Omega))$ will be associated. We write the symbol Ω as the argument to stress that the solution $u(\Omega)$ (and all others data) is assumed to be a function of Ω. Finally, let $I : (\Omega, y) \rightarrow R^1$, with $\Omega \in \Theta$, $y \in V(\Omega)$ be a *cost functional* the explicit choice of which depends on optimization targets. A large class of optimal shape design problems can be stated as follows

$$(\mathbf{P}) \qquad \begin{cases} \text{Find } \Omega^* \in \Theta \text{ such that} \\ I(\Omega^*, u(\Omega^*)) \leq I(\Omega, u(\Omega)) \quad \forall\, \Omega \in \Theta, \end{cases}$$

with $u(\Omega) \in V(\Omega)$ being the solution to $(P(\Omega))$. Problem $(\mathbf{P})$ is a variational type problem. Conditions, under which $(\mathbf{P})$ admits at least one solution are now very well established (see Pironneau (1984), Haslinger, Neittaanmäki(1988,1996)). As mentioned above, in contact shape optimization state problem $(P(\Omega))$ is given by a variational inequality.

In this presentation we restrict ourselves to the so called *Signorini problem*, i.e. to the contact problem for a linearly elastic body, unilaterally supported by a rigid foundation. For the sake of simplicity, only the plane and frictionless case will be considered. However, a simple model of friction (the so called given friction) can be analysed as well (see Haslinger, Neittaanmäki (1996), Haslinger (1999)). Another simplification used here concerns of the geometry of deformable bodies represented by domains Ω and of a rigid foundation S. Next,we shall consider that

$$S = R^2_- = \{(x_1, x_2) \in R^2 |\; x_2 \leq 0\}$$

and Ω is a 'curved" rectangle:

$$\Omega = \{(x_1, x_2) \in R^2 |\; x_1 \in (a, b),\; \alpha(x_1) < x_2 < \gamma\},$$

where a, b, γ are given real numbers and $\alpha : [a, b] \rightarrow R^1$ is a non-negative Lipschitz continuous function, describing the contact part Γ_C of Ω, coming in contact with S. Next, we shall consider all such α's belonging to the set U_{ad}:

$$U_{ad} = \{\alpha \in C^{0,1}([a, b]) |\; 0 \leq \alpha \leq C_0,\; |\alpha'| \leq C_1, \text{ a.e. in } (a, b),\; \int_a^b \alpha dx_1 = C_2\},$$

i.e. U_{ad} contains all non-negative Lipschitz continuous functions, which are uniformly bounded and uniformly Lipschitz continuous and satisfy an integral type constraint. With any such $\alpha \in U_{ad}$, domain $\Omega(\alpha)$ of the above mentioned shape will be associated. The family of admissible domains is now given by

$$\Theta = \{\Omega(\alpha) |\ \alpha \in U_{ad}\}.$$

Remark 1.1 It is easy to see that Θ consists of the domains with uniformly Lipschitz continuous boundaries and satisfying a constant volume constraint.

On any $\Omega(\alpha) \in \Theta$, the following sets of functions are introduced:

$$V(\alpha) = \{v \in (H^1(\Omega(\alpha)))^2 |\ v = 0 \text{ on } \Gamma_u(\alpha)\},$$

$$K(\alpha) = \{v = (v_1, v_2) \in V(\alpha) |\ v_2(x_1, \alpha(x_1)) \geq -\alpha(x_1),\ x_1 \in (a, b)\},$$

i.e. $V(\alpha)$ is the space of *virtual displacements* and $K(\alpha)$ is its closed, convex subset of *kinematically admissible displacements*. Next, we shall suppose that a portion $\Gamma_u(\alpha)$, where the zero displacements are prescribed is *non-empty and open* in $\partial\Omega(\alpha)$, $\text{meas}_1\Gamma_u(\alpha) \geq \delta$ for any $\alpha \in U_{ad}$, where the symbol meas_1 stands for the one-dimensional Lebesgue measure of sets and $\delta > 0$ is given. Moreover, $\Gamma_u(\alpha)$ will depend continuously on variations of $\alpha \in U_{ad}$. Finally, let

$$J_\alpha(v) = \frac{1}{2}(\Lambda\varepsilon(v), \varepsilon(v))_{0,\Omega(\alpha)} - L_\alpha(v)$$

be the *total potentional energy functional* with the following meaning of notations:

$$\Lambda \in \mathcal{L}((R^2)^2_{sym}, (R^2)^2_{sym})$$

is a linear, symmetric and positive definite mapping from the space of (2×2) symmetric matrices into itself, defining a linear relation between stress tensors $\sigma = (\sigma_{ij})^2_{i,j=1}$ and linearized strain tensors $\varepsilon(v) = (\varepsilon_{ij}(v))^2_{i,j=1}$ corresponding to $v \in V(\alpha)$. Further

$$L_\alpha(v) = (F_i, v_i)_{0,\Omega(\alpha)} + (P_i, v_i)_{0,\Gamma_p(\alpha)}.$$

Here $F \in (L^2_{loc}(R^2))^2$, $P \in (H^1_{loc}(R^2))^2$ are given, $\Gamma_p(\alpha) \equiv \partial\Omega(\alpha)\backslash(\Gamma_C(\alpha) \cup \Gamma_u(\alpha))$ is a part of $\partial\Omega(\alpha)$ submitted to surface tractions P. Recall that $\Gamma_C(\alpha)$ is given by the graph of $\alpha \in U_{ad}$.

The weak formulation of the Signorini problem on $\Omega(\alpha)$ reads as follows:

$(P(\alpha))$ Find $u(\alpha) \in K(\alpha) : J_\alpha(u(\alpha)) \leq J_\alpha(v) \ \ \forall\, v \in K(\alpha)$

or equivalently

$$(P(\alpha)) \qquad \begin{cases} \text{Find } u(\alpha) \in K(\alpha) \text{ such that} \\ (\Lambda\varepsilon(u(\alpha)), \varepsilon(v - u(\alpha)))_{0,\Omega(\alpha)} \geq L_\alpha(v - u(\alpha)) \quad \forall\, v \in K(\alpha). \end{cases}$$

Under the assumptions, introduced above, $(P(\alpha))$ has a unique solution $u(\alpha)$ for any $\alpha \in U_{ad}$.

Now, we are ready to define a class of contact shape optimization problems. Taking into account the geometry of $\Omega(\alpha)$'s, we see that the function α, describing $\Gamma_C(\alpha)$ is the only design variable. So we are led to the following definition:

$$(\mathbf{P}) \qquad \begin{cases} \text{Find } \alpha^* \in U_{ad} \text{ such that} \\ I(\alpha^*, u(\alpha^*)) \leq I(\alpha, u(\alpha)) \quad \forall\, \alpha \in U_{ad}, \end{cases}$$

where $I : (\alpha, y) \rightarrow R^1$ with $\alpha \in U_{ad}$, $y \in V(\alpha)$ is a cost functional and $u(\alpha) \in K(\alpha)$ solves $(P(\alpha))$. Provided that **(P)** has a solution $\alpha^* \in U_{ad}$, the corresponding $\Omega(\alpha^*) \in \Theta$ will be called an *optimal shape* (with respect to U_{ad} and I). The existence of solutions to **(P)**, when the state is given by the Signorini problem, together with practical aspects of the numerical realization is analyzed in details in Haslinger, Neittaanmäki (1988,1996).

As mentioned at the beginning of this contribution, the control of contact stresses is of primal interest in many problems. One tries to find such a shape of $\Gamma_C(\alpha)$ along which the contact stresses are evenly distributed. How to choose I to achieve such a distribution? A usual way, widely used in an ingeneering community is to minimize the maximum, i.e. the L^∞-norm of $T_n(u)$ along $\Gamma_C(\alpha)$. This choice of the cost is meaningful in a discrete case only, while in continuous setting is useless due to lack of regularity of $u(\alpha)$. There is no reason to expect that $T_n(u) \in L^\infty(\Gamma_C(\alpha))$. In 1981, Benedict and Taylor proposed another I, namely

$$I(\alpha, u(\alpha)) \equiv J_\alpha(u(\alpha)), \tag{1.1}$$

i.e. I is equal to the value of J_α evaluated at the equilibrium state $u(\alpha)$. Problem **(P)** with (1.1) is mathematically analyzed in Haslinger, Neittaanmäki (1988,1996). It was shown that **(P)** has at least one solution α^* in this case. Moreover it was numerically verified that shape optimization with (1.1) yields a constant stress distribution along $\Gamma_C(\alpha)$. The mathematical justification of this fact was done later in Klarbring, Haslinger (1993) by a detailed interpretation of the corresponding optimality conditions. The fact that $T_n(u)$ is constant (or better "almost" constant) is a consequence of the constant volume constraint appearing in the definition of Θ. Another very important fact is that the mapping

$$\alpha \rightarrow I(\alpha, u(\alpha)), \quad \alpha \in U_{ad}$$

is *once continuously differentiable*, regardless the fact that the inner mapping: control $\rightarrow$ state is not, in general. Thus, classical gradient type optimization methods can be used. On the other hand there are some drawbacks when using (1.1):

(a) if there is no constant volume constraint imposed on Ω's then one can not expect the constant distribution of contact stresses;

(b) the control of $T_n(u)$ is *passive*: We do not have any influence on its magnitude. The value of $T_n(u)$ is related to the Lagrange multiplier associated with the constant volume constraint.

In order to drive contact stresses in an active way, a least square approach seems to be natural: one tries to adjust the shape of $\Gamma_C(\alpha)$ in such a way that the resulting contact stress is as close as possible to an a-priori given desired distribution z_d. To avoid difficulties, related to the regularity of $u(\alpha)$, one has to be careful with the choice of a norm, defining the least square functional. This was done in Haslinger, Klarbring (1993), see also Haslinger, Neittaanmäki (1996). Below we briefly describe the approach used there in.

Let $M(\alpha) \subset \partial\Omega(\alpha)$ be a non-empty, open and connected part of $\partial\Omega(\alpha)$ such that $M(\alpha) \cap \Gamma_C(\alpha) \neq \emptyset$ and $M(\alpha) \cap \Gamma_u(\alpha) = \emptyset$. With any such $M \equiv M(\alpha)$, $\alpha \in U_{ad}$ the following space will be associated:

$$V_M(\alpha) = \{v \in (H^1(\Omega(\alpha)))^2 \mid v = 0 \text{ on } \partial\Omega(\alpha))\backslash\overline{M}\}$$

equipped with the energy norm

$$\|v\|_\alpha \equiv (\Lambda\varepsilon(v), \varepsilon(v))^{1/2}_{0,\Omega(\alpha)}, \quad v \in V_M(\alpha).$$

Let $W_M(\alpha)$ be the corresponding space of traces, i.e.

$$\phi \in W_M(\alpha) \text{ iff } \exists\, v \in V_M(\alpha) \text{ and } v = \phi \text{ on } \partial\Omega(\alpha).$$

It is well-known that $W_M(\alpha)$ is a Banach space with the quotient norm:

$$\|\phi\|_{1/2,M,\alpha} = \inf_{\substack{v \in V_M(\alpha) \\ v=\phi \text{ on } \partial\Omega(\alpha)}} \|v\|_\alpha. \tag{1.2}$$

Denote by $W'_M(\alpha)$ the dual space to $W_M(\alpha)$ and by $\|\ \|_{-1/2,M,\alpha}$ the corresponding dual norm.

As a cost functional in (**P**) we now take

$$I(\alpha, u(\alpha)) \equiv \frac{1}{2}\|T_n(u) - z_d\|^2_{-1/2,M,\alpha}, \tag{1.3}$$

where $z_d \in H^1_{loc}(R^2)$ is given. It is readily seen that the couple $(T_n(u), 0) \in W'_M(\alpha)$ (do not forget that the frictionless case in considered). Here $T_n(u)$ stands for the normal stress along $\partial\Omega(\alpha)$ and it includes the contact normal stress on the intersection $M(\alpha) \cap \Gamma_C(\alpha)$. Since the dual norm is not very practical for numerical computations, an equivalent expression of (1.3) is necessary. Indeed, it is not difficult to show that the functional given by (1.3) is equal to

$$I(\alpha, u(\alpha)) = \frac{1}{2}\|z(\alpha)\|^2_\alpha, \tag{1.4}$$

where $z(\alpha) \in V_M(\alpha)$ is the unique solution of the auxiliary linear elasticity problem

$$(\Lambda\varepsilon(z(\alpha)), \varepsilon(\phi))_{0,\Omega(\alpha)} = \langle T_n(u(\alpha)) - z_d, \phi\rangle_\alpha \quad \forall\, \phi \in V_M(\alpha), \tag{1.5}$$

where the symbol $\langle\ ,\ \rangle_\alpha$ stands for the duality pairing between $W'_M(\alpha)$ and $W_M(\alpha)$. From the Green's theorem if follows that the right hand side of (1.5) can be written as follows:

$$\langle T_n(u) - z_d, \phi\rangle_\alpha = (\Lambda\varepsilon(u(\alpha)), \varepsilon(\phi))_{0,\Omega(\alpha)} - L_\alpha(\phi) - \langle z_d, \phi\rangle_\alpha. \tag{1.6}$$

The existence of at least one solution to (**P**) with I given by (1.3) is mathematically analysed in Haslinger, Klarbring (1993). Problem (**P**) with (1.3) has the following advantages:

(c) *the active* control of $T_n(u)$ is possible;

(d) choosing $M(\alpha) = \partial\Omega(\alpha)\backslash\overline{\Gamma}_u(\alpha)$, $z_d|_{\Gamma_p(\alpha)} = P_n$ (the normal component of P) and $z_d = 0$ on $\Gamma_C(\alpha)$ then the mapping $\alpha \to I(\alpha, u(\alpha))$ is again *once continuously differentiable* so that gradient type minimization methods can be used, again.

In both examples however, the following discrepancy appears: in order to control contact stresses, i.e. quantities defined on a part of $\partial\Omega(\alpha)$, the knowledge of $u(\alpha)$ in the whole $\Omega(\alpha)$ is necessary (see (1.1),(1.6)). A natural idea arises, namely to use another variational formulation expressed solely in terms of stresses. To this end we use the so called *reciprocal variational formulation* (see Haslinger, Panagiotopoulos (1984)). Below we recall briefly its derivation. We use the form which is convenient for the shape optimization.

Let $\hat{\Omega} = (a,b) \times (0,\gamma)$ be the reference domain and $F_\alpha : \hat{\Omega} \to \Omega(\alpha)$ the mapping defined by

$$\begin{cases} (x_1, x_2) = F_\alpha(\hat{x}_1, \hat{x}_2), \quad (\hat{x}_1, \hat{x}_2) \in \hat{\Omega} \\ x_1 = \hat{x}_1 \\ x_2 = \frac{\gamma - \alpha(\hat{x}_1)}{\gamma}\hat{x}_2 + \alpha(\hat{x}_1). \end{cases} \tag{1.7}$$

The boundary $\partial\hat{\Omega}$ is split into three non-overlaping parts $\hat{\Gamma}$, $\hat{\Gamma}_u$ and $\hat{\Gamma}_p$, where $\hat{\Gamma} = (a,b) \times \{0\}$. Moreover, we shall suppose that the partition of $\partial\Omega(\alpha)$ into $\Gamma_u(\alpha)$, $\Gamma_C(\alpha)$ and $\Gamma_p(\alpha)$ is derived from this one of $\partial\hat{\Omega}$ as follows:

$$\Gamma_C(\alpha) = F_\alpha(\hat{\Gamma}), \quad \Gamma_u(\alpha) = F_\alpha(\hat{\Gamma}_u), \quad \Gamma_p(\alpha) = F_\alpha(\hat{\Gamma}_p).$$

Futher let

$$V(\hat{\Omega}) = \{v \in H^1(\hat{\Omega}) |\ v = 0 \text{ on } \hat{\Gamma}_u\},$$

be the space of scalar functions defined in $\hat{\Omega}$ and vanishing on $\hat{\Gamma}_u$.

We define

$$H^{1/2}(\hat{\Gamma}) \equiv \text{trace}_{\hat{\Gamma}}\ V(\hat{\Omega})$$

$$H^{-1/2}(\hat{\Gamma}) \equiv (H^{1/2}(\hat{\Gamma}))'$$

the trace space on $\hat{\Gamma}$, the dual space, respectively with the duality denoted by $\langle\ ,\ \rangle$. Finally, Λ stands for the non-negative cone in $H^{-1/2}(\hat{\Gamma})$:

$$\Lambda = \{\mu \in H^{-1/2}(\hat{\Gamma}) | \langle \mu, \phi \rangle \geq 0 \quad \forall\, \phi \in H^{1/2}(\hat{\Gamma}),\ \phi \geq 0\}.$$

Using the duality approach, the kinematical constraint $v \in K(\alpha)$ can be released by using Lagrange multipliers and problem $(P(\alpha))$ can transformed into the following saddle-point formulation:

$$(M(\alpha)) \qquad \begin{cases} \text{Find } (w, \lambda) \in V(\alpha) \times \Lambda \text{ such that} \\ \mathcal{L}_\alpha(w, \mu) \leq \mathcal{L}_\alpha(w, \lambda) \leq \mathcal{L}_\alpha(v, \lambda) \quad \forall\, (v, \mu) \in V(\alpha) \times \Lambda, \end{cases}$$

where the Lagrangian $\mathcal{L}_\alpha$ is given by

$$\mathcal{L}_\alpha(v,\mu) \equiv J_\alpha(v) - \langle \mu, v_2 + \alpha \rangle, \quad v \in V(\alpha),\ \mu \in \Lambda.$$

The value $\langle \mu, v_2 \rangle$ has to be taken in the following sense

$$\langle \mu, v_2 \rangle \equiv \langle \mu, v_2 \circ F_\alpha \rangle, \quad \alpha \in U_{ad},$$

with $F_\alpha : \hat{\Omega} \to \Omega(\alpha)$ given by (1.7).

It is easy to show that the solutions $u(\alpha)$ of $(P(\alpha))$ and (w, λ) of $(M(\alpha))$ are related as follows:

$$u(\alpha) = w(\alpha) \ \text{ in } \Omega(\alpha)$$
$$\lambda = T_n(u)\sqrt{1 + (\alpha')^2} \ \text{ on } \hat{\Gamma}.$$

Elimination of the displacement field w in $(M(\alpha))$ leads to the reciprocal variational formulation, expressed solely in terms of contact stresses. This variational formulation reads as follows:

$$(R(\alpha)) \qquad \begin{cases} \text{Find } \lambda(\alpha) \in \Lambda \text{ such that} \\ S_\alpha(\lambda(\alpha)) \le S_\alpha(\mu) \quad \forall\, \mu \in \Lambda, \end{cases}$$

where

$$S_\alpha(\mu) = \frac{1}{2}\langle \mu, G_\alpha(\mu) \rangle + \langle \mu, G_\alpha(L_\alpha) \rangle + \langle \mu, \alpha \rangle.$$

Here $G_\alpha : V'(\alpha) \to V(\alpha)$ is the Green's operator defined as follows

$$G_\alpha(f) = u(f) \in V(\alpha), \quad f \in V'(\alpha),$$

where $u(f)$ is the unique solution of

$$(\Lambda\varepsilon(u(f)), \varepsilon(\phi))_{0,\Omega(\alpha)} = f(\phi) \quad \forall \phi \in V(\alpha).$$

Let us recall that $(R(\alpha))$ has a unique solution for any $\alpha \in U_{ad}$ and from the duality theory it follows that this solution equals to the second component of the saddle-point in $(M(\alpha))$. Let us mention also that the reciprocal variational formulation presents one of the most efficient ways of the numerical realization of contact problems. It makes possible (among others) to parallelize the computational process. For practical applications however, the approximation of G_α is necessary. A standart way is to use the inverse of the corresponding stiffness matrix.

Now, let us consider a class of optimal shape design problems (**P**) in which the cost functional I depends on the design variables $\alpha \in U_{ad}$ and on the contact stress $T_n(u)$, which will be now approximated directly by means of $\lambda(\alpha)$ being the solution of $(R(\alpha))$. Therefore (**P**) reads as follows:

$$(\mathbf{P}) \qquad \begin{cases} \text{Find } \alpha^* \in U_{ad} \text{ such that} \\ I(\alpha^*, \lambda(\alpha^*)) \le I(\alpha, \lambda(\alpha)) \quad \forall\, \alpha \in U_{ad} \end{cases}$$

with $\lambda(\alpha)$ solving $(R(\alpha))$.

In order to guarantee the existence of solutions to (**P**), the following *lower semicontinuity* of I will be needed:

$$(A) \qquad \left\{ \begin{array}{l} \alpha_n \Rightarrow \alpha \text{ (uniformly) in } [a,b],\ \alpha_n, \alpha \in U_{ad} \\ \lambda(\alpha_n) \rightharpoonup \lambda(\alpha) \text{ in } \Lambda \end{array} \right\} \Longrightarrow$$

$$\Longrightarrow \liminf_{n\to\infty} I(\alpha_n, \lambda(\alpha_n)) \geq I(\alpha, \lambda(\alpha)).$$

Here $\lambda(\alpha_n), \lambda(\alpha)$ stands for the solutions of $(R(\alpha_n))$, $(R(\alpha))$, respectively and the weak convergence used in (A) is defined as follows:

$$\lambda(\alpha_n) \rightharpoonup \lambda(\alpha) \text{ in } \Lambda \text{ iff}$$

$$\langle \lambda(\alpha_n), (\text{trace}_{\Gamma(\alpha_n)}\phi) \circ F_{\alpha_n} \rangle \longrightarrow \langle \lambda(\alpha), (\text{trace}_{\Gamma(\alpha)}\phi) \circ F_\alpha \rangle,$$

holds for any function $\phi \in H^1(\hat{\Omega})$ vanishing in some neighborhood of $F_\alpha(\hat{\Gamma}_u)$. Recall that $F_{\alpha_n}(\hat{\Omega}) = \Omega(\alpha_n)$ and $F_\alpha(\hat{\Omega}) = \Omega(\alpha)$ are given by (1.7).

Then one can prove the following result:

Theorem 1.1 Let (A) be satisfied. Then (**P**) has a solution $\alpha^* \in U_{ad}$.

As an example let as consider the cost functional

$$I(\alpha, \lambda(\alpha)) = \frac{1}{2}\|\lambda(\alpha)\|^2_{-1/2,\alpha}, \tag{1.8}$$

where $\|\ \|_{-1/2,\alpha}$ equals to $\|\ \|_{-1/2,M,\alpha}$ (the norm in $W'_M(\alpha)$ introduced before) with $M(\alpha) = \partial\Omega(\alpha)\backslash\overline{\Gamma}_u(\alpha)$ and $z_d \equiv 0$. Then it is not difficult to verify that I given by (1.8) satisfies (A). Problem (**P**) with such a choice of I reduces the dual norm of the contact stresses. Moreover it is also very easy to show that I can be expressed as follows:

$$I(\alpha, \lambda(\alpha)) = \frac{1}{2}\langle \lambda(\alpha), v_2(\alpha) \rangle,$$

where $v(\alpha) = (v_1(\alpha), v_2(\alpha)) \in V(\alpha)$ is the unique solution to

$$(\Lambda\varepsilon(v(\alpha)), \varepsilon(\phi))_{0,\Omega(\alpha)} = \langle \lambda(\alpha), \phi_2 \rangle \quad \forall\, \phi \in V(\alpha).$$

It can be shown (see Haslinger (1999)) that the trace of $v_2(\alpha)$ on $\Gamma(\alpha)$, necessary for the evaluation of I can be written in the form:

$$\text{trace}_{\Gamma(\alpha)} v_2(\alpha) = \text{trace}_{\Gamma(\alpha)} u_2(\alpha) - \text{trace}_{\Gamma(\alpha)} w_2(\alpha),$$

where $u = (u_1, u_2) \in K(\alpha)$ solves $(P(\alpha))$ and $w = (w_1, w_2) \in V(\alpha)$ solves the linear elasticity problem

$$(\Lambda\varepsilon(w), \varepsilon(\phi))_{0,\Omega(\alpha)} = L_\alpha(\phi) \quad \forall\, \phi \in V(\alpha). \tag{1.9}$$

$\text{Trace}_{\Gamma(\alpha)} u_2(\alpha)$ is usually available as a by product, when conjugate gradient type methods are used for the numerical realization of $(R(\alpha))$, since $\text{trace}_{\Gamma(\alpha)} u_2(\alpha)$ is the Lagrange multiplier associated with the constraint $\lambda \in \Lambda$. Problem (1.9) is a classical linear elasticity problem, which can be solved independently on the Signorini problem $(P(\alpha))$.

Acknowledgments

This work was supported by grant 101/98/0535 of the grant agency of Czech Republic.

References

[1] R.L. Benedict and J.E. Taylor: Optimal design for elastic bodies in contact, Optimization of Distributed-Parameter Structures, E.J. Haug and J. Céa(eds), *Sijthoff and Noordhoff aan den Rijn*, Holland, pp. 1553-1599 (1981).

[2] J. Haslinger and A. Klarbring: Shape Optimization in Unilateral Contact Problems using Generalized Reciprocal Energy as Objective Functional, *Nonlinear Analysis*, Methods & Appl. 21(1993), No11, pp.815-834.

[3] J. Haslinger and P. Neittaanmäki: *Finite Element Approximation for Optimal Shape Design: Theory and Applications*, J. Wiley, Chichester-New York (1988).

[4] J. Haslinger and P. Neittaanmäki: Finite Element Approximation for Optimal Shape, Material and Topology Design, 2nd Edition, J. Wiley, Chichester-New York (1996).

[5] J. Haslinger and P.D. Panagiotopoulos: Approximation of Contact Problems with Friction by Reciprocal Variational Formulation, Proc. Roy. Soc. Edingburgh, 98A (1984), pp.365-383.

[6] A. Klarbring and J. Haslinger: On almost constant contact stress distributions by shape optimization, *Struct. Opt.*, 5(1993), pp. 213-216.

[7] O. Pironneau: Optimal Shape Design for Elliptic Systems, Springer series in Computational Physics, Springer-Verlag, New York (1984).

[8] G. Fichera: Existence Theorems in Elasticity. Boundary Value Problems with Unilateral Constraints. Springer Verlag, Berlin-Heidelberg-New York (1972).

[9] J. Haslinger: Contact shape optimization based on the reciprocal variational formulation, Appl. Math. 44 (1999), pp. 321-358.

11 RECENT RESULTS IN CONTACT PROBLEMS WITH COULOMB FRICTION

Jiří Jarušek and Christof Eck

Math. Inst., Acad. Sci. of the Czech Republic
and Inst. Appl. Math., University of Erlangen-Nuernberg
jarusek@math.cas.cz, eck@uni-erlangen.de

Abstract: A survey about existence theorems both for static and for dynamic Signorini contact problems with Coulomb–type friction is presented.

11.1 INTRODUCTION

Contact problems with friction represent an important task of applied mathematics. Usually, they are formulated for a contact of an elastic or viscoelastic body with a rigid obstacle (support). The static version of such a problem was formulated in Duvaut and Lions (1972) as an open problem. The first proof of a theorem about the existence of a solution appeared in Nečas, Jarušek and Haslinger for a strip in $\mathbb{R}^2$ and a rigid obstacle. It was based on the regularity of solutions to the auxiliary problem with given friction and on the Tichonov fixed–point theorem. For the proof of the regularity, the shift technique published in Fichera (1972) was essentially employed. The existence proof was extended for a general domain in $\mathbb{R}^2$ or $\mathbb{R}^3$ in Jarušek (1983) and Jarušek (1984), where also results for a contact of two elastic bodies are included. Then for a longer time no new results appeared. In the meantime several concepts modifying and simplifying the problem were published as nonlocal friction and the normal compliance approach. While the first idea had no connection with the further development of the existence results, the penalization of the contact condition included into the second idea appeared as an efficient tool to extend the above cited results and to solve also dynamical problems.

In the presented contribution we survey our recent progress in the field. In Sec. 2 we sum up the results for the static problem. In Sec. 3 we mention the result for the dynamic problem with given friction, where the contact condition is formulated in displacements. In Sec. 4 the dynamic contact problem

R.P. Gilbert et al. (eds.), From Convexity to Nonconvexity, 147–160.

with Coulomb friction is treated, but with the contact condition formulated in velocities. Its possible generalizations including also thermal aspects of the friction are mentioned in Sec. 5.

Throughout the paper we assume $\Omega \subset \mathbb{R}^N$ to be a domain with a Lipschitz boundary Γ divided into three measurable and pairwisely disjoint parts: Γ_U, Γ_T and the contact part Γ_C which is assumed to be of the class $C_{1,1}$ and to have a positive measure. Moreover, the following notation is employed: For a domain $M \subset \mathbb{R}^N$, $k > 0$, and $p \in [1, +\infty)$, $W_p^k(M)$ denotes the isotropic Sobolev–Slobodetskii space, where k is the maximal order of differentiability and p the power of integrability. If $k \in \mathbb{R}_+^2 \equiv [0, +\infty)^2$, then its first coordinate indicates the existence of the appropriate time derivative and the second the existence of the appropriate space derivatives such that all derivatives mentioned belong to $L_p(M)$. Moreover, $H^k(M) \equiv W_2^k(M)$ stands for the spaces of Hilbert type. $H^{k*}(M)$ are their duals with respect to the L_2–pairing, $\mathring{H}^k(\Omega)$ the corresponding spaces with zero traces and $H^{-k}(\Omega)$ their duals, $k \geq 0$. The same notation is used for the analogous Bochner–type spaces. The spaces with range in $\mathbb{R}^N$, however, are denoted by bold letters, e.g. $H^1(\Omega; \mathbb{R}^N) = \boldsymbol{H}^1(\Omega)$. B_0 is the space of bounded functions with the sup–norm. The strong and weak convergence in the appropriate spaces is denoted by "$\to$" and "$\rightharpoonup$", respectively. For a displacement v: $\Omega \to \mathbb{R}^N$, $e_{ij}(v) \equiv \frac{1}{2}(v_{i,j} + v_{j,i})$, $i, j = 1, \ldots, N$, are the components of the linear strain tensor $\widetilde{e}(v)$. Here and in the sequel $v_{i,j} \equiv \partial v_i / \partial x_j$. The time derivatives are denoted by dots. The summation convention is employed consequently. For the stress tensor $\widetilde{\sigma} \equiv (\sigma_{ij};\ i, j = 1, \ldots, N)$ and the outer normal vector n at a point of Γ we denote $T(u) \equiv (\sigma_{ij}(u) n_j;\ i = 1, \ldots, N)$. Moreover, for any vector w defined there, w_n and w_t are its normal and tangential components, respectively. To avoid any confusion with this notation, we use the letter τ for the time variable in the dynamic problems. For a set M, $\operatorname{Int} M$ signifies its interior and $\operatorname{cl} M$ its closure. For a number $o \in \mathbb{R}$, $o^+ = \max(o, 0)$ is its nonnegative part. δ denotes the Kronecker symbol.

11.2 THE STATIC CASE

We start with the classical formulation: for a displacement u we assume the linear strain–stress relation

$$\sigma_{ij}(u) = a_{ijk\ell} e_{k\ell}(u) \tag{11.1}$$

in the general case. The coefficients in (11.1) shall fulfil the conditions of symmetry, $a_{ijk\ell} = a_{jik\ell} = a_{k\ell ij}$, and of boundedness and ellipticity, this means for all symmetric tensors $(\xi_{ij})_{i,j=1}^N, (\varrho_{ij})_{i,j=1}^N \in \mathbb{R}^{N^2}$ and all $x \in \Omega$ there holds

$$a_{ijk\ell}(x) \xi_{ij} \varrho_{k\ell} \leq A_0 \sqrt{\xi_{ij}\xi_{ij}\varrho_{k\ell}\varrho_{k\ell}} \quad \text{and} \quad a_{ijk\ell}(x)\xi_{ij}\xi_{k\ell} \geq a_0 \xi_{ij}\xi_{ij} \tag{11.2}$$

with real constants $0 < a_0 \leq A_0$. In the particular homogeneous isotropic case

$$a_{ijk\ell} = \frac{E}{1+\nu}\left[\frac{1}{2}(\delta_{ik}\delta_{j\ell} + \delta_{i\ell}\delta_{jk}) + \frac{\nu}{1-2\nu}\delta_{ij}\delta_{kl}\right], \tag{11.3}$$

where $E > 0$ is the Young modulus of elasticity and $\nu \in (-1, \frac{1}{2})$ the Poisson ratio. We look for a displacement u such that

$$-\sigma_{ij,j}(u) = f_i \quad \text{in } \Omega,\ i = 1, \ldots, N, \tag{11.4}$$

$$u = U \quad \text{on } \Gamma_U, \tag{11.5}$$

$$T(u) = T_0 \quad \text{on } \Gamma_T, \tag{11.6}$$

$$u_n \le g,\ T_n(u) \le 0,\ T_n(u)\,(u_n - g) = 0 \quad \text{on } \Gamma_C, \tag{11.7}$$

$$\left.\begin{array}{l} u_t = 0 \;\Rightarrow\; |T_t(u)| \le \mathfrak{F}(0)|T_n(u)| \\ u_t \ne 0 \;\Rightarrow\; T_t(u) = \mathfrak{F}(u_t)\, T_n(u) \frac{u_t}{|u_t|} \end{array}\right\} \text{ on } \Gamma_C. \tag{11.8}$$

If we multiply the equilibrium of forces (11.4) by $v - u$, where v is a suitable test function, integrate the divergence of the stress tensor by parts and insert the boundary conditions, we obtain the variational inequality:

Find $u \in \mathcal{K} := \{v \in \boldsymbol{H}^1(\Omega);\ v = U$ *on* Γ_U *and* $v_n \le g$ *on* $\Gamma_C\}$ *such that for all* $v \in \mathcal{K}$

$$\int_\Omega \sigma_{ij}(u) e_{ij}(v-u)\,dx + \int_{\Gamma_C} \mathfrak{F}(u_t)\,|T_n(u)|\,(|v_t| - |u_t|)\,dx_s \ge \mathfrak{L}(v-u) \tag{11.9}$$

with $\mathfrak{L}(v) = \int_\Omega f_i v_i\,dx + \int_{\Gamma_T} (T_0)_i v_i\,dx_s$.

Let us denote by $\mathfrak{R}$ the kernel of the linearized stress operator and let $R \equiv \mathfrak{R} \cap (\mathcal{K} - g)$. The problem is solved under the following assumptions:

Assumption 1. *Under above mentioned assumption about* Ω, Γ *and its parts let, moreover, one of the following conditions about* Γ_U *be satisfied:*

(i) mes $\Gamma_U > 0$,

(ii) *R does not contain a nontrivial subspace and* $\mathfrak{L} > 0$ *on* $R \setminus \{0\}$,

(iii) *the nontrivial maximal subspace R_0 of R contains only constants,* $\mathfrak{L}(R_0) = \{0\}$ and $\mathfrak{L} > 0$ *on* $R \cap R_0^\perp \setminus \{0\}$.

Besides the mentioned ellipticity, boundedness and symmetry conditions, the coefficients $a_{ijk\ell}$ of the strain–stress relation are Lipschitz with respect to x. The given data fulfil $U \in \boldsymbol{H}^1(\Omega)$, $f \in \boldsymbol{H}^{\frac{1}{2}*}(\Omega)$, $T_0 \in \boldsymbol{H}^{\frac{1}{2}*}(\Gamma_T)$ *and* $g \in H^1(\Gamma_C)$, $g \ge 0$ *a.e. in* Γ_C. *The compatibility condition* $U = 0$ *a.e. on* Γ_C *shall be valid. Moreover,* $\mathfrak{F} \in L_\infty(\Gamma_C \times \mathbb{R}^N)$ *such that* $\mathfrak{F}(x, \cdot)$ *is continuous for a.e.* $x \in \Gamma_C$. *For* $\widetilde{supp}\,\mathfrak{F} := \mathrm{cl}\{x \in \Gamma_C; \exists y \in \mathbb{R}^N$ *such that* $\mathfrak{F}(x,y) \ne 0\}$ *it holds* $\mathrm{dist}(\widetilde{supp}\,\mathfrak{F},\ \Gamma \setminus \Gamma_C) > 0$, *and* $\|\mathfrak{F}\|_{L_\infty} < C_{\mathfrak{F}}$ *with*

$$C_{\mathfrak{F}} = \begin{cases} \dfrac{\sqrt{3-4\nu}}{2-2\nu},\ N = 2, \qquad \sqrt{\dfrac{3-4\nu}{4-4\nu}},\ N = 3, \\ \sqrt{\dfrac{3a_0}{4A_0}},\ N = 2, 3, \end{cases} \tag{11.10}$$

for the homogeneous isotropic case and the general material, respectively.

Under above listed assumptions the following theorem was proved in Eck and Jarušek (1997[1]) and (1997[2]):

Theorem 11.1. *There exists a solution to (11.9).*

To prove the theorem, we replace the contact condition (11.7) by the Newton–type condition

$$T_n(u) = -\frac{1}{\varepsilon}[u_n - g]^+,\ \varepsilon > 0 \tag{11.11}$$

and we smooth the Euclidian norm in the friction term by smooth and convex approximations K_η, e.g.

$$K_\eta(x) = \left(\begin{array}{ll} |x|, & x \in \mathbb{R}^N \quad \text{and } |x| \geq \eta, \\ -\frac{|x|^4}{8\eta^3} + \frac{3|x|^2}{4\eta} + \frac{3}{8}\eta, & x \in \mathbb{R}^N \quad \text{and } |x| < \eta. \end{array} \right. \tag{11.12}$$

Due to the properties of convex functions, such a modified variational problem obtains a form of a variational equation, see Eck and Jarušek (1997[1]) and Eck and Jarušek (1997[2]). If, moreover, $\operatorname{mes}\Gamma_U = 0$ (the so called semicoercive case), then we create an artificial global coercivity of the problem by adding a penalty term $\gamma \int\limits_\Omega (\pi_{\mathfrak{R}} u)^2\, dx$. A standard a–priori estimate independent of the parameters ε, η and possible γ is derived and the existence of solutions to this problem for every value of ε, η, γ is proved employing an appropriate theorem about pseudomonotone operators. For this result, the fact that for the described modification and fixed $\varepsilon > 0$ the friction term is of lower order than the energy term seems to be the most important. In the semicoercive case the assumption (ii) or (iii) in Assumption 1 is essential, too. Then the limit procedure with $\eta \to 0$ and $\gamma \to 0$ for the case (ii) of Assumption 1 is easy and we arrive to the following variational problem:
Find $u_\varepsilon \in \boldsymbol{H}^1(\Omega)$ such that for all $v \in \boldsymbol{H}^1(\Omega)$

$$\int\limits_\Omega \sigma_{ij}(u_\varepsilon)e_{ij}(v - u_\varepsilon)\, dx + \int\limits_{\Gamma_C} \frac{1}{\varepsilon}\left[(u_\varepsilon)_n - g\right]^+ (v_n - (u_\varepsilon)_n)\, dx_s \tag{11.13}$$

$$+ \int\limits_{\Gamma_C} \frac{1}{\varepsilon}\left[(u_\varepsilon)_n - g\right]^+ \mathfrak{F}(u_\varepsilon)\left(|v_t| - |(u_\varepsilon)_t|\right) dx_s \;\geq\; \mathfrak{L}(v - u_\varepsilon).$$

It is not difficult to prove that there exists a solution to (11.13) for any $\varepsilon > 0$.

The procedure for the particular case (iii) of Assumption 1 is a little bit more complicated, we refer to Eck and Jarušek (1997[2]) for details.

In the limit procedure for $\varepsilon \to 0$ the relation (11.11) looses its regularizing effect to $T_n(u)$. Then only its dual estimate (by means of the Green theorem) is available. Due to this fact and to the nonmonotone character of the friction term, the whole term can be dominated in some important estimations by the energy norm only quantitatively and a certain smallness of $\|\mathfrak{F}\|_{L_\infty}$ is required therefore. Moreover, to carry out the limit procedure, a certain regularity of the friction term is necessary, i.e

$$\|T_n(u_\varepsilon)\|_{L_2(\widetilde{supp\,\mathfrak{F}})} \leq c \quad \text{and} \quad \|u\|_{\boldsymbol{H}^1(\widetilde{supp\,\mathfrak{F}})} \leq c, \tag{11.14}$$

with the constant c independent of $\varepsilon > 0$.

The proof of this regularity is rather cumbersome and technical. It exploits essentially the localization technique described in Fichera (1972) and the local straightening of the boundary. To obtain (11.14) we use also the technique of shifts in arguments and the appropriate renormation results proved with the help of the Fourier transformation. To obtain the lowest possible estimate of the friction term, we use both the trace theorem and the inverse trace theorem based on the direct calculation of the solution to the homogeneous isotropic Lamé system with a Dirichlet boundary value condition an a strip or a half-space. We performed this cumbersome calculation only for $N = 2, 3$.

After proving (11.14) we obtain for a certain sequence $\varepsilon_k \to 0$ besides the usual convergence of the volume term and the linear term in (11.13) that $T_n(u_{\varepsilon_k}) \rightharpoonup T_n(u)$ in $L_2(\widetilde{supp\,\mathfrak{F}})$, $(u_{\varepsilon_k})_t \to u_t$ in $L_p(\widetilde{supp\,\mathfrak{F}})$, $p \in [1, +\infty)$ (from the Sobolev imbedding theorem), and $\mathfrak{F}(u_{\varepsilon_k}) \to \mathfrak{F}(u)$ in $L_p(\widetilde{supp\,\mathfrak{F}})$, $p \in [1, +\infty)$ (from the Lebesgue dominant convergence theorem). Thus the convergence of the friction term of u_{ε_k} to that of the limit u is proved and u is a solution of (11.9). Theorem 1 is proved.

In the static case we have solved also the contact of two elastic bodies. Let they occupy two domains $\Omega_\iota \subset \mathbb{R}^N$, $\iota = 1, 2$, such that the properties of the boundaries and their parts are analogous to those formulated for Ω in the case of one body and a rigid support. However, the contact part Γ_C is a common part of both boundaries such that $g = 0$ holds. On Γ_C we define the outer normal vector with respect to Ω_1, the relative displacement $u_{\rm rel} = u_1 - u_2$, where u_ι is the displacement on Ω_ι, $\iota = 1, 2$, and $T_n(u) \equiv T_n(u_1) = -T_n(u_2)$. Moreover, we define $\Omega = \Omega_1 \cup \Omega_2$ and analogously Γ_X, $X = U, T$. Then the classical formulation (11.4–11.8) remains valid and an appropriate adaptation of (11.9) is possible. Adapting suitably Assumption 1, too, we obtain the existence of solution of this problem. Here, *in the homogeneous isotropic case*

$$C_{\mathfrak{F}} = \begin{cases} \frac{1}{2\sqrt{S_1 S_2}} \frac{R_1 S_1 + R_2 S_2}{R_1\sqrt{S_1} + R_2\sqrt{S_2}}, & 4R_2^2 \geq \frac{(R_1 S_1 - R_2 S_2)^2}{S_1 S_2}, \\ \sqrt{\frac{R_2}{S_1}} \frac{1}{\sqrt{R_1 + R_2} + \sqrt{R_2}}, & 4R_2^2 \leq \frac{(R_1 S_1 - R_2 S_2)^2}{S_1 S_2}, \end{cases} \tag{11.15}$$

if $R_2 S_2 \geq R_1 S_1$, where

$$\text{for } N = 2 \quad R_\iota = \frac{E_\iota}{1 - \nu_\iota^2} \quad \text{and} \quad S_\iota = \frac{(1 - \nu_\iota)^2}{3 - 4\nu_\iota}, \ \iota = 1, 2,$$

$$\text{for } N = 3 \quad R_\iota = \frac{E_\iota}{1 + \nu_\iota} \quad \text{and} \quad S_\iota = \frac{1 - \nu_\iota}{3 - 4\nu_\iota}, \ \iota = 1, 2.$$

For the case $R_2 S_2 \leq R_1 S_1$ the symmetry (the possibility to reindex the bodies) is exploited.

In the general case, if $a_0^{(\iota)}$ and $A_0^{(\iota)}$ denote the corresponding ellipticity and boundedness constants of the respective bilinear forms, then

$$C_{\mathfrak{F}} = \begin{cases} \sqrt{\dfrac{3a_0^{(1)}a_0^{(2)}}{4A_0^{(1)}A_0^{(2)}}}\dfrac{A_0^{(1)}+A_0^{(2)}}{\sqrt{a_0^{(1)}A_0^{(1)}}+\sqrt{a_0^{(2)}A_0^{(2)}}}, & \dfrac{a_0^{(2)}}{a_0^{(1)}} \geq \dfrac{\left(A_0^{(1)}-A_0^{(2)}\right)^2}{4A_0^{(1)}A_0^{(2)}}, \\ \sqrt{\dfrac{3a_0^{(1)}a_0^{(2)}}{A_0^{(1)}}}\dfrac{1}{\sqrt{a_0^{(1)}+a_0^{(2)}}+\sqrt{a_0^{(2)}}}, & \dfrac{a_0^{(2)}}{a_0^{(1)}} \leq \dfrac{\left(A_0^{(1)}-A_0^{(2)}\right)^2}{4A_0^{(1)}A_0^{(2)}}, \end{cases} \tag{11.16}$$

if $A_0^{(2)} \geq A_0^{(1)}$. For the case $A_0^{(2)} \leq A_0^{(1)}$ we exploit again the symmetry of the problem.

11.3 DYNAMIC PROBLEM WITH CONTACT CONDITION IN DISPLACEMENT AND GIVEN FRICTION

We assume that the problem is defined for a time interval $I_{\mathfrak{T}} \equiv (0, \mathfrak{T})$ and we define $Q_{\mathfrak{T}} \equiv I_{\mathfrak{T}} \times \Omega$, $S_{\mathfrak{T}} \equiv I_{\mathfrak{T}} \times \Gamma$ and $S_{X\mathfrak{T}} \equiv I_{\mathfrak{T}} \times \Gamma_X$, $X = C, T, U$. Let us introduce the formulation of the problem. We look for a displacement u on $Q_{\mathfrak{T}}$ such that

$$\ddot{u}_i - \frac{\partial}{\partial x_j}\sigma_{ij}(u) = f_i \quad \text{on } Q_{\mathfrak{T}},\ i = 1, \ldots, N\ , \tag{11.17}$$

$$u = U \quad \text{on } S_{U\mathfrak{T}}, \tag{11.18}$$

$$T(u) = T_0 \quad \text{on } S_{T\mathfrak{T}}, \tag{11.19}$$

$$u_n \leq 0,\ T_n(u) \leq 0,\ T_n(u)u_n = 0 \quad \text{on } S_{C\mathfrak{T}}, \tag{11.20}$$

$$\left.\begin{array}{l} \dot{u}_t = 0 \ \Rightarrow\ |T_t(u)| \leq G \\ \dot{u}_t \neq 0 \ \Rightarrow\ T_t(u) = -G\dfrac{\dot{u}_t}{|\dot{u}_t|} \end{array}\right\} \text{ on } S_{C\mathfrak{T}}, \tag{11.21}$$

$$u(0, \cdot) = u_0, \qquad \dot{u}(0, \cdot) = u_1 \quad \text{on } \Omega. \tag{11.22}$$

Here, the given friction force $-G \leq 0$ stands for $\mathfrak{F}T_n(u)$ in the problems with Coulomb friction. The stress tensor is composed from an inviscid and a viscous part

$$\sigma_{ij}(u) = \sigma_{ij}^I(u) + \sigma_{ij}^V(u),\ i, j = 1, \ldots, N \tag{11.23}$$

and the strain–stress relation has the form

$$\sigma_{ij}^I(u) = \frac{\partial W}{\partial e_{ij}}(\cdot, \tilde{e}(u)) \quad \text{and} \quad \sigma_{ij}^V(u) = a_{ijk\ell}e_{k\ell}(\dot{u}),\ i, j = 1, \ldots, N. \tag{11.24}$$

The coefficients of the form A: $(v, w) \to a_{ijk\ell}e_{ij}(v)e_{k\ell}(w)$ satisfy the same symmetry conditions as the analogous coefficients in Sec. 2 and the validity of relations (11.2) is also assumed. The space–dependent stored energy function W: $\mathbb{R}^{N+N^2} \to \mathbb{R}$ is assumed to be C_2–smooth on $\overline{\Omega} \times \mathbb{R}^{N^2}$, to satisfy $W(\cdot, 0) = 0$, $(\partial W/\partial \tilde{e})(\cdot, 0) = 0$ and to have the partial Hesse matrix $\partial^2 W/\partial \tilde{e}^2$ uniformly strongly elliptic with the ellipticity constant β_0 and uniformly bounded (with constant β_1) for almost every $x \in \Omega$.

The mathematical difficulty of the problem, which has a somewhat parabolized character by the viscosity, consists in the Signorini boundary condition formulated in displacements. This leads to an opposite sign at the time derivative in the variational formulation of the problem which is for the purely elastic material hard to be overcome (the only satisfactory results obtained in the literature for such settings are those for strings, i.e. $N = 1$). The occurence of a certain physically well–based viscosity in the model solved seems to be almost necessary to enable the proof of the existence of solutions and to find a mathematically well–based numerical method for the approximation of these solutions.

For $w \in \boldsymbol{H}^{\frac{1}{2}}(\partial\Omega)$ denote $\mathcal{C}_w \coloneqq \{v \in \boldsymbol{H}^1(\Omega);\ v = w \text{ on } \Gamma_U,\ v_n \leq 0 \text{ a.e. in } \Gamma_C\}$. We introduce the variational formulation of the problem:

A weak solution to (11.17–11.22) *will be a function* $u \in B_0(I_{\mathfrak{T}};\boldsymbol{H}^1(\Omega))$ *for which* $u(\tau,\cdot) \in \mathcal{C}_{U(\tau,\cdot)}$ *for a.e.* $\tau \in I_{\mathfrak{T}}$, $\dot{u} \in B_0(I_{\mathfrak{T}};\boldsymbol{L}_2(\Omega)) \cap L_2(I_{\mathfrak{T}};\boldsymbol{H}^1(\Omega))$, $\dot{u}(\mathfrak{T},\cdot) \in \boldsymbol{L}_2(\Omega)$ *(therefore* $\ddot{u} \in \boldsymbol{H}^{1*}(Q_{\mathfrak{T}};\mathbb{R}^N)$*) and for all* $v \in \boldsymbol{H}^1(Q_{\mathfrak{T}})$ *such that* $v(\tau,\cdot) \in \mathcal{C}_{U(\tau,\cdot)}$ *a.e. in* $I_{\mathfrak{T}}$ *the following inequality holds:*

$$\int\limits_{Q_{\mathfrak{T}}} (\sigma_{ij}(u)e_{ij}(v-u) - \dot{u}_i(\dot{v}_i - \dot{u}_i))\,dx\,d\tau + \int\limits_{\Omega} (\dot{u}_i(v_i - u_i))\,(\mathfrak{T},\cdot)\,dx + \tag{11.25}$$

$$\int\limits_{S_{c,\mathfrak{T}}} G\,(|v_t + \dot{u}_t - u_t| - |\dot{u}_t|)\,dx_s\,d\tau \geq \int\limits_{\Omega} (u_1)_i(v_i(0,\cdot) - (u_0)_i)\,dx + \mathfrak{L}(v-u)$$

with $\mathfrak{L}\colon v \mapsto \int\limits_{Q_{\mathfrak{T}}} f_i v_i\,dx\,d\tau + \int\limits_{S_{T\mathfrak{T}}} (T_0)_i\, v_i\,dx_s\,d\tau$. The inequality (11.25) clearly follows from (11.17) by multiplying the equilibrium of forces by $v - u$, by integrating the result over $Q_{\mathfrak{T}}$, using the Green theorem both in the time and space variables and the boundary value conditions (11.18–11.21) as well as the initial conditions in (11.22).

The introduced problem will be solved under the following set of assumptions:

Assumption 2. *Let the assumptions concerning* Ω*, its boundary,* $I_{\mathfrak{T}}$*, the form* A *and the stored energy function* W *be satisfied. Let, moreover,* $u_0 \in \mathcal{C}_{U(0,\cdot)}$, $u_1 \in \boldsymbol{H}^1(\Omega)$, $U \in \boldsymbol{H}^2(Q_{\mathfrak{T}})$ *such that* $U(0,\cdot)|_{\Gamma_U} = u_0|_{\Gamma_U}$, $\frac{\partial U}{\partial \tau}(0,\cdot)|_{\Gamma_U} = u_1|_{\Gamma_U}$ *and* $U = 0$ *a.e. in* $S_{C\mathfrak{T}}$, $T_0 \in L_2\Big(I_{\mathfrak{T}};\boldsymbol{H}^{\frac{1}{2}*}(\Gamma_T)\Big)$, $f \in L_2(I_{\mathfrak{T}};\boldsymbol{H}^{1*}(\Omega))$ *and* $0 \leq G \in L_2\Big(I_{\mathfrak{T}};H^{\frac{1}{2}*}(\Gamma_C)\Big)$.

It holds (cf. Jarušek (1996)):

Theorem 11.2. *Let Assumption 2 be valid. Then there exists at least one solution to (11.25) (a weak solution to (11.17–11.22).*

For the proof of Theorem 2 the following property of W, which follows straightforwardly from the assumption, is important:

$$W(\cdot,\tilde{e}(w)) = \int\limits_0^1 (1-\theta)\frac{\partial^2 W}{\partial e_{ij}\partial e_{kl}}(\cdot,\tilde{e}(\theta w))\,e_{ij}(w)\,e_{kl}(w)\,d\theta$$

$$\in \Big(\frac{1}{2}\beta_0 e_{ij}(w)\,e_{ij}(w), \frac{1}{2}\beta_1 e_{ij}(w)\,e_{ij}(w)\Big). \tag{11.26}$$

Moreover, $\partial W/\partial\widetilde{e}$ is strongly monotone with the constant β_0. We exploit the penalty method changing the Signorini condition (11.20) by (11.11) for any $\tau \in \mathfrak{T}$. We smooth the norm of u_t with the help of (11.12) again. We put $\mathcal{H}:= \{w \in \boldsymbol{H}^1(\Omega);\ w = 0 \text{ a.e. in } \Gamma_U\}$. Then we obtain the variational equation:

For $\varepsilon > 0$ and $\eta > 0$ we define that $u_{\varepsilon,\eta}$ is the weak solution of the penalized and smoothed problem, iff $u_{\varepsilon,\eta} \in U + B_0(I_\mathfrak{T};\mathcal{H})$ for which $\dot{u}_{\varepsilon,\eta} \in B_0(I_\mathfrak{T};\boldsymbol{L}_2(\Omega)) \cap L_2(I_\mathfrak{T};\boldsymbol{H}^1(\Omega))$ and $\ddot{u}_{\varepsilon,\eta} \in L_2(I_\mathfrak{T};\mathcal{H}^*)$, the initial condition (11.22) is satisfied and the equation

$$\int_{Q_\mathfrak{T}} ((\ddot{u}_{\varepsilon,\eta})_i v_i + \sigma_{ij}(u_{\varepsilon,\eta})e_{ij}(v))\, dx\, d\tau + \int_{S_{C\mathfrak{T}}} \left(G(\nabla K_\eta)((\dot{u}_{\varepsilon,\eta})_t)v_t + \frac{1}{\varepsilon}(u_{\varepsilon,\eta})_n^+ v_n \right) dx_s\, d\tau = \mathfrak{L}(v) \tag{11.27}$$

holds for every $v \in L_2(I_\mathfrak{T};\mathcal{H})$.

Using (11.26) and the occurance of the acceleration term in (11.27), we arrive by a more or less standard way to the a–priori estimate

$$\sup_{\tau \in I_\mathfrak{T}} \left(\|\dot{u}_{\varepsilon,\eta}(\tau,\cdot)\|^2_{\boldsymbol{L}_2(\Omega)} \right) + \|\nabla \dot{u}_{\varepsilon,\eta}\|^2_{L_2(Q_\mathfrak{T};\mathbb{R}^{N^2})} \leq c, \tag{11.28}$$

where c depends on the input data, but does not depend on $\varepsilon > 0$ and $\eta > 0$. The estimate (11.28) shows that the effect of the semicoercivity can not occur in the dynamic problems. The crucial point for our approach is to exploit the dual estimate

$$\|\ddot{u}_{\varepsilon,\eta}\|^2_{L_2(I_\mathfrak{T};\boldsymbol{H}^{-1}(\Omega))} \leq c\|\nabla \dot{u}_{\varepsilon,\eta}\|^2_{L_2(Q_\mathfrak{T};\mathbb{R}^{N^2})} + \|f\|^2_{L_2(I_\mathfrak{T};\boldsymbol{H}^{-1}(\Omega))} \tag{11.29}$$

with c independent of ε, η and the boundary data. To derive (11.29) we put test functions $w \in \mathring{\boldsymbol{H}}^1(Q_\mathfrak{T})$ into the variational equation and use (11.17). By means of a standard interpolation technique and a certain trace theorem we derive from (11.28) and (11.29) the following very important estimates

$$\|\dot{u}_{\varepsilon,\eta}\|^2_{H^{\frac{1}{2}}(I_\mathfrak{T};\boldsymbol{L}_2(\Omega))} \leq c\|\nabla \dot{u}_{\varepsilon,\eta}\|^2_{L_2(Q_\mathfrak{T};\mathbb{R}^{N^2})} + \kappa\|f\|^2_{L_2(I_\mathfrak{T};\boldsymbol{H}^{-1}(\Omega))}, \tag{11.30}$$

$$\|\dot{u}_{\varepsilon,\eta}\|_{\boldsymbol{H}^{\frac{1}{4},\frac{1}{2}}(S_\mathfrak{T})} \leq c\|\dot{u}_{\varepsilon,\eta}\|_{\boldsymbol{H}^{\frac{1}{2},1}(Q_\mathfrak{T})} \tag{11.31}$$

with c, κ independent of ε, η and the boundary data.

The variational equation (11.27) is solved via the Galerkin method. The estimates (11.28-11.31) can be repeated for the finite–dimensional approximation to (11.27) and then the weak convergences of some sequences of the approximate solutions to an element which belongs to all spaces mentioned in the estimates can be proved. The compact imbedding theorem $H^{\frac{1}{4},\frac{1}{2}}(Q_\mathfrak{T}) \hookrightarrow L_2(Q_\mathfrak{T})$, however, proves the crucial strong convergence of the velocities. Moreover, the monotonicity of the operator $\widetilde{e} \mapsto \widetilde{\sigma}^I$ yields the strong convergence of the space gradients. Thus the limit is a solution of (11.27). Then we successively pass two limit procedures: for $\eta \to 0$ and $\varepsilon \to 0$. They are based on the same ideas as the procedure for the Galerkin approximation. This proves Theorem 2.

By the localization and shift technique some tangential regularity of the solution to (11.25) along $S_{C\mathfrak{T}}$ can be proved. Inside Ω a better regularity of u is obvious. It is not clear, how to prove any better time regularity of u along $S_{C\mathfrak{T}}$ and therefore also a better normal regularity there seems to be an open question. To prove the existence of a solution to the appropriate problem with Coulomb friction, however, we need to prove some better regularity of $\dot{u}$. For the contact condition in displacements, this does not seem to be possible via the shift technique.

11.4 DYNAMIC PROBLEM WITH COULOMB FRICTION AND CONTACT CONDITION IN VELOCITIES

The classical formulation of the problem is similar to the preceding one: we assume that the assumptions about Ω, the parts of its boundary and the time interval as well as the conditions (11.17–11.19) and (11.22) hold. On $S_{C\mathfrak{T}}$, however, we assume the validity of the following relations

$$\dot{u}_n \leq 0,\ T_n(u) \leq 0,\ T_n(u)\,\dot{u}_n = 0 \quad \text{on } S_{C\mathfrak{T}}, \tag{11.32}$$

$$\left.\begin{array}{lll} \dot{u}_t = 0 & \Rightarrow & |T_t(u)| \leq -\mathfrak{F}(u)T_n(u) \\ \dot{u}_t \neq 0 & \Rightarrow & T_t(u) = \mathfrak{F}(u)T_n(u)\dfrac{\dot{u}_t}{|\dot{u}_t|} \end{array}\right\} \text{ on } S_{C\mathfrak{T}}. \tag{11.33}$$

The stress is composed like in (11.23) from an elastic and a viscous part. The strain–stress relation assumed in Jarušek and Eck (1997) is completely linear

$$\sigma^I_{ij}(u) = a^{(0)}_{ijk\ell}e_{k\ell}(u) \quad \text{and} \quad \sigma^V_{ij}(u) = a^{(1)}_{ijk\ell}e_{k\ell}(\dot{u})\ i,j = 1,\ldots,N. \tag{11.34}$$

Some nonlinearity like e.g. in (11.24) seems to be possible. We assume that the coefficients $a^{(\iota)}_{ijk\ell}$, $i,j,k,\ell = 1,\ldots,N$, are space–dependent and Lipschitz. Moreover, they are symmetric and the ellipticity and continuity condition like in (11.2) are satisfied with positive real constants a_ι, A_ι, respectively, for $\iota = 0,1$.

The mathematical difficulty of the problem consists in the Coulomb–friction term. We define the cone $\mathcal{C}_w := \{v \in L_2(I_\mathfrak{T}; \boldsymbol{H}^1(\Omega))\,;\ v = w \text{ on } S_{U\mathfrak{T}},\ v_n \leq 0$ a.e. in $S_{C\mathfrak{T}}\}$. Then the variational formulation is the following:

A weak solution of the problem shall be a function $u \in B_0(I_\mathfrak{T}; \boldsymbol{H}^1(\Omega))$ for which $\dot{u} \in (B_0 \cap H^{\frac{1}{2}})(I_\mathfrak{T}; \boldsymbol{L}_2(\Omega)) \cap \mathcal{C}_{\dot{U}}$ and $\ddot{u} \in L_2(I_\mathfrak{T}; \boldsymbol{H}^{-1}(\Omega)) \cap H^{\frac{1}{2}*}(I_\mathfrak{T}; \boldsymbol{L}_2(\Omega))$ such that the initial conditions (11.22) hold and for each $v \in \mathcal{C}_{\dot{U}} \cap H^{\frac{1}{2}}(I_\mathfrak{T}; \boldsymbol{L}_2(\Omega))$ the following inequality is valid:

$$\int_{Q_\mathfrak{T}} (\ddot{u}_i(v_i - \dot{u}_i) + \sigma_{ij}(u)e_{ij}(v - \dot{u}))\,dx\,d\tau + \int_{S_{C\mathfrak{T}}} \mathfrak{F}\,|\sigma_n(u)|\,(|v_t| - |\dot{u}_t|)\,dx_s\,d\tau \geq \mathfrak{L}(v - u) \tag{11.35}$$

with the same $\mathfrak{L}$ as in (11.25).

The necessity to use the regularization technique like in Sec. 2 leads to the following set of assumptions which are somewhat stronger than those in Assumption 2:

Assumption 3. *Under the usual assumptions about the domain, the parts of its boundary, the time interval and the coefficients from* (11.34) *let* $u_0 \in \boldsymbol{H}^{\frac{3}{2}}(\Omega)$, $u_1 \in \boldsymbol{H}^{\frac{3}{2}}(\Omega)$ *such that* $u_1|_{\Gamma_C} = 0$, , *let* $U \in \boldsymbol{H}^2(Q_{\mathfrak{T}})$ *such that* $U(0,\cdot)|_{\Gamma_U} = u_0|_{\Gamma_U}$, $\dot{U}(0,\cdot)|_{\Gamma_U} = u_1|_{\Gamma_U}$ *and* $U = 0$ *a.e. in* $S_{C\mathfrak{T}}$, *let* $f \in H^{\frac{1}{4}}(I_{\mathfrak{T}}; \boldsymbol{H}^{1*}(\Omega)) \cap L_2\left(I_{\mathfrak{T}}; \boldsymbol{H}^{\frac{1}{2}*}(\Omega)\right)$, $T_0 \in L_2\left(I_{\mathfrak{T}}; \boldsymbol{H}^{\frac{1}{2}*}(\Gamma_T)\right)$, *and let* $\mathfrak{F}$ *be space-dependent and satisfy all the conditions from Assumption 1 with the following* $C_{\mathfrak{F}}$:

$$C_{\mathfrak{F}} = \sqrt{\frac{a_1}{2A_1}} \cdot \begin{cases} \dfrac{\sqrt{z}}{\sqrt{\left(1+\sqrt{8z}\right)(z+1)}}, & z \geq 1, \\ \dfrac{\sqrt[4]{z}}{\sqrt{2\left(1+\sqrt{8z}\right)}}, & z \leq 1 \end{cases} \tag{11.36}$$

for

$$z = \frac{\pi\sqrt{2a_1A_1}}{8\left(\int\limits_{\mathbb{R}} \frac{\sin^2 s}{s^2} ds \int\limits_{\mathbb{R}^{N-2}} (1+|s|^2)^{-\frac{N}{2}} ds\right)^2}.$$

Theorem 11.3. *Under Assumption 3 there exists a solution to (11.35).*

The proof of the theorem is based on a combination of the methods employed in the preceding sections. We penalize the contact condition and smooth the Euclidian norm in the friction term and we arrive at a variational equation similar to (11.27) which is solved via the Galerkin method. For the variational equation, the estimates (11.28–11.31) are derived. We perform the limit procedure for $\eta \to 0$ and prove the existence of a solution of the penalized variational inequality:

Look for $u_\varepsilon \in U + B_0(I_{\mathfrak{T}}; \mathcal{H})$ *such that* $\dot{u}_\varepsilon \in B_0(I_{\mathfrak{T}}; \boldsymbol{L}_2(\Omega)) \cap L_2(I_{\mathfrak{T}}; \boldsymbol{H}^1(\Omega))$ *and* $\ddot{u}_\varepsilon \in L_2(I_{\mathfrak{T}}; \mathcal{H}^*)$, *the initial condition (11.22) is satisfied and the inequality*

$$\int\limits_{Q_{\mathfrak{T}}} \left((\ddot{u}_\varepsilon)_i(v_i - (\dot{u}_\varepsilon)_i) + \sigma_{ij}(u_\varepsilon)e_{ij}(v - \dot{u}_\varepsilon)\right) dx\, d\tau \tag{11.37}$$

$$+ \int\limits_{S_{C\mathfrak{T}}} \frac{1}{\varepsilon}(\dot{u}_\varepsilon)_n^+ \left(v_n - (\dot{u}_\varepsilon)_n + \mathfrak{F}\left(|v_t| - |(\dot{u}_\varepsilon)_t|\right)\right) dx_s\, d\tau \geq \mathfrak{L}(v - \dot{u}_\varepsilon)$$

holds for each $v \in L_2\left(I_{\mathfrak{T}}; \dot{U} + \mathcal{H}\right)$.

Since the nonmonotone friction term is in $L_1(S_{C\mathfrak{T}})$ and therefore of a lower order than the volume energy for a fixed ε, the solution is also unique, if $\mathfrak{F}$ is solution–independent.

Like in Sec. 2, however, we need to prove a higher regularity of the friction term to perform the limit procedure $\varepsilon \to 0$, namely we prove

$$\|T_n(u_\varepsilon)\|_{L_2(I_{\mathfrak{T}} \times \widehat{supp\,\mathfrak{F}})} \leq c \quad \text{and} \quad \|u\|_{\boldsymbol{H}^{\frac{1}{2},1}(I_{\mathfrak{T}} \times \widehat{supp\,\mathfrak{F}})} \leq c \tag{11.38}$$

with c independent of $\varepsilon > 0$. To prove this we employ again the localization and shift technique. The shift technique must be used simultaneously for the time and space shifts here. We need again both the trace and the inverse trace theorems as exact as possible to obtain $C_{\mathfrak{F}}$ as large as possible. The well–known unpotential character of the evolution problem disables us to use the Lamé system as an auxiliary problem in the general case. For the inverse trace theorem we use the equation $\dot{w} = \mu \Delta w$ and we optimize with respect to μ. In the trace theorem we estimate the fractional time and space derivative separately and we optimize the ratio between the time- and the normal component by the appropriate estimate of the fractional time derivative.

However, the occurence of the acceleration for which only the dual estimate of the type (11.29) is available inflicts that even after all these optimization the value of $C_{\mathfrak{F}}$ is so low compared to that in the static case. A suitable application of the Gronwall lemma shows that the inviscid part has no influence on its magnitude. In the homogeneous isotropic case the direct calculation of the solution to the appropriate Lamé system on a halfspace with a Dirichlet boundary value condition can be employed. Even for $N = 2$, however, the direct calculation for the trace and the inverse trace theorems leads to a very complicated although one-dimensional optimization problem (the variable is the ratio between the dual time and the dual tangential space variable after the use of a partial Fourier transformation). For an earlier stage of these estimates cf. Eck, Thesis. Up–to–now we have still no sufficiently good estimates of the optimal values there.

After proving (11.38) the final limit procedure is quite analogous to that in Sec. 2. The only difference is that from the Sobolev imbedding theorem we obtain $p = 2 + 4/(N-1)$ for the convergence of $|(u_{\varepsilon_k})_t|$. Thus Theorem 3 is proved.

11.5 APPENDIX: THERMAL ASPECTS OF FRICTION

We assume that all relations of the classical formulation of the problem in the preceding section remain valid with the stress tensor satisfying a linear thermoviscoelastic law of the Kelvin–Voight type,

$$\sigma_{ij}(u) = a^{(0)}_{ijk\ell} e_{k\ell}(u) + a^{(1)}_{ijk\ell} e_{k\ell}(\dot{u}) - b_{ij}\Theta. \tag{11.39}$$

The coefficients $a^{(\iota)}_{ijk\ell}, b_{ij}$ satisfy the usual symmetries and the Lipschitz continuity condition with respect to the space variable and generate the usual elliptic and bounded bilinear (or quadratic) forms. The temperature field Θ satisfies the heat equation

$$\dot{\Theta} - (c_{ij}\Theta_{,j})_{,i} + b_{ij}\Theta\, \dot{u}_{i,j} = 0 \tag{11.40}$$

with the tensor of thermal conductivity (c_{ij}) being symmetric. Its further properties will be stated below. The boundary value conditions for the temperature are given by

$$\begin{aligned}\Theta &= 0 \quad \text{on } S_{U\mathfrak{T}},\\ c_{ij}\Theta_{,j}n_i &= K(\Upsilon-\Theta) \text{ on } S_{T\mathfrak{T}},\\ c_{ij}\Theta_{,j}n_i &= J(\cdot,\mathfrak{F}|T_n|,|\dot{u}_t|)+K(\Upsilon-\Theta) \text{ on } S_{C\mathfrak{T}}\end{aligned}$$

and the initial condition is

$$\Theta = \Theta_0.$$

In Eck and Jarušek (1999) the existence of a weak solution to such a system is proved under the assumption that J is monotone in the second and third argument, satisfies the Carathéodory conditions and the linear growth condition

$$J(x,y,z) \le \check{c}(1+|y|+|z|). \tag{11.41}$$

Then for the tensor (c_{ij}) the usual ellipticity and boundedness is sufficient. However, in the deformation heat (the last term in (11.40)) we must replace the original temperature Θ by a reference temperature $\tilde{\Theta}$.

In Jarušek and Eck (1997) the existence of a weak solution is proved for $J(x,y,z) = yz$ in (11.41), i.e. for the physically best based condition. The quadratic–type growth here must be compensed by a particular growth of the energy norm. The following model introduced by Nečas was employed here: $c_{ij} \equiv c_{ij}(\nabla\Theta)$ satisfies the growth condition

$$\check{c}_1\left(1+|\nabla\Theta|^2\right)\xi_i\xi_i \le c_{ij}(\nabla\Theta)\xi_i\xi_j \le \check{c}_2\left(1+|\nabla\Theta|^2\right)\xi_i\xi_i,\ \xi\in\mathbb{R}^N, \tag{11.42}$$

the strong monotonicity

$$\begin{aligned}&\langle c_{ij}(\nabla\Theta)\Theta_{,j}-c_{ij}(\nabla\Xi)\Xi_{,j},\Theta_{,i}-\Xi_{,i}\rangle_{Q_{\mathfrak{T}}}\\ &\ge\ \check{c}_3\|\Theta-\Xi\|^4_{L_4(I_{\mathfrak{T}};W^1_4(\Omega))}+\check{c}_4\|\Theta-\Xi\|^2_{L_2(I_{\mathfrak{T}};H^1(\Omega))}\end{aligned} \tag{11.43}$$

for each $\Theta,\Xi\in L_4\big(I_{\mathfrak{T}};W^1_4(\Omega)\big)$, and the continuity relation

$$c_{ij}\left(\nabla\Theta^{(k)}\right)\Theta^{(k)}_{,j}\to c_{ij}(\nabla\Theta)\Theta_{,j} \text{ in } L_{\frac{4}{3}}(Q_{\mathfrak{T}}),\ i=1,\dots,N, \tag{11.44}$$

for $\Theta^{(k)}\to\Theta$ strongly in $L_4\big(I_{\mathfrak{T}};W^1_4(\Omega)\big)$. The constants $\check{c}_1,\dots,\check{c}_4$ in (11.42) and (11.43) are positive and completely independent of all other employed variables. An example for such a matrix–valued function is $c_{ij}(x,\Xi)=\delta_{ij}(b_0(x)+b_1(x)|\Xi|^2)$ with measurable functions b_0 and b_1 such that $b_i\in[q_1,q_2]$, $i=0,1$, on Ω for some positive real constants q_1,q_2. Under such assumptions the deformation heat in (11.40) can preserve its original form.

The scheme of proofs of these result is the following: in the first step we treat the contact problem with the penalized Signorini contact condition. The solution Θ is replaced by an artificial given Θ_1. Using the appropriate result mentioned in Sec. 4 we prove that the operator $\Theta_1\mapsto u$ is well defined. Then we consider the heat equation with given displacement u and solution Θ. The operator $u\mapsto\Theta$ is also well defined. The existence of an appropriate solution to the penalized problem is proved via the Schauder fixed–point theorem after some estimations based on the interpolation technique.

In the second step new a–priori estimates must be derived. Here, Theorem 3 and its proof is employed together with the interpolation technique again. The result of Jarušek and Eck (1997) e.g. exploits the fact that there is a constant c independent of ε such that $\|\Theta_\varepsilon\|_{L_p(S_\mathfrak{T})} \leq c$ for any $p \in (4,6)$. Then the limit procedure for $\varepsilon \to 0$ can be performed.

11.6 CONCLUSION

We hope that the series of results surveyed here represents a remarkable progress in the field. However, several basic problems remain still unsolved. The existence of a solution to the dynamic contact problem with Coulomb friction with the contact condition in displacements is certainly the most important. Moreover, it is necessary to find a suitable and solvable dynamic formulation of the contact condition for two bodies respecting the fact that with the slip the pairs of points for which the relative normal displacement is defined change in time. In all mentioned problems the uniqueness remains still an open problem whose solution is expected to be negative in many cases. The hope to prove some existence theorems for the mentioned dynamic problems with purely elastic materials does not seem to be too high.

Since 1998 when the contribution was submitted some progress in the last topic has been done. The references are updated to the year 2000.

Acknowledgments

The work presented here was partially supported by the Czech Academy of Sciences under grant A 1075707 and by the the German Research Foundation (DFG) via Sonderforschungsbereich 404 and grant We 659/30-1.

References

Duvaut, Georges, and Lions, Jacques L., (1972). *Les inéquations en mécanique et en physique.* Dunod, Paris.

Eck, Christof, (1996). *Existenz und Regularität der Lösungen für Kontaktprobleme mit Reibung.* Thesis, University of Stuttgart.

Eck, Christof, and Jarušek, Jiří, (1997[1]). Existence results for the static contact problem with Coulomb friction. *SFB 404 Bericht* Nr. 97/2, Universität Stuttgart. *Math. Meth. & Models Appl. Sci.*8, 445-468,(1998).

Eck, Christof, and Jarušek, Jiří, (1997[2]). Existence results for the semicoercive static contact problem with Coulomb friction. *SFB 404 Bericht* Nr. 97/45, Universität Stuttgart *Nonlinear Anal. TMA* 42, 961-976, (2000).

Eck, Christof, and Jarušek, Jiří,. The solvability of a coupled thermoviscoelastic contact problem with small Coulomb friction and linearized growth of frictional heat. Math. Meth. App. Sci. 22, 1221-, (1999).

Fichera, Gaetano, (1972). *Existence Theorems in Elasticity. Boundary Value Problems of Elasticity with Unilateral Constraints.* Springer V., Berlin–Heidelberg–New York.

Jarušek, Jiří, (1983). Contact problems with bounded friction. Coercive case. *Czech. Math. J.* 33(108):237–261.

Jarušek, Jiří, (1984). Contact problems with bounded friction. Semicoercive case. *Czech. Math. J.* 34(109): 619–629.

Jarušek, Jiří, (1996). Dynamical contact problems with given friction for viscoelastic bodies. *Czech. Math. J.* 46:475–487.

Jarušek, Jiří, and Eck, Christof, (1996). Dynamic contact problems with friction in linear viscoelasticity. *Comp. Rend. Acad. Sci. Paris* Ser. I 332:497–502.

Jarušek, Jiří, and Eck, Christof, (1997). Dynamic contact problems with small Coulomb friction for viscoelastic bodies. Existence of solutions. *SFB 404 Bericht* Nr. 97/1, Universität Stuttgart. *Math. Meth. & Models Appl. Sci.*, 9, 11-34,(1999).

Jarušek, Jiří, and Eck, Christof, (1997). Existence of solutions to a nonlinear coupled thermo-viscoelastic contact problem with small Coulomb friction. In Applied Nonlinear Analysis (A. Sequeira et al. eds.) Kluwer/Plenum Press, 49-62.

Nečas, Jindřich, Jarušek, Jiří, and Haslinger, Jaroslav (1980). On the solution of the variational inequality to the Signorini problem with small friction. *Boll. Unione Mat. Ital.* 5(17 B): 796–811.

12 POLARIZATION FIELDS IN THE VARIATIONAL ANALYSIS OF THE LINEAR PIEZOELECTRIC PROBLEM

P. Bisegna and F. Maceri

Department of Civil Engineering
University of Rome "Tor Vergata"
00133 Rome - Italy

Abstract: Some new variational formulations, based on the use of polarization fields, are developed for the fundamental problem of the linear theory of piezoelectricity.

The effectiveness of the proposed formulations is shown by solving in closed-form a simple problem of interest in applications.

12.1 INTRODUCTION

Piezotransducers and piezocomposites exhibit widespread applications in technology. Therefore, how to find the effect of inhomogeneities and defects in piezoelectric bodies and how to evaluate the effective moduli of piezoelectric composites is a central problem for the design and manufacturing of piezodevices.

The main difficulty of these analyses is the strong coupling of the elastic and electric fields; thus, new analytical solutions for typical problems, like a single inhomogeneity in an infinite medium, are needed. Wang [15] and Dunn and Taya [4] found the piezoelectric analogous of Eshelby's tensor [5] and estimated the effective moduli of fiber-reinforced piezoelectric composites.

R.P. Gilbert et al. (eds.), From Convexity to Nonconvexity, 161–175.

Liang at al. [11] and Chung and Ting [3] developed an accurate analysis of an elliptic inclusion in an infinite piezoelectric medium, thus generalizing the results of Muskhelishvili [13].

Bisegna and Luciano [1, 2], when estimating the effective moduli of piezocomposites, adopted a Hashin-Shtrikman type approach. The physical idea of the Hashin-Shtrikman variational principles [7, 8], originally introduced in the homogenization problem of multiphase composite materials, is to substitute the given body with a homogeneous reference body subject to polarization fields, used to exactly simulate the actual constitutive behaviour of the original body.

The Hashin-Shtrikman principles are characterized by two nested stationary problems: the first one, the auxiliary problem, is the elastostatic problem of the homogeneous reference body subject to a given polarization field; the second one is an optimization problem which has as unknown the optimal polarization field.

In this paper, comprehensive variational formulations of the linear piezoelectric problem are established and a generalization of the original Hashin-Shtrikman approach is made, providing flexible and powerful methods of analysis of nonhomogeneous piezoelectric bodies under general loading.

A simple application to the problem of a single ellipsoidal piezoelectric inhomogeneity in an infinite isotropic medium shows how the proposed methodology supplies all the needed concentration effects in a very simple way.

12.2 THE LINEAR PIEZOELECTRIC PROBLEM

Let $\mathbb{R}$ be the real field, $\mathcal{V}$ the three-dimensional Euclidean vector space and Sym the Euclidean vector space of symmetric linear applications of $\mathcal{V}$ into itself.

Let a body $\mathcal{B}$ be given, whose reference configuration is the regular region $\mathcal{R}$ of the three-dimensional Euclidean point space. In the framework of the quasi-static linear theory of piezoelectricity, the following fields defined over $\mathcal{R}$ are introduced:

(*i*) the $\mathcal{V}$-valued displacement $\mathbf{s}$;

(*ii*) the $\mathbb{R}$-valued electric potential ϕ;

(*iii*) the Sym-valued strain $\boldsymbol{\varepsilon}$;

(*iv*) the $\mathcal{V}$-valued electric field $\mathbf{e}$;

(*v*) the Sym-valued stress $\boldsymbol{\sigma}$;

(*vi*) the $\mathcal{V}$-valued electric displacement $\mathbf{d}$.

In order to formulate the piezoelectric problem for the body $\mathcal{B}$, the assignment of the following data is needed:

(*i*) volume forces $\mathbf{b}$;

(*ii*) volume charge ρ;

(*iii*) surface forces $\mathbf{p}$ on a part $\partial_f\mathcal{R}$ of the boundary $\partial\mathcal{R}$;

(*iv*) surface charge ω on a part $\partial_\omega\mathcal{R}$ of $\partial\mathcal{R}$;

(*v*) prescribed displacement $\mathbf{s}_0$ on $\partial_s\mathcal{R} := \partial\mathcal{R} \setminus \partial_f\mathcal{R}$;

(*vi*) prescribed electric potential ϕ_0 on $\partial_\phi\mathcal{R} := \partial\mathcal{R} \setminus \partial_\omega\mathcal{R}$.

The piezoelectric problem consists in searching for the fields $\mathbf{s}$, ϕ, $\boldsymbol{\varepsilon}$, $\mathbf{e}$, $\boldsymbol{\sigma}$, $\mathbf{d}$ which satisfy the following equations.

Compatibility equations

$$\begin{cases} \boldsymbol{\varepsilon} = \hat{\nabla}\mathbf{s} & \text{in } \mathcal{R} \\ \mathbf{s} = \mathbf{s}_0 & \text{on } \partial_s\mathcal{R}, \end{cases} \tag{12.1}$$

and

$$\begin{cases} \mathbf{e} = -\nabla\phi & \text{in } \mathcal{R} \\ \phi = \phi_0 & \text{on } \partial_\phi\mathcal{R}, \end{cases} \tag{12.2}$$

where ∇ and $\hat{\nabla}$ denote, respectively, the gradient and the symmetric part of the gradient.

Equilibrium equations

$$\begin{cases} \operatorname{div}\boldsymbol{\sigma} + \mathbf{b} = \mathbf{0} & \text{in } \mathcal{R} \\ \boldsymbol{\sigma}\mathbf{n} = \mathbf{p} & \text{on } \partial_f\mathcal{R}, \end{cases} \tag{12.3}$$

and

$$\begin{cases} \operatorname{div}\mathbf{d} - \rho = 0 & \text{in } \mathcal{R} \\ \mathbf{d}\cdot\mathbf{n} = -\omega & \text{on } \partial_\omega\mathcal{R}, \end{cases} \tag{12.4}$$

where the assumption that $\mathbf{d}$ vanishes outside $\mathcal{R}$ is made, div denotes the divergence operator and $\mathbf{n}$ is the outward normal field to $\partial\mathcal{R}$.

Constitutive equations

The constitutive equations may be given one of the following equivalent formulations (e.g., [9]):

***h*-form.** $\boldsymbol{\varepsilon}$ and $\mathbf{d}$ are assumed as independent variables:

$$\begin{cases} \boldsymbol{\sigma} = \partial\varphi/\partial\boldsymbol{\varepsilon} \\ \mathbf{e} = \partial\varphi/\partial\mathbf{d}, \end{cases} \tag{12.5}$$

where

$$\varphi(\boldsymbol{\varepsilon},\mathbf{d}) = \tfrac{1}{2}\mathcal{C}^{\mathbf{d}}\boldsymbol{\varepsilon}\cdot\boldsymbol{\varepsilon} - h\boldsymbol{\varepsilon}\cdot\mathbf{d} + \tfrac{1}{2}\boldsymbol{\beta}^{\boldsymbol{\varepsilon}}\mathbf{d}\cdot\mathbf{d}. \tag{12.6}$$

Here $\mathcal{C}^{\mathbf{d}}$ is the elasticity tensor at constant electric displacement ("open-circuit"), $\boldsymbol{\beta}^{\boldsymbol{\varepsilon}}$ is the impermeability tensor at fixed strain ("clamped") and h is the open-circuit/clamped piezoelectric tensor.

The quadratic function $\varphi(\boldsymbol{\varepsilon},\mathbf{d})$ has the physical meaning of free-energy density [10]. It is supposed to be positive definite, in order to guarantee a stable behaviour of the material. Hence, $\varphi(\boldsymbol{\varepsilon},\mathbf{d})$ is strictly convex.

***d*-form.** $\boldsymbol{\sigma}$ and $\mathbf{e}$ are assumed as independent variables:

$$\begin{cases} -\boldsymbol{\varepsilon} = \partial\psi/\partial\boldsymbol{\sigma} \\ -\mathbf{d} = \partial\psi/\partial\mathbf{e}, \end{cases} \tag{12.7}$$

where

$$\psi(\boldsymbol{\sigma},\mathbf{e}) = -\tfrac{1}{2}\mathcal{S}^{\mathbf{e}}\boldsymbol{\sigma}\cdot\boldsymbol{\sigma} - d\boldsymbol{\sigma}\cdot\mathbf{e} - \tfrac{1}{2}\boldsymbol{\kappa}^{\boldsymbol{\sigma}}\mathbf{e}\cdot\mathbf{e}. \tag{12.8}$$

Here $\mathcal{S}^{\mathbf{e}}$ is the compliance tensor at fixed electric field ("short-circuit"), $\boldsymbol{\kappa}^{\boldsymbol{\sigma}}$ is the dielectric tensor at fixed stress ("free") and d is the short-circuit/free piezoelectric tensor.

The quadratic function $\psi(\boldsymbol{\sigma},\mathbf{e})$ is the opposite of the Legendre conjugate of $\varphi(\boldsymbol{\varepsilon},\mathbf{d})$ and hence it is strictly concave. The function $-\psi(\boldsymbol{\sigma},\mathbf{e})$ has the meaning of complementary energy density.

***e*-form.** $\boldsymbol{\varepsilon}$ and $\mathbf{e}$ are assumed as independent variables:

$$\begin{cases} \boldsymbol{\sigma} = \partial\xi/\partial\boldsymbol{\varepsilon} \\ -\mathbf{d} = \partial\xi/\partial\mathbf{e}, \end{cases} \tag{12.9}$$

where

$$\xi(\boldsymbol{\varepsilon},\mathbf{e}) = \tfrac{1}{2}\mathcal{C}^{\mathbf{e}}\boldsymbol{\varepsilon}\cdot\boldsymbol{\varepsilon} - e\boldsymbol{\varepsilon}\cdot\mathbf{e} - \tfrac{1}{2}\boldsymbol{\kappa}^{\boldsymbol{\varepsilon}}\mathbf{e}\cdot\mathbf{e}. \tag{12.10}$$

Here $\mathcal{C}^{\mathbf{e}}$ is the "short-circuit" elasticity tensor, $\boldsymbol{\kappa}^{\boldsymbol{\varepsilon}}$ is the "clamped" dielectric tensor, and e is the short-circuit/clamped piezoelectric tensor.

The quadratic function $\xi(\boldsymbol{\varepsilon},\mathbf{e})$ accounts for the mechanical energy density and the complementary electric energy density, since it is the opposite of the partial conjugate of $\varphi(\boldsymbol{\varepsilon},\mathbf{d})$ with respect to $\mathbf{d}$. Hence, $\xi(\boldsymbol{\varepsilon},\mathbf{e})$ is strictly convex with respect to $\boldsymbol{\varepsilon}$ and strictly concave with respect to $\mathbf{e}$.

***g*-form.** $\boldsymbol{\sigma}$ and $\mathbf{d}$ are assumed as independent variables:

$$\begin{cases} -\boldsymbol{\varepsilon} = \partial\upsilon/\partial\boldsymbol{\sigma} \\ \mathbf{e} = \partial\upsilon/\partial\mathbf{d}, \end{cases} \tag{12.11}$$

where

$$\upsilon(\boldsymbol{\sigma},\mathbf{d}) = -\tfrac{1}{2}\mathcal{S}^{\mathbf{d}}\boldsymbol{\sigma}\cdot\boldsymbol{\sigma} - g\boldsymbol{\sigma}\cdot\mathbf{d} + \tfrac{1}{2}\boldsymbol{\beta}^{\boldsymbol{\sigma}}\mathbf{d}\cdot\mathbf{d}. \tag{12.12}$$

Here $\mathcal{S}^{\mathbf{d}}$ is the "open-circuit" compliance tensor, $\boldsymbol{\beta}^{\boldsymbol{\sigma}}$ is the "free" impermeability tensor, and g is the open-circuit/free piezoelectric tensor.

The quadratic function $\upsilon(\boldsymbol{\sigma},\mathbf{d})$ is strictly concave with respect to $\boldsymbol{\sigma}$ and strictly convex with respect to $\mathbf{d}$, since it is the opposite of the partial conjugate of $\varphi(\boldsymbol{\varepsilon},\mathbf{d})$ with respect to $\boldsymbol{\varepsilon}$. It accounts for the electric energy density and the complementary mechanical energy density.

Needless to say that the elasticity and compliance tensors are symmetric second-order tensors over Sym; the dielectric and impermeability tensors are symmetric second-order tensors over $\mathcal{V}$; the piezoelectric tensors are linear applications of Sym into $\mathcal{V}$. Of course, if these tensors do depend upon the point, the body is inhomogeneous.

A straightforward consequence, via the divergence theorem, of the previous formulation is Betti's reciprocal theorem in linear piezoelectricity:

$$L_{12} = L_{21}, \tag{12.13}$$

where L_{12} and L_{21}, defined by:

$$L_{12} = \int_{\mathcal{R}} \mathbf{b}^{(1)} \cdot \mathbf{s}^{(2)}\, dv + \int_{\partial_f \mathcal{R}} \mathbf{p}^{(1)} \cdot \mathbf{s}^{(2)}\, da + \int_{\partial_s \mathcal{R}} \boldsymbol{\sigma}^{(1)} \mathbf{n} \cdot \mathbf{s_0}^{(2)}\, da$$

$$+ \int_{\mathcal{R}} \rho^{(2)} \phi^{(1)}\, dv + \int_{\partial_\omega \mathcal{R}} \omega^{(2)} \phi^{(1)}\, da - \int_{\partial_\phi \mathcal{R}} (\mathbf{d}^{(2)} \cdot \mathbf{n}) {\phi_0}^{(1)}\, da, \tag{12.14}$$

$$L_{21} = \int_{\mathcal{R}} \mathbf{b}^{(2)} \cdot \mathbf{s}^{(1)}\, dv + \int_{\partial_f \mathcal{R}} \mathbf{p}^{(2)} \cdot \mathbf{s}^{(1)}\, da + \int_{\partial_s \mathcal{R}} \boldsymbol{\sigma}^{(2)} \mathbf{n} \cdot \mathbf{s_0}^{(1)}\, da$$

$$+ \int_{\mathcal{R}} \rho^{(1)} \phi^{(2)}\, dv + \int_{\partial_\omega \mathcal{R}} \omega^{(1)} \phi^{(2)}\, da - \int_{\partial_\phi \mathcal{R}} (\mathbf{d}^{(1)} \cdot \mathbf{n}) {\phi_0}^{(2)}\, da, \tag{12.15}$$

are the mutual works of two different electric and mechanical loading conditions, distinguished by the superscripts (1) and (2).

12.3 WEAK FORMULATIONS OF THE LINEAR PIEZOELECTRIC PROBLEM

It is a simple matter to verify that the piezoelectric problem recalled in the previous section can be recast by adopting either one of the following natural variational formulations, each one involving one field of mechanical type and one field of electrical type:

(*i*) to find the infimum with respect to $\boldsymbol{\varepsilon}$ and $\mathbf{d}$, over the constraint manifold (12.1) and (12.4), of the functional

$$\Phi = \int_{\mathcal{R}} \varphi(\boldsymbol{\varepsilon}, \mathbf{d})\, dv - \int_{\mathcal{R}} \mathbf{b} \cdot \mathbf{s}\, dv - \int_{\partial_f \mathcal{R}} \mathbf{p} \cdot \mathbf{s}\, da + \int_{\partial_\phi \mathcal{R}} (\mathbf{d} \cdot \mathbf{n}) \phi_0\, da; \tag{12.16}$$

(*ii*) to find the supremum with respect to $\boldsymbol{\sigma}$ and $\mathbf{e}$, over the constraint manifold (12.3) and (12.2), of the functional

$$\Psi = \int_{\mathcal{R}} \psi(\boldsymbol{\sigma}, \mathbf{e})\, dv + \int_{\mathcal{R}} \rho \phi\, dv + \int_{\partial_\omega \mathcal{R}} \omega \phi\, da + \int_{\partial_s \mathcal{R}} \boldsymbol{\sigma} \mathbf{n} \cdot \mathbf{s}_0\, da; \tag{12.17}$$

(*iii*) to find the infimum with respect to $\boldsymbol{\varepsilon}$ and the supremum with respect to $\mathbf{e}$, over the constraint manifold (12.1) and (12.2), of the functional

$$X = \int_{\mathcal{R}} \xi(\boldsymbol{\varepsilon}, \mathbf{e})\, dv - \int_{\mathcal{R}} \mathbf{b} \cdot \mathbf{s}\, dv + \int_{\mathcal{R}} \rho \phi\, dv - \int_{\partial_f \mathcal{R}} \mathbf{p} \cdot \mathbf{s}\, da + \int_{\partial_\omega \mathcal{R}} \omega \phi\, da; \tag{12.18}$$

(*iv*) to find the supremum with respect to $\boldsymbol{\sigma}$ and the infimum with respect to $\mathbf{d}$, over the constraint manifold (12.3) and (12.4), of the functional

$$\Upsilon = \int_{\mathcal{R}} \upsilon(\boldsymbol{\sigma}, \mathbf{d})\, dv + \int_{\partial_s \mathcal{R}} \boldsymbol{\sigma}\mathbf{n} \cdot \mathbf{s}_0\, da + \int_{\partial_\phi \mathcal{R}} (\mathbf{d} \cdot \mathbf{n})\phi_0\, da. \tag{12.19}$$

In the purely elastic case, the formulations (*i*) and (*iii*) [respectively, (*ii*) and (*iv*)] reduce to the classical total potential energy principle [respectively, complementary energy principle].

Furthermore, by stating in weak form the essential constraint manifolds of the foregoing variational principles, many other variational formulations of mixed type can be obtained. Among them, the generalizations to the piezoelectric problem of the well-known Hu-Washizu and Hellinger-Prange-Reissner principles of linear elasticity are especially useful in applications.

In order to complete the mathematical structure of the formal statements given above, an appropriate functional equipment is needed. As a matter of fact, it is clear that the natural functional framework of all the weak formulations (*i*)-(*iv*) is the space of square integrable functions. In this framework, well-known existence, uniqueness and regularity results (e.g., [6]) can be easily extended.

As far as existence is concerned, some necessary conditions can be derived by the variational formulations (*i*)-(*iv*). In particular, when $\partial_\phi \mathcal{R}$ is empty, a solution of the piezoelectric problem exists only if the total electric charge on the body vanishes:

$$\int_{\mathcal{R}} \rho\, dv + \int_{\partial_\omega \mathcal{R}} \omega\, da = 0. \tag{12.20}$$

Indeed, if equation (12.20) is not satisfied, then the constraint manifolds of the variational formulations (*i*) and (*iv*) are empty, while the functionals (12.17) and (12.18) are not bounded from above. Analogously, when $\partial_s \mathcal{R}$ is empty, a necessary condition for the existence of a solution is the equilibrium of the external forces:

$$\int_{\mathcal{R}} \mathbf{b}\, dv + \int_{\partial \mathcal{R}} \mathbf{p}\, da = \mathbf{0} \quad \text{and} \quad \int_{\mathcal{R}} \mathbf{x} \times \mathbf{b}\, dv + \int_{\partial \mathcal{R}} \mathbf{x} \times \mathbf{p}\, da = \mathbf{0}, \tag{12.21}$$

where $\mathbf{x}$ is the position vector with respect to a fixed origin and $\times$ denotes the vector product. Indeed, if equation (12.21) is not satisfied, then the constraint manifolds of the variational formulations (*ii*) and (*iv*) are empty, while the functionals (12.16) and (12.18) are not bounded from below.

12.4 HASHIN-SHTRIKMAN TYPE VARIATIONAL PRINCIPLES

In this section, alternative variational formulations of the linear piezoelectric problem are obtained by generalizing the Hashin-Shtrikman approach, originally introduced to analyze elastic composites. Therefore, the piezoelectric problem is reformulated by substituting the given body with a reference body comprised by homogeneous material and subject to suitable polarization fields, to be chosen in such a way as to exactly simulate the constitutive equations of the material comprising the original body. This approach proves to be especially effective if the given body is nonhomogeneous (in particular, composite).

The free-energy density of the material comprising the reference body is denoted by $\varphi_0(\boldsymbol{\varepsilon}, \mathbf{d})$ and is supposed to be strictly convex. The opposite of the conjugate of $\varphi_0(\boldsymbol{\varepsilon}, \mathbf{d})$ with respect to both $\boldsymbol{\varepsilon}$ and $\mathbf{d}$, or with respect to $\mathbf{d}$ only, or with respect to $\boldsymbol{\varepsilon}$ only are respectively denoted by ψ_0, ξ_0 or υ_0. Moreover, let Φ_0, Ψ_0, X_0 and Υ_0 denote the functionals obtained by substituting φ_0, ψ_0, ξ_0 and υ_0 respectively to φ, ψ, ξ and υ at the right-hand terms of equations (12.16)-(12.19). Of course, Φ_0, Ψ_0, X_0 and Υ_0 would govern the piezoelectric problem introduced in the previous section, if the body $\mathcal{B}$ would be comprised by the reference material.

Four different variational principles useful to determine the unknown polarization fields can be obtained. These principles correspond to the four different formulations of the constitutive equations of a piezoelectric material presented above. The h-form formulation is fully developed in what follows.

***h*-form.** The h-form constitutive equations are rewritten by introducing the unknown polarization fields $\boldsymbol{\sigma}^*$ and $\mathbf{e}^*$, which respectively represent the difference between the stress and the electric field arising in the given body and in the reference body, under the same strain and electric displacement field:

$$\begin{cases} \boldsymbol{\sigma} = \partial\varphi/\partial\boldsymbol{\varepsilon} = \partial\varphi_0/\partial\boldsymbol{\varepsilon} + \boldsymbol{\sigma}^* \\ \mathbf{e} = \partial\varphi/\partial\mathbf{d} = \partial\varphi_0/\partial\mathbf{d} + \mathbf{e}^*. \end{cases} \tag{12.22}$$

¿From these equations it is a simple matter to compute $\boldsymbol{\varepsilon}$ and $\mathbf{d}$ as functions of $\boldsymbol{\sigma}^*$ and $\mathbf{e}^*$:

$$\begin{cases} \boldsymbol{\varepsilon} = \partial\psi_\varphi/\partial\boldsymbol{\sigma}^* \\ \mathbf{d} = \partial\psi_\varphi/\partial\mathbf{e}^*, \end{cases} \tag{12.23}$$

where ψ_φ is the conjugate potential of $\varphi - \varphi_0$. It is worth to note that ψ_φ may be a convex, concave or saddle functional, depending on the choice of φ_0.

The piezoelectric problem introduced in the previous section is then recast as the piezoelectric problem for the body $\mathcal{B}$ comprised by the reference material and subject to the polarization fields $\boldsymbol{\sigma}^*$ and $\mathbf{e}^*$, which are constrained to satisfy equations (12.23). Hence, the latter problem can be formulated as follows: to find the infimum with respect to $\boldsymbol{\varepsilon}$ and $\mathbf{d}$, over the constraint manifold (12.1) and (12.4) and the stationary conditions with respect to $\boldsymbol{\sigma}^*$ and $\mathbf{e}^*$ of the functional

$$\mathrm{HS}_\Phi(\boldsymbol{\varepsilon}, \mathbf{d}, \boldsymbol{\sigma}^*, \mathbf{e}^*) = \Phi_0(\boldsymbol{\varepsilon}, \mathbf{d}) + \int_{\mathcal{R}} [\boldsymbol{\sigma}^* \cdot \boldsymbol{\varepsilon} + \mathbf{e}^* \cdot \mathbf{d} - \psi_\varphi(\boldsymbol{\sigma}^*, \mathbf{e}^*)]\, dv. \tag{12.24}$$

It is pointed out that the above stated Hashin-Shtrikman type principle is characterized by two nested stationary problems: the first one (i.e., to take the infimum with respect to $\boldsymbol{\varepsilon}$ and $\mathbf{d}$) may be called auxiliary problem and is relevant to the homogeneous reference body subject to given polarization fields $\boldsymbol{\sigma}^*$, $\mathbf{e}^*$; the second one (i.e., to make the functional stationary with respect to $\boldsymbol{\sigma}^*$,

$\mathbf{e}^*$) is an optimization problem which has as unknowns the optimal polarization fields which exactly simulate the actual constitutive behaviour of the original body. Accordingly, the functional $\mathbf{HS}_\Phi$ has a mixed (or Lagrangian) structure: the stationary conditions with respect to $\boldsymbol{\varepsilon}$ and $\mathbf{d}$ accounts for the mechanical equilibrium and electric compatibility, while the stationary conditions with respect to the polarization fields $\boldsymbol{\sigma}^*$, $\mathbf{e}^*$ accounts for the constraint (12.23) (or, equivalently, (12.22)).

Finally, it is emphasized that, if the reference material is chosen such that $\varphi-\varphi_0$ (and hence ψ_φ) is *concave*, the foregoing variational formulation is indeed a minimum principle: this issue may be useful in searching for approximate (e.g., numerical) solutions.

An illustrative example

In this section a simple application of the proposed variational principle is presented. In particular, explicit formulae are derived for the exact solution of the problem of a piezoelectric ellipsoidal inhomogeneity Ω in an isotropic, linearly elastic and dielectric, infinitely extended matrix subject to uniform fields at infinity. The whole space $\mathbb{R}^3$ filled with the matrix material is taken as reference body.

Let μ_0, λ_0 and β_0 be, respectively, the Lamé constants and the dielectric impermeability of the matrix. Let $\boldsymbol{\varepsilon}_\infty$, $\boldsymbol{\sigma}_\infty$, $\mathbf{e}_\infty$ and $\mathbf{d}_\infty$ denote, respectively, the strain, stress, electric and electric-displacement fields at infinity, related by the equations:

$$\boldsymbol{\sigma}_\infty = \mathcal{C}_0\boldsymbol{\varepsilon}_\infty \quad \text{and} \quad \mathbf{e}_\infty = \boldsymbol{\beta}_0\mathbf{d}_\infty, \tag{12.25}$$

where

$$\mathcal{C}_0 = 2\mu_0\mathcal{I} + \lambda_0\mathbf{I}\otimes\mathbf{I} \quad \text{and} \quad \boldsymbol{\beta}_0 = \beta_0\mathbf{I}, \tag{12.26}$$

$\mathcal{I}$ is the identity over Sym, $\mathbf{I}$ is the identity over $\mathcal{V}$ and $\otimes$ is the tensor product.

In order to frame the present problem in square integrable function spaces, it is convenient to assume as unknowns the fields:

$$\hat{\boldsymbol{\varepsilon}} = \boldsymbol{\varepsilon} - \boldsymbol{\varepsilon}_\infty, \quad \hat{\boldsymbol{\sigma}} = \boldsymbol{\sigma} - \boldsymbol{\sigma}_\infty, \quad \hat{\mathbf{e}} = \mathbf{e} - \mathbf{e}_\infty \quad \text{and} \quad \hat{\mathbf{d}} = \mathbf{d} - \mathbf{d}_\infty. \tag{12.27}$$

Therefore, the present problem amounts to finding the infimum with respect to compatible $\hat{\boldsymbol{\varepsilon}}$ and self-equilibrated $\hat{\mathbf{d}}$, and the stationary conditions with respect to $\boldsymbol{\sigma}^*$ and $\mathbf{e}^*$ of the functional

$$\int_{\mathbb{R}^3}[\tfrac{1}{2}\mathcal{C}_0\hat{\boldsymbol{\varepsilon}}\cdot\hat{\boldsymbol{\varepsilon}} + \tfrac{1}{2}\boldsymbol{\beta}_0\hat{\mathbf{d}}\cdot\hat{\mathbf{d}}]\,dv + \int_\Omega[\boldsymbol{\sigma}^*\cdot\hat{\boldsymbol{\varepsilon}} + \mathbf{e}^*\cdot\hat{\mathbf{d}} - \psi_\varphi(\boldsymbol{\sigma}^* - \Delta\boldsymbol{\sigma}_\infty, \mathbf{e}^* - \Delta\mathbf{e}_\infty)]\,dv, \tag{12.28}$$

arising from (12.24) when the constraints (12.22) and (12.23) are rewritten in terms of $\hat{\boldsymbol{\varepsilon}}$ and $\hat{\mathbf{d}}$. Here

$$\Delta\boldsymbol{\sigma}_\infty = \mathcal{C}^{\mathbf{d}}\boldsymbol{\varepsilon}_\infty - h^t\mathbf{d}_\infty - \boldsymbol{\sigma}_\infty \quad \text{and} \quad \Delta\mathbf{e}_\infty = -h\boldsymbol{\varepsilon}_\infty + \boldsymbol{\beta}^{\varepsilon}\mathbf{d}_\infty - \mathbf{e}_\infty \tag{12.29}$$

denote the difference of the stress and electric fields, in the matrix and in the inhomogeneity, under the same strain and electric-displacement fields $\boldsymbol{\varepsilon}_\infty$ and $\mathbf{d}_\infty$ and the superscript t denotes the transpose.

It is worth noting that the infimum problem with respect to compatible $\hat{\varepsilon}$ and the infimum problem with respect to self-equilibrated $\hat{\mathbf{d}}$ are decoupled, due to the present choice of an isotropic (thus, nonpiezoelectric) reference material. Moreover, it is well known [12, 14] that, when $\boldsymbol{\sigma}^*$ [respectively, $\mathbf{e}^*$] is constant inside Ω, then the corresponding optimal $\hat{\varepsilon}$ [respectively, $\hat{\mathbf{d}}$] is constant inside Ω, too. Analogously, a null-average $\boldsymbol{\sigma}^*$ [respectively, $\mathbf{e}^*$] implies a null-average optimal $\hat{\varepsilon}$ [respectively, $\hat{\mathbf{d}}$].

On the other hand, the density functional ψ_φ is quadratic, and the data $\Delta\boldsymbol{\sigma}_\infty$ and $\Delta\mathbf{e}_\infty$ are constant inside Ω.

The foregoing observations trivially imply that the optimal polarization fields $\boldsymbol{\sigma}^*$ and $\mathbf{e}^*$ are indeed constant inside Ω: hence, in the remainder of this section, $\boldsymbol{\sigma}^*$ and $\mathbf{e}^*$ denote unknown constant fields inside Ω.

As a consequence, explicit formulae are available for the solutions of the infimum problems with respect to compatible $\hat{\varepsilon}$ and self-equilibrated $\hat{\mathbf{d}}$ [12, 14]:

$$\inf_{\hat{\varepsilon}\text{ compatible}} \int_{\mathbf{R}^3} \tfrac{1}{2}\mathcal{C}_0\hat{\varepsilon}\cdot\hat{\varepsilon}\,dv + \int_\Omega \boldsymbol{\sigma}^*\cdot\hat{\varepsilon}\,dv = \int_\Omega \tfrac{1}{2}\mathcal{T}\boldsymbol{\sigma}^*\cdot\boldsymbol{\sigma}^* \tag{12.30}$$

$$\inf_{\hat{\mathbf{d}}\text{ self-equilibrated}} \int_{\mathbf{R}^3} \tfrac{1}{2}\boldsymbol{\beta}_0\hat{\mathbf{d}}\cdot\hat{\mathbf{d}}\,dv + \int_\Omega \mathbf{e}^*\cdot\hat{\mathbf{d}}\,dv = \int_\Omega \tfrac{1}{2}\mathbf{V}\mathbf{e}^*\cdot\mathbf{e}^*, \tag{12.31}$$

where the fourth-order tensor $\mathcal{T}$ and the second-order tensor $\mathbf{V}$ are reported in the first appendix.

The variational problem for the functional (12.28) is thus reduced to search for the stationary conditions with respect to constant fields $\boldsymbol{\sigma}^*$ and $\mathbf{e}^*$ of:

$$\int_\Omega [\tfrac{1}{2}\mathcal{T}\boldsymbol{\sigma}^*\cdot\boldsymbol{\sigma}^* + \tfrac{1}{2}\mathbf{V}\mathbf{e}^*\cdot\mathbf{e}^* - \psi_\varphi(\boldsymbol{\sigma}^* - \Delta\boldsymbol{\sigma}_\infty, \mathbf{e}^* - \Delta\mathbf{e}_\infty)]\,dv. \tag{12.32}$$

After simple algebra, the optimal values of $\boldsymbol{\sigma}^*$ and $\mathbf{e}^*$ turn out to be:

$$\begin{pmatrix}\boldsymbol{\sigma}^*\\ \mathbf{e}^*\end{pmatrix} = \left[\mathbb{M} - \begin{pmatrix}\mathcal{T} & 0\\ 0 & \mathbf{V}\end{pmatrix}\right]^{-1} \mathbb{M}\begin{pmatrix}\Delta\boldsymbol{\sigma}_\infty\\ \Delta\mathbf{e}_\infty\end{pmatrix}, \tag{12.33}$$

where

$$\mathbb{M} = \begin{pmatrix}\mathcal{C}^{\mathbf{d}} - \mathcal{C}_0 & -h\\ -h & \boldsymbol{\beta}^{\varepsilon} - \boldsymbol{\beta}_0\end{pmatrix}^{-1}. \tag{12.34}$$

¿From the optimal polarization fields $\boldsymbol{\sigma}^*$ and $\mathbf{e}^*$, the strain, stress, electric and electric-displacement fields inside the inhomogeneity are computed as linear functions of the far fields ε_∞ $\mathbf{d}_\infty$ (the corresponding linear applications are called concentration tensors) [15]:

$$\begin{pmatrix}\varepsilon\\ \mathbf{d}\end{pmatrix} = \mathbb{M}\begin{pmatrix}\boldsymbol{\sigma}^* - \Delta\boldsymbol{\sigma}_\infty\\ \mathbf{e}^* - \Delta\mathbf{e}_\infty\end{pmatrix} + \begin{pmatrix}\varepsilon_\infty\\ \mathbf{d}_\infty\end{pmatrix}, \tag{12.35}$$

$$\begin{pmatrix}\boldsymbol{\sigma}\\ \mathbf{e}\end{pmatrix} = \begin{pmatrix}\mathcal{C}^{\mathbf{d}} & -h\\ -h & \boldsymbol{\beta}^{\varepsilon}\end{pmatrix}\begin{pmatrix}\varepsilon\\ \mathbf{d}\end{pmatrix}. \tag{12.36}$$

The conceptual scheme shown above can be also applied to augment the variational formulations based on the *d*-, *e*- and *g*-form constitutive equations. The corresponding steps and the results of the analysis of the same simple application are collected in the second appendix.

12.5 CONCLUSIONS

In this paper, several variational formulations of the linear piezoelectric problem were presented. Besides the natural formulations, involving one mechanical field and one electrical field, Hashin-Shtrikman type weak formulations, based on the concept of polarization fields, were presented. The latter ones are especially suitable in the analysis of problems involving heterogeneous bodies.

An application of the proposed formulations to the classical problem of a piezoelectric ellipsoidal inhomogeneity in an infinite isotropic medium shown the effectiveness of the present approach to determine the concentration effects.

Acknowledgements

This research was developed within the framework of Lagrange Laboratory, an European research group between CNRS, CNR, University of Rome "Tor Vergata", University of Montpellier II, ENPC and LCPC.

The financial supports of the Italian National Research Council (CNR), of the Italian Ministry of University and Research (MURST) and of the Italian Space Agency (ASI) are gratefully acknowledged.

References

[1] Bisegna, P. and Luciano, R. (1996). Variational bounds for the overall properties of piezoelectric composites. *J. Mech. Phys. Solids,* **44**:583-602.

[2] Bisegna, P. and Luciano, R. (1997). On methods for bounding the overall properties of periodic piezoelectric fibrous composites. *J. Mech. Phys. Solids,* **45**:1329-1356.

[3] Chung, M. Y. and Ting, T. C. T. (1996). Piezoelectric solid with an elliptic inclusion or hole. *Int. J. Solids Structures,* **33**:3343-3361.

[4] Dunn, M. L. and Taya, M. (1993). Micromechanics predictions of the effective piezoelectric moduli of piezoelectric composites. *Int. J. Solids Structures,* **30**:161-175.

[5] Eshelby, J. D. (1957). The determination of the elastic field of an ellipsoidal inclusion, and related problems. *Proc. R. Soc. London,* A**241**:376-396.

[6] Fichera, G. (1972). Existence theorems in elasticity. *Handbuch der Physik,* VIa/2. Flügge, S. ed. Springer-Verlag, Berlin.

[7] Hashin, Z. and Shtrikman, S. (1962). On some variational principles in anisotropic and nonhomogeneous elasticity. *J. Mech. Phys. Solids,* **10**:335-342.

[8] Hashin, Z. and Shtrikman, S. (1962). A variational approach to the theory of the elastic behavior of multiphase materials. *J. Mech. Phys. Solids,* **11**:127-140.

[9] Ikeda, T. (1990). *Fundamentals of Piezoelectricity.* Oxford University Press, Oxford.

[10] Landau, L. D. and Lifshitz E. M. (1981). *Electrodynamics of Continuous Media.* Course of Theoretical Physics, Vol. 8. Pergamon Press, Oxford.

[11] Liang, J., Jiecai H., Wang, B. and Du S. (1995). Electroelastic modelling of anisotropic piezoelectric materials with an elliptic inclusion. *Int. J. Solids Structures,* **32**:2989-3000.

[12] Mura, T. (1982). *Micromechanics of Defects in Solids.* Martinus Nijhoff, London.

[13] Muskhelishvili N. I. (1953). *Some basic problems of the mathematical theory of elasticity.* Noordhoff, Groningen.

[14] Stratton J. A. (1952). *Electromagnetic theory.* McGraw-Hill, New York.

[15] Wang, B. (1992). Three-dimensional analysis of an ellipsoidal inclusion in a piezoelectric material. *Int. J. Solids Structures,* **29**:293-308.

Appendix. Eshelby's tensors of an ellipsoidal inclusion

Eshelby's tensors [5] $\mathcal{T}$, $\mathcal{U}$, $\mathbf{V}$ and $\mathbf{W}$ are defined by the equations:

$$\inf_{\hat{\varepsilon} \text{ compatible}} \int_{\mathbf{R}^3} \tfrac{1}{2}\mathcal{C}_0\hat{\varepsilon}\cdot\hat{\varepsilon}\,dv + \int_\Omega \sigma^*\cdot\hat{\varepsilon}\,dv = \int_\Omega \tfrac{1}{2}\mathcal{T}\sigma^*\cdot\sigma^*, \tag{12.37}$$

$$\inf_{\hat{\sigma} \text{ self-equilibrated}} \int_{\mathbf{R}^3} \tfrac{1}{2}\mathcal{S}_0\hat{\sigma}\cdot\hat{\sigma}\,dv + \int_\Omega \varepsilon^*\cdot\hat{\sigma}\,dv = \int_\Omega \tfrac{1}{2}\mathcal{U}\varepsilon^*\cdot\varepsilon^*, \tag{12.38}$$

$$\inf_{\hat{\mathbf{d}} \text{ self-equilibrated}} \int_{\mathbf{R}^3} \tfrac{1}{2}\boldsymbol{\beta}_0\hat{\mathbf{d}}\cdot\hat{\mathbf{d}}\,dv + \int_\Omega \mathbf{e}^*\cdot\hat{\mathbf{d}}\,dv = \int_\Omega \tfrac{1}{2}\mathbf{V}\mathbf{e}^*\cdot\mathbf{e}^*, \tag{12.39}$$

$$\inf_{\hat{\mathbf{e}} \text{ compatible}} \int_{\mathbf{R}^3} \tfrac{1}{2}\boldsymbol{\kappa}_0\hat{\mathbf{e}}\cdot\hat{\mathbf{e}}\,dv + \int_\Omega \mathbf{d}^*\cdot\hat{\mathbf{e}}\,dv = \int_\Omega \tfrac{1}{2}\mathbf{W}\mathbf{d}^*\cdot\mathbf{d}^*, \tag{12.40}$$

where $\sigma^* \in \text{Sym}$, $\varepsilon^* \in \text{Sym}$, $\mathbf{e}^* \in \mathcal{V}$, $\mathbf{d}^* \in \mathcal{V}$ are constant fields defined inside the ellipsoidal inclusion Ω and $\mathcal{S}_0 = \mathcal{C}_0^{-1}$, $\boldsymbol{\kappa}_0 = \boldsymbol{\beta}_0^{-1}$.

These tensors are related by the identities:

$$\mathcal{S}_0\mathcal{U} + \mathcal{I} = -\mathcal{T}\mathcal{C}_0, \quad \text{and} \quad \boldsymbol{\kappa}_0\mathbf{W} + \mathbf{I} = -\mathbf{V}\boldsymbol{\beta}_0. \tag{12.41}$$

In the particular case of a spherical inhomogeneity, they are known to be:

$$\mathcal{T} = \frac{-1}{30\mu_0(1-\nu_0)}[2(4-5\nu_0)\mathcal{I} - \mathbf{I}\otimes\mathbf{I}], \tag{12.42}$$

$$\mathcal{U} = \frac{-2\mu_0}{15(1-\nu_0)}[(7-5\nu_0)\mathcal{I} + (1+5\nu_0)\mathbf{I}\otimes\mathbf{I}], \tag{12.43}$$

$$\mathbf{V} = -\frac{2}{3\beta_0}\mathbf{I}, \tag{12.44}$$

$$\mathbf{W} = -\frac{1}{3}\beta_0\mathbf{I}, \tag{12.45}$$

where $\nu_0 = \lambda_0/(2\mu_0 + 2\lambda_0)$ is the Poisson's ratio of the reference material.

More complex expressions hold in the case of an ellipsoidal inclusion [12, 14]. For the sake of simplicity, the elastic and dielectric components of the reference material are reported without the subscript 0. A Cartesian reference frame is chosen with the x_1, x_2 and x_3 axes parallel to the principal diameters of the ellipsoid, whose lengths are denoted by $2a_1$, $2a_2$ and $2a_3$. Some of the Cartesian components of $\overline{\mathcal{U}} = 8\pi(1-\nu)\mathcal{U}/(2\mu)$, $\overline{\mathcal{T}} = 16\pi(1-\nu)\mu\mathcal{T}$, $\mathbf{V}$ and $\mathbf{W}$ are reported in what follows. All other non-zero components are obtained by the cyclic permutation of $(1,2,3)$. The components which cannot be obtained in this way are zero.

(*i*) Components of $\overline{\mathcal{U}} = 8\pi(1-\nu)\mathcal{U}/(2\mu)$:

$$\begin{aligned}\overline{\mathcal{U}}_{1111} &= [3{a_1}^2(1-\nu)I_{11} + \nu {a_1}^2(I_{12}+I_{13}) - 8\pi(1-\nu)^2]/(1-2\nu) \\ &\quad +(1-\nu)I_1 - \nu(I_2+I_3) &(12.46)\\ \overline{\mathcal{U}}_{1122} &= [{a_2}^2(1-\nu)I_{12} + \nu {a_2}^2(3I_{22}+I_{23}) - 8\pi\nu(1-\nu)]/(1-2\nu) \\ &\quad -(1-\nu)I_1 + (I_2-I_3) &(12.47)\\ \overline{\mathcal{U}}_{1133} &= [{a_3}^2(1-\nu)I_{13} + \nu {a_3}^2(3I_{33}+I_{23}) - 8\pi\nu(1-\nu)]/(1-2\nu) \\ &\quad -(1-\nu)I_1 + (I_3-I_2) &(12.48)\\ \overline{\mathcal{U}}_{1212} &= ({a_1}^2+{a_2}^2)I_{12} + (I_1+I_2)(1-2\nu) - 8\pi(1-\nu). &(12.49)\end{aligned}$$

(*ii*) Components of $\overline{\mathcal{T}} = 16\pi(1-\nu)\mu\mathcal{T}$:

$$\begin{aligned}\overline{\mathcal{T}}_{1111} &= [-3a_1^2 I_{11} + \nu a_2^2 I_{12} + \nu a_3^2 I_{13} - (1-4\nu^2)I_1]/(1+\nu) &(12.50)\\ \overline{\mathcal{T}}_{1122} &= [3\nu a_1^2 I_{11} + \nu a_3^2 I_{13} - a_2^2 I_{12} + (1-2\nu)I_1]/(1+\nu) &(12.51)\\ \overline{\mathcal{T}}_{1133} &= [3\nu a_1^2 I_{11} + \nu a_2^2 I_{12} - a_3^2 I_{13} + (1-2\nu)I_1]/(1+\nu) &(12.52)\\ \overline{\mathcal{T}}_{1212} &= -(a_1^2+a_2^2)I_{12} - (1-2\nu)(I_1+I_2). &(12.53)\end{aligned}$$

(*iii*) Components of $\mathbf{V}$:

$$V_{11} = (I_1/(4\pi) - 1)/\beta. \tag{12.54}$$

(*iv*) Components of $\mathbf{W}$:

$$W_{11} = -\beta I_1/(4\pi). \tag{12.55}$$

The integrals:

$$\begin{aligned}I_i &= 2\pi a_1 a_2 a_3 \int_0^\infty \frac{ds}{(a_i^2+s)\Delta(s)} &(12.56)\\ I_{ij} &= 2\pi a_1 a_2 a_3 \int_0^\infty \frac{ds}{(a_i^2+s)(a_j^2+s)\Delta(s)}, &(12.57)\end{aligned}$$

with $\Delta(s) = \sqrt{(a_1^2+s)(a_2^2+s)(a_3^2+s)}$ can be expressed by the standard elliptic integrals [12].

Appendix. The remaining three Hashin-Shtrikman type variational principles

***d*-form.** Constitutive equations (here φ_ψ is the conjugate potential of $\psi - \psi_0$):

$$\begin{cases} -\varepsilon = \partial\psi/\partial\sigma = \partial\psi_0/\partial\sigma - \varepsilon^* \\ -\mathbf{d} = \partial\psi/\partial\mathbf{e} = \partial\psi_0/\partial\mathbf{e} - \mathbf{d}^* \end{cases} \quad \text{or} \quad \begin{cases} -\sigma = \partial\varphi_\psi/\partial\varepsilon^* \\ -\mathbf{e} = \partial\varphi_\psi/\partial\mathbf{d}^*. \end{cases} \tag{12.58}$$

HS_Ψ functional:

$$\mathrm{HS}_\Psi(\sigma,\mathbf{e},\varepsilon^*,\mathbf{d}^*) = \Psi_0(\sigma,\mathbf{e}) + \int_{\mathcal{R}} [-\varepsilon^*\cdot\sigma - \mathbf{d}^*\cdot\mathbf{e} - \varphi_\psi(\varepsilon^*,\mathbf{d}^*)]\,dv. \tag{12.59}$$

Solution of the illustrative example:

$$\begin{pmatrix}\boldsymbol{\varepsilon}^*\\ \mathbf{d}^*\end{pmatrix} = \left[\mathbb{N} - \begin{pmatrix}\mathcal{U} & 0\\ 0 & \mathbf{W}\end{pmatrix}\right]^{-1} \mathbb{N}\begin{pmatrix}\Delta\boldsymbol{\varepsilon}_\infty\\ \Delta\mathbf{d}_\infty\end{pmatrix}, \qquad (12.60)$$

$$\begin{pmatrix}\boldsymbol{\sigma}\\ \mathbf{e}\end{pmatrix} = \mathbb{N}\begin{pmatrix}\boldsymbol{\varepsilon}^* - \Delta\boldsymbol{\varepsilon}_\infty\\ \mathbf{d}^* - \Delta\mathbf{d}_\infty\end{pmatrix} + \begin{pmatrix}\boldsymbol{\sigma}_\infty\\ \mathbf{e}_\infty\end{pmatrix}, \qquad (12.61)$$

$$\begin{pmatrix}\boldsymbol{\varepsilon}\\ \mathbf{d}\end{pmatrix} = \begin{pmatrix}\mathcal{S}^{\mathbf{e}} & d\\ d & \boldsymbol{\kappa}^{\boldsymbol{\sigma}}\end{pmatrix}\begin{pmatrix}\boldsymbol{\sigma}\\ \mathbf{e}\end{pmatrix}, \qquad (12.62)$$

where

$$\mathbb{N} = \begin{pmatrix}\mathcal{S}^{\mathbf{e}} - \mathcal{S}_0 & d\\ d & \boldsymbol{\kappa}^{\boldsymbol{\varepsilon}} - \boldsymbol{\kappa}_0\end{pmatrix}^{-1}. \qquad (12.63)$$

e-form. Constitutive equations (here υ_ξ is the conjugate potential of $\xi - \xi_0$):

$$\begin{cases}\boldsymbol{\sigma} = \partial\xi/\partial\boldsymbol{\varepsilon} = \partial\xi_0/\partial\boldsymbol{\varepsilon} + \boldsymbol{\sigma}^*\\ -\mathbf{d} = \partial\xi/\partial\mathbf{e} = \partial\xi_0/\partial\mathbf{e} - \mathbf{d}^*\end{cases} \quad \text{or} \quad \begin{cases}\boldsymbol{\varepsilon} = \partial\upsilon_\xi/\partial\boldsymbol{\sigma}^*\\ -\mathbf{e} = \partial\upsilon_\xi/\partial\mathbf{d}^*.\end{cases} \qquad (12.64)$$

HS_X functional:

$$\mathrm{HS}_X(\boldsymbol{\varepsilon}, \mathbf{e}, \boldsymbol{\sigma}^*, \mathbf{d}^*) = X_0(\boldsymbol{\varepsilon}, \mathbf{e}) + \int_{\mathcal{R}} [\boldsymbol{\sigma}^* \cdot \boldsymbol{\varepsilon} - \mathbf{d}^* \cdot \mathbf{e} - \upsilon_\xi(\boldsymbol{\sigma}^*, \mathbf{d}^*)]\, dv. \qquad (12.65)$$

Solution of the illustrative example:

$$\begin{pmatrix}\boldsymbol{\sigma}^*\\ \mathbf{d}^*\end{pmatrix} = \left[\mathbb{P} - \begin{pmatrix}\mathcal{T} & 0\\ 0 & -\mathbf{W}\end{pmatrix}\right]^{-1} \mathbb{P}\begin{pmatrix}\Delta\boldsymbol{\sigma}_\infty\\ \Delta\mathbf{d}_\infty\end{pmatrix}, \qquad (12.66)$$

$$\begin{pmatrix}\boldsymbol{\varepsilon}\\ -\mathbf{e}\end{pmatrix} = \mathbb{P}\begin{pmatrix}\boldsymbol{\sigma}^* - \Delta\boldsymbol{\sigma}_\infty\\ \mathbf{d}^* - \Delta\mathbf{d}_\infty\end{pmatrix} + \begin{pmatrix}\boldsymbol{\varepsilon}_\infty\\ -\mathbf{e}_\infty\end{pmatrix}, \qquad (12.67)$$

$$\begin{pmatrix}\boldsymbol{\sigma}\\ -\mathbf{d}\end{pmatrix} = \begin{pmatrix}\mathcal{C}^{\mathbf{e}} & -e\\ -e & -\boldsymbol{\kappa}^{\boldsymbol{\varepsilon}}\end{pmatrix}\begin{pmatrix}\boldsymbol{\varepsilon}\\ \mathbf{e}\end{pmatrix}, \qquad (12.68)$$

where

$$\mathbb{P} = \begin{pmatrix}\mathcal{C}^{\mathbf{e}} - \mathcal{C}_0 & -e\\ -e & -\boldsymbol{\kappa}^{\boldsymbol{\varepsilon}} + \boldsymbol{\kappa}_0\end{pmatrix}^{-1}. \qquad (12.69)$$

***g*-form.** Constitutive equations (here ξ_υ is the conjugate potential of $\upsilon - \upsilon_0$):

$$\begin{cases}-\boldsymbol{\varepsilon} = \partial\upsilon/\partial\boldsymbol{\sigma} = \partial\upsilon_0/\partial\boldsymbol{\sigma} - \boldsymbol{\varepsilon}^*\\ \mathbf{e} = \partial\upsilon/\partial\mathbf{d} = \partial\upsilon_0/\partial\mathbf{d} + \mathbf{e}^*\end{cases} \quad \text{or} \quad \begin{cases}-\boldsymbol{\sigma} = \partial\xi_\upsilon/\partial\boldsymbol{\varepsilon}^*\\ \mathbf{d} = \partial\xi_\upsilon/\partial\mathbf{e}^*.\end{cases} \qquad (12.70)$$

HS_Υ functional:

$$\mathrm{HS}_\Upsilon(\boldsymbol{\sigma}, \mathbf{d}, \boldsymbol{\varepsilon}^*, \mathbf{e}^*) = \Upsilon_0(\boldsymbol{\sigma}, \mathbf{d}) + \int_{\mathcal{R}} [-\boldsymbol{\varepsilon}^* \cdot \boldsymbol{\sigma} + \mathbf{e}^* \cdot \mathbf{d} - \xi_\upsilon(\boldsymbol{\varepsilon}^*, \mathbf{e}^*)]\, dv. \qquad (12.71)$$

Solution of the illustrative example:

$$\begin{pmatrix} \varepsilon^* \\ \mathbf{e}^* \end{pmatrix} = \left[\mathbb{B} - \begin{pmatrix} -\mathcal{U} & 0 \\ 0 & \mathbf{V} \end{pmatrix} \right]^{-1} \mathbb{B} \begin{pmatrix} \Delta\varepsilon_\infty \\ \Delta\mathbf{e}_\infty \end{pmatrix}, \tag{12.72}$$

$$\begin{pmatrix} -\sigma \\ \mathbf{d} \end{pmatrix} = \mathbb{B} \begin{pmatrix} \varepsilon^* - \Delta\varepsilon_\infty \\ \mathbf{e}^* - \Delta\mathbf{e}_\infty \end{pmatrix} + \begin{pmatrix} -\sigma_\infty \\ \mathbf{d}_\infty \end{pmatrix}, \tag{12.73}$$

$$\begin{pmatrix} -\varepsilon \\ \mathbf{e} \end{pmatrix} = \begin{pmatrix} -\mathcal{S}^{\mathbf{d}} & -g \\ -g & -\beta^{\sigma} \end{pmatrix} \begin{pmatrix} \sigma \\ \mathbf{d} \end{pmatrix}, \tag{12.74}$$

where

$$\mathbb{B} = \begin{pmatrix} -\mathcal{S}^{\mathbf{d}} + \mathcal{S}_0 & -g \\ -g & \beta^{\sigma} - \beta_0 \end{pmatrix}^{-1}. \tag{12.75}$$

13 SURVEY OF THE METHODS FOR NONSMOOTH OPTIMIZATION

M. M. Mäkelä

Department of Mathematical Information Technology, University of Jyväskylä
P.O. Box 35, FIN-40351 Jyväskylä, Finland

13.1 INTRODUCTION

In nonsmooth (nondifferentiable) optimization the considered functions are not required to have continuous derivatives in classical sense. Nonsmooth optimization problems arise in very many field of applications, for example in economics, mechanics (see [38]), chemical engineering and optimal control (see [12]). The source of nonsmoothness can be divided into four classes: physical, technological, methodological and numerical nonsmoothness.

In physical class the original phenomenon to be considered itself contains different kind of discontinuities and irregularities. Typical examples of physical nonsmoothness are the phase changes of material in continuous casting of steel (see e.g. [33]) and piecewise linear tax models in economics. Although the objective and constraints functions originally are continuously differentiable some elements from outside may bring with nonsmooth components to the problem. The technological nonsmoothness in model is usually caused by some extra technological constraints, which affect a nonsmooth dependence between the variables and functions. That kind of examples are so-called obstacle problems in optimal shape design (see [12]) and discrete feasible sets in product planning. On the other hand some solution algorithms for constrained optimization may also lead to a nonsmooth problem. The examples of methodological nonsmooth-

R.P. Gilbert et al. (eds.), From Convexity to Nonconvexity, 177–191.

ness are the exact penalty function method and the Lagrange decomposition method. Finally, the problems may be analytically smooth but numerically nonsmooth. That is the case with, for instance, noisy input data or so-called stiff problems, which are numerically unstable and behave like nonsmooth problems.

There exist also several approaches to solve nonsmooth optimization problems. The application of smooth gradient-based methods is a simple manner, but may lead to the failure in convergence, in optimality conditions or in gradient approximation as was stated in [23]. On the other hand the direct methods, for example Powel's method and polytope search (see [8]), employing no derivative information are quite unreliable and becomes inefficient when the size of the problem is getting large. The different kind of regularization techniques introduced, for instance, in [12] or in [46] may give satisfactory results in some cases but are not, in general, as efficient as the direct nonsmooth approach as was noticed in [31].

The methods for nonsmooth optimization can be divided into two main classes: subgradient methods and bundle methods. Both of them are based on the assumption that only the objective function value and its arbitrary subgradient at each point are available. The basic idea behind the subgradient methods is to generalize the smooth methods by replacing the gradient by an arbitrary subgradient. Due to this simple structure they are widely used, but suffer from some drawbacks in the form of poor convergence rate and the lack of an implementable stopping criteria. For an excellent overview of the subgradient methods we refer to [43].

In what follows we concentrate on the bundle methods, which are at the moment the most efficient and promising methods for nonsmooth optimization. We give a short review concerning the historival development of the bundle methods. For simplicity, we restrict first ourselves to the convex case, but later on the considered methods are extended also to the nonconvex optimization.

13.2 CONVEX OPTIMIZATION

To point out the basic ideas and differences between the methods we consider in this section an unconstrained optimization problem of the form

$$\min_{x \in \mathbb{R}^n} f(x), \tag{13.1}$$

where the objective function $f : \mathbb{R}^n \to \mathbb{R}$ is convex (not necessarily differentiable), in other words

$$f(\lambda x + (1 - \lambda)y) \leq \lambda f(x) + (1 - \lambda) f(y)$$

for all $x, y \in \mathbb{R}^n$ and $\lambda \in [0, 1]$. We suppose, that at every point $x \in \mathbb{R}^n$ we can compute the function value $f(x)$ and an arbitrary subgradient $\xi(x)$ from the subdifferential

$$\partial f(x) = \{\xi \in \mathbb{R}^n \mid f(y) \geq f(x) + \xi^T(y - x) \quad \text{for all } y \in \mathbb{R}^n\}.$$

For convex function we have the following necessary and sufficient optimality condition (see for example [31]).

Theorem 13.1. *Convex $f : \mathbb{R}^n \to \mathbb{R}$ attains its global minimum at x^*, if and only if*

$$0 \in \partial f(x^*).$$

General bundle algorithm

Firstly we present a general bundle method that produces a sequence $\{x_k\}_{k=1}^{\infty} \subset \mathbb{R}^n$ converging to a global minimum of f, if it exists. We suppose that in addition to the current iteration point x_k we have some trial points $y_j \in \mathbb{R}^n$ (from past iterations) and subgradients $\xi_j \in \partial f(y_j)$ for $j \in J_k$, where the index set J_k is a nonempty subset of $\{1, \dots, k\}$.

The idea behind bundle methods is to approximate f below by a piecewise linear function, in other words, we replace f by so-called *cutting-plane model*

$$\hat{f}_k(x) := \max_{j \in J_k} \{f(y_j) + \xi_j^T (x - y_j)\}, \tag{13.4}$$

which equivalently can be written in the form

$$\hat{f}_k(x) = \max_{j \in J_k} \{f(x_k) + \xi_j^T (x - x_k) - \alpha_j^k\},$$

with the *linearization error*

$$\alpha_j^k := f(x_k) - f(y_j) - \xi_j^T (x_k - y_j) \quad \text{for all} \quad j \in J_k. \tag{13.6}$$

Note, that the convexity implies

$$\hat{f}_k(x) \le f(x) \quad \text{for all } x \in \mathbb{R}^n \quad \text{and} \quad \alpha_j^k \ge 0 \quad \text{for all } j \in J_k. \tag{13.7}$$

The next iterate candidate is then defined by

$$y_{k+1} := x_k + d_k$$

where the search direction d_k is calculated by

$$d_k := \arg \min_{d \in \mathbb{R}^n} \{\hat{f}_k(x_k + d) + \frac{1}{2} d^T M_k d\}. \tag{13.9}$$

The role of the stabilizing term $\frac{1}{2} d^T M_k d$ is to guarantee the existence of the solution d_k and keep the approximation local enough. The $n \times n$ matrix M_k is intend to accumulate information about the curvature of f around x_k.

A *serious step* $x_{k+1} := y_{k+1}$ is taken if y_{k+1} is significantly better than x_k, in other words

$$f(y_{k+1}) \le f(x_k) + m_L v_k,$$

where $m_L \in (0, \frac{1}{2})$ is a line search parameter and

$$v_k = \hat{f}_k(y_{k+1}) - f(x_k)$$

is the predicted descent of f at x_k.

Otherwise, a *null step* $x_{k+1} := x_k$ is taken to improve the cutting plane model $\hat{f}_{k+1}$, since we set $J_{k+1} := J_k \cup \{k+1\}$. The iteration is terminated if $v_k \geq -\varepsilon_s$, where $\varepsilon_s > 0$ is a final accuracy tolerance supplied by the user.

Notice, that the problem (13.9) still is a nonsmooth optimization problem. However, due to piecewise linear nature it can be rewritten as a (smooth) quadratic programming subproblem

$$\begin{cases} \text{minimize} & v + \frac{1}{2} d^T M_k d \\ \text{subject to} & -\alpha_j^k + \xi_j^T d \leq v \quad \text{for all} \quad j \in J_k. \end{cases} \tag{QP}$$

For computational reasons it might be, in some cases, more effective to solve the dual problem of (QP), i.e. we find multipliers λ_j^k for $j \in J_k$ solving the problem

$$\begin{cases} \text{minimize} & \frac{1}{2}\left[\sum_{j\in J_k} \lambda_j \xi_j\right]^T M_k^{-1} \left[\sum_{j\in J_k} \lambda_j \xi_j\right] + \sum_{j\in J_k} \lambda_j \alpha_j^k \\ \text{subject to} & \sum_{j\in J_k} \lambda_j = 1 \\ \text{and} & \lambda_j \geq 0. \end{cases} \tag{DP}$$

Theorem 13.2. *Problems (QP) and (DP) are equivalent, and they have unique solutions (d_k, v_k) and λ_j^k for $j \in J_k$, respectively, such that*

$$d_k = -\sum_{j\in J_k} \lambda_j^k M_k^{-1} \xi_j, \tag{13.12}$$

$$v_k = -d_k^T M_k d_k - \sum_{j\in J_k} \lambda_j^k \alpha_j^k. \tag{13.13}$$

In what follows we shortly present several versions of bundle methods, which all conform the above ideas. To avoid technical details we focus on their main differences in the choice of the cutting plane approximation $\hat{f}_k$, the linearization error α_j^k or the stabilizing matrix M_k.

Cutting plane methods

The history of bundle methods originates from the cutting plane idea, which was developed independently in [6] and [13]. In the original cutting plane method the curvature matrix was chosen to be

$$M_k \equiv 0.$$

For this reason the direction finding subproblems (QP) and (DP) are linear programming problems and thus simpler to solve. It is also obvious, that if the original objective function f is piecewise linear or almost piecewise linear, then the cutting plane method may convergence reliable way and rapidly to the exact global optimum.

However, the lack of stabilizing quadratic term affects that (QP) or (DP) may not have a finite optimum. In addition, the convergence of the cutting plane algorithm can, in some cases, be really slow (see [47]). This is mainly due to fact that $||d_k||$ may be so large that the new trial point y_{k+1} is far from the previous ones y_j, $j = 1, \ldots, k$. Then y_{k+1} lies in the region where the model $\hat{f}_k$ poorly approximates the original objective f. The method also stores all the past subgradients, in other words $J_k = \{1, \ldots, k\}$, which may cause serious computational difficulties in the form of unbounded storage requirement.

Conjugate subgradient methods

The next step towards bundle methods was the method of conjugate subgradients, which was developed in two journal articles [47] and [19] of the same issue. The main idea of the conjugate subgradient method was to treat each past subgradient $\xi_j \in \partial f(y_j)$ for $j \in J_k$ as it were a subgradient at the current point x_k, in other words the linearization error (13.6) was neglected. On the other hand, no curvature of the objective function were taken account, thus the search direction finding problems (QP) and (DP) were applied by

$$M_k \equiv I \qquad \text{and} \qquad \alpha_j^k \equiv 0.$$

With these choices the opposite direction of d_k can be found by projecting the origin onto the set

$$\operatorname{conv}\{\xi_j \mid j \in J_k\}$$

approximating the set $\partial f(x_k)$. The neglection of α_j^k affects that the cutting-plane model $\hat{f}_k$ is an usable approximation to f only if the trial points y_j are close enough to x_k. For this reason the choice of the index set $J_k \subset \{1, \ldots, k\}$ is a crucial point of the algorithm. Several *subgradient selection strategies* to chooce J_k were proposed, for instance, in [47], [19], [34] and [39] to keep the approximation local enough and limit the number of stored subgradients. In [47] and [19] the authors introduce first time the *subgradient aggregation strategy*, which requires storing only finite number of subgradients. The numerical experiments has shown, that the convergence of the conjugate subgradient method is rather slow in practice (see [22]).

ε-steepest descent methods

The first method really called "bundle method" was the ε-steepest descent method introduced in [20]. The method was further developed in [26], where the subgradient aggregation strategy was utilized to limit the number of stored subgradients, and more lately in [44]. Contrary to the conjugate subgradient method some attention to the linearization error (13.6) was paid. Differing from the standard treatment as a term in objective function of (DP) the linearization error was controlled by an extra constraint. Since no curvature of the objective function were still taken account, in other words

$$M_k \equiv I,$$

the dualized search direction finding problem (DP) has the form

$$\begin{cases} \text{minimize} & \frac{1}{2}\|\sum_{j\in J_k}\lambda_j\xi_j\|^2 \\ \text{subject to} & \sum_{j\in J_k}\lambda_j = 1 \\ & \lambda_j \geq 0 \\ \text{and} & \sum_{j\in J_k}\lambda_j\alpha_j^k \leq \varepsilon_k, \end{cases} \tag{DP'}$$

where $\varepsilon_k > 0$ is a controlling parameter. This tolerance controls the radius of the ball in which the cutting-plane model is thought to be a good approximation to the objective function. The main difficulty in the ε-steepest descent method is the design of convergent rules for automatic updating schemas of ε_k. However, due to the numerical tests of [22] this method works better in general than the conjugate subgradient method.

Generalized cutting plane methods

To avoid the difficulties related to above methods in handling the linearization error the idea of the generalized cutting plane method was introduced in [21] and further developed in [14]. The starting point was the classical cutting plane algorithm. The stabilizing term $\frac{1}{2}\|d\|^2$ was added to its objective function to guarantee the existence of the solution and to keep the approximation local enough. Thus the search direction was obtained as a solution of (QP) or (DP) with the first order choice

$$M_k \equiv I.$$

In spite of different backgrounds the ε-steepest descent method and the generalized cutting plane method have the following connection: if λ_j^k for $j \in J_k$ are optimal multipliers of (DP) with $M_k \equiv I$, then they solve also the problem (DP') with the choice

$$\varepsilon_k = \sum_{j\in J_k}\lambda_j^k\alpha_j^k.$$

The storage saving subgradient selection and aggregation strategies were introduced in [14]. Especially the aggregation strategy of [14] has been succesfully applied in fortcoming methods. When compared with the ε-steepest descent method the difficulty of selecting the approximation tolerance ε_k was avoided in the generalized cutting plane method. Also the numerical tests of [29] support the superiority of the generalized cutting plane method. However, the tests also disclose its sensitivity to the scaling of the objective function (i.e. multiplication of f by a positive constant).

Diagonal variable metric bundle methods

The next improvement in the history of the bundle methods was the diagonal variable metric idea. A weighting parameter was added to the quadratic term of the objective functions in (QP) and (DP) in order to accumulate some second order information about the curvature of f around x_k. Thus the variable metric matrix M_k took the diagonal form

$$M_k = u_k I,$$

with the weighting parameter $u_k > 0$. Based on the proximal point algorithm of [41] and the work of [1] the ***proximal bundle*** method was derived in [15], where also an adaptive safeguarded quadratic interpolation technique for updating u_k was introduced. Somewhat similar outcome was concluded in [42], where the ***bundle trust region*** method was developed combining the bundle idea with the classical trust region method of [27] and [32]. Furthermore, the ***diagonal quasi-Newton*** method of [25] based on the Moreau-Yosida regularizations (see [37] and [48]) belongs to this class of bundle methods. The methods deviates mainly in the strategies for updating the weight u_k: the bundle trust region method employs the same safeguarded quadratic interpolation technique than the proximal bundle method, while the diagonal quasi-Newton method relys on the curved search technique. The numerical tests in [15], [31] and [42] demonstrates the obvious progress in the convergence speed of the diagonal variable metric bundle methods when compared with the earlier methods.

Variable metric bundle methods

The development of second order method has been facinating the researchers of nonsmooth optimization during its whole history. Although the tools from nonsmooth analysis (see for example [40]) already exist, their influence in the numerical methodology has not been as fruitful as in first order case. However, several attempts to employ

$$M_k \text{ as a full matrix}$$

with some updating scheme has been proposed by various authors. Already in his pioneering work [21] Lemaréchal derived a version of variable metric bundle method utilizing the classical BFGS secant updating formula from smooth optimization (see [8]). Due to the disappointing numerical results in [22] this idea was buried nearly for two decades. In [11] the space dilation updating scheme of [43] was adopted from the subgradient method context. More lately, based on Moreau-Yosida regularization and BFGS update, variants of the variable metric bundle idea were proposed in [3], [25] and [36]. According to very limited numerical experiments (see for example [11]) it seems that the variable metric bundle methods works fairly well. However, when proportion the results to the extra computational efforts needed with the full matrix algebra they do not offer substantial advancement in numerical solution process.

Translated bundle methods

Another strategy to exploit higher order information about f is to keep the stabilizing matrix constant, in other words

$$M_k \equiv I,$$

but move the supporting hyperplane by varying the linearization error (13.6). First attempts into this direction was made in [35], where so-called α-function was introduced, and in [9], where the linearization error α_j^k was replaced by the notion

$$\beta_j^k := f(y_j) - f(x_k). \tag{13.23}$$

This idea was extended in [11], where the combination

$$\rho_j^k := \max\left\{(1-\theta)\alpha_j^k, \min\{\alpha_j^k, \beta_j^k\}\right\},$$

with $\theta \in (0,1)$ was utilized. In the same paper [11] also the quadratically based translation were proposed in the form

$$\rho_j^k := \xi_k^T(x_k - y_j).$$

None of above methods behaves uniformly better than the others in numerical tests of [11]. Due to these limited results the convergence rate of the translated bundle methods places itself between ε-steepest descent method and diagonal variable metric bundle methods.

Tilted bundle methods

Based on the work [45] the tilted bundle method was developed in [16] in order to accumulate some second-order information and some interior point features to proximal bundle method of [15]. The matrix M_k was handled in the original diagonal form

$$M_k = u_k I,$$

with safeguarded quadratic interpolation technique for updating u_k. The cutting-plane model (13.4) was replaced by so-called *tilted cutting-plane model*

$$\breve{f}_k(x) := \max_{j\in J_k}\{f(x_k) + (1-\theta_j^k)\xi_j^T(x-x_k) - \rho_j^k\}$$

with some tilting parameters $\theta_j^k \in [0, 1-\kappa]$ for $j \in J_k$ and $\kappa \in (0,1]$. The linearization error (13.6) was replaced by the combination

$$\rho_j^k := \max\{\breve{\alpha}_j^k, \kappa\cdot\alpha_j^k,\},$$

where the tilted linearization error is

$$\breve{\alpha}_j^k := f(x_k) - f(y_j) - (1-\theta_j^k)\xi_j^T(x_k - y_j).$$

Note, that if $\kappa = 1$, then $\theta_j^k = 0$ for all $j \in J_k$ and the method restores to the proximal bundle method of [15]. The tilted cutting plane is a real "cutting plane", since it cuts the parts of the epigraph of f while the standard cutting planes cuts nothing. The numerical tests in [16] appear rather promising, but the question how to chooce the tilting parameters κ and θ_j^k is still open.

Level bundle methods

Since the level-sets of the cutting plane model (13.4) behave rather regular way a new bundle method variant based on minimizing the stabilizing quadratic term subject to some level set of $\hat{f}_k$ was proposed in [24]. In other words, the search direction finding problem (13.9) was replaced by

$$d_k := \arg \min_{d \in \mathbb{R}^n} \{\frac{1}{2} d^T M_k d \mid \hat{f}_k(x_k + d) \le f_k^{\text{lev}}\},$$

where the target level $f_k^{\text{lev}} < f(x_k)$ is chosen to ensure $f_k^{\text{lev}} \uparrow \inf_{\mathbb{R}^n} f$ as $k \to \infty$. The stabilizing matrix was chosen to be constant

$$M_k \equiv I.$$

Bounded storage versions of the level bundle idea were introduced in [4], [17] and more lately in [5], where the global convergence were proved without any compactness assumptions of [24]. In numerical tests the level bundle methods worked in reliable way (see [24] and [5]) but they lost, for example, to diagonal variable metric bundle methods.

Bundle-Newton method

The recent advance in the development of the second-order bundle method was made in [28] where the bundle-Newton method was derived. Instead of piecewise linear cutting-pane model (13.4) they introduced a quadratic model of the form

$$\tilde{f}_k(x) := \max_{j \in J_k} \{f(y_j) + \xi_j^T (x - y_j) + \frac{1}{2} \varrho_j (x - y_j)^T M_j (x - y_j)\},$$

where $\varrho_j \in [0, 1]$ is some damping parameter. The search direction finding problem (13.9) was then replaced by the problem

$$d_k := \arg \min_{d \in \mathbb{R}^n} \{\tilde{f}_k(x_k + d)\}. \tag{13.31}$$

When compared with the earlier variable metric bundle methods we can state that bundle-Newton method is a "real" second-order method, since every piece in the model contains second-order information in the form of the stabilizing matrix M_j. For the approximation

$$M_j \approx \nabla^2 f(y_j)$$

the authors proposed optionally analytic or finite difference approximations. The numerical experiments in [28] seem to be very promizing, especially in piecewise quadratic test cases (see [30]).

13.3 NONCONVEX OPTIMIZATION

In this chapter we consider the unconstrained optimization problem (13.1) of the form where the objective function $f : \mathbb{R}^n \to \mathbb{R}$ is supposed to be locally Lipschitz continuous. This means that for any bounded set $S \subset \mathbb{R}^n$ there exists a (Lipschitz) constant $K > 0$ such that

$$|f(x) - f(y)| \le K\|x - y\| \quad \text{for all } x, y \in S.$$

Note, that f needs not to be convex nor differentiable. The subdifferential of f is now defined (see [7]) as

$$\partial f(x) = \text{conv}\{\lim_{i\to\infty} \nabla f(x^i) \mid x^i \to x \text{ and } \nabla f(x^i) \text{ exists}\}.$$

In order to guarantee the convergence of the methods f is supposed to be upper semidifferentiable due to [2], in other words, for any $x \in \mathbb{R}^n$, $d \in \mathbb{R}^n$ and sequences $\xi_i \subset \mathbb{R}^n$ and $t_i \subset (0, \infty)$ satisfying $\xi_i \in \partial f(x + t_i d)$ and $t_i \downarrow 0$, one has

$$\limsup_{i\to\infty} \xi_i^T d \geq \liminf_{i\to\infty} [f(x + t_i d) - f(x)]/t_i.$$

For locally Lipschitz continuous function we have the following necessary optimality condition.

Theorem 13.3. *If locally Lipschitz continuous $f : \mathbb{R}^n \to \mathbb{R}$ attains its local minimum at x^*, then*

$$0 \in \partial f(x^*). \tag{13.36}$$

Since the optimality condition is not sufficient without convexity assumption, the methods can not guarantee even local optimality of the solutions. Only some candidates, called *substationary* points, satisfying the condition (13.36) are to be looked for.

In convex case the cutting-plane model (13.4) was an under estimate for the objective function and the nonnegative linearization error (13.6) measured how good an approximation the model was to the original problem (see (13.7)). In nonconvex case these facts are not valid anymore: α_j^k may have a tiny (or even negative!) value, although the trial point y_j lies far away from the current iteration point x_k and thus the corresponding subgradient ξ_j is useless.

For the reasons mention above, the main modifications in the methods needed in nonconvex case concern the linearization error α_j^k. Since the problem is now more complicated, some includes, for instance so-called resetting strategies, and some changes, for example in line search procedure, have to be done in order to guarantee the convergence properties (see [14]). For simplicity we concentrate on the changes made in the form of the linearization error in what follows.

Subgradient deletion rules

As mentioned above in the conjugate subgradient methods the linearization error (13.6) was neglected and some deletion rules was needed to reduce the past subgradient information in order to localize the approximations. For this reason they were used also for nonconvex problems in [34] and [39], where, for example, the following type deletion rules were proposed

$$J_k := \{1 \leq j \leq k \mid \|x_k - y_j\| \leq \delta_k\},$$

where $\delta_k > 0$ tends to zero. Without any special deletion rules the translated bundle method introduced in [9] was proposed to be suitable also for nonconvex case in [10], since the substitute linearization error (13.23) is always nonnegative, in other words

$$\beta_j^k = f(y_j) - f(x_k) \geq 0.$$

More complicated deletion rules was derived in [14] for generalized cutting plane methods. The linearization error (13.6) was replaced by its absolute value, in other words

$$\beta_j^k := |\alpha_j^k| = |f(x_k) - f(y_j) - \xi_j^T(x_k - y_j)| \quad \text{for all} \quad j \in J_k \tag{13.39}$$

and the algorithm was resetted (deleting the old subgradient information) whenever

$$\|d_k\| \leq m_S \cdot \max_{j \in J_k} \{s_j^k\}, \tag{13.40}$$

where $m_S > 0$ is a reset tolerance supplied by the user and

$$s_j^k := \|x_j - y_j\| + \sum_{i=j}^{k-1} \|x_{i+1} - x_i\| \tag{13.41}$$

is the distance measure estimating $\|x_k - y_j\|$ without the need to store the trial points y_j.

More lately, in [18] Kiwiel introduces his *restricted step* proximal bundle method, where the search direction finding problem (13.9) was replaced by

$$d_k := \arg\min_{d \in \mathbb{R}^n} \{\hat{f}_k(x_k + d) + \frac{1}{2} d^T M_k d \mid \|d\| \leq \delta_k\}, \tag{13.42}$$

where $\delta_k > 0$ tends to zero and

$$M_k = u_k I,$$

with the weighting parameter $u_k > 0$. The resetting test (13.40) was replaced by

$$u_k \|d_k\| \leq m_S \cdot \max_{j \in J_k} \{s_j^k\} - \sum_{j \in J_k} \lambda_j^k \beta_j^k,$$

where λ_j^k for $j \in J_k$ are the optimal multipliers of (DP) and β_j^k defined by (13.39).

Subgradient locality measures

In order to add some localizing information to the model, the linearization error (13.6) was replaced by so-called subgradient locality measure

$$\beta_j^k =: \max\{\alpha_j^k, \gamma\|x_k - y_j\|^2\} \tag{13.45}$$

in [26] for the ε-steepest descent method. The distance measure parameter $\gamma \geq 0$ can be set to zero when f is convex. The authors also proposed to use the distance measure (13.41) avoiding to store the trial points y_j, in other words to replace (13.45) by

$$\beta_j^k =: \max\{\alpha_j^k, \gamma(s_j^k)^2\}.$$

In [35] the subgradient locality measure in the form (13.45) and in [14] in the form

$$\beta_j^k =: \max\{|\alpha_j^k|, \gamma(s_j^k)^2\} \tag{13.47}$$

were introduced for generalized cutting plane method. Furthermore, in [42] the subgradient locality measure in the form (13.45) and in [31] in the form (13.47) were proposed for a diagonal variable metric bundle method (bundle trust region and proximal bundle method, respectively). In [28] the generalization of (13.47) was introduced in the form

$$\beta_j^k =: \max\{|\alpha_j^k|, \gamma(s_j^k)^\omega\}$$

with $\omega \geq 1$ for the bundle-Newton method.

The latest modification of the subgradient locality measure was proposed in [18] of the form

$$\beta_j^k =: \max\{|\alpha_j^k|, \gamma_1(s_j^k)^2, \gamma_2\|\xi_j\|s_j^k\}$$

with $\gamma_1, \gamma_2 \geq 0$. The numerical tests in [18] comparing this Levenberg-Marquardt type method with the restricted step method (13.42) shows the slight superiority of the latter subgradient deletion rules based method. Especially in worst cases the restricted step method did not need as much objective function and subgradient calculations as the Levenberg-Marquardt type method.

Acknowledgments

The author acknowledges the support of the grant SA-8583 of the Academy of Finland.

References

[1] A. Auslender. Numerical Methods for Nondifferentiable Convex Optimization. *Mathematical Programming Study*, 30:102–126, 1987.

[2] A. Bihain. Optimization of Upper Semidifferentiable Functions. *Journal of Optimization Theory and Applications*, 4:545–568, 1984.

[3] J.F. Bonnans, J.C. Gilbert, C. Lemaréchal and C. Sagastizábal. A Family of Variable Metric Proximal Methods. *Mathematical Programming*, 68:15–47, 1995.

[4] U. Brännlund. A Descent Method with Relaxation Type Step. In: J. Henry and J.P. Yvon, *Lecture Notes in Control and Information Sciences*, pp. 177–186. Springer-Verlag, New York, 1994.

[5] U. Brännlund, K. C. Kiwiel and P.O. Lindberg. Preliminary Computational Experience with a Descent Level Method for Convex Nondifferentiable Optimization. In: J. Doležal and J. Fidler, *System Modelling and Optimization*, pp. 387–394. Chapman & Hall, London, 1996.

[6] E.W. Cheney and A.A. Golstein. Newton's Method for Convex Programming and Tchebycheff Approximation. *Numerische Mathematik*, 1:253–268, 1959.

[7] F.H. Clarke. *Optimization and Nonsmooth Analysis.* J. Wiley, New York, 1983.

[8] R. Fletcher. *Practical Methods of Optimization.* J. Wiley, Chichester, 1987.

[9] M. Gaudioso and M.F. Monaco. A Bundle Type Approach to the Unconstrained Minimization of Convex Nonsmooth Functions. *Mathematical Programming*, 23:216–226, 1982.

[10] M. Gaudioso and M.F. Monaco. *Some Techniques for Finding the Search Direction in Nonsmooth Minimization Problems.* Dipartimento di Sistemi, Universita' della Calabria, Report 75, 1988.

[11] M. Gaudioso and M.F. Monaco. Variants to the Cutting Plane Approach for Convex Nondifferentiable Optimization. *Optimization*, 25:65–75, 1992.

[12] J. Haslinger and P. Neittaanmäki. *Finite Element Approximation of Optimal Shape Design: Theory and Applications.* J. Wiley, Chichester, 1988.

[13] J.E. Kelley. The Cutting Plane Method for Solving Convex Programs. *SIAM J.*, 8:703–712, 1960.

[14] K.C. Kiwiel. *Methods of Descent for Nondifferentiable Optimization.* Springer-Verlag, Berlin, 1985.

[15] K.C. Kiwiel. Proximity Control in Bundle Methods for Convex Nondifferentiable Optimization. *Mathematical Programming*, 46:105–122, 1990.

[16] K.C. Kiwiel. A Tilted Cutting Plane Proximal Bundle Method for Convex Nondifferentiable Optimization. *Operations Research Letters*, 10:75–81, 1991.

[17] K.C. Kiwiel. Proximal Level Bundle Methods for Convex Nondifferentiable Optimization, Saddle-point Problems and Variational Inequlities. *Mathematical Programming*, 69:89–109, 1995.

[18] K.C. Kiwiel. Restricted Step and Levenberg-Marquardt Techniques in Proximal Bundle Methods for Nonconvex Nondifferentiable Optimization. *SIAM Journal on Optimization*, 6:227–249, 1996.

[19] C. Lemaréchal. An Extension of Davidon Methods to Non Differentiable Problems. *Mathematical Programming Study*, 3:95–109, 1975.

[20] C. Lemaréchal. Combining Kelley's and Conjucate Gradient Methods. In: *Abstracts of IX International Symposium on Mathematical Programming*, Budapest, Hungary, 1976.

[21] C. Lemaréchal. *Nonsmooth Optimization and Descent Methods.* IIASA-report, Laxemburg, Austria, 1978.

[22] C. Lemaréchal. Numerical Experiments in Nonsmooth Optimization. In: E.A. Nurminski, *Progress in Nondifferentiable Optimization*, pp. 61–84. IIASA-report, Laxemburg, Austria, 1982.

[23] C. Lemaréchal. Nondifferentiable Optimization. In: G.L. Nemhauser, A.H.G. Rinnooy Kan, M.J. Todd, *Optimization*, pp. 529–572. North-Holland, Amsterdam, 1989.

[24] C. Lemaréchal, A. Nemirovskii and Yu. Nesterov. New Variants of Bundle Methods. *Mathematical Programming*, 69:111–147, 1995.

[25] C. Lemaréchal and C. Sagastizábal. An Approach to Variable Metric Bundle Methods. In: J. Henry and J.P. Yvon, *Lecture Notes in Control and Information Sciences*, pp. 144–162. Springer-Verlag, New York, 1994.

[26] C. Lemaréchal, J.-J. Strodiot and A. Bihain. On a Bundle Algorithm for Nonsmooth Optimization. In: O.L. Mangasarian, R.R. Mayer and S.M. Robinson, *Nonlinear Programming*, pp. 245–281. Academic Press, New York, 1981.

[27] K. Levenberg. A Method for the Solution of Certain Nonlinear Problems in Least Squares. *Quart. Appl. Math.*, 2:164–166, 1944.

[28] L. Lukšan and J. Vlček. A Bundle-Newton Method for Nonsmooth Unconstrained Minimization. *Mathematical Programming*, 83: 373-391, 1998.

[29] M.M. Mäkelä. *Methods and Algorithms for Nonsmooth Optimization*. Reports on Applied Mathematics and Computing, No. 2, University of Jyväskylä, Department of Mathematics, 1989.

[30] M.M. Mäkelä, M. Miettinen, L. Lukšan and J. Vlček. Comparing Nonsmooth Nonconvex Bundle Methods in Solving Hemivariational Inequalities. *J. Global Optimization*, 14: 117-135, 1999.

[31] M.M. Mäkelä and P. Neittaanmäki. *Nonsmooth Optimization: Analysis and Algorithms with Applications to Optimal Control.* World Scientific Publishing Co., Singapore, 1992.

[32] D.W. Marquardt. An Algorithm for Least-Squares Estimation of Nonlinear Parameters. *SIAM Journal on Applied Mathematics*, 11:431–441, 1963.

[33] K. Miettinen, M.M. Mäkelä and T. Männikkö. Optimal Control of Continuous Casting by Nondifferentiable Multiobjective Optimization. *Computational Optimization and Applications*, 11: 177-194, 1998.

[34] R. Mifflin. An Algorithm for Constrained Optimization with Semismooth Functions. *Mathematics of Operations Research*, 2:191–207, 1977.

[35] R. Mifflin. A Modification and an Extension of Lemaréchal's Algorithm for Nonsmooth Minimization. *Mathematical Programming Study*, 17:77–90, 1982.

[36] R. Mifflin. A Quasi-second-order Proximal Bundle Algorithm. *Mathematical Programming*, 73:51–72, 1996.

[37] J.J. Moreau. Proximité et dualité dans un espace hilbertien. *Bulletin de la Société Mathématique de France*, 93:273–299, 1965.

[38] J.J. Moreau, P.D. Panagiotopoulos and G. Strang (Eds.). *Topics in Nonsmooth Mechanics.* Birkhäuser Verlag, Basel, 1988.

[39] E. Polak, D.Q. Mayne and Y. Wardi. On the Extension of Constrained Optimization Algorithms from Differentiable to Nondifferentiable Problems. *SIAM Journal on Optimal Control and Optimization*, 21:179–203, 1983.

[40] R.A. Poliquin and R.T. Rockafellar. Generalized Hessian Properties of Regularized Nonsmooth Functions. *SIAM Journal on Optimization*, 6:1121–1137, 1996.

[41] R.T. Rockafellar. Monotone Operators and the Proximal Point Algorithm. *SIAM Journal on Optimal Control and Optimization*, 14:877–898, 1976.

[42] H. Schramm and J. Zowe. A Version of the Bundle Idea for Minimizing a Nonsmooth Functions: Conceptual Idea, Convergence Analysis, Numerical Results. *SIAM Journal on Optimization*, 2:121–152, 1992.

[43] N.Z. Shor. *Minimization Methods for Non-differentiable Functions.* Springer-Verlag, Berlin, 1985.

[44] J.-J. Strodiot, V.H. Nguyen and N. Heukemes. ε-Optimal Solutions in Nondifferentiable Convex Programming and Some Related Questions. *Mathematical Programming*, 25:307–328, 1983.

[45] V.N. Tarasov and N.K. Popova. A Modification of the Cutting-plane Method with Accelerated Convergence. In: V.F. Demyanov and D. Pallaschke, *Nondifferentiable Optimization: Motivations and Applications*, pp. 284–290. Springer-Verlag, Berlin, 1984.

[46] K.L. Teo and C.J. Goh. On Constrained Optimization Problems with Nonsmooth Cost Functionals. *Applied Mathematics and Optimization*, 18:181–190, 1988.

[47] P. Wolfe. A Method of Conjugate Subgradients for Minimizing Nondifferentiable Functions. *Mathematical Programming Study*, 3:145–173, 1975.

[48] K. Yosida. *Functional Analysis.* Springer-Verlag, 1964.

14 HEMIVARIATIONAL INEQUALITIES AND HYSTERESIS

M. Miettinen

Department of Mathematics, University of Jyväskylä
P.O. Box 35, FIN-40351 Jyväskylä, Finland

14.1 INTRODUCTION

Hemivariational inequalities introduced by P.D. Panagiotopoulos are generalizations of variational inequalities. This type of inequality problems arises, e.g. in variational formulation of mechanical problems whenever nonmonotone and multivalued relations or nonconvex energy functions are involved. Typical examples of such kind of phenomena are nonmonotone friction laws and adhesive contact laws. Mathematically these nonmonotone relations are described by means of generalized gradients (in sense of F.H. Clarke) of nonconvex potential functions. For applications and for their mathematical treatment we refer to [9],[10],[13]–[18].

Hysteresis is defined as a rate independent memory effect. It occurs typically in ferromagnetism, phase transitions, thermostats and mechanical damage. The basic feature of hysteresis is that it involves relations, e.g. loops depending on the history of the system, which cannot be expressed by means of functions, not even multivalued ones. Therefore, in order to model this complex phenomenon it has been introduced numerous different hysteresis operators, e.g. Preisach, Prandtl, Duhem operators. Furthermore, the mathematical theory of hystere-

R.P. Gilbert et al. (eds.), From Convexity to Nonconvexity, 193–206.

sis has been developed rapidly during the last three decade. We refer to the following four monographs [1],[7],[8],[19], which give an extensive view of this branch of applied mathematics.

In this paper we first give a short insight to the basic features of superpotential relations being the core of the theory of hemivariational inequalities and to hysteretic relations. This is the task of the two following sections. The main aim is to enlighten the complementary nature of these two concepts originating from [11]: hysteresis operators are inevitable tools for describing phenomena involving memory effects and, similarly, superpotentials are essential tools for describing complex relations in which the response is immediate. At the end of this paper we give an overview of the results established in [11],[12] for parabolic B.V.P.s including the both phenomena.

14.2 RELATIONS EXPRESSED VIA SUPERPOTENTIALS

We start with the definitions of the generalized directional derivative and the generalized gradient in the sense of F.H. Clarke [2]:

Definition 14.1. *Suppose that $j : \mathbf{R} \to \mathbf{R}$ is a locally Lipschitz function.*

(i) The generalized directional derivative of j at ξ in the direction η, denoted $j^\circ(\xi;\eta)$, is defined as follows:

$$j^\circ(\xi;\eta) = \limsup_{\xi'\to\xi,\tau\to 0+} \frac{j(\xi'+\tau\eta)-j(\xi')}{\tau}.$$

(ii) The generalized gradient of j at ξ, denoted $\partial j(\xi)$, is the subset of $\mathbf{R}$ given by

$$\partial j(\xi) = \{\tau \in \mathbf{R} : j^\circ(\xi;\eta) \geq \tau\eta \quad \forall \eta \in \mathbf{R}\}.$$

We list some basic properties of the generalized gradient and the generalized directional derivative which will be applied later.

Proposition 14.1. *Let $j : \mathbf{R} \to \mathbf{R}$ be a locally Lipschitz function. Then it holds:*

(i) For each $\xi \in \mathbf{R}$ the generalized gradient $\partial j(\xi)$ is a nonempty, closed and bounded interval of $\mathbf{R}$.

(ii) The function $(\xi,\eta) \mapsto j^\circ(\xi;\eta)$ is upper semicontinuous, i.e.

$$\limsup_{\substack{\bar\xi\to\xi\\ \bar\eta\to\eta}} j^\circ(\bar\xi;\bar\eta) \leq j^\circ(\xi;\eta).$$

(iii) The multivalued function ∂j is closed, i.e. if $\{\xi_k\}$ and $\{\eta_k\}$ are sequences such that $\{\xi_k\}$ converges to ξ and $\eta_k \in \partial j(\xi_k)$, then there is a cluster point η of $\{\eta_k\}$ satisfying $\eta \in \partial j(\xi)$.

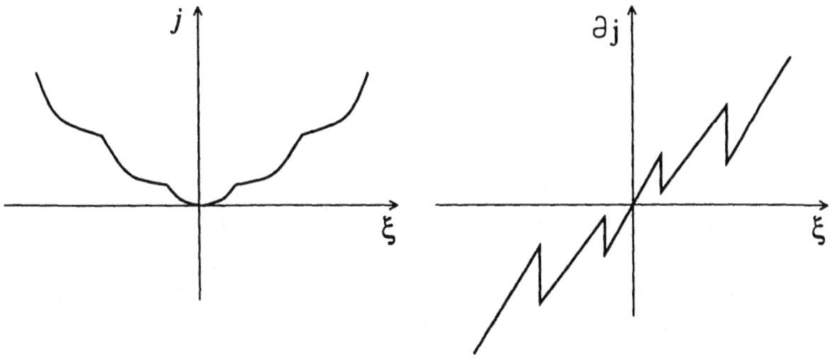

Figure 14.1 Example of a locally Lipschitz function j and its generalized gradient ∂j.

By *superpotential laws* we mean the relations of the form

$$w \in \partial j(u) \tag{14.1}$$

where $j : \mathbf{R} \to \mathbf{R}$, called a *superpotential*, is a locally Lipschitz function (see an example in Fig.14.1). Note that the relation (14.1) can be rewritten equivalently in a form

$$j^{\circ}(u; v) \geq wv \qquad \forall v \in \mathbf{R}. \tag{14.2}$$

If j is convex the generalized gradient ∂j reduces to the subdifferential of the convex analysis and the generalized directional derivative to the classical directional derivative. Moreover, then (14.2) is equivalent to

$$j(v) - j(u) \geq wv \qquad \forall v \in \mathbf{R}. \tag{14.3}$$

The above superpotential relations (14.1)–(14.3) constitute the core of the theory of variational (convex superpotential) and hemivariational inequalities (nonconvex superpotential). They enable us to describe and treat mathematically complicated multivalued, monotone or nonmonotone relations between the studied quantities, e.g. in mechanics the relations between forces and displacements.

14.3 HYSTERESIS OPERATORS

Let us recall from [1],[7],[19] some basic concepts of *hysteresis operators*. For extensive treatment we refer to [1],[7],[19].

Let $[0, T]$ be a given time interval. We denote by $C_{pm}([0, T])$ the set of all continuous and *piecewise monotone* functions on $[0, T]$. A function $v : [0, T] \to \mathbf{R}$ is a piecewise monotone if there exists a *monotonicity partition* $\Delta = \{t_i\}_{i=0}^{n}$, $0 = t_0 < t_1 < ... < t_n = T$ such that $v|_{[t_{i-1}, t_i]}$ is monotone for all $i = 1, ..., n$. The monotonicity partition Δ of v is called the *standard monotonicity partition* of v if the number of the subintervals n is minimal.

We denote by S the set of all strings of real numbers and by S_A the set of all *alternating strings* of real numbers, i.e.

$$S_A = \{(s_0, ..., s_n) : (s_{i+1} - s_i)(s_i - s_{i-1}) < 0, \ 1 \leq i \leq n-1 \text{ and } n \in \mathbf{N}\}.$$

We define the *restriction operator* $\rho_A : C_{pm}([0,T]) \to S_A$ as follows:

$$\rho_A(v) = (v(t_0), ..., v(t_n)), \tag{14.4}$$

where $\{t_i\}_{i=0}^n$ is the standard monotonicity partition of v. Further, the so called *prolongation operator* $\pi_A : S_A \to C_{pm}([0,T])$ maps the string $s = (s_0, ..., s_n)$ to the linear interpolate function $v : [0,T] \to \mathbf{R}$ of the points $(\frac{iT}{n}, v(\frac{iT}{n}) \equiv s_i)$, $i = 0, ..., n$.

In order to be able to formulate the definition of the hysteresis operator we introduce the *rate independent functionals.* A functional $\mathcal{H} : C_{pm}([0,T]) \to \mathbf{R}$ is said to be rate independent if it satisfies

$$\mathcal{H}[v \circ \phi] = \mathcal{H}[v] \tag{14.5}$$

for all $v \in C_{pm}([0,T])$ and for all continuous increasing functions $\phi : [0,T] \to \mathbf{R}$ satisfying $\phi(0) = 0$ and $\phi(T) = T$. This implies that only the local extremal values of v are important for $\mathcal{H}[v]$. Thus, it is straightforward to see that the following bijective correspondence ([1, Proposition 2.2.5])

$$\mathcal{H} = \tilde{\mathcal{H}} \circ \rho_A, \qquad \text{with } \tilde{\mathcal{H}} = \mathcal{H} \circ \pi_A \tag{14.6}$$

holds between the functions $\tilde{\mathcal{H}} : S_A \to \mathbf{R}$ and the rate independent functionals $\mathcal{H} : C_{pm}([0,T]) \to \mathbf{R}$.

For simplicity (as above) we constantly assume that input functions of hysteresis operators are continuous. But, on the other hand, output functions can also be discontinuous; we divide our consideration to two cases according to that if output functions are continuous or not. From the mathematical point of view an appropriate function space for discontinuous output functions is the space of functions of bounded variations, denoted by $BV(0,T)$ (see the definition e.g. in [4]).

Definition 14.2. *(continuous hysteresis operator) An operator $\mathcal{W} : C_{pm}([0,T]) \to C([0,T])$ is said to be a continuous hysteresis operator on $C_{pm}([0,T])$ if there exists a rate independent functional $\mathcal{H}$ called a generating functional of $\mathcal{W}$ such that*

$$\mathcal{W}[v](t) = \mathcal{H}[v_t], \qquad \textit{for all } t \in [0,T] \textit{ and } v \in C_{pm}([0,T]), \tag{14.7}$$

in which

$$v_t(\xi) = \begin{cases} v(\xi), & 0 \le \xi \le t, \\ v(t), & t < \xi \le T. \end{cases}$$

Further, an operator $\tilde{\mathcal{W}} : S_A \to S$ is called a hysteresis operator on S_A if

$$\tilde{\mathcal{W}}(s) = (\tilde{\mathcal{H}}(s_0), \tilde{\mathcal{H}}(s_0, s_1), ..., \tilde{\mathcal{H}}(s)), \qquad \textit{for all } s = (s_0, ..., s_n) \in S_A, \tag{14.8}$$

where $\tilde{\mathcal{H}} = \mathcal{H} \circ \pi_A$ called a generating functional of $\tilde{\mathcal{W}}$ and $\mathcal{H}$ is a rate independent functional on $C_{pm}([0,T])$.

The discontinuous hysteresis operator is defined in a similar way except the following obvious modifications have to be made:

$$\begin{aligned}
&\mathcal{W} : C_{pm}([0,T]) \to P(BV(0,T)),\\
&\tilde{\mathcal{W}} : S_A \to P(S),\\
&\mathcal{H} : C_{pm}([0,T]) \to P(\mathbf{R}),\\
&\tilde{\mathcal{H}} : S_A \to P(\mathbf{R}),
\end{aligned}$$

in which P is a power set, i.e. the mappings above are multivalued.

In the both cases these unique generating functionals $\mathcal{H}$ and $\tilde{\mathcal{H}}$ are often called the *final value mappings* and are denoted by $\mathcal{W}_f$ and $\tilde{\mathcal{W}}_f$, respectively. Due to (14.6) we also have a bijective correspondence between the hysteresis operators $\mathcal{W}$ defined on $C_{pm}([0,T])$ and $\tilde{\mathcal{W}}$ on S_A. Therefore, in the sequel we can use the same notation $\mathcal{W}$ to both $\mathcal{W}$ and $\tilde{\mathcal{W}}$ and, consequently, the notation $\mathcal{W}_f$ to both $\mathcal{W}_f$ and $\tilde{\mathcal{W}}_f$ without any danger of confusion. In the end, we remark that the hysteresis operators defined on $C_{pm}([0,T])$ can be extended to the set of all continuous functions $C([0,T])$ by using the density of $C_{pm}([0,T])$ in $C([0,T])$ (see [1],[7],[19]).

Next we illustrate the above concepts by simple examples, a *completed delayed relay operator* and its slight modification. Let (r_1, r_2) be a fixed couple, called threshold values, from $\mathbf{R}^2$ with $r_1 < r_2$ and $w_{-1} \in [-1,1]$ a given initial state for the hysteresis operator. The completed delayed relay operator (see Fig.14.2 a)

$$\mathcal{R}_{r_1,r_2,w_{-1}} : C([0,T]) \to P(BV(0,T)) \tag{14.9}$$

is defined as follows: for any $u \in C([0,T])$ a function $w \in \mathcal{R}_{r_1,r_2,w_{-1}}[u]$ if and only if

$$w(0) \in \begin{cases} \{-1\}, & u(0) < r_1\\ [-1, w_{-1}], & u(0) = r_1\\ \{w_{-1}\}, & r_1 < u(0) < r_2\\ [w_{-1}, 1], & u(0) = r_2\\ \{1\}, & u(0) > r_2, \end{cases} \tag{14.10}$$

$$w(t) \in \begin{cases} \{-1\}, & u(t) < r_1\\ [-1, 1], & r_1 \le u(t) \le r_2\\ \{1\} & u(t) > r_2, \end{cases} \tag{14.11}$$

and

$$\begin{cases} w \text{ is constant in a neighborhood of } t, & \text{if } u(t) \neq r_1, r_2\\ w \text{ is nonincreasing in a neighborhood of } t, & \text{if } u(t) = r_1\\ w \text{ is nondecreasing in a neighborhood of } t, & \text{if } u(t) = r_2. \end{cases} \tag{14.12}$$

It appears that $\mathcal{R}_{r_1,r_2,w_{-1}}$ really constitutes a discontinuous hysteresis operator from $C([0,T])$ to $P(BV(0,T))$. Indeed, u as a function from $C([0,T])$ can have only a finite number of oscillations between r_1 and r_2. Hence, for any $w \in \mathcal{R}_{r_1,r_2,w_{-1}}[u]$ its total variation in $[0,T]$ is finite implying $w \in BV(0,T)$. The rate independence of $\mathcal{R}_{r_1,r_2,w_{-1}}$ is obvious.

We define a continuous hysteresis operator $\bar{\mathcal{R}}_{r_1,r_2,\rho,w_{-1}}$ from $\mathcal{R}_{r_1,r_2,w_{-1}}$ by slightly perturbing it: Let $0 < \rho < \frac{r_2-r_1}{2}$ be given. Then we define

$$\bar{\mathcal{R}}_{r_1,r_2,\rho,w_{-1}} : C([0,T]) \to C([0,T]) \tag{14.13}$$

as follows: for any $u \in C([0,T])$ a function $w \in \bar{\mathcal{R}}_{r_1,r_2,\rho,w_{-1}}[u]$ if and only if

$$w(0) = \begin{cases} \{-1\}, & u(0) \le r_1 - \rho \\ \min\{w_{-1}, \frac{1}{\rho}(u(0) - (r_1+\rho)) + 1\}, & r_1 - \rho < u(0) < r_1 + \rho \\ w_{-1}, & r_1 + \rho \le u(0) \le r_2 - \rho \\ \max\{w_{-1}, \frac{1}{\rho}(u(0) - (r_2-\rho)) - 1\}, & r_2 - \rho < u(0) < r_2 + \rho \\ \{1\}, & u(0) \ge r_2 + \rho, \end{cases}$$

$$w(t) \in \begin{cases} \{-1\}, & u(t) \le r_1 - \rho \\ [-1, \frac{1}{\rho}(u(t) - (r_1+\rho)) + 1], & r_1 - \rho < u(t) < r_1 + \rho \\ [-1,1], & r_1 + \rho \le u(t) \le r_2 - \rho \\ [\frac{1}{\rho}(u(t) - (r_2-\rho)) - 1, 1], & r_2 - \rho < u(t) < r_2 + \rho \\ \{1\} & u(t) \ge r_2 + \rho, \end{cases}$$

and

$$\begin{cases} w \text{ is nonincreasing in a neighborhood of } t, & \text{if } (u(t),w(t)) \in J_1 \\ w \text{ is nondecreasing in a neighborhood of } t, & \text{if } (u(t),w(t)) \in J_2 \\ w \text{ is constant in a neighborhood of } t, & \text{otherwise} \end{cases}$$

where J_1, J_2 are segments of line defined by

$$J_1 = \{(x, \frac{1}{\rho}(x - (r_1+\rho)) + 1) \in \mathbf{R}^2 \mid x \in [r_1-\rho, r_1+\rho]\},$$
$$J_2 = \{(x, \frac{1}{\rho}(x - (r_2-\rho)) - 1) \in \mathbf{R}^2 \mid x \in [r_2-\rho, r_2+\rho]\},$$

i.e. J_1 and J_2 are the left and right edges of the parallelogram of Fig.14.2 b.

Now it is straightforward to see that $\bar{\mathcal{R}}_{r_1,r_2,\rho,w_{-1}}$ for a fixed ρ generates a continuous hysteresis operator. Furthermore, clearly $\bar{\mathcal{R}}_{r_1,r_2,\rho,w_{-1}}$ tends $\mathcal{R}_{r_1,r_2,w_{-1}}$ as $\rho \to 0_+$. Also we easily see that both hysteresis operators are monotone, i.e.

$$\big(w(t_2) - w(t_1)\big)\big(u(t_2) - u(t_1)\big) \ge 0 \qquad \forall t_1, t_2 \in [0,T] \tag{14.14}$$

where w belongs to $\mathcal{R}_{r_1,r_2,w_{-1}}[u]$ or $\bar{\mathcal{R}}_{r_1,r_2,\rho,w_{-1}}[u]$. Multiplying $\mathcal{R}_{r_1,r_2,w_{-1}}$ and $\bar{\mathcal{R}}_{r_1,r_2,\rho,w_{-1}}$ by -1 we get the corresponding elementary nonmonotone hysteresis operators (see in Fig.14.4 a the nonmonotone completed delayed relay operator).

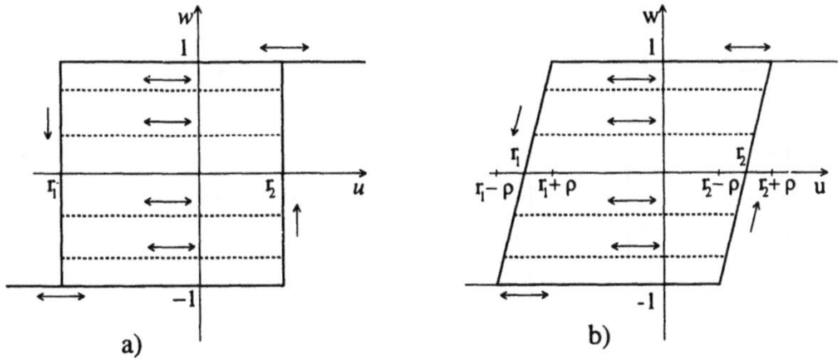

Figure 14.2 Example of discontinuous and continuous hysteresis operator.

In order to form more complicated hysteresis operators one can proceed as follows: Define a linear superposition operator $\mathcal{P}$ from the completed delayed relay operators through a formula

$$\mathcal{P}[u] = \int_0^{+\infty} \int_{-\infty}^{+\infty} \omega(r,s)\mathcal{R}_{s-r,s+r,w_{-1}}[u](t)\, ds\, dr, \tag{14.15}$$

where ω is a density function satisfying $\omega \in L^1_{loc}(\mathbf{R}_+ \times \mathbf{R})$ (see an example of a hysteresis curve given by a Preisach operator in Fig.14.5 a). It is possible to show that $\mathcal{P}$ is identical to the so called (continuous) *scalar Preisach operator*, although usually the ***delayed relay operator*** not the completed delayed relay operator is used in the definition of Preisach operator (see details e.g. [19]). Further, if more general density functions like measures are used one can create discontinuous scalar Preisach operators (again we refer to [19]).

14.4 COMPARISON OF TWO MODELS AND THEIR EXTENSION

In the previous sections we saw that terms induced by convex or nonconvex superpotentials are able to describe very complicated relations like vertical jumps to upwards or downwards between the studied quantities. The same also holds true for hysteresis operators. The basic difference between these models is illustrated in Fig.14.3 and 14.4, i.e. the latter model has memory effects (jumps with delay) and in the former model the response is immediate (jumps without delay). In principle one could say that the latter model includes the former model: if $r_2 = r_1$ then Fig.14.3 a is the same as Fig.14.3 b and, similarly, Fig.14.4 a and Fig.14.4 b. But as we have already seen this is not a practical approach; complex monotone or nonmonotone relations without delays can be defined in a much simpler way through convex or nonconvex superpotentials than hysteresis operators. Furthermore, the mathematical treatment of discontinuous hysteresis is more complicated than terms induced by superpotentials, and actually as we shall see below from the mathematical point of view it is not possible to include jumps without delays in a hysteresis operator.

The basic mathematical question in the theory of (dynamic) variational and hemivariational inequalities and the theory of hysteresis is the following one:

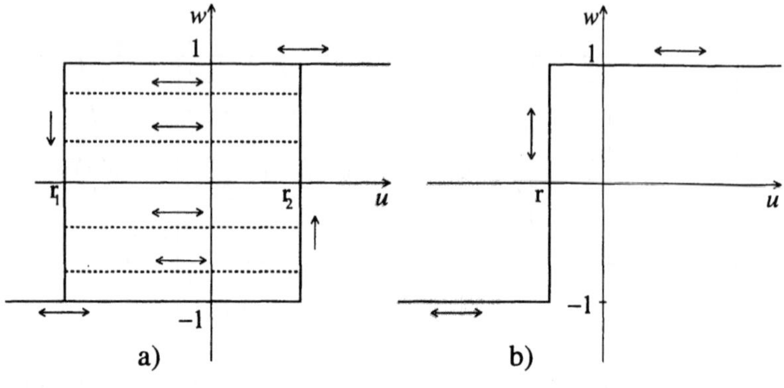

Figure 14.3 Example of monotone hysteretic and superpotential relations

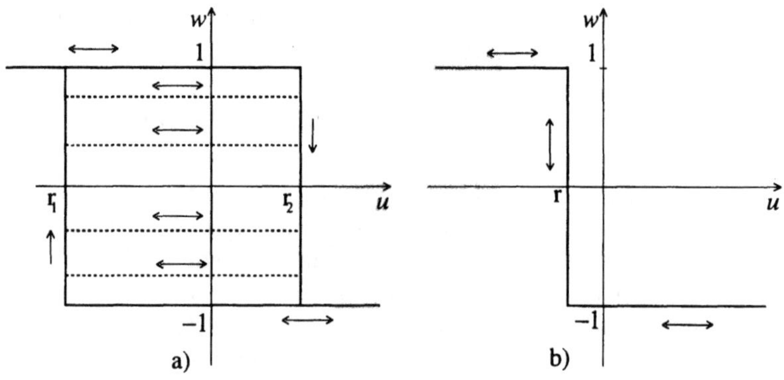

Figure 14.4 Example of nonmonotone hysteretic and superpotential relations

Suppose that we have sequences $\{u_k\}$ and $\{w_k\}$ converging to u and w in some norms and satisfying either $w_k(t) \in \partial j(u_k(t))$ in $[0, T]$, j a superpotential, or $w_k(t) \in \mathcal{W}[u_k](t)$ in $[0, T]$, $\mathcal{W}$ a hysteresis operator. Is it true that $w(t) \in \partial j(u(t))$ in $[0, T]$ in the case of variational or hemivariational inequalities and $w(t) \in \mathcal{W}[u](t)$ in $[0, T]$ in the case of hysteresis problems?

Let us shortly sketch the solution procedures for the both cases. In the former one it is enough to have that $\{u_k\}$ converges strongly to u in $L^2(0, T)$. This implies the pointwise convergence of $\{u_k\}$ (by passing to a subsequence if necessary) almost everywhere in $(0, T)$, and then from Egoroff's theorem it follows the uniform convergence for $\{u_k\}$ except in a set having a arbitrary small measure. Applying Proposition 14.1 (i) we get that $\{w_k\}$ is bounded in a proper sense, and then either from Proposition 14.1 (ii) or (iii) we can deduce that $w(t) \in \partial j(u(t))$ a.e. in $[0, T]$.

In the latter case we consider only the discontinuous hysteresis operator $\mathcal{R}_{r_1, r_2, w_{-1}}$, since the treatment of the continuous hysteresis operators is straightforward. Now it is necessary to assume that $\{u_k\}$ converges strongly to u in $C([0, T])$. Then as in the previous section we note that the sequence $\{w_k\}$

belongs to $BV(0,T)$. Further, we see that it is bounded in $BV(0,T)$ and, thus a subsequence $\{w_k\}$ (denoted by the same symbol as the original sequence) converges to w with respect to the weak$*$–topology of $BV(0,T)$. It means that w_k converges strongly to w in $L^1(0,T)$ and "its derivative" converges weak$*$ in $(C([0,T]))^*$, the dual space of $C([0,T])$. The key point is now that the relation (14.12) can be written by means of the dual pairing between $C([0,T])$ and $(C([0,T]))^*$. And then the strong convergence of $\{u_k\}$ in $C([0,T])$ and the weak$*$–converge of $\{w_k'\}$ in $(C([0,T]))^*$ implies that the limits u and w satisfy (14.12) (for details see [19]). Note that the treatment of the other relations (14.10) and (14.11) is obvious.

As a result of the discussions above it should be clear that if the discontinuous hysteresis operator contains jumps without delay the boundedness of $\{w_k\}$ in $BV(0,T)$ is lost and, consequently, there is not any possibility to prove that the limits u and w satisfy the relation $w(t) \in \mathcal{W}[u](t)$. Here it is necessary to point out that the mathematical theory of more general discontinuous hysteresis operators is not yet well established. However, in spite of this shortage the above conclusion of the complementary nature of superpotentials and hysteresis operators is universal.

Let us shortly discuss a multidimensional case, i.e. the input function u and the output function w are vector-valued functions. The extension for the vector-valued superpotentials is quite straightforward: as a potential function we use a locally Lipschitz function $j : \mathbf{R}^M \to \mathbf{R}$, $M \geq 2$, and, then the generalized gradient ∂j induces a multivalued relation from $\mathbf{R}^M$ to $\mathbf{R}^M$. Further, Proposition 14.1 holds true. Therefore, by means of convex and nonconvex superpotentials it has been developed an extensive mathematical theory which enables the treatment of complicated multidimensional relations, e.g. in mechanics (see e.g. [16]). On the other hand, for the hysteresis operators there do not exist so obvious generalizations in the multidimensional case; although there exists some models for multidimensional hysteresis (see [7],[8],[19]) one can say that the theory is not yet well established. Anyway, also in this case superpotentials and hysteresis operators can be considered to be complementary.

As a result of the above consideration it is natural to introduce the following extension to the theory of hysteresis and the theory of hemivariational inequalities: Let us study relations of the form

$$w \equiv w_1 + w_2, \qquad w_1(t) \in \partial j(u(t)), \quad w_2(t) \in \mathcal{W}[u](t). \tag{14.16}$$

The advantage of this model is very obvious: we can treat more general relations (see Fig.14.5). And also from the mathematical point of view this decomposition is practical, for the term w_1 one can apply the techniques of hysteretic relations and for the term w_2 the techniques of superpotential relations.

Furthermore, here it is important to point out that in many situations due to the complexity of hysteresis operators (or the nonexistence of appropriate hysteresis operators) it can be worthwhile to describe hysteretic phenomena by means of sequences of hemivariational inequalities, e.g. loading and unloading processes (see [18, p.121]) or classical hysteresis phenomena with closed and/or open loops (see [17, p.209]).

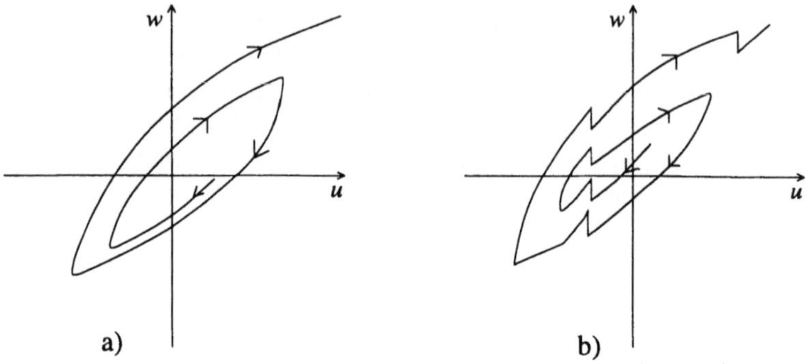

Figure 14.5 Example of a hysteretic curve of the Preisach operator and the same curve added with a superpotential law.

14.5 PARABOLIC B.V.P.S WITH HYSTERETIC AND SUPERPOTENTIAL RELATIONS

In this section we give a short overview of results established in [11] and [12] for parabolic B.V.P.s having relations expressed both via nonconvex superpotentials and hysteresis operators. The problem (P1) considered in [11] is defined as follows:

$$u'(t) + w(t) + Au(t) + \Xi(t) = f(t) \qquad \text{a.e. in } (0,T), \tag{14.17}$$
$$w(x,t) = \mathcal{W}[u(x,\cdot);x](t) \qquad \forall t \in [0,T], \quad \text{a.e. } x \in \Omega, \tag{14.18}$$
$$\Xi(x,t) \in \partial j(x, u(x,t)) \qquad \text{a.e. } (x,t) \in Q_T = \Omega \times (0,T), \tag{14.19}$$
$$u(0) = u_0, \tag{14.20}$$

where ∂j stands for the generalized gradient of j with respect to the second variable. Thus, the sum $\Xi + w$ is exactly of the form (14.16).

The problem (P2) studied in [12] is formulated in a similar way as (P1) except the first condition (14.17) is replaced by

$$u'(t) + w'(t) + Au(t) + \Xi(t) = f(t) \qquad \text{a.e. in } (0,T). \tag{14.21}$$

Note that now the hysteretic term w and the superpotential term Ξ are not of the same order: w appears in the principal part of the above B.V.P. and Ξ is a lower order term. Due to the possible nonmonotonicity of Ξ we are not able to treat the situation in which Ξ depends on u'.

Next we give mathematical formulations of the problems above and state the assumptions needed for the existence results. Let $\Omega \subset \mathbf{R}^N$ be a bounded domain with Lipschitz boundary $\partial\Omega$. Let V be a Hilbert space such that the imbedding $V \subset H^1(\Omega)$ is dense and continuous. We denote by $\|\cdot\|_V$ and $|\cdot|_H$

the norms of V and $H \equiv L^2(\Omega)$, respectively. The duality pairing between V and V^* is denoted by $\langle \cdot, \cdot \rangle_V$ and the inner product in H by $(\cdot, \cdot)_H$.

For a space dependent hysteresis operator $\mathcal{W}[\cdot; x]$ we introduce the following conditions:

(H1) (continuity and boundedness) The hysteresis operator $\mathcal{W}[\cdot; x]$ is continuous on $C([0,T])$ for every $x \in \Omega$ and the parametrized final value mapping

$$(s; x) \mapsto \mathcal{W}_f(s; x)$$

is measurable for all $s = (s_0, ..., s_n) \in S$ and $n \in \mathbf{N}$ and satisfies

$$|\mathcal{W}_f(s; x)| \leq \gamma(x) + c_1 \max_{i=0,\dots,n} |s_i|, \qquad \text{for all } x \in \Omega, \quad s \in S \text{ and } n \in \mathbf{N},$$

where $\gamma \in L^2(\Omega)$ and c_1 a positive constant.

(H2) (monotonicity) The level functions $l_n : \mathbf{R} \to \mathbf{R}$ defined by

$$l_n(\xi) = \mathcal{W}_f(s_0, ..., s_{n-1}, \xi; x)$$

are increasing functions for all $x \in \Omega$, $n \in \mathbf{N}$ and $(s_0, ..., s_{n-1}) \in S$.

(H3) The hysteresis operator $\mathcal{W}[\cdot; x]$ is a mapping from $H^1(0,T)$ into itself for every $x \in \Omega$ and it holds for every $v \in H^1(0,T)$

$$|\mathcal{W}[v; x]'(t)| \leq \gamma_2(x) + c_2|v'(t)|, \qquad \forall x \in \Omega \text{ and a.e. } t \in (0,T),$$

where $\gamma_2 \in L^2(\Omega)$ and c_2 a positive constant.

For a (possibly) nonconvex superpotential $j : \Omega \times \mathbf{R} \to \mathbf{R}$ we impose the following conditions:

(J1) The function satisfies the Caratheodory type conditions

(i) For all $\xi \in \mathbf{R}$ the function $x \mapsto j(x, \xi)$ is measurable on Ω.

(ii) For almost all $x \in \Omega$ the function $\xi \mapsto j(x, \xi)$ is locally Lipschitz on $\mathbf{R}$.

(J2) (linear growth condition) There exists a positive constant c_3 such that

$$\eta \in \partial j(x, \xi) \Longrightarrow |\eta| \leq c_3(1 + |\xi|)$$

for a.e. $x \in \Omega$ and each $\xi \in \mathbf{R}$. Moreover, the function $j(\cdot, 0) \in L^1(\Omega)$.

(J3) (integrability condition) There exists a function $\beta : \Omega \times \mathbf{R}_+ \to \mathbf{R}$ such that

(i) $\beta(\cdot, r) \in L^2(\Omega)$ for each $r \geq 0$,

(ii) If $r' \leq r''$ then for almost all $x \in \Omega$

$$\beta(x, r') \leq \beta(x, r'')$$

and for almost all $x \in \Omega$

$$|j(x,\xi) - j(x,\eta)| \le \beta(x,r)|\xi - \eta| \qquad \forall \xi, \eta \in B(0,r), r \ge 0.$$

Moreover, the function $j(\cdot, 0) \in L^1(\Omega)$.

(J4) (directional growth condition) There exists a nonnegative function α : $\Omega \to \mathbf{R}$ such that $\alpha \in L^2(\Omega)$ and for almost all $x \in \Omega$

$$j^\circ(x,\xi;-\xi) \le \alpha(x)(1+|\xi|) \qquad \forall \xi \in \mathbf{R}.$$

Let A be a linear, bounded and symmetric operator from V to V^* satisfying:

(A1) The operator A is coercive, i.e. there exists constants $c_4 > 0$ and $c_5 \ge 0$ such that

$$\langle Av, v\rangle_V \ge c_4\|v\|_V^2 - c_5|v|_H^2 \qquad \forall v \in V.$$

Furthermore, we suppose that $f \in L^2(0,T;H)$ and the initial value u_0 is regular enough (see [11],[12]).

The *weak formulations* of the problems (P1) and (P2) are as follows:

Definition 14.3. *A function $u \in H^1(0,T;L^2(\Omega)) \cap L^\infty(0,T;V)$ is a solution of the problem (P1) if there exist $w \in L^2(\Omega;C([0,T]))$ and $\Xi \in L^1(Q_T) \cap L^2(0,T;V^*)$ such that*

$$\int_0^T (u'(t),v(t))_H\,dt + \int_0^T (w(t),v(t))_H\,dt + \int_0^T \langle Au(t),v(t)\rangle_V\,dt$$
$$+ \int_0^T \langle \Xi(t),v(t)\rangle_V\,dt = \int_0^T (f(t),v(t))_H\,dt \qquad \forall v \in L^2(0,T;V)$$

and the conditions (14.18)–(14.20) are satisfied.

Definition 14.4. *A function $u \in H^1(0,T;L^2(\Omega)) \cap L^\infty(0,T;V)$ is a solution of the problem (P2) if there exist $w \in L^2(\Omega;C([0,T])) \cap H^1(0,T;V^*)$ and $\Xi \in L^1(Q_T) \cap L^2(0,T;V^*)$ such that*

$$\int_0^T (u'(t),v(t))_H\,dt + \int_0^T \langle w'(t),v(t)\rangle_V\,dt + \int_0^T \langle Au(t),v(t)\rangle_V\,dt$$
$$+ \int_0^T \langle \Xi(t),v(t)\rangle_V\,dt = \int_0^T (f(t),v(t))_H\,dt \qquad \forall v \in L^2(0,T;V)$$

and the conditions (14.18)–(14.20) are satisfied.

The following existence results are proved in [11] and [12]:

Theorem 14.1. *a) Let hypotheses (H1),(J1),(J2),(A1) be satisfied. Then the problem (P1) has at least one solution. Moreover, the function Ξ belongs to $L^\infty(0,T;H)$.*

b) Let hypotheses (H1),(J1),(J3),(J4),(A1) be satisfied. Then the problem (P1) has at least one solution.

Theorem 14.2. *a) Let hypotheses (H1),(H2),(J1),(J2),(A1) be satisfied. Then the problem (P2) has at least one solution. Moreover, the function* Ξ *belongs to* $L^\infty(0,T;H)$.
b) Let hypotheses (H1),(H2),(H3),(J1),(J3),(J4),(A1) be satisfied. Then the problem (P2) has at least one solution. Moreover, the function w *belongs to* $H^1(0,T;H)$.

Acknowledgments

The author acknowledges the support of the grant SA-38962 of the Academy of Finland.

References

[1] M. Brokate and J. Sprekels. *Hysteresis and Phase Transitions.* Springer Verlag, New York, 1996.

[2] F.H. Clarke. *Optimization and Nonsmooth Analysis.* Wiley, New York, 1983.

[3] G. Duvaut and J.L. Lions. *Inequalities in Mechanics and Physics.* Springer Verlag, Berlin, 1976.

[4] L.C. Evans and R.F. Gariepy. *Measure Theory and Fine Properties of Functions.* CRC Press, Boca Raton, 1992.

[5] G. Fichera. Problemi elastostatici con vincoli unilaterali, il problema di Signorini con ambigue condizioni al contorno. *Mem. Accad. Naz. Lincei,* 8(7):91–140, 1964.

[6] G. Fichera. Boundary value problems in elasticity with unilateral constraints. In: C. Truesdell, *Encyclopedia of Physics VI a/2, Mechanics of Solids II,* pp. 391–424. Springer Verlag, Berlin, 1972.

[7] M.A. Krasnoselskii and A.V. Pokrovskii. *Systems with Hysteresis.* Springer Verlag, Heidelberg, 1989.

[8] P. Krejčí. *Hysteresis, Convexity and Dissipation in Hyperbolic Equations* Gakkotosho, Tokyo, 1996.

[9] M. Miettinen. A parabolic hemivariational inequality. *Nonlinear Analysis,* 26:725–734, 1996.

[10] M. Miettinen and P.D. Panagiotopoulos. On parabolic hemivariational inequalities and applications. to appear in *Nonlinear Analysis.*

[11] M. Miettinen and P.D. Panagiotopoulos. Hysteresis and hemivariational inequalities: Semilinear Case. to appear in *J. Global Optimization.*

[12] M. Miettinen and P.D. Panagiotopoulos. Hysteresis and hemivariational inequalities: Quasilinear Case. preprint.

[13] D. Motreanu and P.D. Panagiotopoulos. On the eigenvalue problem for hemivariational inequalities: existence and multiplicity of solutions. *J. Math. Anal. Appl.*, 197:75–89, 1996.

[14] D. Motreanu and P.D. Panagiotopoulos. Double eigenvalue problems for hemivariational inequalities. to appear in *Arch. Rat. Mech. Anal.*

[15] Z. Naniewicz. Hemivariational inequalities with functionals which are not locally Lipschitz. *Nonlinear Analysis*, 25:1307–1320, 1995.

[16] Z. Naniewicz and P.D. Panagiotopoulos. *Mathematical Theory of Hemivariational Inequalities and Applications.* Marcel Dekker, New York, 1995.

[17] P.D. Panagiotopoulos. *Inequality Problems in Mechanics and Applications. Convex and Nonconvex Energy Functions.* Birkhäuser, Basel, Boston, 1985.

[18] P.D. Panagiotopoulos. *Hemivariational Inequalities. Applications in Mechanics and Engineering.* Springer Verlag, New York, Berlin, 1993.

[19] A. Visintin. *Differential Models of Hysteresis.* Springer Verlag, Berlin, 1994.

15 NON CONVEX ASPECTS OF DYNAMICS WITH IMPACT

Lætitia Paoli[1]
and Michelle Schatzman[2]

[1]CNRS UMR 5585 Analyse Numérique
Faculté des Sciences, 23 Rue du Docteur Michelon
Université Jean Monnet – Saint-Etienne
42023 Saint-Etienne Cedex, France
paoli@anum.univ-st-etienne.fr

[2]CNRS UMR 5585 Analyse Numérique
Université Claude Bernard – Lyon 1
69622 Villeurbanne Cedex, France
schatz@iris.univ-lyon1.fr

Abstract: We review several aspects of dynamics with impact with a finite number of degrees of freedom. We recall that the variational formulation has to be supplemented by a constitutive law for the impact. We take this law to be Newton's law, with restitution coefficient $r \in [0, 1]$. Dynamical problems have an intrinsic nonconvexity, due to the presence of second derivatives in time. Several results on the approximation of dynamics with impact are stated when the set of constraints is smooth. When the set of constraints is an angular domain, phenomena of ill-posedness are important. Moreau's rule for inelastic impact is justified by a penalty method.

15.1 INTRODUCTION

In this article we consider a general class of dynamical problems with impact with a finite number of degrees of freedom. More precisely, assume that the free dynamics of the problem is described by

$$\ddot{u} = f(t, u, \dot{u}), \tag{15.1}$$

R.P. Gilbert et al. (eds.), From Convexity to Nonconvexity, 207–222.

where the representative point of the system, u, belongs to $\mathbb{R}^N$ and f is a function from $[0,T] \times \mathbb{R}^N \times \mathbb{R}^N$ to $\mathbb{R}^N$.

Assume that the set of constraints K is a closed subset of $\mathbb{R}^N$. We will assume that either K is a convex set, or the boundary of K is of class C^3. Therefore, the tangent cone at $x \in \partial K$ to K, denoted $T_x K$ is well defined in both cases: in the first case, it is a convex cone, and in the second case, it is always a half-space.

We look for a solution of (15.1) supplemented with the constraint

$$\forall t \in [0,T], \quad u(t) \in K. \tag{15.2}$$

Suppose that at time t_1, $u(t_1)$ belongs to ∂K and that on an interval (t_0, t_1), $u(t)$ belongs to the interior of K; then elementary considerations in convex analysis and ordinary differential equations show that if (15.1) hold whenever u belongs to the interior of K, then $\dot{u}(t_1 - 0)$ belongs to the cone $-T_{u(t_1)}K$. If $\dot{u}(t_1 - 0)$ does not belong to the cone $T_{u(t_1)}K$, then $\dot{u}$ must undergo a discontinuity at time t_1 if the constraint (15.2) is satisfied on some interval starting at t_1.

Therefore, equation (15.1) must be modified as follows:

$$\ddot{u} = f(t, u, \dot{u}) + \mu, \tag{15.3}$$

where μ describes the reaction of the constraints. From the mathematical viewpoint, μ is a measure: in particular, when $\dot{u}$ has a discontinuity at some time t_1, μ has a vector valued Dirac mass at t_1. Let $\xi \in \mathbb{R}^N$ be the coefficient of this Dirac mass: it is equal to $\dot{u}(t_1 + 0) - \dot{u}(t_1 - 0)$; it is clear that $\dot{u}(t_1 + 0)$ belongs to $T_{u(t_1)}K$ and $\dot{u}(t_1 - 0)$ belongs to $-T_{u(t_1)}K$; therefore,

$$\xi \in T_{u(t_1)}K. \tag{15.4}$$

If we assume moreover that the constraints are perfect, then the reaction of the constraints belongs to the normal cone to K at $u(t)$ whenever $u(t) \in \partial K$. There is a more mathematical way of describing the conditions satisfied by μ: in the case of a convex K, we can write, following [16] and [17] that for all $v \in C^0(0,T;\mathbb{R}^N)$ taking its values in K, we must have

$$\langle \mu, v - u \rangle \geq 0. \tag{15.5}$$

In the case of a smooth ∂K, let us denote $\nu(x)$ the interior normal at x to ∂K; write $\mu = g|\mu|$, where $|\mu|$ is the absolute value of the measure μ (known to be a positive measure on $[0,T]$) and g is $|\mu|$ integrable and $|\mu|$-almost everywhere different from 0 on the suport of $|\mu|$. Then the following condition must be satisfied:

$$\begin{aligned} &\operatorname{supp}(|\mu|) \subset \{t \in [0,T] : u(t) \in \partial K\}, \\ &g \in \mathbb{R}^+ \nu \circ u, \ |\mu|\text{-almost everywhere.} \end{aligned} \tag{15.6}$$

If the impact takes place at a point x of ∂K where ∂K is smooth, then (15.5), or more generally (15.6) says that the tangential component of the velocity is conserved, and (15.2) says that the normal component of the velocity (measured by taking the scalar product with the interior normal) is negative before the impact and positive afterwards.

This is all the knowledge on μ that we can infer from the assumption of perfect unilateral constraints. It does not imply uniqueness, since it tells us nothing on the behavior of the kinetic energy at impact times.

Therefore, we need a constitutive law of the impact to complete the model. We choose a Newton's law, i.e. we assume that there is a restitution coefficient r such that the normal velocity after the impact is equal to the opposite of the normal velocity before the impact, multiplied by r. It is important to observe that this constitutive law makes sense only if the impact takes place at a point x of ∂K where ∂K is smooth. If the set of constraints is not smooth complicated phenomena can and will happen: some of them will be described here.

We will also take initial conditions: u_0 is given in K and we choose $\dot{u}_0$ arbitrarily if u_0 belongs to the interior of K and in the tangent cone $T_{u_0}K$ if u_0 belongs to ∂K.

The article is organized as follows: in section 2, we comment on the essentially non convex nature of vibration problems with constraints on the position; we show how a natural numerical scheme contains the implicit assumption that the restitution coefficient vanishes. In section 3, when K is smooth, we describe a penalty method and a numerical scheme which approximate the solutions of (15.2)-(15.3)-(15.5) with restitution coefficient r, and at the end, we comment on the uniqueness of solutions, or rather, the lack of it, even with a given restitution coefficient. In section 4 we describe the behavior of the penalized approximation when K is an angular domain, in two different cases: the first case is when there is no loss of kinetic energy at impact (elastic impact) and the second case is when the normal component of the velocity is totally lost (inelastic impact).

Let us mention here that a very complete review of dynamics with impact and/or friction is given in [4].

15.2 NON CONVEXITY OF DYNAMICAL PROBLEMS

Assume that A is a symmetric positive definite $N \times N$ matrix. Given $g \in \mathbb{R}^N$, the following very classical minimization problem

$$\text{to minimize over } K \quad \frac{x^* A x}{2} - g^* x$$

admits a unique solution for all closed convex subset K of $\mathbb{R}^N$. There is an equivalent variational formulation of this problem, i.e.

$$\text{to find } x \in K \text{ such that for all } y \in K \quad (y - x)^* (Ax - g) \geq 0. \qquad (15.7)$$

If $w = A^{-1} g$, then (15.7) expresses the fact that x is the projection of w on K relatively to the metric defined by the scalar product $\langle x, y \rangle = y^* A x$.

Given a function g with values in $\mathbb{R}^N$, we define a "parabolic" generalization of (15.7) if we find $x(t)$ with values in K such that

$$\int_0^T (y(t) - x(t))^*(\dot{x}(t) + Ax(t) - g(t))\,dt \geq 0 \tag{15.8}$$

for all continuous function y taking its values in K. The problem (15.8) possesses a solution thanks to the theory of semigroups generated by maximal monotone operators, as exposed in [3] for instance.

Define an operator $\mathcal{A}$ in $L^2(0,T;\mathbb{R}^N)$ by

$$D(\mathcal{A}) = \{z \in H^1(0,T;\mathbb{R}^N) : z(0) = 0\}, \quad \mathcal{A}x = \dot{x} + Ax.$$

The operator $\mathcal{A}$ is certainly not self-adjoint in $L^2(0,T;\mathbb{R}^N)$: it is not even symmetric; however, it is nonnegative in the following sense: for all z belonging to the domain of $\mathcal{A}$,

$$(\mathcal{A}z, z)_{L^2} = \frac{1}{2} z(T)^* z(T) + \int_0^T z(t)^* Az(t)\,dt,$$

which is clearly strictly positive for all non zero z. Define also the closed convex subset $\mathcal{K}$ of $L^2(0,T;\mathbb{R}^N)$ by

$$\mathcal{K} = \{z \in L^2(0,T;\mathbb{R}^N) : z(t) \in K \text{ for almost every } t\}.$$

With these definitions, if g is given in $L^2(0,T;\mathbb{R}^N)$ and w is the unique solution of

$$\dot{w} + Aw = g, \quad w(0) = x(0),$$

given by

$$w(t) = e^{-tA}x(0) + \int_0^t e^{-(t-s)A} g(s)\,ds,$$

then (15.8) can be rewritten

$$\text{to find } x \in D(\mathcal{A}) \text{ such that for all } y \in \mathcal{K} \quad (\mathcal{A}(x - w), y - x)_{L^2} \geq 0. \tag{15.9}$$

The formal analogy of (15.9) with (15.7) can be understood more precisely if we consider a discretization of (15.8): we replace the integral by a finite sum, the derivative by a finite difference, and we consider the variational problem

$$\sum_{n=0}^{J-1} (Y^{n+1} - X^{n+1})^* \left(\frac{X^{n+1} - X^n}{\Delta t} + AX^{n+1} - G^{n+1}\right) \Delta t \geq 0. \tag{15.10}$$

Here the test vectors Y^n must belong to K for $n = 1, \ldots J$, and for each n, X^n is sought in K. Therefore, if Y^n coincides with X^n except for $n = k+1$, we can see that X^{k+1} solves the variational problem

$$(Y^{k+1} - X^{k+1})^* \left(\frac{X^{k+1} - X^k}{\Delta t} + AX^{k+1} - G^{k+1}\right) \geq 0,$$

which is exactly (15.7) with g substituted by $X^k + \Delta t G^{k+1}$, and A substituted by $\mathbf{1} + \Delta t A$. If the sequence of piecewise constant functions

$$\sum_{n=1}^{J} G^n 1_{((n-1)\Delta t, n\Delta t]}$$

converges strongly to g in $L^2(0,T;\mathbb{R}^N)$ and if X^0 converges to $x(0)$, it is an exercise in the theory of maximal monotone operators to show that the piecewise affine interpolation of the X^j's converges to the unique solution x of (15.8) as Δt tends to 0, strongly in $H^1(0,T;\mathbb{R}^N)$.

Similarly, one could consider the second order problem: to find $x(t)$ with values in K such that

$$\int_0^T (y(t) - x(t))^* (\ddot{x}(t) + Ax(t) - g(t))\, dt \geq 0 \tag{15.11}$$

for all continuous function y taking its values in K. This is exactly the kind of problem that we were considering in the introduction of this article: if we let $f(t, x, \dot{x}) = -Ax + g(t)$, we find that (15.11) is analogous to (15.3) with the condition (15.5).

However, we have already observed that this formulation does not include information on the constitutive law of the impact, and therefore the solution is not well defined.

Moreover, if $\mathcal{B}$ is the operator in $L^2(0,T;\mathbb{R}^N)$ defined by

$$D(\mathcal{B}) = \{z \in H^2(0,T;\mathbb{R}^N) : z(0) = \dot{z}(0) = 0\}, \quad \mathcal{B}z = \ddot{z} + Az, \tag{15.12}$$

the operator $\mathcal{B}$ has some degree of symmetry: if $z, \dot{z}$ and $w, \dot{w}$ vanish at $t = T$, an integration by parts shows that

$$(\mathcal{B}z, w)_{L^2} = (z, \mathcal{B}w)_{L^2}.$$

However, there is no sense in which $\mathcal{B}$ could be considered positive: an integration by parts also shows that for $z \in D(\mathcal{B})$,

$$(\mathcal{B}z, z)_{L^2} = z(T)^* \dot{z}(T) - \int_0^T \dot{z}(t)^* \dot{z}(t)\, dt + \int_0^T z(t)^* A z(t)\, dt.$$

Let ξ be a unitary eigenvector of A relative to the eigenvalue λ, and let $z(t) = \sin^2(m\pi t/T)\xi$ where m is an integer. Then

$$\int_0^T \dot{z}(t)^* \dot{z}(t)\, dt = \frac{m^2\pi^2}{2T}, \quad \int_0^T z(t)^* A z(t)\, dt \leq \lambda T.$$

Therefore, for large values of m, $(\mathcal{B}z, z)_{L^2}$ can take arbitrarily large negative values.

If we consider a discretized version of (15.11) similar to (15.10), we can write for instance

$$\sum_{n=1}^{J-1}(Y^{n+1}-X^{n+1})^* \left(\frac{X^{n+1}-2X^n+X^{n-1}}{\Delta t^2}+AX^{n+1}-G^{n+1}\right)\Delta t \geq 0. \tag{15.13}$$

The constraints $X^n \in K$ and $Y^n \in K$ must be satisfied for all $n = 1, \dots, J$. The above argument shows that we can solve a sequence of variational problems of the form (15.7), where we substitute g by $2X^n - X^{n-1} + \Delta t^2 G^{n+1}$ and A by $\mathbf{1} + \Delta t^2 A$.

Though (15.13) does not mention a restitution coefficient, it contains an implicit constitutive law of the impact, as we shall see now on the simplest possible example: $N = 1$, $A = 1$, $K = \mathbb{R}^+$, $g = 0$. Denote s^+ the positive part of the real number s. With $G^n = 0$, the discrete solution of (15.13) must satisfy the relation

$$X^{n+1} = \left(\frac{2X^n - X^{n-1}}{1+\Delta t^2}\right)^+. \tag{15.14}$$

We can solve explicitly this recurrence for particular initial data, in order to get a feeling for the behavior of (15.13): define

$$\alpha = \arctan(\Delta t), \quad X^0 = 1, \quad X^1 = \frac{\cos\alpha}{(1+\Delta t^2)^{1/2}}.$$

Then, as long as $2X^{n-1} - X^{n-2}$ is greater than or equal to 0, X^n is given by

$$X^n = \frac{\cos n\alpha}{(1+\Delta t^2)^{n/2}}.$$

If k is the first index for which $2X^{k-1} - X^{k-2}$ is strictly negative, then

$$X^k = 0.$$

For $n = k + 1$, the right hand side of (15.14) contains the positive part of $2X^k - X^{k-1}$: X^k vanishes and X^{k-1} is certainly non negative, since it is the positive part of a real number; therefore, X^{k+1} vanishes, and so do all the subsequent X^n, by an immediate induction.

Therefore (15.13) contains implicitly an inelastic constitutive law of the impact: the normal component of the velocity is cancelled at the impact.

15.3 APPROXIMATION AND UNIQUENESS WHEN K IS SMOOTH

We will consider two different types of approximation: the penalty approximation and the numerical approximation.

Approximation by penalty

The penalty approximation is used very often for computational purposes: we replace the rigid constraint by a very stiff drawback force whenever the representative point leaves the set of constraints: if K is smooth there exists a neighborhood Ω of K such that the projection P_K is uniquely defined in $\Omega \setminus K$. We have also to define a viscous friction term which dissipates the normal component of the velocity in exactly the right proportion. Let

$$H(u,v) = \begin{cases} (v^*(u-P_K u))(u-P_K u)/|u-P_K u|^2, & \text{if } u \in \Omega \setminus K, \\ 0 & \text{if } u \in K. \end{cases} \tag{15.15}$$

With these definitions, the penalized equation for (15.2)-(15.3)-(15.5) with restitution coefficient r is

$$\ddot{u} + 2\alpha\sqrt{k} H(u,\dot{u}) + k(u - P_K u) = f(t,u,\dot{u}). \tag{15.16}$$

The positive number k is assumed to be very large. The number α is greater than or equal to 0: if α belongs to the interval $[0,1)$ we let

$$r = \exp(-\alpha\pi/\sqrt{1-\alpha^2}). \tag{15.17}$$

We have proved in [9] that when K is convex and $\alpha < 1$, the solution of (15.16) with initial conditions u_0, $\dot{u}_0$ converges to a solution of (15.3) with restitution coefficient r given by (15.17).

We conjecture that the same result holds even when K is not convex, but its boundary is smooth enough. We also conjecture that if $\alpha \geq 1$, the solution of (15.16) converges to a solution of (15.3) with zero restitution coefficient.

The basic idea for these results and conjectures is extremely simple: in the case when K is the half-plane $x_N \geq 0$, f vanishes, u_0 is on the boundary of the half plane and $\dot{u}_0$ points outward, the solution of (15.16) is given explicitly by

$$\begin{aligned} u_j(t) &= u_{0,j} + t\dot{u}_{0,j}, \quad 1 \leq j \leq N-1, \text{ if } t \geq 0, \\ u_N(t) &= \frac{1}{\sqrt{k(1-\alpha^2)}} \sin(t\sqrt{k(1-\alpha^2)}) \exp(-\alpha t\sqrt{k})\dot{u}_{0,N} \\ &\qquad \text{if } 0 \leq t \leq \pi/\sqrt{(1-\alpha^2)k}, \\ u_N(t) &= -(t - \pi/\sqrt{(1-\alpha^2)k}) r\dot{u}_{0,N}, \text{ if } t \geq \pi/\sqrt{(1-\alpha^2)k}. \end{aligned}$$

Thus, in this case, the behavior of the solution of (15.16) is very easy to understand. All the work of the proof given in [9] is to prove estimates which enable us to substantiate this intuition; a delicate point is the convergence to a solution with the restitution coefficient r: it entails a precise local study close to the impact times.

Time discretization

The numerical approximation generalizes (15.13) so as to contain all the possible constitutive laws with restitution coefficient between 0 and 1. We defined in [10] and [8] a numerical scheme by

$$\begin{aligned} U^{n+1} &= (1+r)P_K\left(\frac{2U^n - (1-r)U^{n-1} + \Delta t^2 F^n}{1+r}\right) - rU^{n-1}, \\ F^n &= F\big(t^n, U^n, U^{n-1}, (U^{n+1} - U^{n-1})/2\Delta t, \Delta t\big). \end{aligned} \tag{15.18}$$

Here F is a function satisfying the consistency condition

$$\forall u, v \in \mathbb{R}^N, \quad \forall t \in [0,T], \quad F(t,u,u,v,0) = f(t,u,v).$$

We assume moreover that F is continuous with respect to all its arguments and that it is Lipschitz continuous with respect to its second, third and fourth arguments.

We have proved that the solution of the implicit scheme (15.18) converges to a solution of (15.3) with restitution coefficient r, provided that the boundary of K is of class C^3 in the convex case. The analogous statement for the nonconvex case is also true ([11], [12]): the proof is much more technical than in [8], since we have to show that U^n stays in Ω for all n and for all small enough h.

The idea of the proof can be understood by examining a very simple case: if f vanishes, $N = 1$ and $K = \mathbb{R}^+$, if the initial data are $U^0 = 1$ and $U^1 = 1 - \Delta t$, we write explicitly the solution of the recurrence

$$U^{n+1} = \left(2U^n - (1-r)U^{n-1}\right)^+ - rU^{n-1}.$$

We find that as long as $2U^n - (1-r)U^{n-1}$ is nonnegative,

$$U^{n+1} = 1 - (n+1)\Delta t.$$

If k is the first index for which $2U^k - (1-r)U^{k-1}$ is strictly negative, U^{k+1} is equal to $-rU^{k-1}$. The expression $2U^{k+1} - (1-r)U^k$ is equal to $-2rU^{k-1} - (1-r)U^k$. We substitute the value $U^k = 2U^{k-1} - U^{k-2}$ into the expression $-2rU^{k-1} - (1-r)U^k$, and we find that it is equal to $-\left(2U^{k-1} - (1-r)U^{k-2}\right)$ which is nonpositive according to the definition of k. Therefore U^{k+2} is equal to $-rU^k$. Finally, $2U^{k+2} - (1-r)U^{k+1}$ is equal to $-r\left(2U^k - (1-r)U^{k-1}\right)$, which is nonnegative; therefore U^{k+3} is equal to $2U^{k+2} - U^{k+1}$, and an immediate induction shows that for $n \geq k+2$, $U^{n+1} - U^n$ is equal to $r\Delta t$.

Once again all the work in the proof of the convergence of the discretized solution to a solution of the continuous problem depends on getting good estimates and performing the local study of convergence in the neighborhood of impact times: the curvature of ∂K creates serious difficulties, and we have to proceed in two big steps: step one consists in obtaining estimates on the solution of (15.18) for $n\Delta t \leq \tau$, where τ is some strictly positive time, which may be much smaller than T. These estimates enable us to pass to the limit as Δt tends to 0. The second step is a continuity argument: the limit satisfies global estimates, provided that it exists, and a connectivity argument leads to the conclusion.

Let us comment on the numerical difficulties related to penalty methods: first, they do not eliminate the necessity to test whether the representative point belongs to the set of constraints. This is a time consuming process which is necessary in order to know whether the penalization must be switched on. Second, the approximation is good only if k is very large. This means that we have to integrate very stiff equations, so that the time step must be small relatively to $1/\sqrt{k}$. It may also be necessary to use implicit schemes, all of which is very costly. Therefore, we concluded that except in very simple cases where one could calculate solutions by analytical formulae during free flight and seek impact times by standard methods, there was no or little advantage for the penalty method with respect to the numerical scheme (15.18). This conclusion is based on extensive numerical testing on problems with 1 to 19 degrees of freedom, reported in [8] and in [7].

Observe that the proof of convergence of the solution of (15.18) to a solution of (15.3) with restitution coefficient r yields another existence proof, which includes the case of a vanishing restitution coefficient.

Uniqueness and non uniqueness

In the above discussion, we mentioned several times "a solution" instead of "the solution" of our problem. The reason is that, even with restitution coefficient $r \in (0, 1]$, there are some subtle instances of nonuniqueness. Such an example was first given in [2]. Two other examples of the same type were constructed in [18]. The same mechanism is used: flat points of C^∞ functions create serious difficulties. It can be conjectured that if the problem has two distinct solutions, then it has an uncountable number of solutions. This has been proved in the particular case of the second example of [18] using an isoenergetic change of time, [19].

A result of generic uniqueness has been proved in the case of elastic impact in [5], in [14] and in [15].

A result of uniqueness for $N = 1$ when f is an analytic function of its three arguments and $r \in [0, 1]$ has been established in [20]. Moreover this result is generalized for any N when ∂K is analytic and f is analytic in its three arguments in [1].

15.4 NON SMOOTH CONSTRAINTS

The penalization (15.16) and the numerical scheme (15.18) can be as well defined for a convex set of constraints K which has corners. However, the behavior of these approximations is complicated, because the problems under consideration are ill–posed.

We will restrict ourselves to the case when K is an angular domain in $\mathbb{R}^2$ and examine the limit of the penalty method when $\alpha = 0$ or when $\alpha > 1$.

Let us define our coordinate conventions: the angle at the vertex of K is $\bar{\theta} \in (0, \pi)$; the domain K is the intersection of the half-planes $x_1 \leq 0$ and $-x_1 \cos \bar{\theta} + x_2 \sin \bar{\theta} \leq 0$. It will be convenient to consider a second system of euclidean coordinates, with basis vectors $\mathbf{i} = (-\cos \bar{\theta}, \sin \bar{\theta})^*$ and $\mathbf{j} = (-\sin \bar{\theta}, -\cos \bar{\theta})$. In this new basis, the coordinates will be denoted y_1 and y_2. See figure 15.1 for a graphical description.

Elastic impact in an angle

The first case that we shall consider is the case when $\alpha = 0$. We show that if $\bar{\theta} \geq \pi/2$ we can find a sequence of initial data such that the sequence of penalized solution starting at these initial data tends towards $t\hat{v}$ when k tends

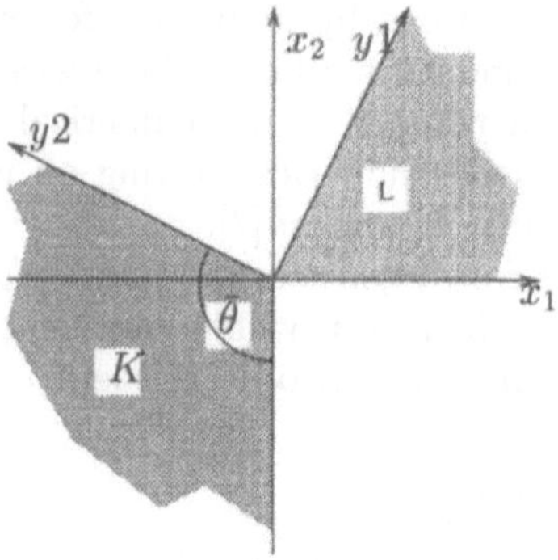

Figure 15.1 The convex set K and the coordinate systems.

to infinity and $\hat{v}$ can be chosen in an uncountable set. Therefore we cannot enforce uniqueness in a natural way, and the problem is ill-posed.

Let $\bar{v} = (\bar{v}_1, \bar{v}_2)^*$ be a vector with strictly positive components. Choose a number β in the interval $[0, \pi]$, and define

$$\bar{u} = \begin{pmatrix} 0 \\ -\beta\bar{v}_2 \end{pmatrix}.$$

We assume moreover that $\bar{v}_1 \cos\bar{\theta} + \bar{v}_2 \sin\bar{\theta} \leq 0$ which implies that $(-\bar{v}_1, \bar{v}_2)^*$ belongs to K.

Consider now the penalized problem

$$\ddot{u}_k + k(u_k - P_K u_k) = 0, \quad u_k(0) = \bar{u}/\sqrt{k}, \quad \dot{u}_k(0) = \bar{v}. \tag{15.19}$$

We perform the change of variables

$$t = \tau/\sqrt{k}, \quad u_k(t) = U(\tau)/\sqrt{k}.$$

In the new variables, (15.19) becomes

$$\ddot{U} + U - P_K U = 0, \quad U(0) = \bar{u}, \quad \dot{U}(0) = \bar{v}. \tag{15.20}$$

If $\beta = \pi$, the solution of (15.20) is given by

$$U(\tau) = \begin{pmatrix} \bar{v}_1 \sin\tau \\ \bar{v}_2(\tau - \pi) \end{pmatrix}, \text{ if } 0 \leq \tau \leq \pi, \quad U(\tau) = \begin{pmatrix} -\bar{v}_1(\tau - \pi) \\ \bar{v}_2(\tau - \pi) \end{pmatrix}, \text{ if } \tau \geq \pi. \tag{15.21}$$

If $\beta = 0$, the solution of (15.20) depends on the sign of $\bar{v}_1 \sin\bar{\theta} + \bar{v}_2 \cos\bar{\theta} = p$. If $p \geq 0$, the solution is given by

$$U(\tau) = \begin{cases} \bar{v}\sin\tau & \text{if } 0 \leq \tau \leq \pi, \\ -(\tau - \pi)\bar{v} & \text{if } \tau \geq \pi. \end{cases}$$

If $p \leq 0$, the solution is given by

$$U(\tau) = \begin{cases} (\bar{v} \cdot \mathbf{i})\mathbf{i} \sin \tau + (\bar{v} \cdot \mathbf{j})\mathbf{j}\tau & \text{if } 0 \leq \tau \leq \pi, \\ -(\bar{v} \cdot \mathbf{i})\mathbf{i}(\tau - \pi) + (\bar{v} \cdot \mathbf{j})\mathbf{j}\tau & \text{if } \tau \geq \pi. \end{cases}$$

When β belongs to $(0, \pi)$, U remains in the set $\{x_1 \geq 0, \quad x_2 \leq 0\}$ on the time interval $[0, \beta]$. We claim that U has to enter $K^\perp$ and to leave it in finite time, whereas it enters in $\{x \cdot \mathbf{i} \geq 0, x \cdot \mathbf{j} \geq 0\}$. To see that this claim is true, we observe that on $[0, \beta]$,

$$U(\tau) = \begin{pmatrix} \bar{v}_1 \sin \tau \\ \bar{v}_2(\tau - \beta) \end{pmatrix}.$$

At time β, we start using polar coordinates: we let

$$U(\tau) = R \begin{pmatrix} \cos(\Theta) \\ \sin(\Theta) \end{pmatrix}.$$

The equations satisfied by Θ and R are

$$\ddot{R} - R\dot{\Theta}^2 + R = 0, \tag{15.22}$$

$$\frac{d}{d\tau}(R^2\dot{\Theta}) = 0. \tag{15.23}$$

The data at time β are

$$R(\beta) = \bar{v}_1 \sin \beta, \quad \dot{R}(\beta) = \bar{v}_1 \cos \beta, \quad \Theta(\beta) = 0, \quad \dot{\Theta}(\beta) = \frac{\bar{v}_2}{\bar{v}_1 \sin \beta}.$$

We integrate once (15.23), we let

$$\Gamma_0 = \bar{v}_1 \bar{v}_2 \sin \beta,$$

and we can see that R satisfies

$$\ddot{R} - \frac{\Gamma_0^2}{R^3} + R = 0. \tag{15.24}$$

Equation (15.24) has the first integral $\dot{R}^2 + R^2 + \Gamma_0^2/R^2$. Thus, R, $1/R$ and $\dot{R}$ remain bounded as long as (15.24) remains valid. Moreover $\dot{\Theta}$ is bounded from below by a strictly positive number. Therefore, there exists a time τ_1 such that $\Theta(\tau_1)$ is equal to $\pi - \bar{\theta}$ for the first time. After time τ_1, U is given by

$$\begin{aligned} U(\tau) = &\left([\dot{U}(\tau_1) \sin(\tau - \tau_1) + U(\tau_1) \cos(\tau - \tau_1)] \cdot \mathbf{i}\right)\mathbf{i} \\ &+ \left([(\tau - \tau_1)\dot{U}(\tau_1) + U(\tau_1)] \cdot \mathbf{j}\right)\mathbf{j}. \end{aligned} \tag{15.25}$$

Of course, $U(\tau_1) \cdot \mathbf{j}$ vanishes and for geometric reasons, $\dot{U}(\tau_1) \cdot \mathbf{j}$ is strictly positive. Therefore, there exists a time τ_2 such that for the first time, the function U given by (15.25) enters K. Elementary arguments show that $U(\tau_2)$, $\dot{U}(\tau_2)$ and τ_2 are continuous functions of β. Let us emphasize the dependence of $\dot{U}(\tau_2)$ by denoting it $\hat{v}(\beta)$.

Morevoer, the energy is conserved: we multiply scalarly (15.20) by $\dot{U}$, we integrate, remembering that

$$\frac{d}{d\tau}\frac{|U - P_K U|^2}{2} = (U - P_K U) \cdot \dot{U},$$

and we find that (15.20) has the first integral $|\dot{U}|^2 + |U - P_K U|^2$. Therefore, $|\dot{U}(\tau_2)| = |\bar{v}|$. A connexity argument enables us to infer that the set $\{\dot{U}(\tau_2) : \beta \in [0, \pi]\}$ contains one of the two arcs which join $\hat{v}(0)$ and $\hat{v}(\pi)$ in the circle of radius $|\bar{v}|$.

Choose now any element in the set $\{\hat{v}(\beta) : \beta \in [0, \pi]\}$, and take initial data as in (15.19). It is immediate that the limit of $u_k(t) = U(t\sqrt{k})/\sqrt{k}$ is equal to $t\hat{v}(\beta)$, which proves our claim.

We can see now that the penalty method will give essentially unpredictable results after two reflexions: small delays can considerably change the outcome of the convergence process, and we consider that this difficulty should be treated in detail, possibly by estimating the measure of the set of initial values for which these phenomena occur in finite time.

We conjecture that in well behaved situation, the set of initial data for which the phenomenon occurs in finite time is negligible in the phase space, in the measure–theoretic sense.

Though a sequence of solutions of penalized problems can have many limit points, it is always possible to extract from it a convergent subsequence whose limit conserves the energy at impact. This fact has been proved in [8].

Inelastic impact in an angle

The second case that we wish to consider is the case $\alpha > 1$, $\theta \in (0, \pi/2) \cup (\pi/2, \pi)$. Let us recall that for inelastic impact in domains with corners, Moreau has proposed the following rule in [6]: if the representative point hits a corner x of K, it stops if the angle is acute; if the angle is obtuse, two cases must be considered: the convex set K is the union of two convex cones K_1 and K_2 with angle $\bar{\theta}/2$. Denote S_j the side of K which is common to K and K_j. If the opposite of the velocity at impact belongs to K_1, then the representative point of the system continues beyond the corner and its new velocity is the projection on S_2 of the old one. The analogous statement holds when K_2 replaces K_1.

In particular, if f vanishes, and if the representative point hits S_1 at time 0, its new velocity after the first impact is the projection of the old velocity on S_1; if the new velocity is oriented toward the vertex of K, then at a later time t_0, the representative point hits the vertex, and its new velocity after impact will be the projection of the velocity before impact on the side S_2.

This phenomenon can be obtained in the limit of a penalty approximation, as we shall see now. The proof is complicated and details are given in [13].

The set K of constraints is the same as above, as well as the coordinates. Now, we consider that all the initial data are fixed and satisfy

$$u_{1,0} = 0, \quad \dot{u}_{1,0} > 0, \quad u_{2,0} < 0, \quad \dot{u}_{2,0} > 0. \tag{15.26}$$

We define

$$\Delta = \alpha^2 - 1, \quad \xi_1 = -\alpha + \sqrt{\Delta}, \quad \xi_2 = -\alpha - \sqrt{\Delta}. \tag{15.27}$$

In the first phase of the motion, i.e. as long as u_1 is strictly positive and u_2 is strictly negative, we have

$$u_1(t) = \frac{\dot{u}_{0,1}}{2\sqrt{\Delta k}} \left(e^{\xi_1 t\sqrt{k}} - e^{\xi_2 t\sqrt{k}} \right), \tag{15.28}$$

$$u_2(t) = u_{0,2} + t\dot{u}_{0,2}. \tag{15.29}$$

A sign analysis shows that for k large enough, u_1 remains positive as long as u_2 remains negative. Moreover, for finite times such that u_2 is negative, the velocity is exponentially close to $(0, \dot{u}_{0,2})^*$ which is the projection of the outgoing velocity on the side $\{0\} \times \mathbb{R}^-$ of the angular domain K.

The second phase of the evolution starts at a time t_0 such that $u_{0,2} + t_0\dot{u}_{0,2} = 0$: at this time u penetrates $K^\perp$, and we shift to polar coordinates. In these coordinates, the following differential equations is satisfied, together with (15.23):

$$\ddot{r} - r\dot{\theta}^2 + 2\alpha\sqrt{k}\dot{r} + kr = 0. \tag{15.30}$$

The initial conditions are

$$\dot{r}(t_0 + 0) = \frac{\dot{u}_{1,0}}{2\sqrt{\Delta}} \left(\xi_1 e^{\xi_1 t_0 \sqrt{k}} - \xi_2 e^{\xi_2 t_0 \sqrt{k}} \right), \tag{15.31}$$

$$\dot{\theta}(t_0 + 0) = \frac{\dot{u}_2(t_0 - 0)}{u_1(t_0 - 0)} = \frac{2\dot{u}_{2,0}\sqrt{\Delta k}}{\dot{u}_{1,0}\left(e^{\xi_1 t_0 \sqrt{k}} - e^{\xi_2 t_0 \sqrt{k}} \right)}. \tag{15.32}$$

Define

$$\Gamma = \frac{\dot{u}_{1,0}\dot{u}_{2,0}}{2\sqrt{\Delta k}} \left(e^{\xi_1 t_0 \sqrt{k}} - e^{\xi_2 t_0 \sqrt{k}} \right). \tag{15.33}$$

An immediate integration of (15.23) yields

$$\ddot{r} - \frac{\Gamma^2}{r^3} + 2\alpha\sqrt{k}\dot{r} + kr = 0. \tag{15.34}$$

We define an exponentially small scale η by

$$\eta = e^{\xi_1 t_0 \sqrt{k}/2}, \tag{15.35}$$

and we perform the following changes of variables:

$$\tau = (t - t_0)\sqrt{k}, \quad r(t) = \eta R(\tau)/\sqrt{k}, \quad \theta(t) = \Theta(\tau). \tag{15.36}$$

In these new variables, (15.34) becomes

$$\ddot{R} - \frac{E(1-\varepsilon)^2}{R^3} + 2\alpha\dot{R} + R = 0, \tag{15.37}$$

where

$$E = \frac{\dot{u}_{1,0}^2 \dot{u}_{2,0}^2}{4\Delta}, \tag{15.38}$$

$$\varepsilon = e^{(\xi_2 - \xi_1) t_0 \sqrt{k}}. \tag{15.39}$$

We are able to prove that the solution of (15.37) is very close to the solution of

$$\ddot{R}_0 - \frac{E}{R_0^3} = 0, \quad R_0(0) = R(0), \quad \dot{R}_0(0) = \dot{R}(0). \tag{15.40}$$

This asymptotic is valid on a time interval $[0, \eta^{\gamma_1}]$ with $\gamma_1 \in (1, 2)$. It is justified by a fixed point theorem based on some very precise estimates.

If $\bar{\theta}$ is strictly larger than $\pi/2$, we show that Θ reaches the value $\pi - \bar{\theta}$ at which the representative point leaves $K^\perp$ at a time which is asymtotic to $t_0 + \Gamma \tan(\pi - \bar{\theta})/\dot{u}_{0,2}^2$.

The outgoing velocity is exponentially close to the projection on the second side of the angle of the velocity before entering $K^\perp$. This justifies Moreau's rule when $\bar{\theta} > \pi/2$.

The case $\bar{\theta} = \pi/2$ is analytically difficult and the reader is referred to [13]. However, if $0 < \bar{\theta} < \pi/2$ we can perform one more asymptotic and a phase portrait analysis. The main ideas are the following: if we choose $\gamma_1 \in (1, 4/3)$ and if we let

$$\tau_3 = \delta \ln(1/\eta) \tag{15.41}$$

where

$$0 < \delta < 1/|\xi_1|, \tag{15.42}$$

we can prove that on the interval $[\tau_1, \tau_3]$, R solution of (15.37) is very close to R_1 which is the solution of the linear differential equation

$$\ddot{R}_1 + 2\alpha \dot{R}_1 + R_1 = 0, \quad R_1(\tau_1) = R(\tau_1), \quad \dot{R}_1(\tau_1) = \dot{R}(\tau_1).$$

The previous step of the analysis had given us the following equivalents for $R(\tau_1)$ and $\dot{R}(\tau_1)$:

$$R(\tau_1) \sim |\dot{u}_{2,0}| \eta^{\gamma_1 - 1}, \quad \dot{R}(\tau_1) \sim |\dot{u}_{2,0}| \eta^{-1}. \tag{15.43}$$

A simple differential inequality shows that for $\tau \geq \tau_1$ and for large enough k we have

$$\dot{R} - \xi_1 R \geq 0. \tag{15.44}$$

On the other hand equation (15.37) can be transformed into a system of two differential equations of first order, which has a single attracting critical point at $(R_c, 0)^*$ in the half plane $\{R > 0\}$.

Thanks to the fact that the nonlinear term $E(1-\varepsilon)^2/R^3$ is negligible for R large, there exists $\tau_4 \in (\tau_3, \tau_3 + O(\ln(1/\eta)))$ such that for all $\tau \geq \tau_4$, $R(\tau)$ remains bounded: this is obtained with the help of a Lyapunov function argument which is valid for large values of R.

Together with (15.44) and a topological argument, we can infer that R is larger than a strictly positive number for $\tau \geq \tau_4$. Thus, we can integrate the angle equation

$$\dot{\Theta} = \frac{\sqrt{E}(1-\varepsilon)}{R^2},$$

and we find that the outgoing velocity is exponentially small; therefore, Moreau's rule is also justified in this case.

15.5 CONCLUSION

We have described several new and old results on the approximation by penalty method and by numerical schemes of dynamical problems with impact. The case when the set of constraints does not have a smooth boundary is still far from being wholly understood: in particular, there are some striking phenomena of ill-posedness: for instance the application of Moreau's rule yields solutions which are discontinuous with respect to the initial data: if the initial velocity is fixed and the initial position varies continuously in such a way that the representative point hits first one side and then the other side of the corner of an angular domain K, the outgoing velocity undergoes a discontinuity. We conjecture that for reasonably well-behaved problems the set of initial data for which these phenomena are encountered is negligible in the measure-theoretic sense, which hopefully renders them harmless in most numerical situations.

References

[1] Ballard, P. (1999). Dynamique des systèmes mécaniques discrets avec liaisons unilatérales parfaites. *C.R. Acad. Sci. Paris, Série IIb*, 327:953-958.

[2] Bressan, A. (1959). Questioni di regolarita e di unicita del moto in presenza di vincoli olonomi unilaterali. *Rend. Sem. Mat. Univ. Padova*, 29:271–315.

[3] Brézis, H. (1973). *Opérateurs maximaux monotones et semi-groupes de contractions dans les espaces de Hilbert.* North-Holland Publishing Co., Amsterdam. North-Holland Mathematics Studies, No. 5. Notas de Matemática (50).

[4] Brogliato, B. (1996). *Nonsmooth impact mechanics*, volume 220 of *Lecture Notes in Control and Information Sciences.* Springer-Verlag London Ltd., London. Models, dynamics and control.

[5] Carriero, M. and Pascali, E. (1982). Uniqueness of the one-dimensional bounce problem as a generic property in $L^1([0, T]; \mathbf{R})$. *Boll. Un. Mat. Ital. A (6)*, 1(1):87–91.

[6] Moreau, J.-J. (1983). Liaisons unilatérales sans frottement et chocs inélastiques. *C. R. Acad. Sci. Paris Sér. II Méc. Phys. Chim. Sci. Univers Sci. Terre*, 296(19):1473–1476.

[7] Panet, M., Paoli, L. and Schatzman, M. (1995). Theoretical and numerical study for a model of vibrations with unilateral constraints. In *Contact*

Mechanics, M. Raous, M.Jean and J.J.Moreau eds, Plenum Press, London. 457–464.

[8] Paoli, L. (1993). *Analyse numérique de vibrations avec contraintes unilatérales.* PhD thesis, Université Claude Bernard – Lyon 1.

[9] Paoli, L. and Schatzman, M. (1993a). Mouvement à un nombre fini de degrés de liberté avec contraintes unilatérales : cas avec perte d'énergie,. *Modél. Math. Anal. Num.(M2AN)*, 27(6):673–717.

[10] Paoli, L. and Schatzman, M. (1993b). Schéma numérique pour un modèle de vibrations avec contraintes unilatérales et perte d'énergie aux impacts, en dimension finie. *C.R. Acad. Sc. Paris Sér. I*, 317:211–215.

[11] Paoli, L. and Schatzman, M. (1999a). Approximation et existence en vibro-impact. *C.R. Acad. Sc. Paris Sér. I*, 329:1103–1105.

[12] Paoli, L. and Schatzman, M. (1999b). A numerical scheme for impact problems. Preprint UMR 5585 MAPLY, Available at http://maply.univ-lyon1.fr/

[13] Paoli, L. and Schatzman, M. (2000). Penalty approximation for non smooth constraints in vibroimpact. Preprint UMR 5585 MAPLY, Available at http://maply.univ-lyon1.fr/

[14] Percivale, D. (1985). Uniqueness in the elastic bounce problem. *J. Differ. Equations*, 56:206–215.

[15] Percivale, D. (1991). Uniqueness in the elastic bounce problem. II. *J. Differ. Equations*, 90:304–315.

[16] Rockafellar, R. T. (1968). Integrals which are convex functionals. *Pacific J. Math.*, 24:525–539.

[17] Rockafellar, R. T. (1971). Integrals which are convex functionals. II. *Pacific J. Math.*, 39:439–469.

[18] Schatzman, M. (1978). A class of nonlinear differential equations of second order in time. *Nonlinear Anal., Theory, Methods and Applications*, 2:355–373.

[19] Schatzman, M. (1981). The isoenergetic change of time for convex irregular hamiltonian systems. Technical report, CEREMADE, Université Paris-Dauphine.

[20] Schatzman, M. (1998). Uniqueness and continuous dependence on data for one dimensional impact problems. *Math. Comput. Modelling*, 28(4-8):1–18.

16 ON GLOBAL PROPERTIES OF D.C.FUNCTIONS

L.N.Polyakova*

Applied Mathematics Dept
St.Petersburg State University
Staryi Peterhof St Petersburg
198904 RUSSIA

Abstract: Necessary and sufficient optimality conditions of the difference of convex functions are derived for unconstrained optimization problems. Connection between extremal properties of the difference of convex functions and the extremal properties of the difference of their conjugate functions is established. Duality theorems are proved. A smooth approximation of a d.c. function is investigated.

Key words: global optimization, convex function, conjugate function, infimal convolution, subdifferential, ε-subdifferential, hypodifferential.

The presentation of the material in the present paper follows the terminology and notation used in [1].

Let $f_1, f_2 : \mathbb{R}^n \to \mathbb{R}$ be continuous convex functions and $f(x) = f_1(x) - f_2(x)$. The function f is the difference of convex functions and is called a d.c. function. The function $f(x)$ is quasidifferentiable and $Df(x) = [\partial f_1(x), -\partial f_2(x)]$ is a quasidifferential of f at a point $x \in \mathbb{R}^n$, where $\partial f_i(x)$, $i = 1, 2$, are the subdifferentials in the sense of Convex Analysis of $f_i(x)$, $i = 1, 2$ at x.

Consider the following optimization problem:

*The research was supported by the Russian Foundation for Fundamental Studies (grant RFFI No. 97-01-00499)

R.P. Gilbert et al. (eds.), From Convexity to Nonconvexity, 223–231.

Find

$$\inf_{x \in \mathbb{R}^n} f(x). \tag{16.1}$$

Theorem 1. [2] *Let a d.c.function f attain its extremal value on $\mathbb{R}^n$ at a point $x^* \in \mathbb{R}^n$. If x^* is a minimum point, then*

$$\partial f_2(x^*) \subset \partial f_1(x^*). \tag{16.2}$$

If x^* is a maximum point, then

$$\partial f_1(x^*) \subset \partial f_2(x^*). \tag{16.3}$$

A point x^* is called an inf-stationary point of f, if condition (16.2) holds. A point x^* is called a sup-stationary point of f, if condition (16.3) holds. Note that conditions (16.2) and (16.3) are only necessary conditions for an extremum of d.c. functions.

J.-B. Hiriart-Urruty [3] was the first to deduce necessary and sufficient conditions for a global minimum of a d.c. function. He used ε - subdifferentials of these convex functions. His conditions will be formulated later on.

In the sequel we shall assume that range $\partial f_1^* =$ dom f_1^*, range $\partial f_2^* =$ dom f_2^*, where f_1^*, f_2^* are the conjugate functions of the convex functions f_1 and f_2, dom f_1^*, dom f_2^* are the effective domains of f_1^*, and f_2^*, respectively.

Lemma 1. *Fix a point $x \in \mathbb{R}^n$, then*

$$f_1(x) - f_2(x) \leq f_2^*(v) - f_1^*(v) \qquad \forall v \in \partial f_1(x), \tag{16.4}$$

$$f_1(x) - f_2(x) \geq f_2^*(v) - f_1^*(v) \qquad \forall v \in \partial f_2(x), \tag{16.5}$$

$$f_1(x) - f_2(x) = f_2^*(v) - f_1^*(v) \qquad \forall v \in \partial f_1(x) \cap \partial f_2(x). \tag{16.6}$$

Lemma 2.

1. *Fix a point $v \in$ dom ∂f_1^*, then*

$$f_2^*(v) - f_1^*(v) \geq f_1(x) - f_2(x) \qquad \forall x \in \partial f_1^*(v). \tag{16.7}$$

2. *Fix a point $v \in$ dom ∂f_2^*, then*

$$f_2^*(v) - f_1^*(v) \leq f_1(x) - f_2(x) \qquad \forall x \in \partial f_2^*(v), \tag{16.8}$$

3. *Fix a point $v \in$ dom $\partial f_1^* \cap$ dom ∂f_2^*, then*

$$f_2^*(v) - f_1^*(v) = f_1(x) - f_2(x) \qquad \forall x \in \partial f_1^*(v) \cap \partial f_2^*(v). \tag{16.9}$$

The proofs of these Lemmas follow from the Young-Fenchel inequality.

A point x^* is called a q - stationary point of the function f on $\mathbb{R}^n$, if

$$\partial f_1(x^*) \cap \partial f_2(x^*) \neq \emptyset.$$

It is clear, that inf $-$ and sup - stationary points of the function f are also q - stationary points of f on $\mathbb{R}^n$.

Denote by

$$f^o(v) = f_2^*(v) - f_1^*(v), \quad v \in \mathbb{R}^n .$$

It is not difficult to observe that if a point $v \notin [\mathrm{dom}\, f_1^* \cup \mathrm{dom}\, f_2^*]$, we collide with the case $+\infty - \infty$. Every such case will be treated separately.

Example 1. Consider the function

$$f(x) = f_1(x) - f_2(x) = |x_1| - |x_2|, \quad x = (x_1, x_2) \in \mathbb{R}^2 .$$

The point $x_0 = (0,0)$ is neither a point of a local minimum of the function f, nor a point of a local maximum of it on $\mathbb{R}^n$. But this point is a q - stationary point of f on $\mathbb{R}^n$. Find the conjugate functions of f_1 and f_2.

$$f_1^*(v) = \begin{cases} 0, & v \in \mathrm{co}\ \{(1,0),(-1,0)\}, \\ +\infty, & v \notin \mathrm{co}\ \{(1,0),(-1,0)\}, \end{cases}$$

$$f_2^*(v) = \begin{cases} 0, & v \in \mathrm{co}\ \{(0,1),(0,-1)\}, \\ +\infty, & v \notin \mathrm{co}\ \{(0,1),(0,-1)\}. \end{cases}$$

Since $\partial f_1(x_0) = \mathrm{co}\ \{(1,0),(-1,0)\}, \quad \partial f_2(x_0) = \mathrm{co}\ \{(0,1),(0,-1)\}$, then

$$\mathrm{co}\ \{(1,0),(-1,0)\} \cap \mathrm{co}\ \{(0,1),(0,-1)\} = (0,0),$$

and $f(x_0) = 0 = f^o(0_2)$.

Example 2. Consider the functions

$$f_1(x) = \begin{cases} x^3, & x \geq 0, \\ 0, & x < 0, \end{cases} \qquad f_2(x) = \begin{cases} 0, & x \geq 0, \\ -x^3, & x < 0, \end{cases} \qquad x \in \mathbb{R}.$$

These functions are continuously differentiable. Then $f(x) = x^3$. There is a unique q-stationary point of f. Namely, $x^* = 0$. At this point we get $f_1'(x^*) = 0, \quad f_2'(x^*) = 0$. Then

$$f_1^*(v) = \begin{cases} \frac{2v\sqrt{v}}{3\sqrt{3}}, & v \geq 0, \\ +\infty, & v < 0, \end{cases}$$

$$f_2^*(v) = \begin{cases} +\infty, & v > 0, \\ -\frac{2v\sqrt{|v|}}{3\sqrt{3}}, & v \leq 0, \end{cases},$$

$$f^o(v) = \begin{cases} +\infty, & v > 0, \\ 0, & v = 0, \\ -\infty, & v < 0. \end{cases}$$

The function f^o is finite only at the point $v^* = 0$. Find the subdifferentials of f_1^* and f_2^* at the point $v^* = 0$:

$$\partial f_1^*(0) = (-\infty, 0] \subset \mathbb{R}, \quad \partial f_2^*(0) = [0, +\infty) \subset \mathbb{R}.$$

It is clear that $\partial f_1^*(0) \cap \partial f_2^*(0) = 0$. Note that if we calculate the conjugate function of f, we obtain $f^*(v) = \infty \quad \forall v \in \mathbb{R}$.

Corollary 1. *It follows from* (16.7) *and* (16.8) *that*
1) *if* dom $f_2^* \not\subset$ dom f_1^*, *then the function* f *is unbounded from below,*
2) *if* dom $f_1^* \not\subset$ dom f_2^*, *then the function* f *is unbounded from above.*

At first consider the minimization problem. In this case it will be assumed that dom $f_2^* \subset$ dom f_1^*. Then dom $f^o =$ dom f_2^*. If a point $v \notin$ dom f_1^*, we have $+\infty - \infty$. Since we consider the problem of minimizing the function f, we define the finction f^o to be equal $+\infty$ outside of the set dom f_1^*. Namely, assume

$$f^o_-(v) = \begin{cases} f^o(v), & v \in \text{dom } f_1^*, \\ +\infty, & v \notin \text{dom } f_1^*. \end{cases}$$

Theorem 2. *Let a point* x^* *be a global minimum point of the function* f *on* $\mathbb{R}^n$, *then any subgradient* $v^* \in \partial f_2(x^*)$ *is a global minimum point of the function* f^o_- *on* $\mathbb{R}^n$.

Proof. Since x^* is a global minimum point of f on $\mathbb{R}^n$, then (see (2))

$$\partial f_2(x^*) \subset \partial f_1(x^*),$$

and

$$f_1(x) - f_2(x) \geq f(x^*) \quad \forall x \in \mathbb{R}^n .$$

It follows from the definition of the conjugate function that

$$f_2^*(v) \geq f_1^*(v) + f(x^*) \quad \forall v \in \mathbb{R}^n, \tag{16.10}$$

or, what is the same $f^o_-(v) \geq f(x^*) \quad \forall v \in \mathbb{R}^n$. (16.10) implies that if $v \in$ dom f_2^*, then $v \in$ dom f_1^*. Therefore, dom $f_2^* \subset$ dom f_1^*. (10) implies also

$$f^o_-(v) \geq f(x^*) = f^o_-(v^*) \quad \forall v \in \mathbb{R}^n, \quad \forall v^* \in \partial f_2(x^*).$$

The theorem is proved.

Theorem 3. *Let a point* x^* *be a* q *- stationary point of the function* f *on* $\mathbb{R}^n$. *If there is a point*

$$v^* \in \partial f_1(x^*) \cap \partial f_2(x^*),$$

which is a global minimum point of the function f^o_- on $\mathbb{R}^n$, *then* x^* *is a global minumum point of the function* f *on* $\mathbb{R}^n$.

Proof. Let x^* be a q -stationary point of f on $\mathbb{R}^n$. Tnen $\partial f_1(x^*) \cap \partial f_2(x^*) \neq \emptyset$. Suppose that $v^* \in \partial f_2(x^*) \cap \partial f_2(x^*)$ is a global minimum point of the function f^o_- on $\mathbb{R}^n$. Then

$$f^o_-(v^*) = f_2^*(v^*) - f_1^*(v^*) \leq f_2^*(v) - f_1^*(v) \quad \forall v \in \mathbb{R}^n .$$

Since $(f^*)^* = f$, then

$$f_2(x) \leq f_1(x) - f^o_-(v^*) \quad \forall x \in \mathbb{R}^n .$$

(16.5) yields

$$f(x) \geq f^o_-(v^*) = f(x^*) \quad \forall x \in \mathbb{R}^n .$$

The theorem is proved.

Corollary 2. *If there exists a global minimum point $x^* \in \mathbb{R}^n$ of the function f on $\mathbb{R}^n$, then there exists a point $v^* \in \mathbb{R}^n$ which is a global minimum point of the function f_-^o on $\mathbb{R}^n$ and the following equality holds*

$$\min_{x\in\mathbb{R}^n} f(x) = \min_{v\in\mathbb{R}^n} f_-^o(v).$$

Corollary 3. *If the function f attains its global minimum on $\mathbb{R}^n$, then*

$$\text{dom } f_2^* \subset \text{dom } f_1^*.$$

Let us illustrate these conditions for some well-known problems.

Example 3. Let the function f_2 be equal to a constant value a on $\mathbb{R}^n$. Then

$$f_2^*(v) = \begin{cases} -a, & \text{if } v = 0_n, \\ +\infty, & \text{otherwise.} \end{cases}$$

Suppose that $0_n \in \text{dom } f_1^*$. Then there is a global minimum point of the function f on $\mathbb{R}^n$. Put

$$f_-^o(v) = \begin{cases} -a - f_1^*(v), & \text{if } v = 0_n, \\ +\infty, & \text{otherwise.} \end{cases}$$

Then the function f_-^o can attain its minimum only at the point $v^* = 0_n$. Therefore a global minimum point of f may be the point x^* for which $0_n \in \partial f_1(x^*)$.

Example 4. Let

$$\begin{aligned} f_1(x) &= \frac{1}{2}\langle x, x\rangle, \\ f_2(x) &= \max_{i\in I}\{\langle a_i, x\rangle - b_i\}, \quad x, a_i \in \mathbb{R}^n, \quad b_i \in \mathbb{R}, \quad I = \{1\ldots m\}. \end{aligned}$$

Consider the following optimization problem:

Find

$$\inf_{x\in\mathbb{R}^n} f(x). \tag{16.11}$$

A global minimum point on $\mathbb{R}^n$ for the given function exists. We can find the global minimum points of the strongly convex functions

$$\psi_i(x) = 1/2\langle x, x\rangle - \langle a_i, x\rangle + b_i, \quad i \in I,$$

and choose points where the value of the functions $\psi_i, i \in I$ is the smallest.

As the function f_2 is polyhedral then the formula for its conjugate function is known

$$f_2^*(v) = \inf\{\lambda_1 b_1 + \cdots + \lambda_m b_m\},$$

where

$$v = \lambda_1 a_1 + \cdots + \lambda_m a_m, \ \lambda_1 + \cdots + \lambda_m = 1, \quad \lambda_i \geq 0 \ \ \forall i \in I.$$

$$f_2^*(v) = \begin{cases} \inf\{\mu \in \mathbb{R}^1 \mid [v, \mu] \in F^* \subset \mathbb{R}^n \times \mathbb{R}^1\}, & v \in \text{ dom } f_2^*, \\ +\infty, & v \notin \text{ dom } f_2^*, \end{cases} \tag{16.12}$$

where

$$F^* = \text{co}\left\{\bigcup_{i\in I}[a_i, b_i]\right\} \subset \mathbb{R}^n \times \mathbb{R}^1 .$$

The infimum in (16.12) is attained. Since in this case dom $f_1^* = \mathbb{R}^n$, then

$$f_-^0(v) = f^o(v) = \begin{cases} f_2^*(v) - 1/2\langle x, x\rangle, & v \in \text{dom } f_2^*, \\ +\infty, & v \notin \text{dom } f_2^*. \end{cases}$$

Find a global minimum point of the function f_-^o on $\mathbb{R}^n$. Let $v \in$ dom f_2^*. Since dom $f_2^* =$ co $\left\{\bigcup_{i\in I} a_i\right\}$, then

$$\begin{aligned} v &= \lambda_1 a_1 + \cdots + \lambda_m a_m, \\ &\lambda \in \Lambda = \left\{\lambda = (\lambda_i, \ldots, \lambda_m) \in \mathbb{R}^m \;\middle|\; \begin{array}{r} \lambda_1 + \cdots + \lambda_m = 1, \\ \lambda_i \geq 0 \quad \forall i \in I \end{array}\right\}. \end{aligned}$$

Put

$$\psi^o(\lambda) = \langle \lambda, b\rangle - 1/2\langle G\lambda, \lambda\rangle,$$

where G is the Grama matrix of the vectors $\{a_i\}$, $i \in I$, $b = (b_1, \ldots, b_m) \in \mathbb{R}^m$. To find the global minimum of the function f^o on $\mathbb{R}^n$ it is necessary to solve the following constrained optimization problem :

$$\min_{\lambda\in\Lambda} \psi^o(\lambda).$$

A concave function attains its minimum at a vertex of the polyhedral Λ, that is at the points $\lambda^1 = (1, 0, \ldots, 0)$, $\lambda^2 = (0, 1, \ldots, 0)$, $\ldots$, $\lambda^m = (0, 0, \ldots, 1) \in \mathbb{R}^m$. Therefore the points which can be global minimum points of the function f^0 are the points a_i, $i \in I$. Let a point a_{i^*} be a global minimum point of the function f^0 on $\mathbb{R}^n$. Then $x^* \in \partial f_1^*(a_{i^*})$ is a global minimum point of the function f and $\partial f_1^*(a_{i^*}) = a_{i^*}$.

Consider the function

$$f(x) = f_1(x) - f_2(x) = 0.5x^2 - |x - 0.5|, \quad x \in \mathbb{R}$$

There are three q-stationary points of f: $x_1 = -1$, $x_2 = 0.5$, $x_3 = 1$. The point x_1 is a global minimum point, the point x_2 is a local maximum point, the point x_3 is a local minimum point. In the given example

$$\mathbb{R} = \text{dom } f_1^* \not\subset \text{dom } f_2^* = [-1, 1].$$

Therefore the function f is unbounded from above. Since

$$f_1^*(v) = 0.5v^2, \quad f_2^*(v) = \begin{cases} 0.5v, & v \in [-1, 1], \\ +\infty, & v \notin [-1, 1]. \end{cases}$$

then

$$f^o(v) = \begin{cases} 0.5v(1 - v), & v \in [-1, 1], \\ +\infty, & v \notin [-1, 1]. \end{cases}$$

The function f^o has also three q-stationary points: $v_1 = -1$, $v_2 = 0.5$, $v_3 = 1$. The point v_1 is a global minimum point, the point v_2 is a local maximum point, the point v_3 is a local minimum point of the function f^o on $\mathbb{R}^1$. We have

$$v_1 = f_1'(x_1) = f_2'(x_1),\ v_2 = f_1'(x_2) \subset \partial f_2(x_2),\ v_3 = f_1'(x_3) = f_2'(x_3).$$

Consider the problem of maximizing the function f on $\mathbb{R}^n$. Suppose that dom $f_1^* \subset$ dom f_2^*, and define the finction f^o to be equal $+\infty$ outside of the set dom f_2^*. Namely, assume

$$f_+^o(v) = \begin{cases} f^o(v), & v \in \text{dom } f_2^*, \\ -\infty, & v \notin \text{dom } f_2^*. \end{cases}$$

Analogously we can prove

Theorem 4. *Let a point x^* be a global maximum point of the function f on $\mathbb{R}^n$, then any subgradient $v^* \in \partial f_1(x^*)$ is a global maximum point of the function f_+^o* on $\mathbb{R}^n$.

Theorem 5. *Let a point x^* be a q - stationary point of the function f on $\mathbb{R}^n$. Then if there exists a point $v^* \in \partial f_1(x^*) \cap \partial f_2(x^*)$, which is a global maximum point of the function f_+^o on $\mathbb{R}^n$, then x^* is a global maximum point of the function f on $\mathbb{R}^n$.*

Corollary 4. *If there exists a global maximum point $x^* \in \mathbb{R}^n$ of the function f on $\mathbb{R}^n$, then there exists a point $v^* \in \mathbb{R}^n$ which is a global maximum point of the function f_+^o on $\mathbb{R}^n$ and the following equality holds*

$$\max_{x \in \mathbb{R}^n} f(x) = \max_{v \in \mathbb{R}^n} f_+^o(v).$$

Corollary 5. *If the function f attains its global minimum on $\mathbb{R}^n$, then*

$$\text{dom } f_1^* \subset \text{dom } f_2^*.$$

Theorem 6. *The following equalities hold*

$$\inf_{x \in \mathbb{R}^n} f(x) = \inf_{v \in \text{dom } f_2^*} f^o(v), \quad \sup_{x \in \mathbb{R}^n} f(x) = \sup_{v \in \text{dom } f_1^*} f^o(v).$$

Example 5. Let $f_1(x) = e^x$, $f_2(x) = 0$, $x \in \mathbb{R}$. Then

$$f_1^*(v) = \begin{cases} v \ln v - v, & v > 0 \\ 0, & v = 0, \\ +\infty, & v < 0, \end{cases}$$

$$f_2^*(v) = \begin{cases} 0, & v = 0, \\ +\infty, & v \neq 0. \end{cases}$$

$$f_-^o(v) = \begin{cases} 0, & v = 0, \\ +\infty, & v \neq 0. \end{cases}$$

It is not difficult to see that dom $f_2^* = 0$, $\inf_{v \in \text{dom } f_2^*} f_-^o(v) = f_-^o(0) = 0$, and

$$\inf_{x \in \mathbb{R}^1} f(x) = 0 = \min_{v \in \text{dom } f_2^*} f^o(v).$$

Now we formulate Hiriart-Urruty's theorem and present another proof of it.

Theorem 7. *A point x^* is a global minimum of the function f, if and only if*

$$\partial_\varepsilon f_2(x^*) \subset \partial_\varepsilon f_1(x^*) \quad \forall \varepsilon \geq 0, \tag{16.13}$$

where $\partial_\varepsilon f_i(x^*)$ are ε-subdifferentials of f_i, $i = 1, 2$, at the point x^*.

Proof. *Necessity.*

Let x^* be a global minimum point of the function f on $\mathbb{R}^n$. Fix any $\varepsilon \geq 0$. Then for any $v \in \partial_\varepsilon f_2(x^*)$ the following inequality holds

$$f_2(x^*) + f_2^*(v) - \langle x^*, v\rangle - \varepsilon \leq 0.$$

Making use of (16.6) and (16.10), we have

$$f_1(x^*) + f_1^*(v) - \langle x^*, v\rangle - \varepsilon \leq 0.$$

Therefore $v \in \partial_\varepsilon f_1(x^*)$. Since $v \in \partial_\varepsilon f_2(x^*)$ is arbitrary then

$$\partial_\varepsilon f_2(x^*) \subset \partial_\varepsilon f_1(x^*) \quad \forall \varepsilon \geq 0.$$

Sufficiency. Let condition (16.13) hold at the point x^*. Take any $v \in \operatorname{dom} f_2^*$. Put

$$\varepsilon(v) = f_2(x^*) + f_2^*(v) - \langle x^*, v\rangle.$$

From the Young-Fenchel inequality we have $\varepsilon(v) \geq 0$. Therefore $v \in \partial_{\varepsilon(v)} f_2(x^*)$. Then $v \in \partial_{\varepsilon(v)} f_1(x^*)$. In this case $\operatorname{dom} f_2^* \subset \operatorname{dom} f_1^*$. This relation yields

$$\begin{aligned} 0 &\geq f_1(x^*) + f_1^*(v) - \langle x^*, v\rangle - \varepsilon(v) \\ &= f_1(x^*) + f_1^*(v) - \langle x^*, v\rangle - f_2(x^*) - f_2^*(v) + \langle x^*, v\rangle \\ &= f(x^*) + f_1^*(v) - f_2^*(v). \end{aligned}$$

Therefore

$$-f_1^*(v) \leq -f_2^*(v) + f(x^*) \quad \forall v \in \operatorname{dom} f_2^*.$$

Then for each $x \in \mathbb{R}^n$

$$\begin{aligned} f_1(x) &= \sup_{v \in \mathbb{R}^n} \{\langle x, v\rangle - f_1^*(v)\} \\ &= \sup_{v \in \operatorname{dom} f_1^*} \{\langle x, v\rangle - f_1^*(v)\} \\ &\geq \sup_{v \in \operatorname{dom} f_2^*} \{\langle x, v\rangle - f_1^*(v)\} \\ &\geq \sup_{v \in \operatorname{dom} f_2^*} \{\langle x, v\rangle - f_2^*(v)\} + f(x^*) \\ &= f_2(x) + f(x^*). \end{aligned}$$

Then

$$f(x) \geq f(x^*) \quad \forall x \in \mathbb{R}^n.$$

The theorem is proved.

References

[1] Rockafellar R.T.(1970). Convex Analysis. Princeton University Press.

[2] Polyakova L.N.(1980). Necessary conditions for an extremum of quasidifferentiable functions, Vestnik of Leningrad Univ., 13, 57-62.

[3] Hiriart-Urruty J.-B. (1985). Generalised differentiability, duality and optimization for problems dealing with differences of convex functions. In: Convexity and duality in Optimizaion, Ed. J. Ponstein, 37-50, Lecture Notes in Economics and Mathematical Systems. Vol.256, Springer.

17 EXISTENCE RESULTS FOR VARIATIONAL-HEMIVARIATIONAL INEQUALITIES: A F.E.BROWDER TECHNIQUE

G.Dinca, G.Pop

Faculty of Mathematics
Bucharest University
Academiei 14, 70109
Bucharest, Romania

Abstract: Existence results for variational-hemivariational inequalities are presented as consequences of a unitary proving method based on a regularization method and on a F.E.Browder technique.

17.1 INTRODUCTION

In this paper, some existence results for variational-hemivariational inequalities (already obtained in [2], [3], [4] and [12]) are presented as consequences of a unitary proving method. The problems studied in the quoted papers have the following general form:

R.P. Gilbert et al. (eds.), From Convexity to Nonconvexity, 233–241.

Find $u \in \mathcal{D}(\Phi)$ such that, for every v in $V \cap L^\infty(\Omega; I\!R^N)$, one has

$$\begin{aligned}(Au, v-u) &+ \int_\Omega j^o(u(x); v(x) - u(x))dx + \Phi(v) - \Phi(u) \\ &\geq (f, v-u), \qquad (17.1)\end{aligned}$$

where $j^o(z;\cdot)$ stands for the Clarke derivative of a locally Lipschitz function j at z.

For j a Lipschitz function on $I\!R$, $N = 1$ and Φ the indicator function of a convex, closed, nonempty set $K \subset V$, in the papers [2], [3] and [4] one proves that the previous inequality is satisfied for every $v \in V$.

In the proof of the above mentioned results, the problem (P) is approximated with a regularized problem (P_ε), namely with the problem obtained from (P) by replacing the term $\int_\Omega j^o(u(x); v(x)-u(x))dx$ by $\int_\Omega j'_\varepsilon(u(x))\cdot(v(x)-u(x))dx$. Here, and in the sequel, $j_\varepsilon : I\!R^N \to I\!R$ is defined by

$$j_\varepsilon(x) = \int_{I\!R^N} j(x - \varepsilon y)p(y)dy,$$

where $p \in C_o^\infty(B_{I\!R^N}(0,1))$ is a positive function such that $\int_{I\!R^N} p(x)dx = 1$.

In order to prove the existence result for the problem (P_ε), a proof inspired by that of the Browder surjectivity theorem for monotone, hemicontinuous and coercive operators has been proposed in [2]. This idea is also used in [3], [4] and [12]. The method which will be described below is different from the method used in [11], where a Galerkin basis for $V \cap L^\infty(\Omega; I\!R^N)$ is used, and from the method used in [6], where nonsmooth variants for theorems coming from the critical point theory are used.

An important role in the proposed proof is played by the following "a priori " boundedness on the regularized problems which appear in the approximation of the problem (P):

(B) There are $v_o \in \mathcal{D}(\Phi)$ and $R > 0$, not depending on $\varepsilon \in (0,1)$ such that:

if $u \in \mathcal{D}(\Phi)$ satisfies the inequality

$$\begin{aligned}(Au, v_o - u) &+ \int_\Omega j'_\varepsilon(u(x)) \cdot (v_o(x) - u(x))dx + \Phi(v_o) - \Phi(u) \\ &\geq (f, v_o - u),\end{aligned}$$

then $||u||_V \leq R$.

The proof of the existence result for the regularized problem (P_ε) will be divided in the next main steps:

1. Let $\mathcal{F} = \{F : F$ finite dimensional subspace of V and $v_o \in F\}$. For every $F \in \mathcal{F}$ and $\varepsilon \in (0,1)$ one proves the existence result for the problem (P_ε) with Φ replaced by $\Phi|_F$. In the followings one denotes this problem by $(\mathrm{P}_{\varepsilon,F})$.

2. For every $F \in \mathcal{F}$ and $\varepsilon \in (0,1)$ one considers $W_{\varepsilon,F}$, the weakly closure in V of the set

$$S_{\varepsilon,F} = \bigcup_{G \supset F, G \in \mathcal{F}} \{u : u \text{ solution for } (\mathrm{P}_{\varepsilon,G})\}.$$

One remarks that all the elements of the family $\{W_{\varepsilon,F} : F \in \mathcal{F}\}$ are contained in the weakly compact set $\overline{B_V(0,R)}$. As the considered family has the finite intersection property, it follows that its intersection is not void. One proves that every element of $\bigcap_{F \in \mathcal{F}} W_{\varepsilon,F}$ is a solution for (P_ε).

3. One passes to the limit with $\varepsilon \to 0_+$ in the regularized problem (P_ε).

In the followings we will indicate how the above sketched methodology, essentially based on an "a priori" estimate, is acting for the particular situations considered in the papers [2], [3], [4] and [12].

17.2 HYPOTHESES

(H1) The reflexive space V is compactly imbedded in $L^p(\Omega; \mathbb{R}^N)$, where Ω is an open, bounded subset of $\mathbb{R}^m$.

(H2) $A : V \to V^*$ is weakly continuous and $u \mapsto (Au, u)$ is weakly lower semicontinuous on V.

(H3) $j : \mathbb{R}^N \to \mathbb{R}$ is a locally Lipschitz function such that

$$j^o(s; t - s) \leq \alpha(|t|)(1 + |s|^\sigma) \text{ for every } t,\ s \in \mathbb{R}^N,$$

where $\alpha : [0, \infty) \to [0, \infty)$ is a nondecreasing function and $\sigma \in [1, p)$.

(H4) $f \in V^*$ and $\Phi : V \to \mathbb{R} \cup \{\infty\}$ is a convex, lower semicontinuous, proper function.

Remark: If hypothesis (H_3) is replaced by a stronger hypothesis, namely:

(H_3') j is a Lipschitz function

then the inequality from the problem (P) will be satisfied for every $v \in V$.

17.3 THE PROBLEM ($\mathrm{P}_{\varepsilon,F}$)

In the paper [12] one uses a variant of the Ky-Fan's theorem ([1]). Let F_1 be $F \cap L^\infty(\Omega; \mathbb{R}^N)$. The function j_ε' is locally Lipschitz and $\|\|_V$, $\|\|_\infty$ are equivalent norms on the finite dimensional space F_1. Thus the operator $T : F_1 \to F_1^*$,

$$(Tu, v) = (Au, v) + \int_\Omega j_\varepsilon'(u(x)) \cdot v(x)dx - (f, v),$$

is continuous. One considers the multivalued application $\psi : \mathcal{D}(\Phi) \cap F_1 \to \mathcal{P}(F_1)$ defined by $\psi(v) = \{u \in F_1 : (Tu, v-u)+\Phi(v)-\Phi(u) \geq 0\}$. Let us assume that $v_o = 0$. (The element v_o has appeared in the "a priori " boundedness (B).) It is easy to prove that:

1) $\psi(v)$ is a closed set which contains v, $\forall v \in \mathcal{D}(\Phi) \cap F_1$,

2) $\psi(0)$ is a bounded set,

3) $\text{co}\{v_1, ..., v_n\} \subset \bigcup_{i=1}^{n} \psi(v_i)$ for every finite subset $\{v_1, ..., v_n\}$ of $\mathcal{D}(\Phi) \cap F_1$.

It follows that $\bigcap_{v \in \mathcal{D}(\Phi)\cap F_1} \psi(v)$ is nonempty, and each one of its elements is a solution for the problem $(\mathrm{P}_{\varepsilon,F})$.

In the papers [2], [3], [4] j satisfies (H_3'), Φ is the indicator function of the convex, closed, nonempty set K. In these hypotheses, for the proof of the existence result one may use Theorem 3.1 from [7]. As j is a Lipschitz function, the operator T is continuous from F to F^*. Thus for every positive integer n there is an $u_n \in K \cap F \cap \overline{B_V(0, n)}$ such that

$$(Tu_n, v - u_n) \geq 0, \ \ \forall v \in K \cap F \cap \overline{B_V(0, n)}.$$

For $v = v_o$, from the "a priori " boundedness (with arbitrary v_o), it follows that $(u_n)_n \subset K \cap \overline{B_V(0, R)}$. One can assume that $u_n \to u$ in V. Using the continuity of T, it easily derives that $u \in K \cap F$ satisfies, for every $v \in K$, the inequality:

$$(Au, v - u) + \int_\Omega j_\varepsilon'(u(x)) \cdot (v(x) - u(x))dx \geq (f, v - u).$$

Remark: As the previous proof shows, the hypothesis $N = 1$ doesn't infer in the arguments. Therefore, by using the method described in the introduction, the existence result for any N can be obtained.

By using different methods, similar results are obtained in [5], [9].

17.4 THE PROBLEM (P_ε)

Let us prove that every u_ε from $\bigcap_{F \in \mathcal{F}} W_{\varepsilon,F}$ is a solution for (P_ε). Let v be in $V \cap L^\infty(\Omega; I\!R^N)$ and let F be the subspace of V generated by v and v_o. Then $u_\varepsilon \in W_{\varepsilon,F}$. Therefore, there exist $(F_n)_n$, $(u_n)_n$ such that $u_n \to u$ weakly in$V$ and

$$F_n \supset F, \ F_n \in \mathcal{F}, \ u_n \text{ is a solution for} (\mathrm{P}_{\varepsilon, F_n}), \ \forall n.$$

We may assume that $u_n \to u$ in $L^p(\Omega; \mathbb{R}^N)$ and $u_n \to u$ a.e. on Ω. For every integer n,

$$(Au_n, v - u_n) + \int_\Omega j'_\varepsilon(u_n(x)) \cdot (v(x) - u_n(x))dx + \Phi(v) - \Phi(u_n)$$
$$\geq \; (f, v - u_n). \tag{17.2}$$

Lemma 1, [12] states that a.e. on Ω:

$$j'_\varepsilon(u_n(x)) \cdot (v(x) - u_n(x)) \leq \alpha(\|v\|_{L^\infty(\Omega;\mathbb{R}^N)} + \varepsilon)[1 + (|u_n(x)| + \varepsilon)^\sigma].$$

Passing to the upper limit in the inequality (17.2), one obtains that u_ε satisfies (17.1). For $v = 0$, one derives

$$\Phi(u) - \alpha(\varepsilon) \int_\Omega [1 + (|u(x)| + \varepsilon)^\sigma]dx \leq (f, u) + (Au, u).$$

Thus, $u \in \mathcal{D}(\Phi)$.

If j is a Lipschitz function then

$$j'_\varepsilon(u_n(x)) \cdot (v(x) - u_n(x)) \leq L(|v(x)| + |u_n(x)|).$$

As $u_n \to u$ in $L^1(\Omega; \mathbb{R}^N)$ too, using the same arguments as before it follows that u_ε satisfies (17.1) for every $v \in V$.

17.5 THE PROBLEM (P)

Let $(\varepsilon_n)_n \subset (0,1)$ be such that $\varepsilon_n \to 0_+$. Let $u_n \in \bigcap_{F \in \mathcal{F}} W_{\varepsilon_n, F}$. As $(u_n)_n \subset \overline{B_V(0, R)}$ we may assume, for some $u \in V$, that

$$u_n \to u \text{ weakly in } V,\; u_n \to u \text{ in } L^p(\Omega; \mathbb{R}^N),\; u_n \to u \text{ a.e.on } \Omega.$$

For every n and for every $v \in V \cap L^\infty(\Omega; \mathbb{R}^N)$ we have

$$(Au_n, v - u_n) + \int_\Omega j'_{\varepsilon_n}(u_n(x)) \cdot (v(x) - u_n(x))dx + \Phi(v) - \Phi(u_n)$$
$$\geq \; (f, v - u_n).$$

It is obvious that

$$\begin{aligned} & (Au, v - u) + \Phi(v) - \Phi(u) - (f, v - u) \\ \geq \; & \limsup_n \left[(Au_n, v - u_n) + \Phi(v) - \Phi(u_n) - (f, v - u_n)\right] \\ \geq \; & -\limsup_n \int_\Omega j'_{\varepsilon_n}(u_n(x)) \cdot (v(x) - u_n(x))dx. \end{aligned}$$

Lemma 1, [12] states that a.e. on Ω

$$j'_{\varepsilon_n}(u_n(x)) \cdot (v(x) - u_n(x)) \leq \alpha(\|v\|_{L^\infty(\Omega;\mathbb{R}^N)} + 1)[1 + (|u_n(x)| + \varepsilon_n)^\sigma]$$

and

$$\limsup_n j'_{\varepsilon_n}(u_n(x)) \cdot (v(x) - u_n(x)) \leq j^o(u(x); (v(x) - u(x)).$$

Then by the Fatou's lemma it follows that u satisfies the inequality (17.1) for every $v \in V \cap L^\infty(\Omega; \mathbb{R}^N)$. As in the previous step it follows that $u \in \mathcal{D}(\Phi)$. Also, if j is a Lipschitz function on $\mathbb{R}^N$, (17.1) is verified for every $v \in V$.

17.6 NECESSARY CONDITIONS FOR THE "A PRIORI" BOUNDEDNESS

Proposition 1 (the coercive case)

Suppose that hypotheses (H_1)-(H_4) hold. If, in addition, there exists a function $c : [0, \infty) \to \mathbb{R}$ such that $\lim_{r\to\infty} c(r)r^{1-\sigma} = \infty$, and

$$(Au, u) \geq c(\|u\|_V)\,\|u\|_V\,, \forall u \in V,$$

then the "a priori" boundedness (B) is satisfied with $v_o = 0$.

Proof

The following inequalities hold:

$$\begin{aligned}
&(Au, u) + \int_\Omega j'_\varepsilon(u(x)) \cdot u(x)dx + \Phi(u) - (f, u) \\
\geq\;& c(\|u\|_V)\,\|u\|_V - \alpha(\varepsilon) \int_\Omega [1 + (|u(x)| + \varepsilon)^\sigma]dx - \|l\|_{V^*} \cdot \|u\|_V \\
& -a - \|f\|_{V^*} \cdot \|u\|_V \geq c(\| u \|_V) \| u \|_V - c_1 - c_2 \| u \|_V^\sigma - c_3 \| u \|_V \\
=\;& \| u \|_V^\sigma\, [c(\| u \|_V) \| u \|_V^{1-\sigma} - c_2 - c_1 \| u \|_V^{-\sigma} - c_3 \| u \|_V^{1-\sigma}].
\end{aligned}$$

(In the previous inequalities, $l \in V^*$, $a \in \mathbb{R}$ are chosen such that $\Phi(u) \geq (l, u) - a$, $\forall u \in V$.)

The last term of the previous inequality doesn't depend on $\varepsilon \in (0, 1)$, and tends to ∞ for $\|u\|_V \to \infty$. Therefore a constant R, independent on $\varepsilon \in (0, 1)$, exists such that

$$(Au, u) + \int_\Omega j'_\varepsilon(u(x)) \cdot u(x)dx + \Phi(u) - (f, u) \geq 0, \;\; \forall u \text{ with } \|u\|_V \geq R.$$

Proposition 2 (the semicoercive case)

Let Ω be an open, bounded subset of $\mathbb{R}^m$ and V a Hilbert space compactly imbedded in $L^2(\Omega; \mathbb{R})$. Let $A \in \mathcal{L}(V, V^*)$ be a positive operator and let $c > 0$ be such that $(Au, u) \geq c\|u\|_V^2$, for every $v \in [\ker a]^\perp$. Let j be a Lipschitz function on $\mathbb{R}$ such that

$$j'(-\infty) \stackrel{def}{=} \lim_{n\to\infty} \operatorname*{ess\,sup}_{(-\infty,-n)} j' < 0 < \lim_{n\to\infty} \operatorname*{ess\,inf}_{(n,\infty)} j'(s) \stackrel{def}{=} j'(\infty).$$

Let Φ be the indicator function of a convex, closed, nonempty subset K of V.

If $\ker a$ is a finite dimensional space, and for every $\theta \in \ker a \setminus \{0\}$ one has

$$\int_{\Omega} \left[j'(\infty)\theta^{+}(x) - j'(-\infty)\theta^{-}(x) \right] dx > (l, \theta),$$

then the "a priori" boundedness (B) is satisfied with $v_o = 0$.

(In the previous statement $\ker a = \{u \in V : (Au, u) = 0\}$, $\theta^{+} = \max\{\theta, 0\}$ and $\theta^{-} = \max\{-\theta, 0\}$.)

Proof

For the proof (see [2], [10] or [11]) one uses a similar method with that used in [8] for $\dim \ker a = 1$.

Proposition 3 (the non-coercive case)

Let Ω be an open, bounded subset of $\mathbb{R}^m$ and V a Hilbert space compactly imbedded in $L^2(\Omega; \mathbb{R})$. Let $A \in \mathcal{L}(V, V^*)$ be such that the application $u \mapsto (Au, u)$ is weakly lower semicontinuous on V. Let K be a closed, convex, nonempty subset of V. Let j be a Lipschitz function on $\mathbb{R}$ such that, for some $s_o \in \mathbb{R}$,

$$\operatorname*{ess\,sup}_{(-\infty, -s_o)} j' \leq 0 \leq \operatorname*{ess\,inf}_{(s_o, \infty)} j'. \tag{17.3}$$

Let Φ be the indicator function of a convex, closed, nonempty subset K of V.

If A is positive plus on K, and every sequence $(x_n)_n \subset V$ such that $\liminf_n (Ax_n, x_n) = 0$ and $\|x_n\|_V = 1, \forall n$ admits a subsequence weakly convergent in V to $x \neq 0$, then for every $f \in A(K) - \overbrace{K^{+}(A, K)}^{o}$ the "a priori" boundedness assumption is satisfied. Here

$$\begin{aligned} K^{\infty} &= \bigcap_{\alpha > 0} \alpha(K - x_o) \text{ with } x_o \in K, \\ N(A, K) &= K^{\infty} \cap \ker(A + A^*) = \{x \in K^{\infty} : (Ax, x) = 0\}, \\ K^{+}(A, K) &= \{g \in V^* : (g, y) \geq 0, \forall y \in N(A, K)\}, \end{aligned}$$

where A^* is the adjoint of A.

Definition

$A \in \mathcal{L}(V, V^*)$ is **positive plus** on K if the following conditions are satisfied:

1. $(Ax, x) \geq 0$, $\forall x \in K \cup K^{\infty}$;
2. if $x \in K^{\infty}$ and $(Ax, x) = 0$, then $x \in \ker(A + A^*)$.

Proof

As $f \in A(K) - \overbrace{K^{+}(A, K)}^{o}$, there is a $v_o \in K$ such that

$$(Av_o - f, y) > 0, \quad \forall y \in N(A, K) \setminus \{0\}.$$

From the proof of Lemma 2.1 , [10] it follows that, if j satisfies (17.3) and j is Lipschitz with a Lipschitz constant L, then there is a constant $c > 0$ such that $j'_\varepsilon(y)(z-y) \leq L|z| + c$, $\forall z, y \in I\!R$ and $\forall \varepsilon \in (0,1)$. We obtain that, for every $u \in V$ and for every $\varepsilon \in (0,1)$:

$$\begin{aligned} & (Au, v_o - u) + \int_\Omega j'_\varepsilon(u(x))(v_o(x) - u(x))dx - (f, v_o - u) \\ \leq \; & (Au, v_o - u) + L\,\|v_o\|_{L^1(\Omega;I\!R)} + c\,\mathrm{meas}(\Omega) - (f, v_o - u) \\ = \; & (Au, v_o - u) + c_1 - (f, v_o - u)\,. \end{aligned}$$

We will prove that there is a $R > 0$ such that if $u \in K$ verifies

$$(Au, v_o - u) + c_1 - (f, v_o - u) \geq 0,$$

then $\|u\|_V \leq R$. If else, there is a sequence $(u_n)_n \subset K$ such that $\|u_n\|_V \to \infty$ and for every n

$$(Au_n, v_o - u_n) + c_1 - (f, v_o - u_n) \geq 0. \tag{17.4}$$

Let $x_n = \dfrac{1}{\|u_n\|_V} u_n$. One can assume that $x_n \to \widetilde{x}$ weakly in V.

Let us show that $\widetilde{x} \in K^\infty$, i.e. $v_o + \alpha\widetilde{x} \in K$ for every $\alpha > 0$. (In the definition of K^∞, one can use any element of K.) Let $\alpha > 0$ be fixed. Due to the convexity of K, for n sufficiently large,

$$\left(1 - \frac{\alpha}{\|u_n\|_V}\right) v_o + \alpha x_n = \left(1 - \frac{\alpha}{\|u_n\|_V}\right) v_o + \frac{\alpha}{\|u_n\|_V} u_n \in K.$$

Passing to the weak limit, we get that $v_o + \alpha\widetilde{x} \in K$.

Let us prove that $(A\widetilde{x}, \widetilde{x}) \leq 0$. Indeed, if one divides by $\|u_n\|_V^2$ in (17.4), then

$$(Ax_n, \frac{v_o}{\|u_n\|_V} - x_n) + \frac{c_1}{\|u_n\|_V^2} - \frac{(f, v_o)}{\|u_n\|_V^2} + \frac{(f, x_n)}{\|u_n\|_V} \geq 0.$$

Passing to the lower limit in the previous inequality, one gets

$$-(A\widetilde{x}, \widetilde{x}) \geq -\liminf_n (Ax_n, x_n) \geq -\limsup_n (Ax_n, x_n) \geq 0.$$

As A is positive plus on K and $\widetilde{x} \in K^\infty$, it follows $(A\widetilde{x}, \widetilde{x}) = 0$ and $\liminf_n (Ax_n, x_n) = 0$. Therefore, $\widetilde{x} \neq 0$ and $\widetilde{x} \in N(A, K)$.

If one proves that $(Av_o - f, \widetilde{x}) \leq 0$, a contradiction will be derived.

As T is positive plus on K, from (17.4) one follows that

$$(Au_n, v_o) + c_1 - (f, v_o - u_n) \geq 0. \tag{17.5}$$

Dividing (17.5) by $\|u_n\|_V$ and passing to the limit, one obtains that $(A\widetilde{x}, v_o) + (f, \widetilde{x}) \geq 0$. As $\widetilde{x} \in N(A, K)$, the proof ends here.

References

[1] C.Baiocchi, A.Capelo, *Variational and Quasivariational Inequalities. Applications to free-boundary problems,* John Wiley, 1984;

[2] G.Dinca, P.D.Panagiotopoulos, G.Pop, *Coercive and semicoercive hemivariational inequalities on convex sets,* Vestnik RUDN 2 (1995), 96-110;

[3] G.Dinca, P.D.Panagiotopoulos, G.Pop, *Inegalités hémivariationnelles semi-coercive sur des ensembles convexes,* C.R.Acad.Sci.Paris, t.320, Serié I (1995), 1183-1186;

[4] G.Dinca, P.D.Panagiotopoulos, G.Pop, *An existence result on noncoercive hemivariational inequalities,* Ann. Fac. Sci. Toulouse, 4(1997);

[5] M.Fundo, *An existence result on a general type of hemivariational inequalities,*Appl.Anal., 65(1997), 373-394;

[6] D.Goeleven, D.Motreanu, P.D.Panagiotopoulos, *Semicoercive variational -hemivariational inequalities,* Appl.Anal., 65(1997), 119-134;

[7] D.Kinderlehrer, G.Stampacchia, *An Introduction to Variational Inequalities and Their Applications,* Academic Press, New York, 1980;

[8] P.J.McKenna, J.Rauch, *Strongly nonlinear perturbations of nonnegative boundary value problems with kernel,* J.Diff.Equations, 28(1978), pp.253-265;

[9] Z.Naniewicz , P.D.Panagiotopoulos, *Mathematical Theory of Hemivariational Inequalities and Applications*, Marcel Dekker, New York 1995;

[10] P.D.Panagiotopoulos , *Coercive and semicoercive hemivariational inequalities,* Nonlinear Analysis T.M.A ., 16(1991), 209- 231;

[11] P.D.Panagiotopoulos, *Hemivariational Inequalities. Applications in Mechanics and Engineering,* Springer-Verlag, 1993;

[12] G.Pop, P.D.Panagiotopoulos, Z.Naniewicz, *Variational-hemivariational inequalities for multidimensional superpotential laws*, Int. J. Num. Funct. Anal. Opt, 18(1997), 827-856 ;

18 PERTURBATIONS OF EIGENVALUE PROBLEMS WITH CONSTRAINTS FOR HEMIVARIATIONAL INEQUALITIES

Vicentiu D. Rădulescu

Department of Mathematics
University of Craiova, 1100 Craiova
Romania

radules@ann.jussieu.fr

The study of hemivariational inequalities began with the pioneering work of Panagiotopoulos (*see* [21], [22], [23], [24], [25], [26]). This theory extends the framework of variational inequalities to the study of nonconvex energy functionals associated to concrete problems arising in Mechanics, Hysterezis, Phase Transition, Liquid Cristals etc. Panagiotopoulos also defined the notion of nonconvex superpotential, by replacing the subdifferential of a convex function (as in [18]) with the generalized gradient (in the sense of Clarke) of a locally Lipschitz function. Even if the hemivariational inequalities appear as a generalization of the variational inequalities, they are much more general, in the sense that they are not equivalent to minimum problems but, however, they give rise to substationarity problems. We extend some classical perturbation methods for the framework of hemivariational inequalities and we show how existence results can be obtained for the new problems.

18.1 THE ABSTRACT FORMULATION

Let V be a real Hilbert space with the scalar product $(\cdot,\cdot)$ and the associated norm $\|\cdot\|$. Assume that V is densely and compactly embedded in $L^p(\Omega;\mathbf{R}^N)$, for some $1<p<+\infty$ and $N\geq 1$, where Ω is a bounded domain in $\mathbf{R}^m$, $m\geq 1$. Throughout, the duality pairing between $V^\star$ and V (resp., between $(\mathbf{R}^N)^\star$ and $\mathbf{R}^N$) will be denoted by $\langle\cdot,\cdot\rangle_V$ (resp., $\langle\cdot,\cdot\rangle$).

R.P. Gilbert et al. (eds.), From Convexity to Nonconvexity, 243–253.

Let $a : V \times V \to \mathbf{R}$ be a continuous, symmetric and bilinear form, which is not necessarily coercive. Let $A : V \to V^\star$ be the self-adjoint bounded linear operator which corresponds to a, that is, for every $u, v \in V$,

$$\langle Au, v\rangle_V = a(u, v)\,.$$

For $r > 0$, set S_r the sphere of radius r in V centered at the origin, i.e.

$$S_r = \{u \in V;\ \|u\| = r\}\,.$$

If $C : S_r \times V \to \mathbf{R}$ is a general mapping and $j : \Omega \times \mathbf{R}^N \to \mathbf{R}$ is a Carathéodory function which is locally Lipschitz with respect to the second variable, we shall consider the following eigenvalue hemivariational inequality with constraints:

(P1) Find $(u, \lambda) \in V \times \mathbf{R}$ such that, for all $v \in V$,

$$\begin{cases} a(u,v) + C(u,v) + \displaystyle\int_\Omega j^0(x, u(x); v(x))dx \geq \lambda(u,v)\,; \\ \|u\| = r\,. \end{cases}$$

Motreanu and Panagiotopoulos imposed to this problem some "symmetric" natural hypotheses and they proved that (P1) admits infinitely many solutions $(\pm u_n, \lambda_n)$, with all u_n distinct (*see* [20], Thm. 4). Our first aim is to establish what happens if (P1) is perturbed in two different manners:

i) if we add a "non-symmetric" perturbation under the integral sign, as well as an arbitrary linear perturbation;

ii) if we study the same symmetric hemivariational inequality, but with respect to spaces which converge to V in the sense of Mosco (*see* [19]).

In both cases we prove that the number of solutions becomes greater and greater: for the case i), if the perturbations are "small enough", in a sense which will be described in Theorem 1, and for the case ii), if the indice of convergence is sufficiently large (*see* Theorem 2).

In Section 4 we also consider a symmetric hemivariational inequality with infinitely many solutions. We have seen in the case i) that, heuristically, if we perturb the problem in a quite general non-symmetric way, then the number of solutions of the perturbed problem goes to infinity, as the perturbation tends to disappear. The novelty now is that in our case the new problem has still infinitely many solutions. We have restricted our attention to the case $N \leq 2$, because in this situation the critical values of the unperturbed problem have a suitable growth. Our arguments do not work if the dimension is at least three.

In Section 5 we take into account a periodic problem studied recently by Goeleven, Motreanu and Panagiotopoulos (*see* [13]). Our result in this case shows that a similar behaviour occurs, if the perturbations are not Z-invariant, but sufficiently small.

The results stated in the paper can be extended to double eigenvalue hemivariational inequalities and with respect to more general types of constraints.

As an illustration, we treat in the last part only the case corresponding to that taken into account in Section 2. We obtain in Theorem 6 a result similar to Theorem 1, but for this new type of hemivariational problems on a spherelike type manifold.

18.2 NON-SYMMETRIC PERTURBATIONS OF THE PROBLEM (P1)

In [20] Motreanu and Panagiotopoulos have imposed to the quantities appearing in the problem (P1) the following assumptions:

(H_1) There exists a locally Lipschitz function $f : V \to \mathbf{R}$, even and bounded on S_r, satisfying

$$C(u,v) \geq f^0(u;v), \qquad \text{for all } (u,v) \in S_r \times V\,, \text{ with } (u,v)=0\,,$$

and such that the set

$$\{\zeta \in V^\star;\ \zeta \in \partial f(u),\ u \in S_r\}$$

is relatively compact in $V^\star$. Here $f^0(u;v)$ stands for the Clarke derivative (*see* [6]) of f at $u \in V$ with respect to the direction $v \in V \setminus \{0\}$, that is

$$f^0(u;v) = \limsup_{\substack{w \to u \\ \lambda \searrow 0}} \frac{f(w+\lambda v) - f(w)}{\lambda}\,.$$

Accordingly, Clarke's generalized gradient $\partial f(u)$ of f at u is defined by

$$\partial f(u) = \{\zeta \in V^\star; f^0(u;v) \geq \langle \zeta, v\rangle_V, \text{ for all } v \in V\}\,.$$

We assume that the functional j satisfies the symmetry condition

(H_2) $j(x,y) = j(x,-y)$, for a.e. $x \in \Omega$ and every $y \in \mathbf{R}^N$,

as well as the growth condition

(H_3) there exist $a_1 \in L^{p/(p-1)}(\Omega)$ and $b \in \mathbf{R}$ such that

$$|\, w \,| \leq a_1(x) + b \,|\, y\,|^{p-1}, \qquad \text{for a.e. } (x,y) \in \Omega \times \mathbf{R}^N \text{ and all } w \in \partial j(x,y)\,.$$

We have denoted by $\partial j(x,y)$ Clarke's generalized gradient of the locally Lipschitz mapping $\mathbf{R}^N \ni y \longmapsto j(x,y) \in \mathbf{R}$, for some fixed $x \in \Omega$.

Let $\Lambda : V \to V^\star$ be the duality isomorphism

$$(\Lambda u, v)_V = (u,v), \qquad \text{for all } u,v \in V\,.$$

The link between the quantities we have introduced is given by

(H_4) Let $(u_n) \subset S_r$ be an arbitrary sequence which converges weakly in V to some u. Assume $\zeta_n \in \partial f(u_n)$ such that

$$a(u_n,u_n) + \langle \zeta_n, u_n\rangle_V \to \alpha_0$$

and, for every $w \in L^{p/(p-1)}(\Omega; \mathbf{R}^N)$ verifying

$$w(x) \in \partial j(x, u(x)), \qquad \text{for a.e. } x \in \Omega,$$

the sequence $\{(A - \lambda_0 \Lambda)u_n\}$ is convergent. Then there exists a strongly convergent subsequence of (u_n) in V.

Here λ_0 is defined by

$$\lambda_0 = r^{-2}(\alpha_0 + \int_\Omega \langle w(x), u(x) \rangle dx).$$

The main result of Motreanu and Panagiotopoulos in [20] is that under the above hypotheses the problem (P1) has infinitely many pairs of solutions.

Our question is to see what happens if we perturb (P1) in a suitable manner. We consider the following non-symmetric perturbed hemivariational inequality:

(P2) Find $(u, \lambda) \in V \times \mathbf{R}$ such that

$$\begin{cases} a(u,v) + C(u,v) + \displaystyle\int_\Omega \Big(j^0(x, u(x); v(x)) + g^0(x, u(x); v(x)) \Big) dx + \\ \langle \varphi, v \rangle_V \geq \lambda(u,v), \quad \text{for all } v \in V; \\ \|u\| = r, \end{cases}$$

where $\varphi \in V^\star$ and $g : \Omega \times \mathbf{R}^N \to \mathbf{R}$ is a Carathéodory function which is locally Lipschitz with respect to the second variable. Fix $\delta > 0$. We make no symmetry assumption on g, but we impose only the growth condition

(H$_5$) $| w | \leq a_2(x) + \delta | y |^{p-1}$, for a.e. $(x, y) \in \Omega \times \mathbf{R}^N$ and for all $w \in \partial g(x, y)$, where $a_2 \in L^{p/(p-1)}(\Omega)$.

We also assume

(H$_6$) The mappings a_2 and φ satisfy

$$\|a_2\|_{L^{p'}} \leq \delta \qquad \text{and} \qquad \|\varphi\|_{V^\star} \leq \delta.$$

As a compactness condition we assume the following variant of **(H$_4$)**:

(H$_7$) Let $(u_n) \subset S_r$ be an arbitrary sequence which converges weakly in V to some u. Assume $\zeta_n \in \partial f(u_n)$ such that

$$a(u_n, u_n) + \langle \zeta_n, u_n \rangle_V \to \alpha_0$$

and, for every $w, z \in L^{p/(p-1)}(\Omega; \mathbf{R}^N)$ verifying

$$w(x) \in \partial j(x, u(x)) \quad \text{and} \quad z(x) \in \partial g(x, u(x)) \qquad \text{for a.e. } x \in \Omega,$$

the sequence $\{(A - \widehat{\lambda_0} \Lambda)u_n\}$ is convergent. Then (u_n) is relatively compact in V.

Here $\widehat{\lambda_0}$ is defined by

$$\widehat{\lambda_0} = r^{-2}(\alpha_0 + \int_\Omega \langle w(x) + z(x), u(x)\rangle dx).$$

The next result shows that the number of solutions of (P2) becomes greater and greater, as $\delta \to 0$. More precisely, we have

Theorem 1. *Assume that the hypotheses* $(\mathbf{H}_1)$-$(\mathbf{H}_7)$ *hold. Then, for every* $k \geq 1$, *there exists* $\delta_k > 0$ *such that, for each* $\delta \leq \delta_k$, *the problem (P2) admits at least* k *distinct solutions.*

18.3 PERTURBATIONS OF THE PROBLEM (P1) WITH RESPECT TO SPACES CONVERGING IN THE SENSE OF MOSCO

Let (V_n) be a sequence of convex closed subsets of V which converges to V in the sense of Mosco [19]. As an illustration with a particular case, if $(e_n)_{n\geq 1}$ is a Schauder basis of V, we can take $V_n = \text{span}\{e_1, \cdots, e_n\}$.

We consider the problem

(P3) Find $(u, \lambda) \in V_n \times \mathbf{R}$ such that, for all $v \in V_n$,

$$\begin{cases} a(u,v) + C(u,v) + \int_\Omega j^0(x, u(x); v(x))dx \geq \lambda(u,v); \\ \|u\| = r. \end{cases}$$

Our main result on this problem is

Theorem 2. *Assume that the hypotheses* $(\mathbf{H}_1)$-$(\mathbf{H}_4)$ *hold. Then, for every* $k \geq 1$, *there exists some* $N = N(k) \in \mathbf{N}$ *such that, for every* $n \geq N$, *the problem (P3) has at least* k *solutions* (u_1, λ_1),..., (u_k, λ_k) *with* $u_1, \cdots, u_k$ *all distinct.*

In the case of equations, results of this kind have been obtained by Ambrosetti [1], Bahri-Berestycki [2] and Struwe [29]. For variational inequalities, situations of this type have been considered in bifurcation problems (*see* Canino-Perri [5], Degiovanni [8]-[9], Degiovanni-Marino [11], Lancelotti [13], Miersemann [17], Schuricht [28]).

18.4 PERTURBED HEMIVARIATIONAL INEQUALITIES WITH INFINITELY MANY SOLUTIONS

Throughout this section we shall assume that V is a real Hilbert space which is densely and compactly embedded in $L^p(\Omega; \mathbf{R}^N)$, for some $N \leq 2$. We start with the following problem

(P4) Find $(u, \lambda) \in V \times \mathbf{R}$ such that

$$\begin{cases} a(u,v) + \int_\Omega j^0(x, u(x); v(x))dx \geq \lambda(u,v), \quad \text{for all } v \in V; \\ \|u\| = r. \end{cases}$$

Here we assume that $j : \Omega \times \mathbf{R}^N \to \mathbf{R}$ is a Carathéodory function which is locally Lipschitz with respect to the second variable and satisfying

$$(\mathbf{H}_8) \qquad \begin{cases} j(x,y) = j(x,-y), & \text{for a.e. } x \in \Omega \text{ and every } y \in \mathbf{R}^N, \\ yj(x,y) \geq 0, & \text{for a.e. } x \in \Omega \text{ and every } y \in \mathbf{R}^N, \\ \sup\limits_{|y| \leq t} \mid j(x,y) \mid \in L^1(\Omega), & \forall t > 0. \end{cases}$$

Our first result is

Theorem 3. *Under the hypotheses* ($\mathbf{H}_8$), *the problem (P4) has infinitely many distinct pairs of solutions* $(\pm u, \lambda)$.

We want to study a perturbation of (P4) of the form

(P5) Find $(u, \lambda) \in V \times \mathbf{R}$ such that

$$\begin{cases} a(u,v) + \displaystyle\int_\Omega \Big(j^0(x,u(x);v(x)) + g^0(x,u(x);v(x)) \Big) dx + \\ \langle \varphi, v \rangle_V \geq \lambda(u,v)\,, \quad \text{for all } v \in V\,; \\ \|u\| = r\,. \end{cases}$$

Here $\varphi \in V^\star$, while $g : \Omega \times \mathbf{R}^N \to \mathbf{R}$ is a Carathéodory function such that, for a.e. $x \in \Omega$, $g(x, \cdot)$ is locally Lipschitz, and satisfies the growth condition

$$(\mathbf{H}_9) \qquad \mid g(x,y) \mid \leq a_1(x) + a_2 \mid y \mid^\theta \qquad \text{with } a_1 \in L^1(\Omega) \text{ and } 0 < \theta < 3.$$

We point out that the functional g does not satisfy any symmetry assumption.

The second result of this section is

Theorem 4. *Assume that the hypotheses* ($\mathbf{H}_8$) *and* ($\mathbf{H}_9$) *hold. Then the problem (P5) admits a sequence of solutions* (u_n, λ_n) *with* $\lambda_n \to +\infty$.

18.5 PERTURBATIONS OF HEMIVARIATIONAL INEQUALITIES INVOLVING PERIODIC ENERGY FUNCTIONALS

Let Z be a discrete group of a Banach space X. A functional $f : X \to \mathbf{R}$ is said to be Z-invariant if

$$f(u+z) = f(u), \qquad \forall u \in X, \ \ \forall z \in Z.$$

In [13] Goeleven, Motreanu and Panagiotopoulos proved a multiplicity result if the quantities appearing in the problem (P4) are Z-invariant, in the sense that they satisfy

($\mathbf{H}_{10}$) $a(zu, zv) = a(u,v)$ and $j(x,(zu)(x)) = j(x,u(x))$, $\forall u, v \in X$, $\forall z \in Z$ and for a.e. $x \in \Omega$.

They also imposed

($\mathbf{H}_{11}$) Whenever a sequence $(v_n) \subset X$ with $\|v_n\| = r$, a number $b \in [-r^2\|A\|, r^2\|A\|]$ and a measurable map $s : \Omega \to \mathbf{R}^N$ are such that $v_n \to v$ in $L^p(\Omega, \mathbf{R}^N)$, $a(v_n, v_n) \to \alpha$ and

$$s(x) \in \partial j(x, v(x)) \qquad \text{for a.e. } x \in \Omega,$$

then

$$\inf_{\|w\|=1} a(w,w) - r^{-2}\left(\alpha + \int_\Omega \langle s(x), v(x)\rangle dx\right) > 0.$$

Goeleven, Motreanu and Panagiotopoulos proved in [13] that, under the hypotheses ($\mathbf{H}_3$), ($\mathbf{H}_{10}$) and ($\mathbf{H}_{11}$), the problem (P4) admits infinitely many pairs of distinct solutions (Zu_n, λ_n) and, moreover, the energy associated to the corresponding problems is bounded. The most natural way to perturb in this case the problem (P4) is to add a non Z-invariant functional under the integral sign and, of course, a linear perturbation $\varphi \in V^\star$. Let $g : \Omega \times \mathbf{R}^N \to \mathbf{R}$ be a Carathéodory function which is locally Lipschitz with respect to the second variable and such that $g(\cdot, 0) \in L^1_{\mathrm{loc}}(\Omega, \mathbf{R})$. We impose to g no Z-invariance property, but the growth condition ($\mathbf{H}_5$). The result is similar to that found in Section 2 and it states that the number of solutions is sufficiently large if the perturbations are small enough.

Theorem 5. *Assume that the hypotheses* ($\mathbf{H}_3$), ($\mathbf{H}_5$), ($\mathbf{H}_6$), ($\mathbf{H}_{10}$) *and* ($\mathbf{H}_{11}$) *are fulfilled. Then, for every $k \geq 1$, there exists $\delta_k > 0$ such that, for every $\delta < \delta_k$ (δ is the quantity which appears in* ($\mathbf{H}_5$)*), the problem (P5) has at least k distinct solutions.*

18.6 PERTURBATIONS OF DOUBLE EIGENVALUE HEMIVARIATIONAL INEQUALITIES

Let $a_1, a_2 : V \times V \to \mathbf{R}$ be two continuous symmetric bilinear forms on V and let $B_1, B_2 : V \to V$ be two bounded self-adjoint linear operators which are coercive in the sense that $(B_i v, v)_V \geq b_i \cdot \|v\|_V^2$, for all $v \in V$, $i = 1, 2$, for some constants $b_1, b_2 > 0$. For fixed positive numbers a, b, r we consider the submanifold $S_r^{a,b}$ of $V \times V$ defined by

$$S_r^{a,b} = \{(v_1, v_2) \in V \times V \ : \ a(B_1 v_1, v_1)_V + b(B_2 v_2, v_2)_V = r^2\}$$

and the tangent space associated to this manifold

$$T_{(u_1,u_2)} S^r_{a,b} := \{(v_1, v_2) \in V \times V \ : \ a(B_1 u_1, v_1)_V + b(B_2 u_2, v_2)_V = 0\}.$$

Let us consider a real function $C : S_r^{a,b} \times V \times V \to \mathbf{R}$ to which we impose no continuity assumption. Denote by π_i ($i = 1, 2$) the canonical projections of $V \times V$ onto V. We assume

($\mathbf{H}_{12}$) There exist two locally Lipschitz maps $f_i : V \to \mathbf{R}$, bounded on

$$\pi_i(S_r^{a,b}), \ (i = 1, 2)$$

respectively, and such that $C((u_1,u_2),v_1,v_2) \geq f_1^0(u_1;v_1)+f_2^0(u_2;v_2)$, for each $(u_1,u_2) \in S_r^{a,b}$ and each $(v_1,v_2) \in T_{(u_1,u_2)}S_r^{a,b}$. In addition we suppose that the sets $\{z \in V^* \ : \ z \in \partial f_i(u_i), u_i \in \pi_i(S_r^{a,b})\}$ are relatively compact in V^*, for $i = 1,2$.

Define the map $(A_1,A_2) : V \times V \to V^* \times V^*$ by the relation

$$\langle (A_1,A_2)(u_1,u_2),(v_1,v_2)\rangle_{V\times V} = a_1(u_1,v_1) + a_2(u_2,v_2)$$

and the duality map $\Lambda : V \times V \to V^* \times V^*$ given by

$$\langle \Lambda(u_1,u_2),(v_1,v_2)\rangle_{V\times V} = a(B_1u_1,v_1)_V + b(B_2u_2,v_2)_V.$$

We also assume

($\mathbf{H}_{13}$) For every sequence $\{(u_n^1,u_n^2)\} \subset S_r^{a,b}$ with $u_n^i \rightharpoonup u_i$ weakly in V, for any $z_n^i \in \partial f_i(u_n^i)$, with $a_i(u_n^i,u_n^i) + \langle z_n^i,u_n^i\rangle_V \to \alpha_i \in \mathbf{R}$, $i = 1,2$, and for all $w \in L^{\frac{p}{p-1}}(\Omega;\mathbf{R}^N)$ which satisfies the relation

$$w(x) \in \partial j(x,(u_1-u_2)(x)) \ for \ a.e. \ x \in \Omega,$$

such that $\left[(A_1,A_2) - \widetilde{\lambda_0}\cdot\Lambda\right](u_n^1,u_n^2)$ converges in $V^* \times V^*$, then there exists a convergent subsequence of (u_n^1,u_n^2) in $V \times V$ (thus, in $S_r^{a,b}$). Here,

$$\widetilde{\lambda_0} = r^{-2}(\alpha_1+\alpha_2+\int_\Omega \langle w(x),(u_1-u_2)(x)\rangle\, dx),$$

($\mathbf{H}_{14}$) The functions f_i are even on $\pi_i(S_r^{a,b})$ in the sense that $f_i(-u_i) = f_i(u_i)$, for all $(u_1,u_2) \in S_r^{a,b}$, for $i = 1,2$.

Consider the problem: find $u_1,u_2 \in V$ and $\lambda_1,\lambda_2 \in \mathbf{R}$ such that

$$(P6) \qquad \begin{cases} a_1(u_1,v_1) + a_2(u_2,v_2) + C((u_1,u_2),v_1,v_2) + \\ \int_\Omega j_y^0(x,(u_1-u_2)(x);(v_1-v_2)(x))dx \geq \\ \geq \lambda_1(B_1u_1,v_1)_V + \lambda_2(B_2u_2,v_2)_V,\ \forall\, v_1,v_2 \in V; \\ u_1,u_2 \in S_r^{a,b}. \end{cases}$$

Under the hypotheses ($\mathbf{H}_2$), ($\mathbf{H}_3$), ($\mathbf{H}_{12}$)-($\mathbf{H}_{14}$), it is proved in [3] that the double eigenvalue problem ($P6$) admits infinitely many pairs of solutions

$$\{\pm(u_n^1,u_n^2),(\lambda_n^1,\lambda_n^2)\}.$$

Let us now consider an arbitrary element φ in V^* and $g : \Omega \times \mathbf{R}^N \to \mathbf{R}$ a Carathéodory function which is locally Lipschitz with respect to the second variable and such that $g(\cdot,0) \in L^1(\Omega)$. We assume that

($\mathbf{H}_{15}$) For every sequence $\{(u_n^1, u_n^2)\} \subset S_r^{a,b}$ with $u_n^i \rightharpoonup u_i$ weakly in V, for any $z_n^i \in \partial f_i(u_n^i)$, with

$$a_i(u_n^i, u_n^i) + \langle z_n^i, u_n^i \rangle_V + \langle \varphi, u_n^i \rangle_V \to \alpha_i \in \mathbf{R}, \ i = 1, 2,$$

and for all $w, z \in L^{\frac{p}{p-1}}(\Omega; \mathbf{R}^N)$ which satisfy the relations

$$w(x) \in \partial_y j(x, (u_1 - u_2)(x)), \text{ and } z(x) \in \partial_y g(x, (u_1 - u_2)(x)),$$

for a.e. $x \in \Omega$, if $\left[(A_1, A_2) - \overline{\lambda_0} \cdot \Lambda\right](u_n^1, u_n^2)$ converges in $V^* \times V^*$, then there exists a convergent subsequence of (u_n^1, u_n^2) in $V \times V$. Here, $\overline{\lambda_0} = r^{-2}(\alpha_1 + \alpha_2 + \int_\Omega \langle w(x) + z(x), (u_1 - u_2)(x) \rangle \, dx)$.

We consider the problem: find $u_1, u_2 \in V$ and $\lambda_1, \lambda_2 \in \mathbf{R}$ such that

$$(P7) \quad \begin{cases} a_1(u_1, v_1) + a_2(u_2, v_2) + C((u_1, u_2), v_1, v_2) + \\ \int_\Omega \{ j_y^0(x, (u_1 - u_2)(x); (v_1 - v_2)(x)) + \\ + g_y^0(x, (u_1 - u_2)(x); (v_1 - v_2)(x)) \} dx + \langle \varphi, v_1 \rangle_V + \langle \varphi, v_2 \rangle_V \geq \\ \geq \lambda_1 (B_1 u_1, v_1)_V + \lambda_2 (B_2 u_2, v_2)_V, \ \forall \, v_1, v_2 \in V; \\ u_1, u_2 \in S_r^{a,b}. \end{cases}$$

In the following perturbation result, a_2 and δ are the quantities appearing in ($\mathbf{H}_5$).

Theorem 6. *Assume that the hypotheses* ($\mathbf{H}_2$), ($\mathbf{H}_3$), ($\mathbf{H}_5$), ($\mathbf{H}_{12}$)-($\mathbf{H}_{15}$) *are fulfilled. Then, for every* $k \geq 1$, *there exists* $\delta_k > 0$ *such that, if* $\|\varphi\|_{V^*} \leq \delta_k$, $\|\theta_1\|_{L^{p'}} \leq \delta_k$, *and* $c \leq \delta_k$, *the problem (P7) admits at least* k *distinct solutions.*

References

[1] A. Ambrosetti, *A perturbation theorem for superlinear boundary value problems,* Math. Res. Center, Univ. Wisconsin-Madison, Tech. Sum. Report 1446, 1974.

[2] A. Bahri and H. Berestycki, A perturbation method in critical point theory and applications, *Trans. Amer. Math. Soc.,* 267(1981), 1-32.

[3] M. Bocea, D. Motreanu and P.D. Panagiotopoulos, Multiple Solutions for a Double Eigenvalue Hemivariational Inequality on a Spherelike Type Manifold, *Nonlinear Anal., T.M.A.,* in press.

[4] M. Bocea, P.D. Panagiotopoulos and V.D. Rădulescu, Perturbations of Double Eigenvalue Hemivariational Inequalities and Applications, *J. Global Optimiz.,* 14(1999), 137-156.

[5] A. Canino and U. Perri, Constrained Problems in Banach Spaces with an Application to Variational Inequalities, *Nonlinear Anal., T.M.A.,* in press.

[6] F.H. Clarke, *Optimization and Nonsmooth Analysis,* Wiley, New York, 1983.

[7] J.N. Corvellec, M. Degiovanni and M. Marzocchi, Deformation Properties for Continuous Functionals and Critical Point Theory, *Top. Meth. Nonl. Anal.*, 1(1993), 151-171.

[8] M. Degiovanni, Bifurcation for odd nonlinear elliptic variational inequalities, *Ann. Fac. Sci. Toulouse,* 11(1990), 39-66.

[9] M. Degiovanni, Variational methods in bifurcation problems for variational inequalities, *Autumn School on Variational Inequalities (Paseky, 1992), 27-51, Faculty of Mathematics and Physics, Charles University, Prague, 1993.*

[10] M. Degiovanni and S. Lancelotti, Perturbations of even nonsmooth functionals, *Differential Integral Equations,* 8(1995), 981-992.

[11] M. Degiovanni and A. Marino, Nonsmooth variational bifurcation, *Atti Accad. Naz. Lincei Rend. Cl. Sci. Fis. Mat. Natur.*, 8(1987), 259-269.

[12] M. Degiovanni and M. Marzocchi, A critical point theory for nonsmooth functionals, *Ann. Mat. Pura Appl.,* 167(1994), 73-100.

[13] D. Goeleven, D. Motreanu and P.D. Panagiotopoulos, Multiple Solutions for a Class of Hemivariational Inequalities Involving Periodic Energy Functionals, *Mathematical Methods in the Applied Sciences,* 20(1997), 547-568.

[14] S. Lancelotti, Perturbations of symmetric constraints in eigenvalue problems for variational inequalities, *Nonlinear Anal., T.M.A.*, 27(1996), 633-644.

[15] S. Lancelotti, Infinitely many solutions for one-dimensional eigenvalue problems for variational inequalities, *Università Cattolica del Sacro Cuore di Brescia,* Preprint 6(1995).

[16] J.L. Lions and G. Stampacchia, Variational Inequalities, *Comm. Pure Appl. Math.*, 20(1967), 493-519.

[17] E. Miersemann, On higher eigenvalues of variational inequalities, *Comment. Math. Univ. Carolin.*, 24(1983), 657-665.

[18] J.J. Moreau, La notion de sur-potentiel et les liaisons unilatérales en élastostatique, *C.R. Acad. Sci. Paris,* 267A(1968), 954-957.

[19] U. Mosco, Convergence of convex sets and of solutions of variational inequalities, *Advances in Mathematics,* 3(1969), 510-585.

[20] D. Motreanu and P.D. Panagiotopoulos, Nonconvex Energy Functions, Related Eigenvalue Hemivariational Inequalities on the Sphere and Applications, *J. Global Optimiz.*, 6(1995), 163-177.

[21] Z. Naniewicz and P.D. Panagiotopoulos, *Mathematical Theory of Hemivariational Inequalities and Applications,* Marcel Dekker, New York, 1995.

[22] P.D. Panagiotopoulos, *Inequality Problems in Mechanics and Applications. Convex and Nonconvex Energy Functionals,* Birkhäuser-Verlag, Boston, Basel, 1985 (Russian translation MIR Publ., Moscow, 1989).

[23] P.D. Panagiotopoulos, Coercive and semicoercive hemivariational inequalities, *Nonlinear Anal., T.M.A.,* 16(1991), 209-231.

[24] P.D. Panagiotopoulos, *Hemivariational Inequalities: Applications to Mechanics and Engineering,* Springer-Verlag, New York/Boston/Berlin, 1993.

[25] P.D. Panagiotopoulos, Hemivariational inequalities and their applications, in *Topics in Nonsmooth Mechanics* (J.J. Moreau, P.D. Panagiotopoulos, and G. Strang, Eds.), Birkhäuser-Verlag, Basel, 1988.

[26] P.D. Panagiotopoulos, Nonconvex superpotentials and hemivariational inequalities; Quasidifferentiability in mechanics, in *Nonsmooth Mechanics and Applications* (J.J. Moreau and P.D. Panagiotopoulos, Eds.), CISM Courses and Lectures No. 302, Springer-Verlag, Wien/New York, 1988.

[27] V.D. Rădulescu and P.D. Panagiotopoulos, Perturbations of hemivariational inequalities with constraints and applications, *J. Global Optimiz.*, 12 (1998), 285-297

[28] F. Schuricht, Bifurcation from minimax solutions by variational inequalities in convex sets, *Nonlinear Anal., T.M.A.*, 26(1996), 91-112.

[29] M. Struwe, Infinitely many critical points for functionals which are not even and applications to superlinear boundary value problems, *Manuscripta Math.*, 32(1980), 335-364.

19 IMPLICIT VARIATIONAL INEQUALITIES ARISING IN FRICTIONAL UNILATERAL CONTACT MECHANICS: ANALYSIS AND NUMERICAL SOLUTION OF QUASISTATIC PROBLEMS

Marius Cocu[1] and Michel Raous[2]

[1]Laboratoire de Mécanique et d'Acoustique - C.N.R.S.
and Université de Provence, Marseille
France
cocu@lma.cnrs-mrs.fr

[2]Laboratoire de Mécanique et d'Acoustique - C.N.R.S.
31, Chemin Joseph Aiguier, 13402 Marseille Cedex 20
France
raous@lma.cnrs-mrs.fr

Abstract: This paper is a survey on implicit variational inequalities arising in the study of unilateral contact problems with friction. Recent works on mathematical and numerical approaches of quasistatic problems are presented. The coupling of unilateral contact, friction, and adhesion is considered and previous results are generalized to this case.

19.1 INTRODUCTION

It can be said that the bases of the mathematical formulation of unilateral and frictional contact have been set by a French-Italian school. On the basis of Signorini's formulation (Signorini, 1959), the unilateral contact problems have been studied by variational methods and by an intensive development of the theory of variational inequalities by (Fichera, 1964; 1972), (Lions and Stampacchia, 1967; Duvaut and Lions, 1972; Mosco, 1975), etc.. The

R.P. Gilbert et al. (eds.), From Convexity to Nonconvexity, 255–267.

introduction of friction has complicated the mathematical problems, even in the static case, since the problems are then related to implicit variational inequalities or to quasi variational inequalities (in a dual formulation). The advances during the last twenty years are due to several groups: in Austin (Oden, Kikuchi), for static and dynamic formulations, in Linköping (Klarbring, Andersson), for modelling, mathematical programming and analysis, in Montpellier (Moreau, Jean), for convex analysis and dynamic problems, in Prague (Jarusek, Haslinger, Hlavaček, Nečas), for mathematical and numerical analysis, and in Thessaloniki (Panagiotopoulos), for extensions to nonsmooth mechanics. The names of Curnier (Lausanne), Marques, Martins and Pires (Lisbon), Mróz and Telega (Warsaw), Pfeiffer and Glocker (Munich), Wriggers (Darmstadt) have also to be mentionned. In this paper, we are concerned with a mathematical analysis of quasistatic unilateral contact problems with nonlocal friction law. A generalization to the coupling with adhesion is also considered and various numerical treatments are given.

19.2 QUASISTATIC CONTACT PROBLEMS WITH FRICTION

In this section we present the mathematical and numerical results which have been recently obtained for quasistatic unilateral problems with a nonlocal friction law (Cocu et al., 1996). Unilateral conditions are strictly imposed, we use neither penalization nor normal compliance. Existence results for quasistatic frictional contact problems with a normal compliance law have been proved by (Andersson, 1991; 1995) and, in the presence of a time regularization, by (Klarbring et al., 1991). We consider a variational formulation of these problems, combining an inequality describing the friction and an inequality representing the unilateral contact conditions. A time discretization is adopted and if the friction coefficient is sufficiently small then the incremental formulation that can be derived from this discretization has a unique solution. By using appropriate estimates for the incremental solutions, we are able to prove the existence of a quasistatic solution.

19.2.1 The classical formulation of unilateral contact problems with friction

Consider a linear elastic body occupying a domain Ω of $\mathbf{R}^d$, $d = 2, 3$, with sufficiently regular boundary $\Gamma = \bar{\Gamma}_u \cup \bar{\Gamma}_\varphi \cup \bar{\Gamma}_c$. We assume that on Γ_u (with $\mathrm{mes}(\Gamma_u) > 0$) the displacement vector equals zero. On Γ_φ a surface force density φ is prescribed and on Ω a volume force density f is applied. The solid is initially in contact with a rigid fixed support along Γ_c and the displacements on this part of the boundary satisfy the unilateral constraints of non penetration into the support (Signorini's conditions) and are submitted to nonlocal friction when contact occurs. We suppose that we may neglect the inertial terms. We denote by $u = (u_i)$ the displacement vector, $\sigma = (\sigma_{ij})$ the stress tensor, $\varepsilon = (\varepsilon_{ij})$ the strain tensor and by $A = (A_{ijkl})$ the elasticity tensor. We make the classical decomposition into normal and tangential components for the contact forces and the displacement vector on Γ_c:

$$u = u_N\, n + u_T\,,\ \sigma\, n = \sigma_N\, n + \sigma_T \quad \text{with} \quad u_N = u.n\,,\ \sigma_N = (\sigma\, n).n\,,$$

where n is the outward unit normal vector to Γ. The classical formulation of this problem is as follows.

P$_1$: Find a displacement field $u = u(t,x)$ which satisfies, for all $t \in [0,T]$, the following equations and conditions:

$$div\,\sigma(u) + f = 0 \quad \text{in} \quad \Omega\,, \tag{19.1}$$

$$\sigma(u) = A : \varepsilon(u) \quad \text{in} \quad \Omega\,, \tag{19.2}$$

$$u = 0 \quad \text{on} \quad \Gamma_u\,, \tag{19.3}$$

$$\sigma(u)n = \varphi \quad \text{on} \quad \Gamma_\varphi\,, \tag{19.4}$$

$$u_N \leq 0, \quad \sigma_N(u) \leq 0 \text{ and } \sigma_N(u).u_N = 0 \quad \text{on} \quad \Gamma_c\,, \tag{19.5}$$

$$\begin{array}{l} |\sigma_T(u)| \leq \mu\,|\sigma_N^*(u)| \text{ and} \\ |\sigma_T(u)| < \mu\,|\sigma_N^*(u)| \Rightarrow \dot{u}_T = 0\,, \\ |\sigma_T(u)| = \mu\,|\sigma_N^*(u)| \Rightarrow \\ \quad \exists\lambda \geq 0 \quad \dot{u}_T = -\lambda\sigma_T(u) \text{ on } \Gamma_c\,, \end{array} \tag{19.6}$$

where μ is the friction coefficient and σ_N^* is a regularization of the normal contact force (Duvaut, 1980; Demkowicz and Oden, 1982; Oden and Pires, 1983; Cocu et al., 1996).

19.2.2 *Variational formulation*

We define the space $V = \left\{v \in \left[H^1(\Omega)\right]^d; v = 0 \text{ a.e. on } \Gamma_u\right\}$ and we set $K = \{v \in V; v_N \leq 0 \text{ a.e. on } \Gamma_c\}$. The norm on V is given by $||.||$ and $\langle .,. \rangle$ shall denote the duality pairing on $H^{\frac{1}{2}}(\Gamma_c) \times H^{-\frac{1}{2}}(\Gamma_c)$. We suppose that $f \in W^{1,2}(0,T;\left[L^2(\Omega)\right]^d), \varphi \in W^{1,2}(0,T;\left[L^2(\Gamma_\varphi)\right]^d)$, so that we have $F \in W^{1,2}(0,T;V)$, where

$$(F,v) = \int_\Omega f.v dx + \int_{\Gamma_\varphi} \varphi.v ds \quad \forall v \in V.$$

We shall adopt the following notations :

- $a : V \times V \longrightarrow I\!R$,

  $a(u,v) = \int_\Omega A_{ijkl}\varepsilon_{ij}(u)\varepsilon_{kl}(v)dx \;\; \forall\, u,v \in V,$ where $A_{ijkl} \in L^\infty(\Omega)$,

  $1 \leq i,j,k,l \leq d$, with the usual properties of ellipticity and symmetry.

- $j : V \times V \longrightarrow I\!R$,

  $j(u,v) = \int_{\Gamma_c} \mu|\sigma_N^*(Pu)||v_T|ds$, where

  P is the projection of $W^{1,2}(0,T;V)$ on V_0 with

  $$V_0 = \left\{ w \in W^{1,2}(0,T;V)\,;\, \int_0^T a(w,\psi)\,dt = \int_0^T (f,\psi)_{[L^2(\Omega)]^d}\,dt \right.$$
  $$\left. + \int_0^T (\varphi,\psi)_{[L^2(\Gamma_\varphi)]^d}\,dt, \forall\, \psi \in L^2(0,T;V),\ \psi = 0 \text{ a.e. on } \Gamma_c{\times}]0,T[\right\},$$

  $\mu \in L^\infty(\Gamma_c)$ and $\mu \geq 0$ a.e. on Γ_c,

  $(\,.\,)^* : H^{-\frac{1}{2}}(\Gamma_c) \rightarrow L^2(\Gamma_c)$ is a linear and compact mapping.

Then the problem (19.1)-(19.4), (19.5), (19.6) admits the following variational formulation (Cocu et al., 1996).

$\mathbf{P}_2$: Find $u \in W^{1,2}(0,T;V)$ such that $u(0) = u_0$ and for almost all $t \in [0,T]$ $u(t) \in K$ and

$$a(u, v - \dot{u}) + j(u, v) - j(u, \dot{u}) \geq (F, v - \dot{u}) + \langle \sigma_N(u), v_N - \dot{u}_N \rangle \quad \forall v \in V, \tag{19.7}$$

$$\langle \sigma_N(u), z_N - u_N \rangle \geq 0 \quad \forall z \in K, \tag{19.8}$$

where the initial displacement u_0 belongs to K and satisfies the following compatibility condition :

$$a(u_0, w - u_0) + j(u_0, w - u_0) \geq (F(0), w - u_0) \quad \forall w \in K.$$

A similar variational formulation was proposed in (Telega, 1991). The following well known property of the mapping j involved in the problem $\mathbf{P}_2$ holds with $\bar{\mu} = |\mu|_{L^\infty(\Gamma_c)}$:

$$\exists C > 0 \quad \forall t \in [0,T] \quad \text{and} \quad \forall u, \bar{u}, v, \bar{v} \in V_0$$
$$|j(u,v) - j(u,\bar{v}) - j(\bar{u},v) + j(\bar{u},\bar{v})| \leq \bar{\mu} C ||u - \bar{u}|| \, ||v - \bar{v}|| \, .$$

19.2.3 Incremental formulations

Incremental formulations can be obtained by operating a time discretization of problem $\mathbf{P}_2$, taking $n \in \mathbb{N}^*$ and setting $\Delta t = T/n$, $t_i = i\,\Delta t$ and $F^i = F(t_i)$ for $i = 0, ..., n$. Using an implicit scheme we consider the following sequence of problems $\mathbf{P}_2^{n,i}$, $i = 0, ..., n-1$, defined for $u^0 = u_0$.

$\mathbf{P}_2^{n,i}$: Find $u^{i+1} \in K$ such that :

$$\begin{aligned} &a(u^{i+1}, v - \frac{u^{i+1} - u^i}{\Delta t}) + j(u^{i+1}, v) - j(u^{i+1}, \frac{u^{i+1} - u^i}{\Delta t}) \geq \\ &(F^{i+1}, v - \frac{u^{i+1} - u^i}{\Delta t}) + \langle \sigma_N(u^{i+1}), v_N - \frac{u_N^{i+1} - u_N^i}{\Delta t} \rangle \quad \forall v \in V , \end{aligned} \tag{19.9}$$

$$\langle \sigma_N(u^{i+1}), z_N - u_N^{i+1} \rangle \geq 0 \quad \forall z \in K . \tag{19.10}$$

Problems $\mathbf{P}_2^{n,i}$, $i = 0, ..., n-1$, are equivalent to the following sequence $\mathbf{S}_2^{n,i}$, $i = 0, ..., n-1$, where $\mathbf{S}_2^{n,i}$: Find $u^{i+1} \in K$ such that

$$\begin{aligned} a(u^{i+1}, w - u^{i+1}) + j(u^{i+1}, w - u^i) - j(u^{i+1}, u^{i+1} - u^i) \geq \\ (F^{i+1}, w - u^{i+1}) \qquad \forall w \in K . \end{aligned} \tag{19.11}$$

19.2.4 Existence results

Inequalities such as $\mathbf{S}_2^{n,i}$ are elliptic quasi variational inequalities so that have a unique solution if the friction coefficient is sufficiently small ($\bar{\mu} < m/C$, where m is the coerciveness constant of the bilinear form a) (Cocu, 1984; Capatina and Cocu, 1991). We suppose from now that μ satisfies this condition. In order to pass to the limit in the incremental formulations, the following bounds for the incremental solution u^i and also for the difference $\Delta u^i = u^{i+1} - u^i$ can be obtained (Cocu et al., 1996) :

$$||u^i|| \leq C_1 \, ||F^i|| \;\; \forall i = 0, ..., n \quad \text{and} \quad ||\Delta u^i|| \leq C_1 \, ||\Delta F^i|| \;\; \forall i = 0, ..., n-1 \, .$$

We define the sequences $(u_n)_n$, $(\tilde{u}_n)_n$, such that $u_n \in L^2(0,T;V)$, $\tilde{u}_n \in W^{1,2}(0,T;V) \quad \forall n \in \mathbb{N}^*$, given by $u_n(0) = \tilde{u}_n(0) = u_0$ and

$$u_n(t) = u^{i+1} \quad \forall t \in]t_i, t_{i+1}],$$
$$\tilde{u}_n(t) = u^i + \frac{(t - t_i)}{\Delta t}(u^{i+1} - u^i) \quad \forall t \in]t_i, t_{i+1}],$$

where u^k is the solution of $\mathbf{S}_2^{n,k}$. Then the following existence and approximation result holds (Cocu et al., 1996).

Theorem 19.1. *There exists a subsequence $(\tilde{u}_{n_p})_p$ of $(\tilde{u}_n)_n$ that converges weakly in $W^{1,2}(0,T;V)$ to an element u, such that $\forall t \in [0,T]$ $(u_{n_p}(t))_p$ converges weakly to $u(t)$ in V. Every such weak limit u is a solution of $\mathbf{P}_2$.*

19.3 EXTENSION TO A MODEL COUPLING ADHESION AND FRICTION

This section is concerned with a model where unilateral contact conditions, adhesion and friction between two elastic bodies are strongly coupled. The interface is considered as a material surface and the derivation of this model follows from the principle of virtual power and the principles of thermodynamics (Cangémi, 1997), (Raous et al., 1998; 1997). Adhesion is characterized by a new variable, the intensity of adhesion β introduced by (Frémond, 1987). We propose a variational formulation which generalizes the one already given for the quasistatic friction problems in Section 19.2. The use of an implicit scheme leads us to incremental formulations for which an existence and uniqueness result is given if the friction coefficient is sufficiently small. Algorithms for computing the incremental solutions are presented.

19.3.1 The classical formulation of quasistatic adhesion problems

Let Ω^1 and Ω^2 be two disjoint domains of $\mathbf{R}^d$, $d = 2,3$, occupied by two linear elastic bodies with sufficiently regular boundaries $\Gamma^\alpha = \bar{\Gamma}_u^\alpha \cup \bar{\Gamma}_\varphi^\alpha \cup \bar{\Gamma}_c$, $\alpha = 1,2$, where Γ_c is the contact boundary. Suppose that on $\Omega^1 \cup \Omega^2$ and on Γ_φ^1, Γ_φ^2 the volume force densities f^1, f^2 and the surface force densities φ^1, φ^2 are respectively applied. On Γ_u^1 and Γ_u^2 (with $\text{mes}(\Gamma_u^\alpha) > 0$, $\alpha = 1,2$) the displacements equal zero. We denote by $[u] = u^1 - u^2$ the relative displacement on Γ_c and by n^α the outward unit normal vector to Γ^α, $\alpha = 1,2$.

The equilibrium equations can be deduced from the principle of virtual power to the whole system $\Omega^1 \cup \Omega^2 \cup \Gamma_c$. The constitutive laws are obtained from the state laws , the complementary laws for the bodies and the contact boundary Γ_c which will be considered in what follows as a material boundary with a specific thermodynamic behaviour (Frémond, 1987; Klarbring, 1990). The state variables are: the infinitesimal strain tensor $\varepsilon = (\varepsilon_{ij})$, the jump [u] on Γ_c and the intensity of adhesion β ($\beta = 1$ means that the adhesion is total, $\beta = 0$ means that there is no adhesion and $0 < \beta < 1$ is the case of partial adhesion).

For the interface, we choose the following free energy:

$$\Psi([u_N],[u_T],\beta) = \frac{C_N}{2}\,[u_N]^2\beta^2 + \frac{C_T}{2}\,[u_T]^2\beta^2 - w\,\beta + +I_{\tilde{K}}([u]) + I_P(\beta)\,,$$

with $I_{\tilde{K}}$, I_P the indicator functions of $\tilde{K} = \{v\,;\,v_N \leq 0\}$ and $P = \{\gamma\,;0 \leq \gamma \leq 1\}$. We introduce also a pseudo-potential of dissipation Φ, having the following form (Cocu et al., 1998; Raous et al., 1998):

$$\Phi\left([\dot{u}_T],\dot{\beta};\chi_N\right) = \mu\left|\sigma_N^* + C_N\,[u_N]\beta^2\right|\,\|[\dot{u}_T]\| + \frac{b}{2}\dot{\beta}^2 + I_{C^-}(\dot{\beta}),$$

where $\chi_N = (\sigma_N,[u_N],\beta)$, μ is the friction coefficient, σ_N^* is a regularization of the normal contact force σ_N and $C^- = \{\gamma\,;\gamma \leq 0\}$. The adhesion is governed here by a compliance law, where C_N and C_T are material characteristic coefficients. As shown in (Cocu et al., 1998; Raous et al., 1998; 1997), the state laws and the complementary laws give the following constitutive laws for the interface, coupling unilateral contact, friction and adhesion:

- unilateral contact conditions

$$\sigma_N + C_N\,[u_N]\beta^2 \leq 0\,,\quad [u_N] \leq 0 \text{ and } \left(\sigma_N + C_N\,[u_N]\beta^2\right)\,[u_N] = 0\,, \tag{19.12}$$

- adhesion and friction law

$$\begin{aligned}
&\left|\sigma_T + C_T\,[u_T]\beta^2\right| \leq \mu\left|\sigma_N^* + C_N\,[u_N]\beta^2\right|\,,\\
&\left|\sigma_T + C_T\,[u_T]\beta^2\right| < \mu\left|\sigma_N^* + C_N\,[u_N]\beta^2\right| \;\Rightarrow\; [\dot{u}_T] = 0\,,\\
&\left|\sigma_T + C_T\,[u_T]\beta^2\right| = \mu\left|\sigma_N^* + C_N\,[u_N]\beta^2\right| \;\Rightarrow\\
&\qquad\qquad \exists\lambda \geq 0 \quad \sigma_T + C_T\,[u_T]\beta^2 = -\lambda\,[\dot{u}_T]\,,
\end{aligned} \tag{19.13}$$

- damage evolution equation

$$b\dot{\beta} = -\left[w - (C_N\,[u_N]^2 + C_T\,[u_T]^2)\beta\right]^-\,. \tag{19.14}$$

The conditions (19.12) represent a contact law where no interpenetration between the two bodies occurs. The friction-adhesion law (19.13) expresses the coupling between a (regularized) Coulomb force and an extra compliance force in the tangential direction. When total decohesion occurs ($\beta = 0$), the relations (19.12), (19.13) are the classical unilateral contact (or Signorini's) conditions and friction conditions for the two elastic bodies.

19.3.2 Variational and incremental formulations of adhesion problems

We denote by V^α, $\alpha = 1,2$, the spaces
$V^\alpha = \left\{v^\alpha \in \left[H^1(\Omega^\alpha)\right]^d; v^\alpha = 0 \text{ a.e. on } \Gamma_u^\alpha\right\}$ and we set

$\mathbf{V} = V^1 \times V^2$, $\mathbf{K} = \{v = (v^1, v^2) \in \mathbf{V}; [v_N] \le 0 \text{ a.e. on } \Gamma_c\}$, $H = L^\infty(\Gamma_c)$.
The norm on $\mathbf{V}$ is given by $||v|| = ||v^1||_{V^1} + ||v^2||_{V^2}$ for all $v \in \mathbf{V}$.

We suppose that:

$$(f^1, f^2) \in W^{1,2}(0,T; [L^2(\Omega^1)]^d \times [L^2(\Omega^2)]^d),$$
$$(\varphi^1, \varphi^2) \in W^{1,2}(0,T; [L^2(\Gamma^1_\varphi)]^d \times [L^2(\Gamma^2_\varphi)]^d),$$

which imply $\mathbf{F} \in W^{1,2}(0,T; \mathbf{V})$, where

$$(\mathbf{F}, v) = \sum_{\alpha=1,2} \left[\int_{\Omega^\alpha} f^\alpha.v^\alpha dx + \int_{\Gamma^\alpha_\varphi} \varphi^\alpha.v^\alpha ds \right] \quad \forall v = (v^1, v^2) \in \mathbf{V}.$$

We shall adopt the following notations:

- $\mathbf{a} : \mathbf{V} \times \mathbf{V} \longrightarrow I\!R$,
  $\mathbf{a}(u,v) = a^1(u^1, v^1) + a^2(u^2, v^2) \quad \forall u = (u^1, u^2), v = (v^1, v^2) \in \mathbf{V}$,
  where $\quad a^\alpha(u^\alpha, v^\alpha) = \int_{\Omega^\alpha} A^\alpha_{ijkl} \varepsilon_{ij}(u^\alpha)\varepsilon_{kl}(v^\alpha) dx, \ \alpha = 1,2,$

- $\mathbf{j} : H \times \mathbf{V} \times \mathbf{V} \longrightarrow I\!R$,
  $\mathbf{j}(\beta, u, v) = \int_{\Gamma_c} \mu |\sigma^*_N(P^1 u^1) + C_N \beta^2 [u_N]| |[v_T]| ds, \quad$ where
  P^1 is the projection of V^1 on V^1_0 with

$$V^1_0 = \left\{ w^1 \in W^{1,2}(0,T; V^1) ; \int_0^T a^1(w^1, \psi^1)\, dt = \int_0^T (f^1, \psi^1)_{[L^2(\Omega^1)]^d}\, dt \right.$$
$$\left. + \int_0^T (\varphi^1, \psi^1)_{[L^2(\Gamma^1_\varphi)]^d}\, dt, \forall \psi^1 \in L^2(0,T; V^1),\ \psi^1 = 0 \text{ a.e. on } \Gamma_c \times]0,T[\right\},$$

$\mu \in L^\infty(\Gamma_c)$ and $\mu \ge 0$ a.e. on Γ_c,
$(.)^* : H^{-\frac{1}{2}}(\Gamma_c) \to L^2(\Gamma_c)$ is a linear and compact mapping,

- $b_N, b_T : H \times \mathbf{V} \times \mathbf{V} \longrightarrow I\!R$,
  $b_N(\beta, u, v) = \int_{\Gamma_c} C_N \beta^2 [u_N][v_N] ds$ and $b_T(\beta, u, v) = \int_{\Gamma_c} C_T \beta^2 [u_T][v_T] ds,$

- $g(\beta, u) = -\frac{1}{b} \left[w - (C_N [u_N]^2 + C_T [u_T]^2)\beta \right]^-.$

Then the local problem (19.12)-(19.14) admits the following variational formulation.

$\mathbf{P}_3$: Find $(u, \beta) \in W^{1,2}(0,T; \mathbf{V}) \times W^{1,2}(0,T; H)$ such that $u(0) = u_0$, $\beta(0) = \beta_0$ and for almost all $t \in [0,T]$, $u(t) \in \mathbf{K}$ and

$$\begin{aligned} \mathbf{a}(u, v - \dot{u}) + \mathbf{j}(\beta, u, v) - \mathbf{j}(\beta, u, \dot{u}) + b_T(\beta, u, v - \dot{u}) \ge \\ (\mathbf{F}, v - \dot{u}) + \langle \sigma_N(u^1), [v_N] - [\dot{u}_N] \rangle \quad \forall v \in \mathbf{V}, \end{aligned} \tag{19.15}$$

$$\langle \sigma_N(u^1), [z_N] - [u_N] \rangle + b_N(\beta, u, z - u) \ge 0 \quad \forall z \in \mathbf{K}, \tag{19.16}$$

$$\dot{\beta} = g(\beta, u) \quad \text{a.e. on } \Gamma_c, \tag{19.17}$$

where the initial conditions $u_0 \in \mathbf{K}$, $\beta_0 \in H$ with $\beta_0 \in [0,1[$ a.e. on Γ_c, satisfy the following compatibility condition: for all w $\in \mathbf{K}$

$$\mathbf{a}(u_0, w - u_0) + \mathbf{j}(\beta_0, u_0, w - u_0) + \mathbf{b}(\beta_0, u_0, w - u_0) \geq (\mathbf{F}(0), w - u_0),$$

where $\mathbf{b} = b_N + b_T$. The mapping $\mathbf{j}(.,.,.)$ satisfies the following property:

$$\exists \bar{C} > 0 \quad \forall \beta \in H \text{ with } \beta \in [0,1[\text{ a.e. on } \Gamma_c, \quad \forall u, \bar{u}, v, \bar{v} \in \mathbf{V}_0^1$$
$$|\mathbf{j}(\beta, u, v) - \mathbf{j}(\beta, u, \bar{v}) - \mathbf{j}(\beta, \bar{u}, v) + \mathbf{j}(\beta, \bar{u}, \bar{v})| \leq \bar{\mu}\bar{C}||u - \bar{u}||\,||v - \bar{v}||.$$

In the same way as in Section 19.2, the following sequence of incremental problems $\mathbf{P}_3^{n,i}$, $i = 0, ..., n-1$, can be obtained.

$\mathbf{P}_3^{n,i}$: Find $(u^{i+1}, \beta^{i+1}) \in \mathbf{K} \times H$ such that

$$\begin{aligned} &\mathbf{a}(u^{i+1}, v - u^{i+1}) + \mathbf{j}(\beta^{i+1}, u^{i+1}, v - u^i) - \mathbf{j}(\beta^{i+1}, u^{i+1}, u^{i+1} - u^i) \\ &\qquad + \mathbf{b}(\beta^{i+1}, u^{i+1}, v - u^{i+1}) \geq (\mathbf{F}^{i+1}, v - u^{i+1}) \quad \forall v \in \mathbf{K}, \end{aligned} \tag{19.18}$$

$$\beta^{i+1} - \beta^i = \Delta t\, g(\beta^{i+1}, u^{i+1}) \quad \text{a.e. on } \Gamma_c\,. \tag{19.19}$$

19.3.3 Mathematical results

We may solve the incremental problem $\mathbf{P}_3^{n,i}$ without additional regularity conditions by a fixed point method (Cangémi, 1997; Cocu et al., 1998). For every $\bar{u} \in \mathbf{K}$ let us denote by $s(\bar{u}) = \beta$ the solution of

$$\beta = \Delta t\, g(\beta, \bar{u}) + \beta^i \tag{19.20}$$

and for every $\beta \in H$ we denote by $u = u(\beta) \in \mathbf{K}$ the solution of

$$\begin{aligned} &\mathbf{a}(u, v - u) + \mathbf{j}(\beta, u, v - u^i) - \mathbf{j}(\beta, u, u - u^i) \\ &\qquad + \mathbf{b}(\beta, u, v - u) \geq (\mathbf{F}^{i+1}, v - u) \quad \forall v \in \mathbf{K}. \end{aligned} \tag{19.21}$$

The existence of the solution $s(\bar{u})$ for the problem (19.20) is clear, while inequalities such as (19.21) have a unique solution $u(\beta)$ if the friction coefficient is sufficiently small i.e. $\bar{\mu} < \bar{m}/\bar{C}$, where $\bar{m}$ is the coerciveness constant of the bilinear form $\mathbf{a}$ (Cocu, 1984; Capatina and Cocu, 1991). In the following we shall suppose that μ satisfies this condition.

We define the mapping $\mathbf{T} : \mathbf{K} \longrightarrow \mathbf{K}$ by: $\forall \bar{u} \in \mathbf{K} \quad \mathbf{T}(\bar{u}) = u(s(\bar{u}))$. By a straightforward computation, we can establish that $\exists\, k_1 > 0$ such that for all $\bar{u}_1, \bar{u}_2 \in \mathbf{K}$

$$|s(\bar{u}_1) - s(\bar{u}_2)| \leq k_1 \Delta t\, ||\bar{u}_1| + |\bar{u}_2||\, |[\bar{u}_1 - \bar{u}_2]| \quad \text{a.e. on} \quad \Gamma_c\,. \tag{19.22}$$

Let us now set $u_1 = u(s(\bar{u}_1))$, $u_2 = u(s(\bar{u}_2))$. Adding the inequalities (19.21) with $u = u_1$, $v = u_2$ and $u = u_2$, $v = u_1$, and using the properties of $\mathbf{a}$, $\mathbf{j}$ and $\mathbf{b}$, one can show that $\exists\, k_2 > 0$ such that for all $\bar{u}_1, \bar{u}_2 \in \mathbf{K}$

$$||u_1 - u_2||^2 \leq k_2 \int_{\Gamma_c} |[u_2]|\, |[u_1 - u_2]|\, |s(\bar{u}_1) - s(\bar{u}_2)| ds\,. \tag{19.23}$$

From the inequalities (19.22), (19.23) and the relation $||u|| \leq k_3$ (with k_3 independant of $\bar{u}$), satisfied by $u(s(\bar{u}))$, it follows that $\mathbf{T}$ is a contraction mapping for sufficiently small Δt. In this way the following existence and uniqueness result holds.

Theorem 19.2. *For a sufficiently small Δt the mapping $\mathbf{T}$ has a unique fixed point u and $(u, s(u))$ is the solution of the incremental problem $\mathbf{P}_3^{n,i}$.*

19.4 NUMERICAL METHODS

For dealing with unilateral contact with friction for evolution problems, either quasistatic or dynamic, the key point is to solve a "static" problem. This will be present in the subsection 19.4.1. By using the incremental formulations, algorithms adapted to quasistatic problems and to the coupling with adhesion are given in subsections 19.4.2 and 19.4.3. An extension to dynamic problems is considered in subsection 19.4.4.

19.4.1 Various solvers for the static problem

The static contact problem has been widely studied (Fichera, 1964; 1972; Duvaut and Lions, 1972; Duvaut, 1980; Demkowicz and Oden, 1982; Glowinski et al., 1976; etc.) and leads to variational or implicit variational inequalities. For a unilateral contact with Coulomb's friction law, the following problem is obtained.

$\mathbf{P}_4$: Find $u \in K$ such that

$$a(u, v-u) + j(u,v) - j(u,u) \geq (F, v-u) \quad \forall v \in K . \tag{19.24}$$

We focus in the present paper on methods treating strictly the unilateral and the friction conditions without using a regularization of the unilateral condition. Penalty and augmented lagrangian methods, which are widely used in large finite element codes, are approximations of the contact conditions and need the choice of some computational parameters. When precise analysis is conducted, such as the analysis of instabilities due to friction, see (Martins et al., 1998), or the coupling with adhesion (Raous et al., 1998; 1997), the contact solution has to be independent of computational parameters. Among the various numerical methods which can be used, we can distinguish two classes: projection techniques associated to optimization and mathematical programming. By introducing a fixed point on the sliding limit, the problem (19.24) can be written as a sequence of classical variational inequalities: find $\tilde{p}$, fixed point of the application $p \longrightarrow S(p)$ with $S(p) = \mu\,|\sigma_N(u_p)|$, where u_p is the solution of the following Tresca friction problem (i.e. when the sliding limit is given).

$\mathbf{P}_5$: Find $u_p \in K$ such that

$$a(u_p, v-u_p) + \int_{\Gamma_c} p(|v_T| - |u_{pT}|)ds \geq (F, v-u_p) \quad \forall v \in K . \tag{19.25}$$

Elimination of the implicit character gives the convexity of the problem (19.25). The following minimization problem under constraints is equivalent to $\mathbf{P}_5$.

$\mathbf{P}_6$: Find $u_p \in K$ such that $G(u) \leq G(v) \quad \forall v \in K$, where

$$G(v) = \frac{1}{2}a(v,v) - (F,v) + \int_{\Gamma_c} p|v_T|ds \quad \forall v \in K .$$

The finite element approximation can be easily written and the discrete problem consists in solving a minimization problem under constraints ($v \in K$) for a

functional including a non differentiable term due to friction. As presented in (Licht et al., 1991), an extra projection on the finite element basis of the friction term gives a simple discrete form (for the absolute value). Then various algorithms have been implemented to solve problem $\mathbf{P}_6$, see (Raous et al., 1988; Lebon and Raous, 1992; Raous and Barbarin, 1992) or in (Panagiotopoulos et al., 1998):
- Over-relaxation with projection (very easy to implement, simple treatment of the absolute value and of the unilateral condition),
- Gauss Seidel algorithm with Aitken acceleration (the convergence is equivalent to the previous method but no optimal relaxation parameter is needed),
- Projected conjugate gradient with preconditioning (a regularization of the absolute value is needed and a convenient projection is used to preserve the conjugacy property).

Convergence of the fixed point method is fast (a diagonal process is used to save the CPU time). The two algorithms based on Gauss Seidel methods are robust as they still work for large friction coefficients. A preliminary condensation of the problem to the only contact variables reduces the number of iterations (this is efficient when several loading cases are treated with the same elasticity matrix).

An alternative to these methods is given by mathematical programming techniques applied to the problem written as a linear complementarity problem. By introducing an extra variable (right and left slidings are treated separately), see (Klarbring, 1986), it has been shown that the problem can be written as follows after condensation:

$$\begin{aligned} w - A^* u &= f \\ u &\geq 0 \\ w &\geq 0 \\ u.w &= 0, \end{aligned}$$

where A^* is a non symmetric matrix because of the friction.

Various algorithms can be implemented, see (Cottle et al., 1979). Our Lemke's algorithm implementation can be found in (Raous et al., 1988; Chabrand et al., 1998) and in (Panagiotopoulos et al., 1998). This direct method (pivoting technique) is very fast but needs more programming work. Generalization to 3 D problems is possible by using a Coulomb's cone with facets, see (Klarbring, 1986). All these solvers have been implemented in our finite element code GYPTIS. Some extensions for finite deformations and plasticity problems have been implemented in SIMEM 3.

19.4.2 The quasistatic problem

It has been shown in Section 19.2 that the incremental formulation of a quasistatic contact problem with friction can be written under the form of the implicit variational inequality (19.11). At each time step, by using the same fixed point process on the sliding limit, as presented in subsection 19.4.1, the following problem has to be solved.

$\mathbf{P}_7$: Find $u^{i+1} \in K$ such that $J(u^{i+1}) \leq J(v) \quad \forall v \in K$, where

$$J(v) = \frac{1}{2} a(v,v) - (F^{i+1}, v) + \int_{\Gamma_c} p|v_T - u_T^i| ds .$$

This problem is very similar to problem $\mathbf{P}_6$. The extra term in the part related to the friction characterizes the memory due to the velocity formulation of the friction. Algorithms of subsection 19.4.1 can be used. An alternative solving based on the increments of displacements is given in (Cocu et al., 1996).

19.4.3 Coupling of adhesion and friction

Once the problems has been written under the incremental form $\mathbf{P}_3^{n,i}$, it is easy to show that, by using the fixed point argument, we have to solve for each subiteration the following problem.

$\mathbf{P}_8$: Find $(u^{i+1}, \beta^{i+1}) \in \mathbf{K} \times H$ such that

$$\mathbf{J}(u^{i+1}, \beta^{i+1}) \leq \mathbf{J}(v, \beta^{i+1}) \quad \forall v \in \mathbf{K}, \tag{19.26}$$

$$\beta^{i+1} - \beta^i = \Delta t\, g(\beta^{i+1}, u^{i+1}) \quad \text{a.e. on } \Gamma_c \text{ with} \tag{19.27}$$

$$\mathbf{J}(v, \beta^{i+1}) = \frac{1}{2} \mathbf{a}(v,v) - (\mathbf{F}^{i+1}, v) + \int_{\Gamma_c} p|v_T - u_T^i| ds + \frac{1}{2} \mathbf{b}(\beta^{i+1}, v, v),$$

where p is now the fixed point of the application $p \longrightarrow \bar{S}(p)$ and $\bar{S}(p) = \mu|\sigma_N^*(u_p) + C_N \beta^2 [u_{Np}]|$. Problem $\mathbf{P}_8$ is very similar to problem $\mathbf{P}_7$; the stiffness matrix will be modified by the extra term $\mathbf{b}(\beta^{i+1}, v, v)$ related to the stiffness of the contact. The integration of the differential equation on β is here presented with an implicit Euler method, but a θ-method have been implemented. The implicit character of the integration needs a few loops on β^{i+1} because of the nonlinearity. So, for each time t^{i+1} and for each value of β^{i+1} , the problem is solved by one of the algorithms presented in subsection 19.4.1. Numerical results and application to a fiber/matrix interface can be found in (Raous et al., 1998). Simulation of a micro-indentation experiment is carried out and an excellent agreement between numerical and experimental results has been observed.

19.4.4 An extension to dynamic problems

The LCP algorithms can be extended to the treatment of dynamic problems. LCP formulation of frictional and unilateral contact for a viscoelastic dynamic problem has been given in (Vola et al., 1998). The differential equation is written in terms of differential measures and the integration is conducted by combining two variants of θ-methods. A correct modelling of shocks and impacts is obtained.

References

Andersson, L. E. (1991). A quasistatic frictional problem with normal compliance. *Nonlinear Analysis, Theory, Methods and Applications*, 16:347–369.

Andersson, L. E. (1995). A global existence result for a quasistatic contact problem with friction. *Advances in Mathematical Sciences and Applications*, 5:249–286.

Cangémi, L. (1997). *Frottement et adhérence: modèle, traitement numérique et application à l'interface fibre/matrice.* Thèse, Université d'Aix-Marseille II , Marseille.

Capatina, A., and Cocu, M. (1991). Internal approximation of quasi-variational inequalities. *Numer. Math.*, 59:385–398.

Chabrand, P., Dubois, F., and Raous, M. (1998). Various numerical methods for solving unilateral contact problems with friction. *Mathematical and Computer Modelling*, 28:97–108.

Cocu, M. (1984). Existence of solutions of Signorini problems with friction. *Int. J. Engrg. Sci.*, 22:567–575.

Cocu, M., Cangémi, L., and Raous, M. (1999). Approximation results for a class of quasistatic contact problems including adhesion and friction. In *Proceedings of the IUTAM Symposium on the Variations of Domains and Free-Boundary Problems in Solid Mechanics-1997*, pages 211-218. Kluwer Academic Publishers.

Cocu, M., Pratt, E., and Raous, M. (1996). Formulation and approximation of quasistatic frictional contact. *Int. J. Engrg. Sci.*, 34:783–798.

Cottle, R.W., Giannessi, F., and Lions, J.L., editors (1979). *Variational Inequalities and Complementarity Problems in Mathematical Physics and Economics.* John Wiley, New York.

Demkowicz, L., and Oden, J. T. (1982). On some existence and uniqueness results in contact problems with nonlocal friction. *Nonlinear Analysis, Theory, Methods and Applications*, 6:1075–1093.

Duvaut, G. (1980). Equilibre d'un solide élastique avec contact unilatéral et frottement de Coulomb. *C. R. Acad. Sci. Paris série A*, 290:263–265.

Duvaut, G., and Lions, J. L. (1972). *Les inéquations en mécanique et en physique.* Dunod, Paris.

Fichera, G. (1964). Problemi elastostatici con vincoli unilaterali: il problema di Signorini con ambigue condizioni al contorno. *Mem. Accad. Naz. Lincei Ser. VIII*, 7:91–140.

Fichera, G. (1972). Boundary value problems of elasticity with unilateral constraints. In Flügge, S., editor, *Encyclopedia of Physics, Vol. VI a/2*, pages 391–424. Springer, Berlin.

Frémond, M. (1987). Adhérence des solides. *J. Méc. Théor. et Appl.*, 6:383–407.

Glowinski, R., Lions, J.L., and Trémolières, R. (1976). *Analyse numérique des inéquations variationnelles.* Dunod, Paris.

Klarbring, A. (1986). A mathematical programming approach to three dimensional contact problems with friction. *Comp. Meth. Appl. Mech. Engrg.*, 58:175–200.

Klarbring, A. (1990). Derivation and analysis of rate boundary problems of frictional contact. *European Journal of Mechanics A/Solids*, 9:53–85.

Klarbring, A., Mikelić, A., and Shillor, M. (1991). A global existence result for the quasistatic frictional contact problem with normal compliance. In *Inter-*

national Series of Numerical Mathematics, 101:85–111. Birkhäuser Verlag, Basel.

Lebon, F., and Raous, M. (1992). Multibody contact problems including friction in structure assembly. *Computers and Structures*, 42:925–934.

Licht, C., Pratt, E., and Raous, M. (1991). Remarks on a numerical method for unilateral contact including friction. In *International Series of Numerical Mathematics*, 101:129–144. Birkhäuser Verlag, Basel.

Lions, J.L., and Stampacchia, G. (1967). Variational inequalities. *Comm. Pure Appl. Math.*, 20:493–519.

Martins, J.A.C., Barbarin, S., Raous, M.,and Pinto da Costa, A. (1999). Dynamic stability of finite dimensional linearly elastic systems with unilateral contact and Coulomb friction. *Comp. Meth. Appl. Mech. Engrg.*, 177:289–328.

Mosco, U. (1975). Implicit variational problems and quasi variational inequalities. In *Lecture Notes in Mathematics 543: Nonlinear Operators and the Calculus of Variations, Bruxelles*, pages 83–156. Springer, Berlin.

Oden, J. T., and Pires, E. B. (1983). Nonlocal and nonlinear friction laws and variational principles for contact problems in elasticity. *ASME Journal of Applied Mechanics*, 50:67–76.

Panagiotopoulos, P., Wriggers, P., Frémond, M., Curnier, A., Klarbring, A., and Raous, M. (1998). *Contact problems : theory, methods, applications, CISM Course*. Springer, Wien. to appear.

Raous, M., and Barbarin, S. (1992). Preconditioned conjugate gradient method for a unilateral problem with friction. In Curnier, A., editor, *Contact Mechanics*, pages 423–432. Presses Polytechniques et Universitaires Romandes, Lausanne.

Raous, M., Chabrand, P., and Lebon, F. (1998). Numerical methods for frictional contact problems and applications. *Journal de Mécanique Théorique et Appliquée*, special issue, supplement n°1 to 7:111–128.

Raous, M., Cocu, M., and Cangémi, L. (1997). Un modéle couplant adhérence et frottement pour le contact entre deux solides déformables. *C. R. Acad. Sci. Paris, Série II b*325:503–509.

Raous, M., Cocu, M., and Cangémi, L. (1999). A consistent model coupling adhesion, friction, and unilateral contact. *Comp. Meth. Appl. Mech. Engng.*, 177:383–399.

Signorini, A. (1959). Questioni di elasticità nonlinearizzata et semilinearizzata. *Rend. di Matem. e delle sue appl.*, 18:1–45.

Telega, J.J. (1991). Quasi-static Signorini's contact problem with friction and duality. In *International Series of Numerical Mathematics*, 101:199–214. Birkhäuser Verlag, Basel.

Vola, D., Pratt, E., Jean, M., and Raous, M. (1998). Consistent time discretization for a dynamical frictional contact problem and complementarity techniques. *Revue Européenne des Eléments Finis*, 7:149–162.

20 REGULARITY FOR VARIATIONAL INEQUALITIES - A SURVEY OF RESULTS

Rainer Schumann

Abstract: We give an overview of the methods to prove regularity results for variational inequalities emphasizing application to elasticity.

20.1 INTRODUCTION

Many problems in physics, mechanics and mathematics itself can be stated as boundary value problems (BVP's) for differential equations (or systems of differential equations). Consider for example the Dirichlet problem

$$\begin{aligned} -\Delta u &= f \text{ in } \Omega \\ u &= g \text{ on } \partial\Omega \end{aligned} \tag{20.1}$$

where f and g are given functions and $\Omega \subset \mathbf{R}^n$ is an open bounded set with sufficiently smooth boundary $\partial\Omega$. For an application to mechanics we assume that $\Omega \subset \mathbf{R}^2$ describes a thin elastic membrane in its undeformed state. It is subjected to an outer force f. Let $u : \Omega \to \mathbf{R}$ denote the height function of the equilibrium position of the membrane being fixed ($u = g$) at the boundary $\partial\Omega$. Let us assume further that on a subset $\Omega_1 \subset\subset \Omega$ there is a lower obstacle described by a function $\psi : \Omega_1 \to \mathbf{R}$. This obstacle forces the membrane to stay above it, i.e. the height function must satisfy the unilateral condition $u \geq \psi$ on Ω_1. Another example is an elastic body occupying a domain $\Omega \subset \mathbf{R}^3$. We suppose that the body rests with a part Γ of its boundary $\partial\Omega$ on a rigid support. This support does not permit displacements of the body in the direction of the outer normal to Ω on Γ whereas tangential displacements are possible (without friction). The underlying differential operator is given by Lamé's system of linear elasticity (see 1.2 and 1.3, below) and the problem is called Signorini's

R.P. Gilbert et al. (eds.), From Convexity to Nonconvexity, 269–282.

problem (cf. [36]). This problem was solved in 1963 by Fichera [8]. We refer also to the survey article Fichera [9]).

In the second half of our century the weak formulation of BVP's using Sobolev spaces $W^{k,p}(\Omega)$ became a standard instrument to treat the existence of solutions (Lions [26], Zeidler [39]). In many cases functional analytic arguments (variational principles, theory of monotone operators, e.g.) can be applied to get the existence of solutions of BVP's in a relatively easy way. These so called direct methods of the calculus of variation may also be applied to problems involving unilateral constraints.

Since functions from Sobolev spaces are not smooth in general the following question arises naturally: Are weak solutions also classical (strong) solutions? That means do they belong the spaces C^k ($k \geq 1$)? This part of the investigation of BVP's is usually more complicated than pure existence theory. For the BVP (1.1) the result can roughly be stated as: the smoother the right hand sides f and g are the smoother is the solution u. This statement is not longer true if we consider problems with unilateral constraints.

In the following section problems with unilateral constraints are formulated as variational inequalities. It will be shown in this survey that variational inequalities need their own regularity theory.

20.2 VARIATIONAL INEQUALITIES AND THEIR APPLICATIONS

The membrane problem

Suppose for simplicity $g = 0$. It is well-known that (1.1) arises as the Euler equation of the following variational problem: Find $u \in X_0 = W_0^{1,2}(\Omega)$ such that

$$J(u) \leq J(v) \text{ for all } v \in X_0 \tag{20.2}$$

with $J(u) = \int_\Omega \sum_{i=1}^n |\partial_i u(x)|^2 \, dx - \int_\Omega f(x)u(x)dx$. The functional J is convex and lower semicontinuous on X_0. Furthermore (1.2) is equivalent to the weak formulation of (1.1): Find $u \in X_0$ such that

$$a(u,v) = b(v) \text{ for all } v \in X_0 \tag{20.3}$$

where $a(u,v) = \int_\Omega \sum_{i=1}^n \partial_i u(x)\partial_i v(x)dx$, $\quad b(v) = \int_\Omega f(x)v(x)dx$. (See [39], Vol. 2A, p.56)

In the presence of an obstacle ψ given on $\Omega_1 \subset\subset \Omega$ we consider the restricted minimum problem

$$\min\{J(v) : v \in \mathbf{K}\} \tag{20.4}$$

where $\mathbf{K} = \{v \in X_0 : v \geq \psi \text{ a.e. on } \Omega_1\}$. The set $\mathbf{K}$ is a closed convex subset of X_0. Then a necessary and sufficient condition for $u \in \mathbf{K}$ to be a solution of (1.4) is the variational inequality

$$a(u, v-u) \geq b(v-u) \text{ for all } v \in \mathbf{K}. \tag{20.5}$$

(We remark that in this context (1.3) is sometimes called a variational equation.) Set $X := W^{1,2}(\Omega)$. The bilinear form $a(.,.)$ defines a linear and continuous operator $A : X \to X_0^*$ by

$$\langle Au, v\rangle = a(u,v) \text{ for all } v \in X_0. \tag{20.6}$$

Thus (1.5) can also be written in the equivalent form

$$\langle Au, v-u\rangle \geq b(v-u) \text{ for all } v \in \mathbf{K}. \tag{20.7}$$

Remark that (1.7) is a nonlinear problem even if A is a linear operator. We now may consider abstract variational inequalities of the form (1.7) where $\mathbf{K}$ is a closed convex subset of a Banach space X, $A : X \to X^*$ is a linear or nonlinear operator and $b \in X^*$ can chosen arbitrarily. Many problems of mechanics and other applications of variational methods in science lead to nonconvex functionals J. Such problems are called hemivariational inequalities. Their treatment is more complicated than that of the classical variational inequalities considered here. We refer to Naniewicz and Panagiotopoulos [29].

Obstacles on the boundary

Let us consider the variational inequality (1.5) where a and b are given as above while $\mathbf{K} = \left\{v \in W^{1,2}(\Omega) : v \geq 0 \text{ a.e. on } \partial\Omega\right\}$. It is easy to see that a regular solution of (1.5) satisfies the complementary relations

$$\begin{gathered} -\Delta u = f \text{ in } \Omega \\ u \geq 0,\ \frac{\partial u}{\partial n} \geq 0, \text{ and } u\frac{\partial u}{\partial n} = 0 \text{ on } \partial\Omega. \end{gathered} \tag{20.8}$$

(cf.[26], p. 243) This problem is called an variational inequality with a thin obstacle since the dimension of $\partial\Omega \subset \mathbf{R}^n$ is $n-1$. Sometimes the constraint in 1.2 is called a thick obstacle. The regularity theory for thin obstacles is more involved, in general.

Signorini's problem in linear elasticity

The Signorini problem mentioned in the introduction is also a thin obstacle problem since the constraint is given on $\Gamma \subset \partial\Omega$. In detail the problem can be described as follows. Suppose that $\Omega \subset \mathbf{R}^n$, $\partial\Omega = \overline{\Gamma} \cup \overline{\Gamma}_D$ where the sets Γ and Γ_D are disjoint, open, smooth and simply connected submanifolds of $\partial\Omega$. The linear elastic behavior of the body occupying Ω is described by the elastic coefficients $a_{ij}^{\alpha\beta}$ satisfying

(i) symmetry: $a_{ij}^{\alpha\beta} = a_{\alpha j}^{i\beta} = a_{i\beta}^{\alpha j} = a_{ji}^{\beta\alpha}$ $(i,j,\alpha,\beta = 1,...,n)$

(ii) ellipticity: $a_{ij}^{\alpha\beta}\xi_\alpha^i\xi_\beta^j \geq c_0\left|\xi\right|^2$ for all $\xi \in \mathbf{R}^{n^2}$ with $\xi_\alpha^i = \xi_i^\alpha$ $(i,\alpha = 1,...,n,$ summation convention applied, $\left|\xi\right|^2 = \sum_{i,\alpha=1}^n \left|\xi_\alpha^i\right|^2$, $c_0 > 0)$. The stress tensor is given by $\sigma_{i\alpha}(u) = \sum_{j,\beta=1}^n a_{ij}^{\alpha\beta}e_{j\beta}(u)$ where $e_{j\beta}(u) = \frac{1}{2}(\partial_j u^\beta + \partial_\beta u^j)$ denotes the small strain tensor corresponding to the displacement vector $u : \overline{\Omega} \to \mathbf{R}^n$. Clearly $\sigma_{i\alpha}(u) = \sum_{j,\beta=1}^n a_{ij}^{\alpha\beta}\partial_\beta u^j$. We set

$$a(u,\zeta) = \int_\Omega a_{ij}^{\alpha\beta}\partial_\beta u^j \partial_\alpha \zeta^i dx. \tag{20.9}$$

Then the elastic energy is given by $\frac{1}{2}a(u,u)$. The action of the outer volume forces is $b(\zeta) = \int_\Omega f_i\zeta^i dx$. Denote by $\mathbf{n}$ the outer normal to $\partial\Omega$. Solving Signorini's problem means to find

$$u \in \mathbf{K} = \{\zeta \in W^{1,2}(\Omega, \mathbf{R}^n) : \zeta \cdot \mathbf{n} \leq 0 \text{ a.e. on } \Gamma, \zeta = \tilde{u} \text{ on } \Gamma_D\}$$

such that

$$a(u, v-u) \geq b(v-u) \text{ for all } v \in \mathbf{K}. \tag{20.10}$$

Here we suppose that $\tilde{u} \in W^{1,2}(\Omega, \mathbf{R}^n)$ describes the displacement vector given on Γ_D and satisfies $\tilde{u} \cdot \mathbf{n} \leq \mathbf{0}$ a.e. on Γ. Thus $\mathbf{K} \neq \emptyset$. Clearly (1.10) is the variational formulation of the problem to find an equilibrium configuration of the elastic body under an unilateral constraint on Γ preventing displacements of the body in the direction of the outer normal to Ω on Γ whereas tangential displacements are possible. Furthermore we assume $f \in L_2(\Omega, \mathbf{R}^n)$ for the volume force. We want to stress that the solution u is a vector valued function. This gives additional difficulties for regularity investigations. If the material is homogeneous and isotropic then the elastic coefficients are given by

$$a_{ii}^{ii} = 2\mu + \lambda, a_{ij}^{ij} = \lambda, a_{ii}^{jj} = \mu \text{ for } i \neq j. \tag{20.11}$$

From symmetry we get $a_{ji}^{ij} = \mu$ for all $i \neq j$. In the remaining cases we have $a_{ij}^{\alpha\beta} = 0$. Here $\lambda, \mu > 0$ are the Lamé coefficients. We remark that many other applications may be found in Friedman [13], Kinderlehrer/Stampacchia [?], and Troianiello [37].

20.3 REGULARITY

In contrast to the theory of elliptic BVP's it is not longer true that smooth data imply smooth solutions. The presence of obstacles has the consequence that there is a threshold of smoothness that in general cannot be surpassed even if the data are arbitrarily smooth.

Thick obstacles

We want to present several methods for gaining regularity results. To avoid long formulas in this survey we use $A = -\Delta$ in most cases. It is possible to apply these methods also for more general linear or quasilinear equations. We cite some results later.

A. Penalization. We show that $u \in W^{2,p}(\Omega)$ for all $p \geq 2$. Then from the embedding theorem it follows that $u \in C^{1,\alpha}(\Omega)$ for all $\alpha \in (0,1)$. The idea is to replace the variational inequality with a family of semilinear equations and to demonstrate the convergence of their solutions to the solution of the variational inequality. It is our goal here to prove a-priori estimates in appropriate function spaces to get regularity theorems for the solution of the variational inequality (cf. Brézis [4], Kinderlehrer/Stampacchia [22]). Consider for example the membrane problem of 1.2 where we now suppose $\Omega_1 = \Omega$, $\psi < 0$ on $\partial\Omega$ and f, ψ and $\partial\Omega$ sufficiently smooth. The penalty equation are

$$-\Delta u_\varepsilon + \varepsilon^{-1}[u_\varepsilon - \psi]^- = f \text{ in } \Omega \qquad (20.12)$$
$$u_\varepsilon = 0 \text{ on } \partial\Omega$$

where $\varepsilon > 0$ is the penalty parameter forcing $u \geq \psi$ in the limit $\varepsilon \to 0$.

The weak formulation of this BVP is to find $u_\varepsilon \in X_0 = W_0^{1,2}(\Omega)$ such that

$$\int_\Omega \sum_{i=1}^n \partial_i u_\varepsilon \partial_i v dx + \varepsilon^{-1} \int_\Omega [u_\varepsilon - \psi]^- v dx = \int_\Omega f v dx \text{ for all } v \in X_0. \quad (20.13)$$

Here any function w is decomposed as $w = w^+ + w^-$, $w^+(x) = \max(w(x), 0)$, $w^-(x) = \min(w(x), 0)$. Set $\langle t\rangle^{p-1} = |t|^{p-2}t$ for $p > 1$. Plugging in the test functions $v = \left\langle [u_\varepsilon - \psi]^- \right\rangle^{p-1}$ into (1.13), using Hölder's and Young's inequalities it is not difficult to arrive at the uniform estimate

$$\varepsilon^{-1} \left\| [u_\varepsilon - \psi]^- \right\|_{L_p(\Omega)} \leq const \text{ for all } 0 < \varepsilon \leq \varepsilon_0.$$

This implies $\|-\Delta u_\varepsilon\|_{L_p(\Omega)} \leq const$ for all $0 < \varepsilon \leq \varepsilon_0$. Then the standard elliptic theory yields: $\|u_\varepsilon\|_{W^{2,p}} \leq const$ for all $0 < \varepsilon \leq \varepsilon_0$. The penalization implies $u_\varepsilon \to u$ in X_0 if $\varepsilon \to 0$ where u is the solution of (1.5).This eventually shows $u \in W^{2,p}(\Omega)$, for all $p \geq 2$. Moreover it follows that $\|\Delta u_\varepsilon\|_{L_\infty(\Omega)} \leq const$ for all $0 < \varepsilon \leq \varepsilon_0$.

B. Maximum principles. We are going to show that u has bounded second derivatives. Since the solution of (1.5) is a weak supersolution to the corresponding elliptic equation also methods based on the maximum principle can be applied. Estimating $\left\| [u-k]^+ \right\|_{L_2}$ and using a growth lemma of Hartmann and Stampacchia ([18]) gives the following maximum principle for weak subsolutions (resp. supersolutions) of $-\Delta u = f$.

Lemma 20.1. *Suppose $f = (f_i) \in L_p(\Omega, \mathbf{R}^n)$, $p > n$ and $u \in W^{1,2}(\Omega)$ satisfies*

$$a(u,v) = \int_\Omega \sum_{i=1}^n \partial_i u(x) \partial_i v(x) dx \leq \int_\Omega f_i(x) \partial_i v(x) dx \text{ (resp. } \geq \text{)} \qquad (20.14)$$

for all $0 \leq v \in W_0^{1,2}(\Omega)$.

Then there exists $C > 0$ depending only on p, n and meas(Ω) such that

$$\|u\|_{L_\infty} \leq \max\left(\sup_{\partial\Omega} u, 0\right) + C\|f\|_{L_p}.$$

If "$\geq$" holds in (1.14) then $\|u\|_{L_\infty} \geq \min(\inf_{\partial\Omega} u, 0) - C\|f\|_{L_p}$.

Proof. Rodrigues [31], p.166.

Let us introduce the test function $v = u + \varepsilon\nabla_{-h,\xi}\nabla_{h,\xi}\varphi$, where $0 \leq \varphi \in C_0^\infty(\Omega)$, $u \in \mathbf{K}$ is the solution of (1.5), $\nabla_{h,\xi}u = h^{-1}(u(x+h\xi) - u(x))$ and $\nabla_{-h,\xi}u = h^{-1}(u(x) - u(x-h\xi))$ are the forward and backward difference quotients in the direction of $\xi \in \mathbf{R}^n$, $|\xi| = 1$. Lemma 1 gives a lower bound for the second differences, and $h \to 0$ implies a lower bound for the second derivatives in the ξ-direction: There exists $k > 0$ such that

$$\inf\left\{\frac{\partial^2 u}{\partial \xi^2} : \xi \in \mathbf{R}^n, |\xi| = 1\right\} \geq -k \text{ a.e. on } \Omega.$$

Then a lemma from linear algebra implies the boundedness of D^2u: $u \in W^{2,\infty}(\Omega)$. See Brézis [4], Kinderlehrer/Stampacchia [22], Ch. 4, Rodrigues [31], Ch. 5, Gerhardt [15].

It is easily seen that u need not belong to C^2. Indeed consider for example $\Omega = (-3,3) \subset \mathbf{R}$, $\psi(x) = 1 - x^2$, $a(u,v) = \int_{-3}^{3} u'v'dx$, $f = 0$ and solve (1.5). The solution is $u(x) = 1 - x^2$ for $|x| \leq a = 3 - 2\sqrt{2}$ and is a linear function for $|x| > a$ whose graph is tangent to the parabola at a and satisfies $u(\pm 3) = 0$. Thus $u''(x) = 0$ if $|x| > a$ while $u''(x) = -2$ for $|x| \leq a$. Observe that u is the smallest weak supersolution and thus solves the variational inequality.

The $W^{2,\infty}$-regularity result remains true for elliptic equations with coefficients $a_{ij} \in W^{2,q}(\Omega)$, $q > n$ (Rodrigues [31], Ch. 5, Uralzeva [38]). We remark that other penalty functionals can be used, for example

$$\beta_\varepsilon(u)(v) = \int_\Omega h(x)\theta\left(\frac{u(x) - \psi(x)}{\varepsilon}\right) v(x)dx$$

where $\theta : \mathbf{R} \to [0,1]$ is an nondecreasing Lipschitz function with $\theta(t) = 0$ for $t \leq 0$, $\lim_{t\to+\infty}\theta(t) = 1$ such that $[1-\theta(t)]\,t \leq C$ for all $t > 0$ and $0 \leq h \in L_p(\Omega)$ satisfies $h \geq (A\psi - f)^+$ (Rodrigues [31], p.144). De Giorgi's theorem and an approximation procedure still allow to derive $u \in C^\alpha(\Omega)$ for some $\alpha \in (0,1)$ if $\psi \in C^\beta(\Omega)$ and $a_{ij} \in L_\infty$ (Biroli [3], Frehse [11]). Irregular obstacles were treated by Frehse and Mosco [12]. Hildebrandt and Widman [19] investigated variational inequalities for vector valued functions with obstacles on Ω.

C. Campanato space technique. Another method is proposed by Giaquinta in [16]. The idea is to consider (small) balls B_R and to compare on B_R the solution of the variational inequality with the harmonic function having the same boundary values on ∂B_R. This method allows to prove $C^{1,\alpha}$-regularity for all $\alpha \in (0,1)$ directly without referring to $W^{2,p}$-regularity first. Consider for example the membrane problem from 1.2. Let u denote its solution and suppose $U \in W^{1,2}(B_R(x_0))$ is the solution of

$$\int_{B_R}\sum_{i=1}^{n}\partial_i U\partial_i v dx = 0 \text{ for all } v \in W_0^{1,2}(B_R)$$

$$U - u \in W_0^{1,2}(B_R).$$

For the harmonic function U we get from Campanato's inequality (see Giaquinta [17], p. 78)

$$\int_{B_r} |DU|^2\,dx \le c_1 \left(\frac{r}{R}\right)^n \int_{B_R} |DU|^2\,dx \le c_1 \left(\frac{r}{R}\right)^n \int_{B_R} |Du|^2\,dx$$

and from the variational inequality (1.5)

$$\int_{B_R} \sum_{i=1}^n \partial_i(u-U)\partial_i(u-v)dx \le \int_{B_R} f(u-v)dx$$

for all v with $v \ge \psi$ in $B_R(x_0)$ and $v-u \in W_0^{1,2}(B_R)$. Choosing $v = \max(U,\psi)$ one eventually concludes

$$\int_{B_r} |Du|^2\,dx \le c_2 \left(\frac{r}{R}\right)^n \int_{B_r} |Du|^2\,dx + c_3 R^{n-2+2a} \tag{20.15}$$

where we suppose $\psi \in C^\alpha$. From (1.15) one derives $u \in C^\alpha$(see Giaquinta [17], p. 86). A refinement of this idea shows the following result.

Proposition 20.1. *If $\psi \in C^{1,\alpha}$ then $u \in C^{1,\alpha}$ for the solution of (1.5).*

The method can be extended to elliptic equations with L_∞–coefficients and quasilinear elliptic equations. The p–Laplacian was investigated by Fuchs [14]. He proved that $\psi \in W_{loc}^{2,\infty}(\Omega)$ implies $u \in C_{loc}^{1,\alpha}(\Omega)$ for all $\alpha < 1$ and also showed that the coincidence set has finite perimeter.

Thin obstacles

Let us consider the following simplified geometric situation. Suppose $\Omega \subset \mathbf{R}^n$ $(n \ge 2)$ is a bounded domain with smooth connected boundary $\partial\Omega = \Gamma \cup \Gamma'$ consisting of two disjoint $(n-1)$-dimensional open manifolds Γ and Γ' $(\Gamma, \Gamma' \ne \emptyset)$. We assume $\Gamma = \{x \in \partial\Omega : x_n = 0\}$, $\Gamma' = \{x \in \partial\Omega : x_n > 0\}$ where $x = (x', x_n)$.

We first mention that there is a counterexample due to Shamir [35] that shows that already in dimension $n = 2$ a solution to the thin obstacle problem given in 1.2 may fail to belong to $W_{loc}^{2,4}(\Omega) \cup W_{loc}^{3,2}(\Omega) \cup C^{1,a}(\Omega)$ for $\alpha \in \left(\frac{1}{2}, 1\right)$. Richardson [30] proved that for $n = 2$ the optimal regularity is $C^{1,1/2}(\Omega \cup \Gamma)$. The case $n > 2$ was treated by Frehse [10], Kinderlehrer [23] and [38]. Suppose the data are sufficiently smooth. Consider

$$a(u, v-u) \ge b(v-u) \text{ for all } v \in \mathbf{K} \tag{20.16}$$

where $a(u,v) = \int_\Omega \sum_{i=1}^n \partial_i u(x)\partial_i v(x)dx$, $\mathbf{K} = \{v \in X : v \ge 0 \text{ on } \Gamma\}$ and define $X = \{v \in W^{1,2}(\Omega) : v = 0 \text{ on } \Gamma'\}$. Then we have the following result.

Proposition 20.2. *There exists $\alpha \in (0, \frac{1}{2})$ such that the solution of (1.16) satisfies $u \in C^{1,a}(\Omega)$.*

Let us briefly indicate the technique of the proof. Since regularity is clear on $\Omega \cup \Gamma'$ we consider $x_0 \in \Gamma$. One of the problems when treating variational inequalities is to find appropriate test functions from $\mathbf{K}$. This can be done here by choosing a cut off function $\eta \in C_0^\infty(B_{2r}(x_0))$ with $\eta = 1$ on $B_r(x_0)$, $|D\eta| \le c/r$ and setting for $\varepsilon > 0$

$$v_\varepsilon = u + \varepsilon \nabla_{-h}(\eta^2 \nabla_h u).$$

where $\nabla_h u = h^{-1}(u(x + he_m) - u(x))$, $\nabla_{-h} u = h^{-1}(u(x) - u(x - he_m))$ and e_m denotes the unit vector in the direction of x_m, $m = 1, ..., n-1$. It is easy to see that

$$v_\varepsilon = \left[1 - \frac{\varepsilon}{h^2}(\eta(x)^2 + \eta(x - he_m)^2)\right] u(x) \qquad (20.17)$$
$$+ \frac{\varepsilon}{h^2}\left[\eta(x)^2(u(x + he_m) + \eta(x - te_m)^2 u(x - he_m)\right] \in \mathbf{K}$$

for small fixed $r > 0$, $0 < h < r$ and $0 < \varepsilon < \frac{h^2}{2}$. Plugging in v_ε in (1.16) it follows by standard estimates from elliptic theory that

$$\int_{B_{2r}} \sum_{i=1}^{n} \eta^2 \left|\nabla_h \partial_i u(x)\right|^2 dx \le C \text{ for all } h \in (0, r). \qquad (20.18)$$

Therefore $u \in W^{2,2}_{loc}(\Omega \cup \Gamma)$ (see Giaquinta[17], p. 45). We remark that the method of difference quotients can also be applied in the case of Signorini's problem, see Fichera [9], p.408). Furthermore this test function technique allows to prove estimates for the second derivatives D^2u of the following form

$$\int_{B_r^+} \left|D^2u(x)\right|^2 dx \le \frac{C}{r^2} \int_{B_{2r}^+ - B_r^+} |Du(x) - c|^2 dx + C \int_{B_{2r}^+} \left(|f|^2 + \left|D^2\psi(x)\right|^2\right) dx.$$

where $B_r^+ = B_r^+(x_0) = B_r(x_0) \cap \Omega$ for $x_0 \in \Gamma \subset \partial\Omega$. The complementary relation $u\frac{\partial u}{\partial n} = 0$ on Γ from (1.8) and Poincaré's inequality imply

$$\int_{B_r^+} \left|D^2u(x)\right|^2 dx \le C \int_{B_{2r}^+ - B_r^+} \left|D^2u(x)\right|^2 dx + Cr^n.$$

The hole-filling trick then gives

$$\int_{B_r^+} \left|D^2u(x)\right|^2 dx \le \theta \int_{B_{2r}^+} \left|D^2u(x)\right|^2 dx + Cr^n \text{ where } 0 < \theta < 1.$$

Finally a growth estimate (see [22], p. 81) shows that there exists $\alpha > 0$ such that

$$\int_{B_r(x_0)} \left|D^2u\right|^2 dx \le cr^{2\alpha}$$

and Morrey's lemma can be used to prove $Du \in C^\alpha(\Omega)$ if $n = 2$. Kinderlehrer ([24]) used the same technique to prove the $C^{1,\alpha}$-regularity for Signorini's problem in plane elasticity ($n = 2$). Remark that this procedure fails for the elliptic

Lamé system in dimension $n \geq 3$, since the Morrey lemma states that $u \in C^{1,a}_{loc}$ if $\int_{B_r(x_0)} |D^2u|^2 \, dx \leq cr^{n-2+2\alpha}$. It is not clear how to obtain the additional factor r^{n-2} in Signorini's problem.

Ignoring the case of systems for a moment a generalization for scalar equations is known if $n \geq 3$. Kinderlehrer ([23]) uses Green's function as test function to get the factor r^{n-2}. Frehse ([10]) shows that $Du \in C^1$ with a logarithmic modulus of continuity for quasilinear problems using a variant of Morrey's lemma. Uralzeva ([38]) uses level set techniques to get $C^{1,a}$ for elliptic equations with $a_{ij} \in W^{2,q}(\Omega)$, $q > n$ and two-sided obstacle problems $\varphi \leq u \leq \psi$ on $\partial\Omega$.

Signorini's problem

It was pointed out in the last section that an alternative method is necessary to show $C^{1,a}$-regularity for Signorini's problem in $\mathbf{R}^n$ $(n > 3)$. We consider the case of an homogeneous and isotropic material where the elastic coefficients are given in (1.11). The idea is to transform the variational inequality (1.10) to a problem on the boundary. It turns out that the transformed problem is a scalar problem on $\partial\Omega$, now with a thick obstacle on $\Gamma \subset \partial\Omega$. For the sake of simplicity we assume the geometric situation described at the beginning of 1.3. Also for technical simplicity we suppose that Ω is not rotational symmetric. This implies coerciveness of strains (Korn's inequality, see Nečas and Hlavaček [28], p. 79, 102).

In 1.2 we defined the bilinear form $a(u, v)$ associated to the elastic energy. The equation

$$\langle Au, v\rangle = a(u, v) \text{ for all } v \in X_0 = W_0^{1,2}(\Omega, \mathbf{R}^n)$$

defines a linear and continuous operator $A : X \to X_0^*$ called Lamé's operator. This operator replaces the Laplacian in 1.2.

Since reduction to the boundary is somewhat tedious for Signorini's problem we first demonstrate an analogous result for the Laplacian.

A. A model problem. The key is Green's formula: Suppose $u, v \in C^1(\overline{\Omega})$. Then

$$\int_\Omega \sum_{i=1}^n \partial_i u \partial_i v dx = \int_{\partial\Omega} \frac{\partial u}{\partial \mathbf{n}} v dS - \int_\Omega \Delta u \, v dx.$$

If $\Delta u = 0$ then the formula reads as

$$\left\langle \frac{\partial u}{\partial \mathbf{n}}, v \right\rangle = \int_\Omega Du \cdot Dv dx. \tag{20.19}$$

Here $\langle .,. \rangle$ denotes the duality pairing between $W^{-1/2,2}(\partial\Omega)$ and $W^{-1/2,2}(\partial\Omega)$. This immediately leads to the idea to write the Dirichlet form $a(u, v)$ as a bilinear form on $\partial\Omega$. For simplicity we show this if Ω is the upper half space $\mathbf{R}^n_+$. General domains $\Omega \subset \mathbf{R}^n$ can be reduced to this situation by localization and coordinate transforms. We consider the following classical Dirichlet problem:

$$\begin{aligned} \Delta u &= 0 \text{ in } \mathbf{R}^n_+ \\ u &= g \text{ on } \partial\mathbf{R}^n_+ \cong \mathbf{R}^{n-1} \end{aligned} \tag{20.20}$$

where $g \in C_0^\infty(\mathbf{R}^{n-1})$ is given. We define $P_1 g = \frac{\partial u}{\partial \mathbf{n}}$ on $\partial\mathbf{R}^n_+$. Partial Fourier transform with respect to $x' \in \mathbf{R}^{n-1}$ shows that the bounded solution to (1.20) is

$$u(x', x_n) = \frac{1}{(2\pi)^{n-1}} \int_{\mathbf{R}^{n-1}} e^{ix'\xi'} e^{-x_n|\xi'|} \widehat{g}(\xi') d\xi'.$$

Thus

$$(P_1 g)(x') = -\frac{\partial u}{\partial x_n}(x', 0) = \frac{1}{(2\pi)^{n-1}} \int_{\mathbf{R}^{n-1}} e^{ix'\xi'} \, |\xi'| \, \widehat{g}(\xi') d\xi'. \tag{20.21}$$

Therefore $P_1 : C_0^\infty(\Omega) \to C^\infty(\Omega)$ is a classical proper pseudodifferential operator of order 1 with principal symbol $\sigma_1(P_1)(x', \xi') = |\xi'|$. Now it follows from (1.19) that

$$\langle P_1 g, v \rangle = \int_\Omega Du \cdot Dv dx.$$

B. A contact BVP. For $a(u, v)$ defined in (1.9) integration by parts gives

$$\begin{aligned} a(u, v) &= \int_\Omega Au \cdot v dx + \int_{\partial\Omega} T(\mathbf{n})u \cdot v dS \\ &= \int_\Omega Au \cdot v dx + \int_{\partial\Omega} T_{\mathbf{n}}u \cdot v_{\mathbf{n}} dS + \int_{\partial\Omega} T_{\mathbf{t}}u \cdot v_{\mathbf{t}} dS. \end{aligned} \tag{20.22}$$

For any deformation u the boundary stress vector is defined as $T(\mathbf{n})u$ with ith coordinate $(T(\mathbf{n})u)_i = \sum_{j=1}^n \sigma_{ij}(u) n_j$. Here $\mathbf{n} = (n_1, ..., n_n)$. The components of $T(\mathbf{n})u$ in tangential and normal directions, respectively are given by the vectors $T_{\mathbf{n}}u = ((T(\mathbf{n})u) \cdot \mathbf{n})\, \mathbf{n}$ and $T_{\mathbf{t}}u = T(\mathbf{n})u - T_{\mathbf{n}}u$. Analogously $u_{\mathbf{n}} = (u \cdot \mathbf{n})\mathbf{n}$ and $u_{\mathbf{t}} = u - u_{\mathbf{n}}$ (see Ciarlet [6], Nečas and Hlavaček [28]).

It was pointed out in the introduction and in 1.2 that for Signorini's problem tangential displacements are possible without friction. Therefore in an equilibrium configuration the tangential stress $T_{\mathbf{t}}u$ should be zero. To proceed analogously to (1.20), (1.21) we consider a boundary contact problem now. Suppose we are given $\varphi \in C_0^\infty(\Omega)$. Let $\Phi \in C^\infty(\Omega, \mathbf{R}^n)$ be the unique solution of

$$\begin{aligned} A\Phi &= 0 \text{ in } \Omega \\ -\Phi \cdot \mathbf{n} &= \varphi \text{ on } \partial\Omega \\ T_{\mathbf{t}}\Phi &= 0 \text{ on } \partial\Omega \end{aligned} \tag{20.23}$$

We set

$$P\varphi = -(T(\mathbf{n})\Phi) \cdot \mathbf{n}. \tag{20.24}$$

Equation (1.22) implies

$$a(\Phi, v) = \int_{\partial\Omega} T_{\mathbf{n}}\Phi \cdot v_{\mathbf{n}} dS = -\int_{\partial\Omega} P\varphi\ (v \cdot \mathbf{n})\, dS$$

We remark that many different contact problems in elasticity can be found in Jentsch [21], Kupradze [25]. On the flat part $\Gamma \subset \partial\Omega$ we have $\mathbf{n} = (0, ..., 0, -1)$ and the boundary conditions of (1.23) read as follows:

$$\Phi^n = \varphi,\ \sigma_{1n}(\Phi) = \sigma_{2n}(\Phi) = ... = \sigma_{n-1,n}(\Phi) = 0.$$

Furthermore $P\varphi = -\sigma_{nn}(\Phi)$ and

$$a(\Phi, v) = -\int_{\partial\Omega} P\varphi\ (v \cdot \mathbf{n})\, dS = \int_{\Gamma} (-\sigma_{nn}(\Phi))\ v^n dS - \int_{\Gamma'} P\varphi\ (v \cdot \mathbf{n})\, dS$$

Problem (1.23) can be formulated and solved in weak form in Sobolev spaces. We state properties of the operator P:

Proposition 20.3. *(i) The boundary contact problem (1.23) has exactly one weak solution $\Phi \in W^{1,2}(\Omega, \mathbf{R}^n)$ for any $\varphi \in W^{1/2,2}(\partial\Omega)$.*

(ii) The operator $P : C_0^\infty(\Omega) \to C^\infty(\Omega)$ is a classical proper pseudodifferential operator of order 1. P extends uniquely to a linear, continuous operator from $W^{1/2,2}(\partial\Omega)$ into $W^{-1/2,2}(\partial\Omega)$ which we denote by P, too.

(iii) For any $\varphi_1, \varphi_2 \in W^{1/2,2}(\partial\Omega)$ we have

$$\langle P\varphi_1, \varphi_2 \rangle = a(\Phi_1, \Phi_2) \tag{20.25}$$

where Φ_1 is the solution of (1.23) corresponding to φ_1 and $\Phi_2 \in W^{1,2}(\Omega, \mathbf{R}^n)$ is an arbitrary vector function satisfying $-\Phi_2 \cdot \mathbf{n} = \varphi_2$ on $\partial\Omega$.

(iv) The principal symbol of P on Γ is

$$\sigma_1(P)(x', \xi') = \frac{2\mu(\mu + \lambda)}{(2\mu + \lambda)} |\xi'| \tag{20.26}$$

for $(x', 0) \in \Gamma$, $\xi' \in \mathbf{R}^{n-1} - \{0\}$.

The proof can be found in Schumann ([32]), ([33]). It uses the theory of elliptic BVP's and their reduction to the boundary as given in Hörmander ([20], Ch. 20). Since the fundamental solution of A is known one can construct the Caldéron projector.

C. Reduction for Signorini's problem. The first step is to reduce Signorini's problem (1.10) involving Dirichlet data on the upper part Γ' of $\partial\Omega$ by a cut-off procedure to the following variational inequality with a contact boundary condition on Γ': Find

$$u \in \mathbf{K}_0 = \{\zeta \in W^{1,2}(\Omega, \mathbf{R}^n) : \zeta^n \geq 0 \text{ on } \Gamma,\ \zeta \cdot \mathbf{n} = \tilde{u} \cdot \mathbf{n} \text{ on } \Gamma'\}$$

such that

$$a(u, v-u) \geq b(v-u) \text{ for all } v \in \mathbf{K}_0. \tag{20.27}$$

We suppose that $\widetilde{u} \in W^{1,2}(\Omega, \mathbf{R}^n)$ is a given function describing the boundary values on Γ' and satisfies $\widetilde{u}^n \geq 0$ on Γ. Thus $\mathbf{K}_0 \neq \emptyset$. Define

$$X = \{\zeta \in W^{1,2}(\Omega, \mathbf{R}^n) : \zeta \cdot \mathbf{n} = 0 \text{ on } \Gamma'\}.$$

It follows from our geometric assumptions that $a(.,.)$ is coercive on X. Therefore there exists exactly one solution u of (1.27). It is easy to see that the solution of (1.27) belongs to the same regularity class as the solution of (1.10).

To complete the transformation of (1.27) to a variational inequality on $\partial\Omega$ we have to solve the following mixed BVP

$$\begin{aligned} Av &= f \text{ in } \Omega \\ T(\mathbf{n})v &= 0 \text{ on } \Gamma \\ v \cdot \mathbf{n} &= \widetilde{u} \cdot \mathbf{n} \text{ on } \Gamma' \\ T_{\mathbf{t}} v &= 0 \text{ on } \Gamma'. \end{aligned} \tag{20.28}$$

Again, there is exactly one weak solution v of (1.28) solving

$$a(v, \varphi) = b(\varphi) \text{ for all } \varphi \in X.$$

The difference $w = u - v$ eventually solves

$$a(w, \varphi - w) \geq 0 \text{ for all } \varphi \in \mathbf{K} \tag{20.29}$$

where $\mathbf{K} = \left\{\varphi \in X : \varphi^n \geq -v^n \stackrel{def}{=} \psi \text{ a.e. on } \Gamma\right\}$.

Now we set $\mathbf{K}_1 = \{\zeta \in W^{1/2,2}(\partial\Omega) : \zeta \geq \psi \text{ a. e. on } \Gamma, \zeta = 0 \text{ on } \Gamma'\}$ and consider the problem to find $g \in \mathbf{K}_1$ solving the (scalar) variational inequality on $\partial\Omega$

$$\langle Pg, v - g\rangle \geq 0 \text{ for all } v \in \mathbf{K}_1. \tag{20.30}$$

There exists exactly one solution of (1.30). By (1.25) the unique solution w of (1.23) with φ replaced by g satisfies (1.29). The result discussed in 1.3 shows that the solution of (1.30) satisfies $g \in C^{1,\alpha}_{loc}(\Gamma)$. This result and standard regularity theory for elliptic BVP's imply $w \in C^{1,\alpha}_{loc}(\Omega \cup \Gamma)$. Therefore we finally get the following proposition.

Proposition 20.4. *The solution u of (1.10) belongs to $C^{1,\alpha}_{loc}(\Omega \cup \Gamma)$ for some $\alpha \in (0,1)$.*

References

[1] Archipova, A.A., and N.N. Uralzeva: On the regularity of the solution of a problem with two-sided obstacles on the boundary (Russian). Vestnik Leningr. Univ., Ser. 1, 1986, No. 1, 3-10.

[2] Baiocchi, C., and A. Capelo: Variational and quasivariational inequalities. Wiley 1984.

[3] Biroli, M.: De Giorgi-Nash-Moser results for a variational inequality. Boll. UMI (5)16-A(1979), 598-605.

[4] Brézis, H.: Problèmes unilateraux. J. Math. Pures et appl. 51(1972), 1-168.

[5] Brézis, H., and G. Stampacchia.: Sur la régularité de la solutions d'inéquations elliptiques. Bull. Soc. Math. France. 96(1968),153-180.

[6] Ciarlet, P.: Mathematical elasticity. North Holland , Amsterdam 1983.

[7] Duvaut, G., and J.L. Lions: Les inéquations en mathématique et en physique. Dunod, Paris 1972.

[8] Fichera, G: Problemi elastostatici con vincoli unilaterali: il problema di Signorini con ambigue condizioni al contorno. Atti Accad. Naz. Lincei Mem. Cl. Fis. Mat. Nat. Sez. Ia 7(8)VII (1963/64), 91-140.

[9] Fichera, G: : Existence theorems in elasticity.

Boundary value problems of elasticity with unilateral constraints. In: Handhuch der Physik (S. Flügge, ed.), Vol. 6a/2, Springer-Verlag, Berlin 1972.

[10] Frehse, J.: On Signorini's problem and variational problems with thin obstacles. Ann. Sc. Norm. Sup. Pisa 4(1977), 343-362.

[11] Frehse, J.: On the smoothness of variational inequalities with obstacles. In: Partial differential equations, p. 87-128. Banach Centre Publications. Warsaw 1983.

[12] Frehse, J. and U. Mosco: Variational inequalities with one sided irregular obstacles. Manuscripta math. 28(1979), 219-234.

[13] Friedman, A.: Variational principles and free-boundary problems. Wiley 1982.

[14] Fuchs, M.: Hölder contunuity of the gradient for degenerate variational inequalities. Nonlin. Anal. 15(1990), 85-100.

[15] Gerhard, C.: Regularity of solutions of nonlinear variational inequalities. Arch. Rat. Mech. Anal. 52(1973), 389-393.

[16] Giaquinta, M.: Remarks on the regularity of weak solutions of some variational inequalities. Math.Z. 177(1981), 15-31.

[17] Giaquinta, M.: Multiple integrals in the calculus of variations and nonlinear elliptic systems. Princeton University Press. Princeton 1983.

[18] Hartman, P., and G. Stampacchia: On some non-linear elliptic differential functional equations. Acta Mat. 115(1966), 271-310.

[19] Hildebrandt, S., and K.-O. Widman: Variational inequalities for vector-valued functions. J. reine und angew. Mathematik 309(1979), 181-220.

[20] Hörmander, L.: The analysis of linear partial differential operators, Vol. 3. Springer-Verlag New York, Berlin 1985.

[21] Jentsch, L.: Über Wärmespannungen in Körpern mit stückweise konstanten Laméschen Elastizitätsmoduln. Schriftenreihe des Zentralinstitus für Mathematik und Mechanik. Akademieverlag, Berlin 1972.

[22] Kinderlehrer, D., and G. Stampacchia: An introduction to variational inequalities and their applications. Academic Press, New York 1980.

[23] Kinderlehrer, D.: The smothness of the solution of the boundary obstacle problem. J. Math. Pures Appl. 60(1981), 193-212.

[24] Kinderlehrer, D.: Remarks about Signorini's problem in linear elasticity. Ann. Sc. Norm. Sup. Pisa (IV) 8(1981), 605-645.

[25] Kupradze, V.D.: Three dimensional mathematical theory of elasticity and thermoelasticity. Mir, Moscow 1976.

[26] Lions, J.L.: Quelques méthodes de résolution des problèmes aux limites non linéaires. Dunod, Paris 1969.

[27] Lions, J.L., and Stampacchia, G.: Variational inequalities. Comm. Pure Appl. Math. 20(1967). 439-519

[28] Nečas, J. and I. Hlavaček: Mathematical theory of elastic and elastoplastic bodies. SNTL Prague and Elsevier, Amsterdam 1981.

[29] Naniewicz, Z. and P.D. Panagiotopoulos: Mathematical theory of hemivariational inequalities and applications. Marcel Dekker, New York 1995.

[30] Richardson, D.: The regularity of the solution of a variational inequality. Report No. 5, 1977, Institut Mittag-Leffler

[31] Rodrigues, J.F.: Obstacle problems in mathematical physics. North Holland, Amsterdam 1987.

[32] Schumann, R.: Regularity for Signorini's problem in linear elasticity, Manuscripta math. 63(1989), 255-291.

[33] Schumann, R.: A remark on a boundary contact problem in linear elasticity. Manuscripta math. 63(1989), 455-468.

[34] Schumann, R.: Regularity for a variational inequality with a pseudodifferential operator of negative order. Zeitschrift für Analysis und ihre Anwendungen 15(1996), 357-375.

[35] Shamir, E.: Regularition of mixed second order elliptic problems. Israel Math. J. 6(1968), 150-168.

[36] Signorini, A.: Questioni di elasticità non linearizzata o semilinearizzata. Rend. di Matem. e delle sue appl. 18(1959).

[37] Troianiello, G.M.: Elliptic differential equations and obstacle problems. Plenum, NewYork 1987.

[38] Uralzeva, N.N.: On the regularity of solutions of variational inqualities (Russian). Uspechi mat. Nauk 42(1987)6, 151-174.

[39] Zeidler, E.: Nonlinear functional analysis and applications, Vols. 1-4. Springer-Verlag New York, Berlin 1985-1992.

21 A SURVEY OF 1-D PROBLEMS OF DYNAMIC CONTACT OR DAMAGE

Meir Shillor

Department of Mathematics and Statistics
Oakland University
Rochester, MI USA

shillor@oakland.edu

Abstract: We survey four of our recent results on one-dimensional dynamic contact with or without friction or with damage. We present the classical models, the weak or variational formulations and state our results. The purpose of considering one-dimensional problems is to gain insight into the behavior of dynamic models for contact, without having to address the considerable mathematical complications that arise in two or three dimensions.

21.1 INTRODUCTION

We present four of our recent results on one-dimensional dynamic contact with or without friction or thermal effects, or with damage. The problems are (i) dynamic contact of a thermoviscoelastic rod with an obstacle; (ii) vibrations of a viscoelastic beam between two stops; (iii) frictional contact of a beam with an obstacle; and (iv) evolution of the mechanical damage in a rod.

Currently, there is considerable industrial interest in dynamic contact. Machine or engine parts and components, pistons, crankshafts, brakes and tires, to mention but a few, come into dynamic, often periodic, contact. The dynamic characteristics of a system determine its reliability, the wear of the component, and the perceived smoothness of its operation. The noise and vibration behavior of cars is important to the automotive industry since they directly influence the customer's perception of car quality. But, general treatment of dynamic contact is very recent because of the considerable mathematical and numerical difficulties associated with dynamic models of contact. Indeed, existence of solutions to dynamic contact of an elastic body using Signorini's condition is still an unresolved problem in two or three space dimensions. We considered one-dimensional problems to gain insight into the behavior of such models,

R.P. Gilbert et al. (eds.), From Convexity to Nonconvexity, 283–295.

without having to address the mathematical complications that arise in more dimensions.

The problems presented are steps toward a comprehensive mathematical and computational theory of dynamic thermoelastic contact with or without friction. We model the contact with either Signorini's or the normal compliance conditions. The friction is modeled with appropriate versions of Coulomb's law. The conditions are those currently used in models for contact and friction. The first three problems can be set as quasivariational inequalities, for which existence of solutions has been established. For each of the problems, we present the classical model and the weak or variational formulation. Then we state the existence, possible uniqueness and partial regularity results. Complete details, proofs and references can be found in the surveyed publications.

We also describe the dynamic evolution of damage in a one-dimensional mechanical system. The practical importance of such a model is in its prediction of the material damage and consequently its possible failure modes. The problem is set as a variational equality, for which the existence of a local solution has been established, since the problem exhibits quenching. This represents the breaking of the system, when the load carrying capacity vanishes. In the quasistatic problem we derived estimates on the existence time.

Much remains to be done, now that the existence of solutions has been established. There is a need to investigate the possible types of behavior and further analyze the models. Because of the complexity of the problems, it is important to construct convergent numerical schemes and perform numerical simulations to investigate the solutions. These will eventually lead to a deeper understanding of dynamic contact and damage and will indicate the way to full analysis in two or three dimensions. Some progress has been made recently in [19, 2, 3].

Following Section 2, which is devoted to preliminaries, the problems are described in Sections 3–6. The list of references is limited, and we refer the reader to the original papers for a complete list.

21.2 PRELIMINARIES

In this section we present the notation. Let $E = \{v \in H^1(0,1) : v(0) = 0\}$, $H = L^2(0,1)$, $\mathbb{E} = L^2(0,T;E)$ and $\mathbb{H} = L^2(0,T;H) = L^2(\Omega_T)$. Let $(\cdot,\cdot)_H$ denote the inner product in H, $\|\cdot\|_H$ the induced norm, and let $(\xi,\eta)_E = (\xi_x,\eta_x)_H$ be the inner product on E, with the associated norm $\|\cdot\|_E$, which is equivalent to the usual H^1 norm. Also, $\langle\cdot,\cdot\rangle$ denotes the duality pairing between $\mathbb{E}$ and its dual $\mathbb{E}'$. These will be used in the problems for rods. We shall use the following when we deal with beams: $V = \{w \in H^2(0,1) : w(0) = w'(0) = 0\}$. Clearly we have

$$V \subseteq E \subseteq H = H' \subseteq E' \subseteq V',$$

where E' and V' are the topological duals of E and V, respectively. For definitions of any unexplained notation we refer the reader to [1], [10] or [18].

The models are set in dimensionless form such that the reference configuration (stress free and isothermal) is $0 \leq x \leq 1$. Let $\Omega_T = (0,1) \times (0,T)$ for $T > 0$. We denote by $u = u(x,t)$ the displacements field, horizontal in a rod, and vertical in a beam; $\sigma = \sigma(x,t)$ denotes the stress field, the pressure in a rod, and the shear stress in a beam. We denote by $\theta = \theta(x,t)$ the temperature, rescaled and measured with respect to some convenient zero point. These are rescaled and defined on Ω_T.

Partial derivatives are denoted by subscripts: $u_t = \partial u/\partial t$, etc. Also, we use "$'$" to denote the time derivative. $(f)_+ = \max\{f, 0\}$ denotes the positive part of f.

21.3 DYNAMIC THERMOVISCOELASTIC CONTACT OF A ROD

We follow [16] and consider a long, slender, viscoelastic rod insulated lengthwise, clamped at its left end while the right end is free to expand or contract as a result of the evolution of the temperature and the stress fields. However, the expansion is limited by a rigid obstacle situated to the right. The quasistatic problem was analysed in [15].

The constitutive relation is viscoelastic of the Kelvin-Voigt type

$$\sigma = u_x + cu_{xt} - a\theta. \tag{21.1}$$

Here, a and c are positive material constants related to the coefficient of thermal expansion and the coefficient of viscosity, respectively.

In dimensionless units, we may write the mechanical and energy equations, in Ω_T, in the form

$$u_{tt} - cu_{xxt} - u_{xx} = -a\theta_x, \tag{21.2}$$
$$\theta_t - \theta_{xx} = -au_{xt}. \tag{21.3}$$

The rod is clamped at the left end so $u(0,t) = 0$ and has the (scaled) wall temperature $\theta(0,t) = 1$. A rigid wall is situated at $x = 1 + g$, and at the free end Singnorini's contact condition holds

$$u(1,t) \leq g, \quad \sigma(1,t) \leq 0, \quad \text{and} \quad \sigma(1,t)(g - u(1,t)) = 0. \tag{21.4}$$

The expansion of the free end is restricted by the presence of the wall, $u \leq g$. When the end is free $\sigma = 0$, otherwise the stress is compressive. And, either the end is free $u < g$ and $\sigma = 0$, or it is in contact and $u = g$.

We model the thermal interaction between the free end and the obstacle by

$$-\theta_x(1,t) = k(g - u(1,t))\theta(1,t) \tag{21.5}$$

where $k(s)$ is the heat exchange coefficient, which depends on $s = g - u$, the separation distance between the rod's end and the obstacle. k measures the thermal conductivity (the reciprocal of the thermal resistance) of the gap, so it is natural to assume that it is a decreasing function of the distance for $s > 0$.

k may be either a continuous function or a graph. When $k(\cdot)$ is a graph we rewrite (26.15) as

$$-\theta_x = \xi\theta, \tag{21.6}$$
$$\xi(t) \in [k((g-u(t,1))+), k(g-u(t,1))], \tag{21.7}$$

i.e., ξ is a measurable selection out of the graph $k(\cdot)$. We assume that the graph has a vertical segment $[k_0,\ k^*]$at the origin which represents the heat exchange when the end is in contact with the obstacle. A more realistic condition is when the thermal conductivity depens on the contact stress too. For a thorough discussion of $k(\cdot)$ we refer the reader to [5].

To complete the model, we specify the initial conditions $u(x,0) = u_0(x)$, $u_t(x,0) = u_1(x)$ and $\theta(x,0) = \theta_0(x)$, for $0 \le x \le 1$.

The classical formulation of the *thermoviscoelastic contac problem of a rod* is: Find a pair $\{u,\theta\}$ such that (26.12)–(26.15) and the initial and boundary conditions hold.

The contact condition (26.14) makes it necessary to consider a weak formulation, since it imposes a ceiling on the regularity of solutions. Generally, they will not have all the classical derivatives. Let $\mathbb{K}_u$ be the closed convex set

$$\mathbb{K}_u = \{v \in \mathbb{E} : v' \in \mathbb{H},\ v(1,t) \le g \text{ a.e. } t,\ v(T) = u(T)\}. \tag{21.8}$$

The weak formulation of (26.12)–(26.17) is: Find $\{u,\theta\} \in \mathbb{K}_u \times \mathbb{E}$, such that

$$-\int_0^T (u_t, v_t - u_t)dt + \int_0^T (\sigma, v_x - u_x)dt + (u_1, v(\cdot,0) - u_0) \le 0, \tag{21.9}$$

for each $v \in \mathbb{K}_u$, and for each $\varphi \in C^1(0,T;E)$ such that $\varphi(\cdot,T) = 0$,

$$\begin{aligned} -\int_0^T (\theta,\varphi_t)\,dt \quad &+ \quad \int_0^T (\theta_x,\varphi_x)\,dt + a\int_0^T (u_{tx},\varphi)\,dt - (\theta_0,\varphi(0,\cdot)) \\ &+ \quad \int_0^T k(g-u(1,t))\theta(1,t)\varphi(1,t)\,dt = 0. \end{aligned} \tag{21.10}$$

Note that (21.9) is a hyperbolic quasivariational inequality, because of Signorini's condition (26.14) and the dependence of $\mathbb{K}_u$ on u. (21.10) is a parabolic equality. We obtain these by multiplication of (26.12) by $v-u$, and (26.13) by φ, integration over Ω_T, and use of the initial and boundary conditions.

Our main results in [16] are summarized in the following theorems.

Theorem 21.1. *Assume that k is a continouous and bounded function, that $u_0 \in E$ and satisfyies $u_0(1) \le g$ and $u_1 \in H$. Then, there exists a pair $\{u,\theta\}$ satisfying $u \in \mathbb{K}_u$, $u' \in \mathbb{E}$, $\theta \in \mathbb{E}$, such that $u(0) = u_0$, $\theta(0) = \theta_0$, and* (21.9) *and* (21.10) *hold.*

The proof is based on the regularization of Signorini's condition.

We have a partial regularity result, namely:

Theorem 21.2. *There exists a solution* $\{u, \theta\}$ *satisfying* (21.9) *and* (21.10) *such that the following energy estimate holds:*

$$\frac{1}{2}||\theta(t)||_H^2 + \frac{1}{2}||u(t)||_E^2 + \int_0^t ||\theta(s)||_E^2 ds + c\int_0^t ||u'(s)||_E^2 ds$$
$$\leq \frac{1}{2}||\theta_0||_H^2 + \frac{1}{2}||u_0||_E^2 + \frac{1}{2}||u_1||_H^2. \tag{21.11}$$

Remark. To investigate the related impact problem, we need only replace the boundary conditions at the left end. We assume that the end is free so $\sigma(0,t) = 0$ and is thermally insulated $\theta_x(0,t) = 0$. The variational formulation of the impact problem is (21.9) and (21.10) but $\mathbb{K}_u$ needs to be modified in an obvious way. One can show that the problem has a weak solution by a straightforward modification of the proof of Theorem 1.

We turn to consider the case when k is a graph defined on $[0, \infty)$, decreasing, positive and continuous on $(0, \infty)$ such that

$$k(r) = \begin{cases} [k_0, k^*] & r = 0, \\ k(r) & r > 0, \end{cases}$$

where $k(0+) = k_0 < k^* < \infty$. The proof is based on an approximating sequence of continuous functions to the graph k. The results of Theorem 26.3.1 guarantee the existence of a solution to each of these problems. Once we obtain the necessary a priori estimates we pass to the limit and find:

Theorem 21.3. *Assume that k is a graph as above. Then, there exists a triplet* $\{u, \theta, \xi\}$ *such that* $\theta, u, u' \in \mathbb{E}$, $u \in \mathbb{K}_u$, $\theta' \in \mathbb{E}'$, $\xi \in L^\infty(0,T)$;

$$u(0) = u_0,\ \theta(0) = \theta_0, \tag{21.12}$$
$$\xi(t) \in [k((g - u(t,1))+),\ k(g - u(t,1))] \quad a.e.\ t, \tag{21.13}$$
$$\theta' + L\theta + \gamma * (\xi\gamma\theta) + au'_x = 0 \quad in\ \mathbb{E}', \tag{21.14}$$

and (21.9) *holds for all* $v \in \mathbb{K}_u$.

Here $\gamma : \mathbb{E} \to L^2(0,T)$ is the trace map $\gamma\theta(t) = \theta(t,1)$, γ^* is its inverse, and $\langle L\theta, \eta\rangle = (\theta_x, \eta_x)$.

We conclude this section with the observation that the uniqueness, further regularity and continuous dependence on the data are open questions.

21.4 VIBRATIONS OF A BEAM BETWEEN TWO STOPS

We consider, following [17], the vibrations of a beam that is attached at its left end to a device that may oscillate, while the motion of the right end is constrained by two rigid obstacles – the stops. The setting is depicted in Fig. 1. A simplified version of the problem was investigated in [21] (see also ([20]) where possible chaotic behavior was established.

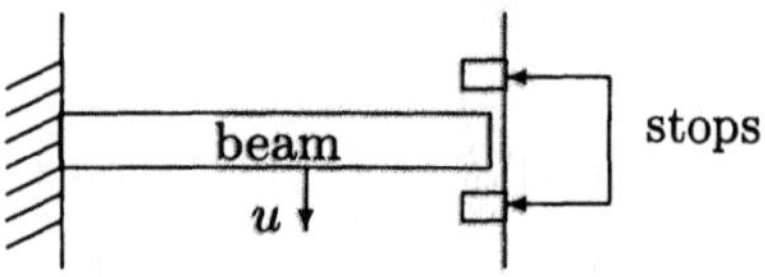

Figure 1 The physical setting.

Let g_1 and g_2 ($g_1 < 0 < g_2$) be the positions of the stops, $u = u(x,t)$ be the vertical displacement and $\sigma = \sigma_T(x,t)$ be the shear stress. We choose the constitutive law to be the Kelvin-Voigt viscoelasticity

$$\sigma(x,t) = -u_{xxx}(x,t) - cu_{xxxt}(x,t), \tag{21.15}$$

where c is a nonnegative viscosity coefficient; when $c = 0$ the beam is elastic.

The equation of motion is $u_{tt} - (\sigma)_x = f$, in Ω_T, where f denotes the (linear) density of the applied force. Initially, $u(x,0) = u_0(x)$ and $u_t(x,0) = v_0(x)$ for $0 \le x \le 1$, where u_0 and v_0 represent the beam's initial deflection and velocity, respectively. The beam is fixed at its left end to a device that may move as $\varphi = \varphi(t)$, thus, $u(0,t) = \varphi(t)$ and $u_x(0,t) = 0$. Indeed, there is a considerable interest when the device oscillates periodically (see [20]).

At the free end we choose either the classical Signorini condition or the *normal compliance* condition. Also, we assume that no moments act on the free end, i.e., $u_{xx}(1,t) + cu_{xxt}(1,t) = 0$.

Signorini's condition describes the idealized case of completely rigid stops: the displacement $u(1,t)$ is allowed only between the stops, thus, $g_1 \le u(1,t) \le g_2$, and either the end is free and $\sigma(1,t) = 0$; or it is in contact and the stress is opposite to the displacement. Let $\chi = \chi_{[g_1,g_2]}(r)$ be the indicator function of the interval $[g_1, g_2]$, i.e. $\chi(r) = 0$ when $r \in [g_1, g_2]$ and $\chi(r) = +\infty$, otherwise. Then, Signorini's condition, in addition to $g_1 \le u(1,t) \le g_2$, states that the stress satisfies

$$-\sigma(1,t) \in \partial\chi(u(1,t)), \tag{21.16}$$

where $\partial\chi$ is the subdifferential of χ, i.e.,

$$\partial\chi(r) = \begin{cases} [0, +\infty) & r = g_2, \\ 0 & g_1 < r < g_2, \\ (-\infty, 0] & r = g_1. \end{cases}$$

The normal compliance condition (see, e.g., [19, 12]) describes flexible stops, with resistance proportional deflection, thus,

$$\sigma(1,t) = -\kappa \left[(u(1,t) - g_2)_+ - (g_1 - u(1,t))_+ \right], \tag{21.17}$$

where κ is a positive constant. When $\kappa \to +\infty$ we obtain Signorini's condition, therefore, we may consider (21.17) as its regularization. We could have used $\sigma(1,t) = -p_2((u(1,t) - g_2)_+) + p_1((g_1 - u(1,t))_+)$, where p_1 and p_2 are given nonnegative functions that vanish for negative argument values.

To incorporate Signorini's condition (21.16) in the weak formulation we set

$$K = \{w \in V : g_1 \leq w(1) \leq g_2\},$$

the convex set of all admissible displacements. Next, we change the dependent variable as $u(\cdot,t) = u(\cdot,t) - (1-x^2)\varphi(t)$, then f changes to $f(\cdot,t) = f(\cdot,t) - (1-x^2)\varphi''(t)$, and $u(0,t) = 0$.

The variational formulation of the problem of *vibrations of a viscoelastic beam between two stops* with Signorini's condition is: **Definition 1**
A function $u \in L^2(0,T;V)$ is said to be a weak solution provided that $u' \in L^2(0,T;V)$; $u(t,\cdot) \in K$, a.e. t; $u(\cdot,0) = u_0 \in K$, and for each $w \in L^2(0,T;V)$, such that $w(\cdot,t) \in K$, a.e. t, $w' \in L^2(0,T;H)$ and $w(\cdot,T) = u(\cdot,T)$,

$$\int_0^T (u', w' - u')\, dt + \int_0^T (u_{xx}, w_{xx} - u_{xx})dt + d\int_0^T (u'_{xx}, w_{xx} - u_{xx})dt$$
$$+ (v_0, w(0) - u_0) \geq \int_0^T (f, w-u)dt. \qquad (21.18)$$

Theorem 21.4. *Let $c > 0$ and assume that $f \in L^2(0,T;H)$, $\varphi \in H^2(0,T)$, $u_0 \in K$ and $v_0 \in H$. Then, for each $T < \infty$, there exists a solution to* (21.18). *Moreover, if $\int_0^\infty |f|_H^2 < \infty$, then the solution exists on $[0,\infty)$.*

The proof is is based on a priori estimates for the solutions of a family of approximate problems with the normal compliance condition, for which we have the following result.

Theorem 21.5. *Let $c \geq 0$, and assume that the data satisfies the above conditions, then the problem with* (21.17) *has a weak solution for each $T < \infty$. When $c > 0$ the solution is unique and satisfies $u \in L^\infty(0,T;V)$, $u' \in L^2(0,T;V) \cap L^\infty(0,T;H)$ and $u'' \in L^2(0,T;V')$. When $c = 0$, $u \in L^\infty(0,T;V)$, $u' \in L^\infty(0,T;H)$ and $u'' \in L^2(0,T;V')$.*

The proof is based on an abstract existence result of Kuttler [13].

Theorem 21.6. *Assume that $c = 0$ and the data is as above. Then, there exists a function u satisfying $u \in L^2(0,T;V)$, $u' \in L^2(0,T;H)$, $u \in K$, $u(0) = u_0 \in K$, and the variational inequality* (21.18) *(with c=0), which holds for all $w \in L^2(0,T;K)$ with $w' \in L^2(0,T;H)$ and $w(\cdot,T) = u(\cdot,T)$.*

Remark It is seen in Theorems 21.4, 21.5 and 21.6, that there is a progressive decrease, which is to be expected, in the solution regularity. When u solves Signorini's problem without viscosity, ($c = 0$), then, $u \in L^2(0,T;V)$, $u' \in L^2(0,T;H)$, and there is a loss of control over u'' altogether. Indeed, the velocity u' may be discontinuous upon impact, and therefore, the acceleration u'' is unbounded, which implies unbounded contact force.

Based on the results of [21] we expect chaotic behavior, thus, in some parameter ranges, the problem has sensitive dependence on initial conditions.

21.5 A BEAM IN FRICTIONAL CONTACT

We follow [4] and present models for dynamic evolution of a viscoelastic beam in frictional contact with a rigid obstacle. The beam is attached to a wall at its left end; its right end may come in frictional contact with a rigid obstacle situated a distance $g > 0$ to the right. Let $u = u(x,t)$ and $v = v(x,t)$ represent the horizontal and vertical displacements of the beam, respectively. Let $\sigma_N = \sigma_N(x,t)$ be the pressure and $\sigma_T = \sigma_T(x,t)$ be the shear stress. Then, the equations of motion take the form

$$u_{tt} - (\sigma_N)_x = f_N, \tag{21.19}$$

$$v_{tt} - (\sigma_T)_x = f, \tag{21.20}$$

where f and f_N denote the vertical and horizontal applied forces, respectively. We assume the Kelvin-Voigt viscoelasticity laws

$$\sigma_N(x,t) = au_x(x,t) + cu_{xt}(x,t), \tag{21.21}$$

$$-\sigma_T(x,t) = v_{xxx}(x,t) + dv_{xxxt}(x,t). \tag{21.22}$$

Here, c and d represent the viscosities and a the elastic modulus. Initially $u(x,0) = u_0(x)$, $u_t(x,0) = u_1(x)$, $v(x,0) = v_0(x)$ and $v_t(x,0) = v_1(x)$, for $0 \le x \le 1$, where u_0, u_1, v_0 and v_1 are prescribed functions. The beam is attached at $x = 0$, thus $u(0,t) = 0$, $v(0,t) = 0$ and $v_x(0,t) = 0$. The moment acting on the free end is zero, thus, $v_{xx}(1,t) + dv_{xxt}(1,t) = 0$.

We consider several alternative conditions at the free end. Horizontally we use Signorini's nonpenetration condition (26.14) with $\sigma = \sigma_N$. We shall also consider the alternative *normal compliance* condition

$$\sigma_N(1,t) = -c_N(u(1,t) - g)_+^{m_N}, \tag{21.23}$$

where c_N and m_N are two positive constants that represent the contact surfaces. (21.23) permits the contacting end to penetrate the obstacle, i.e, $u(1,t) > g$, which may also represent the elastic response of the obstacle. (26.14) may be obtained from (21.23) as $c_N \to \infty$.

The horizontal problem is an isothermal version of the one considered in Section 20.3 thus, we infer the existence of weak solutions from Theorem 1. Therefore, below we deal only with the vertical friction problem.

We employ the friction law

$$|\sigma_T(1,t)| \le h(t), \tag{21.24}$$

$$\text{if } |\sigma_T(1,t)| = h(t) \text{ then } v_t(1,t) = -\lambda\sigma_T(1,t), \tag{21.25}$$

$$\text{if } |\sigma_T(1,t)| < h(t) \text{ then } v_t(1,t) = 0, \tag{21.26}$$

for some $\lambda \ge 0$. The function $h = h(t)$ is the *friction bound* and the conditions (21.24)–(21.26) may be interpreted as follows: when $\sigma_T = \pm h$ then the shear is opposite to the slip, and when $|\sigma_T| < h$ the end sticks to the obstacle. Thus, we may have a stick-slip motion.

We consider three possible choices for the function h. The first is Tresca's condition in which h is given. It was shown in [4] that the problem has a weak solutions provided $h \in H^{-\varepsilon}(0,T)$, for $0 < \varepsilon < 1/8$.

A second way of treating h is the classical Coulomb law of dry friction

$$h(t) = -\mu\sigma_N(1,t), \tag{21.27}$$

where μ is *the coefficient of friction.* We may use it in combination with either the Signorini or the normal compliance conditions. However, in the latter case, we may use a more general h of the form

$$h(t) = c_T(u(1,t) - g)_+^{m_T}, \tag{21.28}$$

where c_T and m_T are two positive constants which may be chosen independently of c_N and m_N. This completes the description of the various boundary conditions that were considered in [4].

A friction law of the form (21.24)– (21.26) imposes a regularity ceiling which, generally, precludes the existence of classical solutions. Thus, it is natural to consider weak, or variational formulations.

Our first formulation incorporates the normal compliance condition (21.23) and the generalized Coulomb condition (21.28) and so we introduce

$$j_N(u,w) = c_N \int_0^T (u(1,t) - g)_+^{m_N} w(1,t)dt, \tag{21.29}$$

the *normal compliance* functional, and the *friction* functional

$$j_T(u,w) = c_T \int_0^T (u(1,t) - g)_+^{m_T} |w(1,t)|dt. \tag{21.30}$$

Note that for $m_N \geq 1$ and $m_T \geq 1$ both functionals are defined and convex on $L^2(0,T;E) \times L^2(0,T;E)$, but $j_T(u,\cdot)$ is not Gâteaux differentiable.

The weak formulation, which is obtained in the usual way, is as follows.

Definition 2

A pair of functions $(u,v) \in L^2(0,T;E) \times L^2(0,T;V)$ is said to be a weak solution to (21.19)–(21.26) and (21.28), provided that $u_t \in L^2(0,T;E)$, $u_{tt} \in L^2(0,T;E')$; $u(\cdot,0) = u_0$, $u_t(\cdot,0) = u_1$, $v_t \in L^2(0,T;V)$, $v_{tt} \in L^2(0,T;V')$; $v(\cdot,0) = v_0$ and $v_t(\cdot,0) = v_1$, and $\forall\varphi \in L^2(0,T;E)$ and $\forall w \in L^2(0,T;V)$

$$\int_0^T \langle u_{tt},\varphi\rangle dt + \int_0^T (\sigma_N,\varphi_x)dt + j_N(u,\varphi) = \int_0^T \langle f_N,\varphi\rangle\, dt, \tag{21.31}$$

$$\int_0^T \langle v_{tt}, w - v_t\rangle dt + \int_0^T (v_{xx}, w_{xx} - v_{xxt})dt + d\int_0^T (v_{xxt}, w_{xx} - v_{xxt})dt$$
$$+ j_T(u,w) - j_T(u,v_t) \geq \int_0^T (f, w - v_t)dt. \tag{21.32}$$

We have the following existence and uniqueness results ([4]) for the problem.

Theorem 21.7. *Let $c \geq 0$, $a > 0$ and $d > 0$. Let $f_N, f_{N,t} \in L^2(0,T;E')$, $f \in L^2(\Omega_T)$, $u_0 \in E$, $v_0 \in V$, u_1, $v_1 \in H$, and $m_N \geq 1$, $m_T \geq 1$. Then, there exists a unique solution to problem* (21.31) *and* (21.32), *provided that in the case when $c = 0$, it is required that $u_t \in L^2(0,T;H)$.*

Our second formulation incorporates Signorini's condition (26.14) and the friction law (21.24)–(21.26). We recall the scale of Hilbert spaces of distributions $H^\varepsilon(0,T)$, $\varepsilon \in (-\infty,\infty)$, as defined in [18] and introduce, for $\varepsilon \geq 0$, the functional $J(z) = \langle h, |z(1,\cdot)|\rangle_{-\varepsilon,\varepsilon}$, where $h \in H^{-\varepsilon}(0,T)$ and $\langle\cdot,\cdot\rangle_{-\varepsilon,\varepsilon}$ denotes the duality pairing between $H^\varepsilon(0,T)$ and $H^{-\varepsilon}(0,T)$. Also let $K = \{f \in E : f(1) \leq g\}$. The variational formulation of the problem with Signorini and Coulomb boundary conditions is as follows. **Definition 3**

Let $\varepsilon \geq 0$, and suppose $h \in H^{-\varepsilon}(0,T)$. A pair of functions $(u,v) \in L^2(0,T;K)\times L^2(0,T;V)$ is said to be a weak solution to (21.19) – (21.22), (26.14), (21.24) – (21.27) provided that $u_t \in L^2(0,T;E)$, $u_{tt} \in L^2(0,T;E')$, $u(\cdot,0) = u_0$, $u_t(\cdot,0) = u_1$, $v_t \in L^2(0,T;V)$, $v_{tt} \in H^{-\varepsilon}(0,T;V'$, $v(\cdot,0) = v_0$, $v_t(\cdot,0) = v_1$, and $\forall\varphi \in L^2(0,T;K)$ and $\forall w \in H^\varepsilon(0,T;V)$,

$$\int_0^T \langle u_{tt}, \varphi - u\rangle dt \quad + \quad \int_0^T (\sigma_N, \varphi_x - u_x)dt \geq \int_0^T (f_N, \varphi - u)dt, \tag{21.33}$$

$$\begin{aligned}\int_0^T \langle v_{tt}, w - v_t\rangle dt \quad &+ \quad \int_0^T (v_{xx}, w_{xx} - v_{xxt})dt + d\int_0^T (v_{xxt}, w_{xx} - v_{xxt})dt \\ &+J(w) - J(v_t) \geq \int_0^T (f, w - v_t)dt. \end{aligned}\tag{21.34}$$

Theorem 21.8. *Let $c \geq 0$, $a > 0$ and $d > 0$. Let $f_N \in L^2(\Omega_T)$, $f \in L^2(\Omega_T)$, $u_0 \in K$, $v_0 \in V$ and u_1, $v_1 \in H$. If $h = -\mu\sigma_N(1,\cdot)$, where $\mu > 0$, then there exists a unique solution of problem* (21.33) *and* (21.34); *in the case when $c = 0$ it is required that $u_t \in L^2(0,T;H)$.*

The solution of the horizontal problem (21.33) can be found in [22, 11] for the case when $c = 0$ and in [16] when $c > 0$. The solvability of (21.34) has been established in [4].

Finally, the existence of weak solutions of the problem with the standard Signorini-Coulomb conditions without Kelvin-Voigt damping, which is the kind of model for frictional contact most frequently employed in engineering applications, despite the difficulties involved, can be found in Theorem 5.4 in [4]. It turns out that in such a setting it was necessary to introduce a positive regularization operator for the stress $\sigma_N(1,t)$.

The proofs can be found in [4] and are rather technical.

A thermoviscoelastic version of the vertical problem which takes into account the frictional heat generation can be found in [14].

21.6 THE ELASTIC ROD WITH DAMAGE

In this section we describe, following [8], a one-dimensional dynamic model for the damage process. We consider a bar or a rod, made of concrete for instance, that is fixed at its left end. A time varying traction acts on its right end and as a result the mechanical state of the bar evolves in time. Our interest lies in the evolution of the damage of the material resulting from tension. The theory we employ was developed by Fremond and his collaborators (see [9] and [7] and references therein) and the new feature is the introduction of the damage field β as an additional dependent variable. If E_Y is Young's modulus of the undamaged material then it is assumed that, as a result of microscopic crack openings, the effective modulus of the rod's material decreases and is βE_Y. In this way β measures the fractional decrease in load bearing capacity of the material. We require that $0 \leq \beta \leq 1$ in the reference configuration of the system $0 < x < 1$. When $\beta = 1$ the material is undamaged; when $\beta = 0$ it is completely damaged. When $0 < \beta < 1$, it measures the fractional density of open cracks; it retains its elastic behavior but with a modified elastic modulus – it becomes softer.

We turn to model the evolution of the state of the bar which consists of the displacement field $u = u(x,t)$ and the damage field $\beta = \beta(x,t)$. We assume the Kelvin-Voigt viscoelastic constitutive relation

$$\sigma(\beta) \equiv c\beta u_{tx} + \beta a u_x, \tag{21.35}$$

where σ is the mechanical stress, c is the viscosity coefficient and a the elastic modulus. The evolution of the state of the system $\{u, \beta\}$ is governed by the equations of mechanical motion and of damage. The damage equation has been derived from a dissipation pseudo-potential in [9]. Here we assume that damage is caused only by tension, for more general assumptions we refer to [9] and [7]. The problem is to find a pair $\{u, \beta\}$ such that:

$$u_{tt} - \sigma(\beta)_x = f_e, \tag{21.36}$$

$$\beta_t - k\beta_{xx} - m\left(\frac{1-\beta}{\beta}\right) = -\frac{b}{2}(u_x)_+^2 + w. \tag{21.37}$$

Here, f_e represents the density of applied forces; k is the coefficient of damage diffusion; w is the damage threshold and m is the displacement factor of the threshold, all positive constants (see, e.g., [9]). The first term on the right-hand side of equation (21.37) represents the source of damage resulting from the tension in the rod.

We require that the damage function does not increase, since in many materials once cracks open they will not close. Thus, β has to satisfy the constraints $0 \leq \beta \leq 1$ and $\beta_t \leq 0$. At its left end the rod is clamped, so $u = 0$ at $x = 0$. A traction $p = p(t)$ acts on the right end; therefore,

$$c\beta v_{tx} + \beta a u_x = p, \qquad x = 1. \tag{21.38}$$

We consider a homogeneous Neumann condition at the left end $\beta_x(0,t) = 0$ and a Dirichlet condition at the right end $\beta(1,t) = 1$. At the right-hand side we could prescribe a homogeneous Neumann condition, which means that no energy is provided to the bar. Initially, $u = u_0$, $u_t = u_1$ and $\beta = \beta_0$ for

$0 \leq x \leq 1$. Here u_0, u_1, and β_0 are the initial displacement, velocity and damage, respectively. Also, β_0 equals one identically if the material is initially undamaged.

The dynamic damage problem for $\{u, \beta\}$ consists of the equations (21.36) and (21.37), the constraints, the initial and the boundary conditions.

We assume that the coefficients a, c, k, m and w are all positive constants. Our local existence result for the problem is:

Theorem 21.9. *Assume that* $f_e \in L^2(0,T;H_0^1(0,1))$, $p \in H^1(0,T;E)$, $u_0 \in H_0^1(0,1)$ *and* $u_1 \in H^1(0,1)$. *Moreover, let* $\beta_0 \in H^2(0,1)$, $\beta_0 - 1 \in H_0^1(0,1)$ *and assume that* $0 < \beta_* \leq \beta_0 \leq 1$ *in* $[0,1]$. *Then, there exists* T^*, $0 < T^* \leq T$, *such that there exists a solution* $\{u, \beta\}$ *to the damage problem on* $[0,T^*)$. *This solution satisfies*

$$\beta - 1 \in H^1(0,T^*;E) \cap L^2(0,T^*;H^2(0,1)) \cap C(\Omega_{T^*}), \tag{21.39}$$

$$u \in H^1(0,T^*;E) \cap L^2(0,T^*;H^2(0,1)). \tag{21.40}$$

The proof is based on regularization of the constraints $\beta_t \leq 0$ and application of the result in [13].

We note that the solution is only guaranteed to exist on an interval $[0,T^*)$ since the damage field can reach the value $\beta(x_*,T^*) = 0$, at a point x_*, which means that the damage at this point is complete; physically the bar snaps, and mathematically the model colapses since two terms in the equation (21.37) become infinite. The solutions is said to *quench*. Some estimates on the quenching time for the quasistatic problem can be found in [7].

References

[1] Adams, R. (1975) *Sobolev Spaces*. Academic Press.

[2] Andrews, K. T., K.L. Kuttler and M. Shillor. (1997) "On the dynamic behavior of a thermoviscoelastic body in frictional contact with a rigid obstacle," *Euro. J. Appl. Math.*, **8** 417–436.

[3] Andrews, K. T., A. Klarbring, M. Shillor and S. Wright. (1997) "A dynamic thermoviscoelastic contact problem with friction and wear," *Int. J. Engng. Sci.*, **35** (14) 1291–1309.

[4] Andrews, K. T., M. Shillor and S. Wright. (1996) "On the dynamic vibrations of an elastic beam in frictional contact with a rigid obstacle," *J. Elasticity*, **42** 1–30.

[5] Andrews, K. T., P. Shi, M. Shillor and S. Wright. (1993) "Thermoelastic contact with Barber's heat exchange condition," *Appl. Math. Optim.*, **28** (1) 11–48.

[6] Duvaut, G. and J. L. Lions. (1976) *Inequalities in Mechanics and Physics*. New York: Springer-Verlag.

[7] Fremond, M., K. L. Kuttler, B. Nedjar and M. Shillor. (1998) "One-dimensional models of damage," *Adv. Math. Sci. Appl.*, 8(2), 541–570

[8] Fremond, M., K. L. Kuttler and M. Shillor. (1999) "A one-dimensional dynamic model of damage," *JMAA*, 229, 271–294.

[9] Fremond, M. and B. Nedjar. (1995) "Damage in concrete: the unilateral phenomenon," *Nuclear Engineering and Design*, **156** (1), 323–335.

[10] Kikuchi, N, and J. T. Oden. (1988) *Contact Problems in Elasticity.* Philadelphia: SIAM.

[11] Kim, J. U. (1990) "A one-dimensional dynamic contact problem in linear visco-elasticity," *Math. Meth. Appl. Sci.*, **13** 55–79.

[12] Klarbring, A., A. Mikelic and M. Shillor. (1988) "Frictional contact problems with normal compliance," *Int. J. Engng. Sci.*, **26** (8) 811–832.

[13] Kuttler, K. L. (1986) "Time dependent implicit evolution equations," *Nonlin. Anal.*, **10** (5) 447–463.

[14] Kuttler, K. L., Y. Renard and M. Shillor. (1999) "Models and simulations of dynamic frictional contact," *Comput. Meth. Appl Mech. Engrg.*, 177, 259–272.

[15] Kuttler, K. L. and M. Shillor. (1994) "A one-dimensional thermoviscoelastic contact problem," *Adv. Math. Sci. Appl.*, **4**(1) 141–159.

[16] Kuttler, K. L. and M. Shillor. (1995) "A dynamic contact problem in one-dimensional thermoviscoelasticity," *J. Nonlin. World*, **2** 355–385.

[17] Kuttler, K. L. and M. Shillor. "Vibrations of a beam between two stops," *Dynamics of Continuous, Discrete and Impulsive Systems*, to appear.

[18] Lions, J. L. and E. Magenes. (1972) *Nonhomogeneous Boundary Value Problems and Applications*, Vol. I&II. New York: Springer.

[19] Martins, J. A. C. and J. T. Oden. (1987) "Existence and uniqueness results for dynamic contact problems with nonlinear normal and friction interface laws," *Nonlin. Anal.* **11** (3) 407–428.

[20] Moon, F. C. (1992) *Chaotic and Fractal Dynamics.* Wiley.

[21] Moon, F. C. and S. W. Shaw. (1983) "Chaotic vibration of a beam with nonlinear boundary conditions," *J. Nonlin. Mech.*, **18** 465–477.

[22] Schatzman, M. and M. Bercovier. (1989) "Numerical approximation of a wave equation with unilateral constraints." *Math. Comput.*, **53** (187) 55–79.

22 NONCONVEXITY IN PLASTICITY AND DAMAGE MODELS: APPEARANCE AND NUMERICAL TREATMENT

Georgios E. Stavroulakis

Institute of Applied Mechanics
Technical University of Braunschweig
D-38106 Braunschweig
Germany*

g.stavroulakis@tu-bs.de

Abstract: Convex analysis provides the mathematical framework for a concise formulation and study of convex elastoplasticity which leads to convex variational inequality problems. Several nonlinear effects, including nonlinear kinematics or damage make the problem nonconvex. The appearance of nonconvexity necessitates the use of appropriate mathematical tools for both the theoretical study and for the numerical treatment of the problem. Fortunatelly convexity and concavity informations can be extracted in some cases of structured nonconvex analysis problems. The theories of the quasidifferentiable and of the difference convex optimization are used to analyse these problems. This way some nonconvex variational and hemivariational inequality problems in mechanics can be analysed by solving a systematic conbination of convex variational inequalities. In this paper this approach is outlined for plasticity and damage models. References to other more general superpotential modelling cases and to appropriate numerical algorithms and applications are also included.

22.1 INTRODUCTION

The use of optimization tools in mechanics has both theoretical and practical advantages. For instance, the classical smooth equations of structural analysis

*Supported by the European Union (Marie Curie Grant ERBFMBICT960987).

R.P. Gilbert et al. (eds.), From Convexity to Nonconvexity, 297–310.

can be produced from the minimization of an appropriately defined potential. Furthermore solution algorithms developed within the framework of mathematical programming can be used for the solution of computational mechanics applications. The introduction of constraints in equation form (i.e., of classical constraints) and in inequality form (for instance, of unilateral contact constraints) is a straightforward task within this general framework. The same is true for the convex nondifferentiable potentials which are required for the modelling of complete vertical branches in constitutive relations or in boundary conditions (e.g., in locking effects). Analogously yield surfaces and plastic strains (or strain rates) in plasticity can be modelled by using mathematical programming techniques.

The limits of this approach are set by the stand of development in mathematical and numerical optimization. Practically this approach is mostly favored for convex problems where both theory (e.g., uniqueness of the solution, lack of local minima of the potential, etc) and numerical algorithms are sufficiently studied. One should mention here that excessive use of mathematical programming techniques has been done for the study of unilateral contact and of plasticity problems.

Working further in this direction one soon realizes that not all kind of nonlinearities arising in mechanics can be extracted from convex potentials or can be described by convex (e.g., yield) sets. Even looking at the classical discretized formulation of a structural analysis problem one realizes that monotonicity characteristics of complex, nonlinear composite relations or the convexity of the corresponding potential functions can not be guaranteed. In some cases theories of structured nonconvex optimization may help. The quasidifferential optimization and in some cases the theory of difference convex optimization have been used to solve this problem.

In this review paper two model applications of structured nonconvex modelling in mechanics are outlined. The first application concerns a standard generalized elastoplasticity problem with a nonconvex quasidifferentiable yield set. In the second application a simple model for damage mechanics is proposed. It can be analysed by a difference convex superpotential formulation. Details on the here outlined applications and on nonconvex and nonsmooth modelling in mechanics can be found in [9], [15].

The paper is organized as follows: first a short introduction in nonsmooth mechanical modeling is given. After that plasticity and damage models are discussed in the next two sections respectively. The paper closes with a preliminary numerical application of an elastic-damage model.

22.2 NONSMOOTH MODELING IN MECHANICS

A function f defined on an open set $X \in \mathbb{R}^n$ is called quasidifferentiable at $x \in X$ if it is directionally differentiable ([6], [8]) at x and its directional derivative $f^{'}(x, g)$ along the direction $g \in \mathbb{R}^n$ can be written as:

$$f^{'}(x,g) = \max_{h_1 \in U} < h_1, g > + \min_{h_2 \in V} < h_2, g > \tag{22.1}$$

Here U and V are convex compact sets in $\mathbb{R}^n$. The ordered pair $Df(x) = [U, V]$ is called the quasidifferential of f at x, U is called the subdifferential of f at x and is denoted by $\underline{\partial} f(x)$ and V is the superdifferential of f at x, denoted by $\overline{\partial} f(x)$.

Practically all convex and nonconvex, possibly nondifferentiable, finite – dimensional potential energy functions which arise in structural analysis are quasidifferentiable ([23], [9]). For the calculation of the quasidifferential of given structured functions, calculus rules have been developed ([7], [8]). In particular the important case of structured nonconvex functions which can be written as a difference fo convex constituents (d.c. functions) is included, as a special case of this theory ([25], [9], [15]).

Accordingly a quasidifferential set $A \subset \mathbb{R}^n$ which is defined by a finite number of equalities and inequalities involving quasidifferential functions. For these kind of generally nonconvex sets with nonsmooth boundaries, normal and tangential vectors are exactly defined by the quasidifferentiability concept. Thus mechanical theories which involve normals and tangentials to these sets can be constructed whithout the classical restriction of convexity.

For quasidifferentiable optimization problems necessary and in some cases sufficient optimality conditions can be written and will be used for the description of appropriate structural analysis problems. First-order necessary (resp. sufficient) conditions for a quasidifferentiable function f to admit a local minimum (resp. a strict local minimum) at point $x^* \in \mathbb{R}^n$ read:

$$-\overline{\partial}\,\overline{f}(x^*) \subset \underline{\partial}\,\underline{f}(x^*) \text{ (resp. int}\underline{\partial}\,\underline{f}(x^*)) \tag{22.2}$$

For constrained optimization we have analogous inclusions (see [8], [9] and the subsequent sections).

A model structural analysis application will clarify the use of quasidifferentiability in mechanics. Let us assume a structure with adhesive contact interfaces and let us assume that the phenomenological elastostatic analysis problem of this structure admits a potential characterisation, i.e. the governing relations of the problem can be derived by appropriate differentiation of a potential energy function. For the modelling of interface adhesive effects we adopt nonmonotone, possibly multivalued (i.e. with complete vertical branches) interface traction - relative displacements laws (see e.g. [19], [15]). A finite element discretization of the direct stiffness method with nodal displacements as the primary variables of the problem is assumed here. Let $\mathbf{u}$ be the n-dimensional vector of displacement degrees of freedom and $\mathbf{e}$ is the m-vector of element deformations. The potential energy optimization problem in elastostatics reads:

$$\min_{\mathbf{u} \in U_{ad}} \{\overline{\Pi}(\mathbf{u}) = \Pi(\mathbf{e}(\mathbf{u})) + \Phi(\mathbf{u}) + p(\mathbf{u})\} \tag{22.3}$$

where $\Pi(\mathbf{e})$ is the elastic energy stored in the system due to deformation, $\Phi(\mathbf{u})$ is the potential that counts for various boundary, interface or skin effects and $p(\mathbf{u})$ is the potential that generates the external loading vector. In general relative interface displacements are involved in the definition of $\Phi(\mathbf{u})$ and, respectively, interface tractions appear as their work dual variables in the virtual work expression which follow. The more economical notation is preferred here. For technical details see [21], [22], [9]. The geometric compatibility transformation is written in the form of a generally nonlinear but differentiable operator $\mathcal{A}(\mathbf{u}) : \mathbb{R}^n \to \mathbb{R}^m$, $\mathbf{e} = \mathcal{A}(\mathbf{u})$. Note that in the case of geometrical nonlinear effects the latter relation is nonlinear and thus the first term in the r.h.s. of (22.3) may become nonconvex.

Variational formulations for the elastostatic analysis problem described by (22.3) will be produced in the sequel by writting down the optimality conditions for this quasidifferentiable minimization problem and using the quasidifferential calculus for the derivation of the quasidifferential of the composite function $\overline{\Pi}(\mathbf{u})$. Variational equalities for classical smooth problems, variational inequalities for nonsmooth, subdifferentiable problems (cf. [16], [19]) and systems of variational inequalities for general quasidifferentiable problems (cf. [21], [18], [19], [9]) are thus derived in a systematic way.

Let $\Pi(\mathbf{e})$ of (22.3) be quasidifferentiable and let $D\Pi(\mathbf{e}) = [\underline{\partial}\Pi(\mathbf{e}), \overline{\partial}\Pi(\mathbf{e})] \in \mathbb{R}^m \times \mathbb{R}^m$. $D\overline{\Pi}(\mathbf{e}) \in \mathbb{R}^m \times \mathbb{R}^m$ can be constructed by using the rules of the quasidifferential calculus (see [8], p.127, [23]) for the composite function $\Pi(\mathbf{e}(\mathbf{u}))$. If moreover $\Phi(\mathbf{u})$ and $p(\mathbf{u})$ are differentiable we get

$$D\overline{\Pi}(\mathbf{u}) = [\underline{\partial}\, \overline{\Pi}(\mathbf{u}), \overline{\partial}\, \overline{\Pi}(\mathbf{u})] = [\underline{\partial}\Pi(\mathbf{u}) + \nabla\Phi(\mathbf{u}) + \nabla p(\mathbf{u}), \overline{\partial}\Pi(\mathbf{u})] \tag{22.4}$$

The optimality conditions for the quasidifferentiable unconstrained optimization problem read: find $\mathbf{u} \in \mathbb{R}^n$ such that

$$-\underline{\partial}\, \overline{\Pi}(\mathbf{u}) \subset \overline{\partial}\, \overline{\Pi}(\mathbf{u}) \tag{22.5}$$

They lead to the system of variational inequalities: find $\mathbf{u} \in \mathbb{R}^n$ such that

$$\mathbf{w} \subset \underline{\partial}\, \overline{\Pi}(\mathbf{u}), \ \forall w \in -\overline{\partial}\, \overline{\Pi}(\mathbf{u}) \tag{22.6}$$

An interesting case arises if the potential energy is a *difference convex (d.c.)* function. For instance let a small displacement problem be considered, i.e., the compatibility transformation reads $\mathcal{A}(\mathbf{u}) = \mathbf{G}^T\mathbf{u}$, with $\mathbf{G}^T$ an $(m \times n)$ matrix and accordingly $p(\mathbf{u}) = \mathbf{p}^T\mathbf{u}$, with $\mathbf{p}$ the n-dimensional loading vector. Let moreover $\Pi(\mathbf{e})$ in (22.3) be convex and differentiable, e.g. consider for instance the linear elasticity problem with $\Pi(\mathbf{e}) = \frac{1}{2}\mathbf{e}^T\mathbf{K}_0\mathbf{e}$, where $\mathbf{K}_0$ is the $(m \times m)$ natural stiffness matrix of the structure. Let the only cause of nonconvexity and nondifferentiability in the problem be introduced by a difference convex boundary potential energy function, i.e. $\Phi(\mathbf{u})$ is written as a difference of convex, possibly nondifferentiable terms ([21], [22])

$$\Phi(\mathbf{u}) = \Phi_1(\mathbf{u}) - \Phi_2(\mathbf{u}) \tag{22.7}$$

In this case by using the quasidifferential calculus we get:

$$\underline{\partial}\, \overline{\Pi}(\mathbf{u}) = \mathbf{K}\mathbf{u} + \mathbf{p} + \partial\Phi_1(\mathbf{u}), \ \overline{\partial}\, \overline{\Pi}(\mathbf{u}) = -\partial\Phi_2(\mathbf{u}) \tag{22.8}$$

Optimality conditions for the d.c. potential lead to the following system of variational inequalities that describe the nonsmooth and nonconvex structural analysis problem: find $\mathbf{u} \in \mathbb{R}^n$ such that

$$\mathbf{0} \in \mathbf{K}\mathbf{u} + \mathbf{p} + \partial\Phi_1(\mathbf{u}) - \mathbf{w} \tag{22.9}$$

for each $\mathbf{w} \in \mathbb{R}^n$ with

$$\mathbf{w} \in \partial\Phi_2(\mathbf{u}) \tag{22.10}$$

Problem (22.9) , (22.10) can be solved iteratively, where in each iteration cycle two convex subproblems are solved (see [21], [22]). The above decomposition is explicitly defined due to the global d.c. representation of the potential energy. In the general quasidifferential case the system of variational inequalities (22.6) is implicitly defined.

22.3 ELASTOPLASTICITY

Convex modelling

A unified approach to standard generalized, small displacement and deformation elastoplasticity is obtained by using the methods of convex analysis. For a discretized problem let the only internal variables in an elastoplastic body be the plastic strains e^p. Let moreover the elastic region in stress space be a closed convex subset C in R^n. We assume the existence of a convex *free energy potential*

$$W(e - e^p) = \frac{1}{2}(e - e^p)^T K_0 (e - e^p) \tag{22.11}$$

and a convex, nonnegative, sublinear and positively homogeneous of order one *internal dissipation potential*

$$\mathcal{D} = I_C^c(\dot{e}^p) \equiv \sup_{\sigma \in C}\{\sigma^T \dot{e}^p\} \tag{22.12}$$

Here K_0 is the natural elastic stiffness matrix and I_C^c is the support function of the convex set C (see e.g. [16], p.55).

The relations that govern the perfect elastoplastic analysis problem are derived by differentiating (resp. subdifferentiating) (22.11) (resp. (22.12)) with respect to e (resp. $\dot{e}^p$), i.e.

$$\sigma = K_0(e - e^p) \tag{22.13}$$

$$\sigma = \partial I_C^c(\dot{e}^p) \tag{22.14}$$

Recall that if the set C is convex then, by using the results of Minkowski duality, the function (22.12) is equal to the gauge function $|.|_{C^0}$ of the polar set of C, denoted by C^0, thus

$$\mathcal{D} = I_C^c(\dot{e}^p) = |\dot{e}^p|_{C^0} \tag{22.15}$$

Here the gauge function is defined by:

$$|\dot{e}^p|_{C^0} \equiv \min\{\lambda > 0 \mid \dot{e}^p \in \lambda C^0\} \tag{22.16}$$

and the polar set C^0 is a convex closed set, including the zero element such that

$$C^0 = \{\dot{e}^p \mid \sigma^T \dot{e}^p \leq 1, \ \forall \sigma \in C\} \tag{22.17}$$

The inverse relation to (22.14) takes the more conventional flow rule form:

$$\dot{e}^p \in \partial I_C(\sigma) = \mathcal{N}_C(\sigma) \tag{22.18}$$

where I_C denotes the indicator function of C and $\mathcal{N}_C$ is the normal cone to the set C. If C is described by the convex *yield locus* $F(\sigma)$ as

$$C = \{\sigma \in R^n | F(\sigma) \leq 0\} \tag{22.19}$$

relation (22.18) is equivalent to the *variational inequality* (c.f. Drucker second postulate)

$$\dot{e}^{p\ T}(\tau - \sigma) \leq 0, \ \forall \tau \in C, \ \sigma \in C, \text{ or } \forall \tau \text{ s.t. } F(\tau) \leq 0 \tag{22.20}$$

A model case of convex multisurface elastoplasticity is the Tresca yield function.

Recall here that for a single yield locus the flow rule (22.18), (22.20) is written as

$$\dot{e}^p = \lambda \nabla F(\sigma) \tag{22.21}$$

where λ is the plastic multiplier which obeys to the complementarity conditions

$$\lambda \geq 0, \ F(\sigma) \leq 0, \ \lambda F(\sigma) = 0 \tag{22.22}$$

For a convex, multisurface plasticity, i.e. when the convex yield locus is defined as an intersection of a number of sets which are defined by (22.19) with different functions $F(\sigma), i \in \mathcal{I}$, we have

$$\dot{e}^p = \sum_{i \in \mathcal{I}} \lambda_i \nabla F_i(\sigma) \tag{22.23}$$

and the complementarity conditions take the form:

$$\lambda_i \geq 0, \ F_i(\sigma) \leq 0, \ i \in \mathcal{I}, \ \sum_{i \in \mathcal{I}} \lambda_i F_i(\sigma) = 0 \tag{22.24}$$

Furthermore time discretization techniques can be used for the numerical solution of the elastoplastic analysis problem, where complementarity problems (cf. (22.22), (22.23)-(22.24)) are solved within each time step.

Structured nonconvex modelling

A nonconvex generalization of the above given approach to elastoplasticity which is based on the quasidifferentiability theory is proposed here. Rougly speaking tangent and normal cones to a generally nonconvex and nonsmooth set are systematically decomposed as unions and intersections of a finite number of convex cones. Thus variational inequalities of convex elastoplasticity are replaced by systems of variational inequalities for the general nonconvex case. A more general approach based on a nonconvex gauge theory proposed in [20], [8] and its connection to star-shaped sets is also given. For more informations on the quasidifferentiability concept and the star-shaped sets in nonsmooth mechanics and for applications the reader is referred to [8], [7], [6], [17], [19], [18].

Let the elasticity set be a general nonconvex, quasidifferentiable set with possible corners, i.e. the yield locus function is assumed to be quasidifferentiable and

$$\Omega = \{\sigma \in \mathbb{R}^n \mid \varphi(\sigma) \leq 0\} \tag{22.25}$$

To give a concrete example we assume a model case where Ω is the union of a finite number of convex subsets $\Omega_i, i \in \mathcal{I} = \{1, \ldots, m\}$, i.e.

$$\Omega = \bigcup_{i \in \mathcal{I}} \Omega_i, \ \Omega_i = \{\sigma \in \mathbb{R}^n \mid \varphi_i(\sigma) \leq 0\} \tag{22.26}$$

where φ_i are classical convex and differentiable yield functions (cf. (22.19)).

Before proceeding with the mechanical problem let us recall that the cone of feasible directions (Bouligand cone) to a set defined by (22.25) is characterized by means of the directional derivatives of $\varphi(.)$ as

$$\Gamma_1(\sigma) = \{g \in \mathbb{R}^n \mid \varphi'(\sigma, g) \leq 0\} \tag{22.27}$$

Recall that a function φ is called ***quasidifferentiable*** in the sense of Demyanov [6], [7], [8], iff it is directionally differentiable and its directional derivative is constructed by means of two convex, closed and compact subsets of $\mathbb{R}^n$. These two sets, which are called the subdifferential $\underline{\partial}\varphi$, and the superdifferential $\overline{\partial}\varphi$ respectively, are always considered as an ordered pair of convex sets, i.e. the quasidifferential is the pair $D\varphi = [\underline{\partial}\varphi, \overline{\partial}\varphi]$. By definition the directional derivative of a quasidifferentiable function is written as:

$$\varphi'(\sigma, g) = \max_{w_1 \in \underline{\partial}\varphi(\sigma)} < w_1, g > + \min_{w_2 \in \overline{\partial}\varphi(\sigma)} < w_2, g > \tag{22.28}$$

For a convex, thus subdifferentiable function φ only the first term in the r.h.s. of (22.28) is used, where $\underline{\partial}\varphi = \partial\varphi$ and $\partial\varphi$ is the convex analysis subdifferential. For a concave, thus superdifferentiable function the second term in the r.h.s. of (22.28) is used, where $\overline{\partial}\varphi = \partial\varphi$ and $\partial\varphi$ is the superdifferential. In view of (22.28), Γ_1 is equivalently expressed by

$$\begin{aligned}\Gamma_1(\sigma) = \{g \in R^n \mid \ & < w_1, g > + < w_2, g > \leq 0, \\ & \forall w_1 \in \underline{\partial}\varphi(\sigma), \text{ for some } w_2 \in \overline{\partial}\varphi(\sigma)\}\end{aligned} \tag{22.29}$$

or

$$\begin{aligned}\Gamma_1(\sigma) &= \bigcup_{w_2 \in \overline{\partial}\varphi(\sigma)} \ \bigcap_{w_1 \in \underline{\partial}\varphi(\sigma)} T_{w_1+w_2}(\sigma) \\ &= \bigcup_{w_2 \in \overline{\partial}\varphi(\sigma)} T_{\underline{\partial}\varphi(\sigma)+w_2}(\sigma)\end{aligned} \tag{22.30}$$

where the following shortcuts have been used:

$$T_{w_1+w_2}(\sigma) = \{g \in R^n \mid \ < w_1, g > + < w_2, g > \leq 0 \ \} \tag{22.31}$$

and

$$T_{\underline{\partial}\varphi(\sigma)+w_2}(\sigma) = \{g \in R^n \mid < w_1, g > + < w_2, g > \leq 0, \ \forall w_1 \in \underline{\partial}\varphi(\sigma) \ \} \quad (22.32)$$

One should observe here that (22.30) represents a systematic convex decomposition of the tangential cone to Ω at the point σ, due to the fact that the sets defined in (22.30), (22.31) are convex. Thus a representation of the possibly nonconvex normal cone to Ω at σ by using convex constituents is possible. It has the form:

$$\mathcal{N}_\Omega(\sigma) = \bigcap_{w_2 \in \overline{\partial}\varphi(\sigma)} \ \bigcup_{w_1 \in \underline{\partial}\varphi(\sigma)} T^0_{w_1+w_2}(\sigma) =$$

$$= \bigcap_{w_2 \in \overline{\partial}\varphi(\sigma)} T^0_{\underline{\partial}\varphi(\sigma)+w_2}(\sigma) = \bigcap_{w_2 \in \overline{\partial}\varphi(\sigma)} \{ \ - T^+_{\underline{\partial}\varphi(\sigma)+w_2}(\sigma) \ \} \quad (22.33)$$

Here the polar cone to T, denoted by T^0, and the conjugate cone of T, denoted by T^+, are defined by:

$$T^0 = -T^+ = \{g \in R^n \mid < g, x > \leq 1, \ \forall x \in T \ \} \quad (22.34)$$

One should mention that not all aspects of the above convex decomposition can be demonstrated by a two–dimensional example, since in two dimensions at a reentrant corner we always have $\mathcal{N}_\Omega(\sigma) = \{0\}$. Recall also that Clarke's generalized normal cone is not equal to the singleton $\{0\}$ in the latter case.

The previously given definitions and the optimality conditions for a quasidifferential optimization problem (see e.g. [8]) give rise to variational expressions for a perfect elastoplastic analysis problem with nonconvex, nondifferentiable yield surface, which are analogous to the convex case. Details can be found in [24], [9].

For instance, for the yield surfaces (22.26) the above general scheme results in the following formulae: Function $\varphi(\sigma)$ of (22.25) can be written as a min-type function by using the yield loci of subsets Ω_i in (22.26), i.e.

$$\Omega = \cup_{i \in \mathcal{I}} \Omega_i = \{\sigma \in R^n \mid \tilde{\varphi}(\sigma) = \min_{i \in \mathcal{I}} \{\varphi_i(\sigma)\} \leq 0\} \quad (22.35)$$

For the min-type function of (22.35), which is quasidifferentiable we may assume that its subdifferential and quasidifferential are given by the relations

$$\underline{\partial}\tilde{\varphi}(\sigma) = \{0\}, \ \overline{\partial}\tilde{\varphi}(\sigma) = \underset{i \in \mathcal{I}_0(\sigma)}{co} \{\nabla \varphi_i(\sigma)\} \quad (22.36)$$

where the following active index set $\mathcal{I}_0(\sigma) = \{i \in \mathcal{I}, \text{ s.t. } \tilde{\varphi}(\sigma) = \varphi_i(\sigma)\}$ is used.

Furthermore a link of the above outlined quasidifferential approach to the nonconvex elastoplasticity with nonconvex dissipation potentials and star-shaped yield sets in the sense of [20], p. 289, [8] has been proposed in [24].

22.4 DAMAGE MECHANICS

For the modelling of the gradual strength degradation of structures at a phenomenological level the continuum damage mechanics theory has been developed. Roughly speaking, the presence of microcracks and other microdefects in a material element at point x is represented by a scalar damage variable $d \in [0,1]$, where $d = 0$ represents the intact material and $d = 1$ represents the fully damaged one. Then the overall strength degradation can be considered by including reduction factors in the elasticity constitutive relations as follows:

$$\bar{\sigma} = \frac{\sigma}{(1-d)}, \quad \bar{\varepsilon} = (1-d)\varepsilon. \tag{22.37}$$

Here the real stresses and strains $\bar{\sigma}, \bar{\varepsilon}$ are referred to a reduced, due to damage, unit volume, area or length, while the nominal quantities σ, ε are referred to the initial structure. The latter quantities are used in the continuum mechanics modelling while the damage parameter(s) d serve as internal variables and require special treatment. For instance, a linear elasticity law can be written, in the presence of damage, as follows:

$$\varepsilon = \frac{F_0}{(1-d)}\sigma, \quad \sigma = (1-d)K_0\varepsilon, \tag{22.38}$$

with the usual notations.

Moreover, if appropriate flow rules for the internal damage parameter d are given, the whole damaging process (possibly coupled with other nonlinear structural effects, e.g., plasticity) can be traced for a given loading history.

In some cases the stress–strain laws of an elastic damaging body can be written as a difference of classical, monotone relations or, in the sense of the previous investigations they can be derived from a difference convex potential function. In these cases d.c. optimization techniques and structural analysis methods based on d.c. and quasidifferential optimization can be used for the solution of the arising nonlinear structural analysis task ([25], [9]).

Here a simple isotropic model damage theory with a scalar damage variable will be outlined.

Let us consider a stress-based isotropic continuum damage model. The complementary free energy potential of the body subjected to damage reads:

$$\Xi(\sigma, d_\sigma) = \frac{1}{2} d_\sigma \sigma^T F_0 \sigma = \frac{1}{2} d_\sigma \Pi_0^c(\sigma), \tag{22.39}$$

where $d_\sigma = \dfrac{1}{(1-d)}$ and the initial complementary elastic potential is denoted by $\Pi_0^c(\sigma)$.

Following analogous reasoning with the case of the elastoplasticity, we get the governing relations of the problem in the form

$$\varepsilon = \frac{\partial \Xi}{\partial \sigma} = d_\sigma \frac{\partial \Pi_0^c(\sigma)}{\partial \sigma}, \tag{22.40}$$

and the nonnegative dissipation inequality

$$\mathcal{D} = \Pi_0^c(\sigma).\dot{d}_\sigma \geq 0. \tag{22.41}$$

Damage initiates and evolves if the damage criterion in stress space holds as equality, where the damage criterion is expressed as:

$$g(\bar{\tau}, r) = \bar{\tau} - r = \sqrt{2\Pi_0^c(\sigma)} - r \leq 0. \tag{22.42}$$

Here r is a damage threshold. Moreover, the complementary energy norm is involved as an equivalent strain measure in (22.42) (the latter is dictated by the fact that this quantity is thermodynamically conjugate to d_σ, as it can be seen by considering the expression $\frac{\partial \Xi}{\partial d_\sigma}$).

Assuming, as in plasticity, that damage evolution follows the principle of maximum dissipation and writing down the optimality conditions for the problem (cf. (22.41))):

$$\max_{\{\bar{\tau}:g(\bar{\tau},r)\leq 0\}} = \bar{\tau}.\dot{d}_\sigma, \tag{22.43}$$

we get the damage evolution law along with the damage loading / unloading conditions:

$$\begin{aligned} \dot{d}_\sigma &= \dot{\mu}\frac{\partial g(\bar{\tau}, r)}{\partial \bar{\tau}}, \\ \dot{\mu} &\geq 0, g(\bar{\tau}, r) \leq 0, \quad \dot{\mu}.g(\bar{\tau}, r) = 0. \end{aligned} \tag{22.44}$$

Analogously one may develope a strain-based theory (with a damage criterion written in terms of strains). Moreover, if the damage evolution $\dot{d}_\sigma$ (resp. $\dot{d}$) is known, the current damage level can be found by integration along the previous loading history. Several specialized versions of a damage model can be formulated, which are appropriate for the treatment of specific applications (for instance coupled elastoplastic and damage interface models).

Besides the classical load-path integration techniques for the solution of damage mechanics problems several other approaches are possible. Here one can define limit analysis problems: find the maximum loading magnitude (or a boundary of all loading vectors) which makes the arising elastic damage problem solvable (or for which the real engineering structure still remains safe or reliable in some sense). Thus coupling of limit analysis theory or of structural reliability theory with the structured nonconvex modelling in mechanics, as it is outlined here, may be an interesting possibility for further investigations.

One possible way fo solving problems in this direction follows the lines of algorithms primarily developed for distributed structural optimization tasks. Here instead of looking e.g. for an optimal material distribution which makes a given structure stiffer, one looks for a worst damage variable distribution which renders a given structure subjected to a given loading more flexible [1]. This approach has been studied in [15], Chapt. 5 and in [26] with respect to design of composite materials. Using an analogous procedure a shear damage mechanism in a twodimensional plain stress plate is analysed in this paper. In fact only the shear stress - shear strain part of the elasticity relation is subjected to damage and is equiped with a scalar damage parameter d. The extend of damage, which is depicted by contour plots of d for all elements of the structure, is given in Figs 1a-d. The plate is supported along the two internal holes near the left boundary and it is loaded by a uniformly distributed vertical loading at the internal hole near the right boundary. Finally about 1000 triangular finite elements have been used for the discretization which is not symmetrical. The total allowable measure of damage is prescribed and the actual distribution of the damage which will appear for a given loading is computed by means of the iterative algorithm described in [15], Chapt. 5 and in [26]. Exact numerical results are not given here, our purpose being to show the qualitative nature of the obtained results. Thorough numerical investigations in this direction will possibly lead to results with engineering interest. These results will be reported elsewhere.

Acknowledgments

The author gratefully acknowledges the financial support of the European Union (Marie Curie Grant ERBFMBICT960987).

References

[1] W. Achtziger and M.P. Bendsøe, Design for maximal flexibility as a simple computational model of damage. Structural Optimization, 10, 258–268, 1995.

[2] P. Aubin, H. Frankowska, *Set-valued analysis*, Birkhäuser Verlag, Basel, Boston 1991.

[3] Z.P. Bazant, L. Cedolin, *Stability of structures. Elastic, inelastic, fracture and damage theories*, Oxford University Press, New York, Oxford 1991.

[4] C. Comi, G. Maier and U. Perego, *Generalized variable finite element modelling and extremum theorems in stepwise holonomic elastoplasticity with internal variables*, Computer Method in Applied Mechanics and Engineering 96, 213-237 (1992).

[5] R.A. Eve, B.D. Reddy and R.T. Rockafellar, *An internal variable theory of plasticity based on the maximum plastic work inequality*, Quarterly of Applied Mathematics 68(1), 59–83 (1990).

[6] V.F. Demyanov and L.V. Vasiliev, L.V., *Nondifferential Optimization*, New York, Optimization Software Inc., Publications Division, 1985.

[7] V.F. Demyanov and A.M. Rubinov, *Quasidifferential Calculus*, New York, Optimization Software Inc., Publications Division, 1986.

[8] V.F. Demyanov and A.M. Rubinov, *Introduction to constructive nonsmooth analysis*, Frankfurt a.M., Peter Lang Verlag, 1995.

[9] V.F. Demyanov, G.E. Stavroulakis, L.N. Polyakova and P.D. Panagiotopoulos, *Quasidifferentiability and nonsmooth modelling in mechanics,*

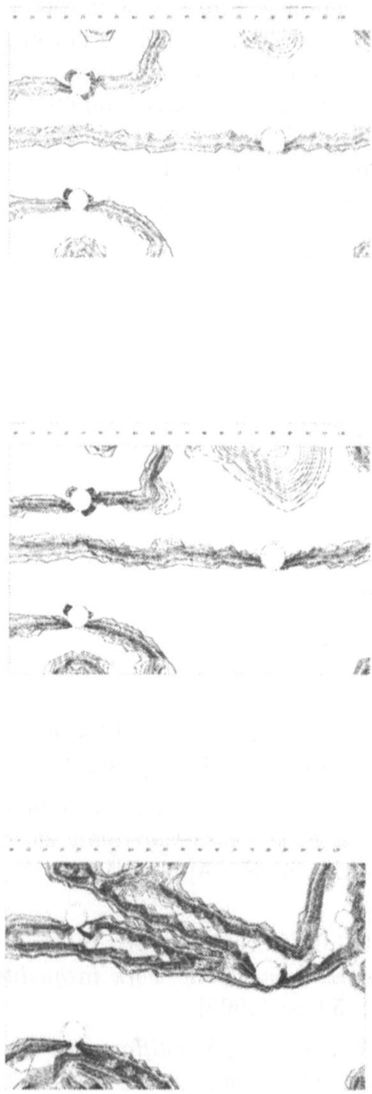

Figure 22.1 Damage distribution. Maximum damage measure is for (a) equal to 0.85, for (b) equal to 0.90 and for (c) equal to 0.95.

engineering and economics, Kluwer Academic, Doldrecht, Boston, London 1996.

[10] R. Glowinski and P. Le Tallec, *Augmented lagrangian and operator-splitting methods for nonlinear mechanics.* Philadelphia, SIAM, 1989.

[11] J.-B. Hiriart-Urruty and C. Lemarechal, *Convex analysis and minimization algorithms I*, Springer-Verlag, Berlin, Heidelberg 1993.

[12] J. Lubliner, *Plasticity Theory*, Macmillan Publ., New York and Collier Macmillan Publ., London 1990.

[13] L.E. Malvern, *Introduction to the mechanics of a continuous medium*, Prentice-Hall Inc., Englewood Cliffs, N.J. 1969.

[14] J.B. Martin and S. Caddemi, *Sufficient conditions for convergence of the Newton-Raphson iterative algorithms in incremental elastic-plastic analysis*, European Journal of Mechanics, A/Solids (3) 13, 351-365 (1994).

[15] E.S. Mistakidis and G.E. Stavroulakis, Nonconvex optimization in mechanics. Algorithms, heuristics and engineering applications by the F.E.M. Kluwer Academic, Dordrecht, Boston, London, 1998.

[16] P.D. Panagiotopoulos, *Inequality problems in mechanics and applications. Convex and nonconvex energy functionals*, Boston - Basel - Stuttgart: Birkhäuser, 1985.

[17] P.D. Panagiotopoulos, *Nonconvex superpotentials and hemivariational inequalities. Quasidifferentiability in mechanics*, In: *Nonsmooth mechanics and applications* J.J. Moreau and P.D. Panagiotopoulos (eds.), CISM Lecture Notes No. 302. Wien - New York: Springer 1988, 82-176.

[18] P.D. Panagiotopoulos, and G.E. Stavroulakis, *New type of variational principles based on the notion of quasidifferentiability*, Acta Mechanica, 94, 171-194 (1992).

[19] P.D. Panagiotopoulos, *Hemivariational inequalities. Applications in mechanics and engineering*, New York - Berlin, Springer Verlag, 1993.

[20] A.M. Rubinov and A.A. Yagubov, *The space of star-shaped sets and its applications in nonsmooth optimization*, In: *Quasidifferential Calculus*, V.F. Demyanov and L.C.W. Dixon, (eds), Mathematical Programming Study 29, Amsterdam, North-Holland, 1986, 176-202.

[21] G.E. Stavroulakis, *Convex decomposition for nonconvex energy problems in elastostatics and applications*, Eur. J. Mechanics, A/Solids, 12(1), 1-20 (1993).

[22] G.E. Stavroulakis and P.D. Panagiotopoulos, *Convex multilevel decomposition algorithms for non-monotone problems*, Int. J. Num. Meths in Engng, 36(11), 1945-1961 (1993).

[23] G.E. Stavroulakis, V.F. Demyanov and L.N. Polyakova, *Quasidifferentiability in nonsmooth, nonconvex mechanics*, Journal of Global Optimization 10(2), 77-87, 1995.

[24] G.E. Stavroulakis, Quasidifferentiability and star-shaped sets. Application in nonconvex, finite dimensional elastoplasticity. Communications on Applied Nonlinear Analysis 2(3), 23-46, 1995.

[25] G.E. Stavroulakis and L.N. Polyakova, Difference convex optimization techniques in nonsmooth computational mechanics. Optimization. Methods and Software 7(1), 55-87, 1996.

[26] P.S. Theocaris and G.E. Stavroulakis, Multilevel optimal design of composite structures including materials with negative Poisson's ratio. Structural Optimization 15(1), 8–15, 1998.

23 AUGMENTED LAGRANGIAN METHODS FOR CONTACT PROBLEMS, OPTIMAL CONTROL AND IMAGE RESTORATION

Józef Joachim Telega and Andrzej Gałka

Polish Academy of Sciences,
Institute of Fundamental Technological Research,
Świętokrzyska 21, 00-049 Warsaw, Poland

Abstract: The aim of this contribution is to review the available results on mathematical aspects of augmented Lagrangian methods and their applications to contact problems without and with friction, parameter estimation, optimal control and image restoration.

23.1 INTRODUCTION

This contribution presents a review of augmented Lagrangian methods both from the point of view of rigorous mathematical approach and application to contact problems without and with friction, parameter estimation in elliptic and parabolic systems and image restoration. On account of limited number of disposable pages our paper is focussed on rather basic aspects of these methods. Thus it may be seen as complementing the excellent review paper by ARORA et all. (1991). In the last paper various augmented Lagrangian methods, called also multiplier methods, are studied with respect to their computational procedures and rate of convergence, cf. also KLARBRING (1992), KIKUCHI and ODEN

R.P. Gilbert et al. (eds.), From Convexity to Nonconvexity, 311–332.

(1988). The reader interested in applications of the augmented Lagrangian methods to elasticity, plasticity and viscoplasticity is referred to GLOWINSKI and LE TALLEC (1989), LE TALLEC (1983, 1986), BLANCHARD and LE TALLEC (1986)

23.2 GENERAL RESULTS

The first section is concerned with the presentation of basic mathematical ideas related to augmented Lagrangian methods.

The multiplier method of Hestens and Powell

The following idea was raised independently by HESTENS (1969) and POWELL (1969). To present this idea we follow ROCKAFELLAR (1973a). Consider the following nonlinear programming problem:

$$\text{minimize} \quad f_0(\mathbf{x}) \quad \text{subject to} \quad f_i(\mathbf{x}) = 0, \; i = 1, ..., n. \tag{23.1}$$

where the functions $f_i : R^n \to R$ are differentiable. The solutions of this problem are not altered if terms of the form $rf_i(\mathbf{x})^2, 1 \leq i \leq m$, are added to the objective function. Consequently, the problem can be represented by the Lagrangian

$$K_r(\mathbf{x}, \mathbf{y}) = f_0(\mathbf{x}) + \sum_{i=1}^{m} [y_i f_i(\mathbf{x}) + r f_i(\mathbf{x})^2], \tag{23.2}$$

where the constant $r > 0$ is the *penalty factor.* The penalty terms suggest the likelihood that, if $\overline{\mathbf{x}}$ is a local optimal solution with the corresponding multiplier vector $\overline{y}$ satisfying

$$\nabla f_0(\overline{\mathbf{x}}) + \sum_{i=1}^{m} \nabla f_i(\overline{\mathbf{x}}) = 0, \tag{23.3}$$

and if r is sufficiently large, then $K_r(\mathbf{x}, \overline{\mathbf{y}})$ will have a local minimum at $\overline{\mathbf{x}}$. Indeed, it can be shown that this is true if $\overline{\mathbf{x}}$ satisfies the standard second-order conditions sufficient for local optimality in 23.1 Consequently, one might attempt to solve 23.1 by solving a sequence of *unconstrained* minimization problems:

$$\text{find } \mathbf{x}^k \text{ minimizing } K_r(\mathbf{x}, \mathbf{y}^k). \tag{23.4}$$

Since $\mathbf{x}^k$ satisfies

$$0 = \nabla_{\mathbf{x}} K_r(\mathbf{x}^k, \mathbf{y}^k) = \nabla f_0(\mathbf{x}^k) + \sum_{i=1}^{m} \left(y_i^k + 2r f_i(\mathbf{x}^k)\right) \nabla f_i(\mathbf{x}^k), \tag{23.5}$$

a natural way to generate the sequence $\{\mathbf{y}^k\}$ is to set

$$y_i^{k+1} = y_i^k + 2r f_i(\mathbf{x}^k). \tag{23.6}$$

The main advantage of the algorithm lies in that it does not require that $r \longrightarrow +\infty$. This algorithm provides a numerical stability that is not found in the usual penalty methods.

In the case of unilateral constraints $f_i(\mathbf{x}) \leq 0$ $(i = 1, ..., m)$, the natural replacement for the Lagrangian K_r is see ROCKAFELLAR (1973b)

$$L_r(\mathbf{x}, \mathbf{y}) = f_0(\mathbf{x}) + \frac{1}{4}r \sum_{i=1}^{m} \left[\theta\left(y_i + 2r f_i(\mathbf{x})\right)^2 - y_i^2\right], \tag{23.7}$$

where

$$\theta(t) = \max\{t, 0\}, \tag{23.8}$$

and

$$\frac{1}{4}r\left[\theta\left(y_i + 2r f_i(\mathbf{x})\right)^2 - y_i^2\right] = \begin{cases} r f_i(\mathbf{x})^2 + y_i f_i(\mathbf{x}) & \text{if} \quad f_i \geq -\frac{y_i}{2r}, \\ -\frac{y_i^2}{4r} & \text{if} \quad f_i(\mathbf{x}) \leq -\frac{y_i}{2r}. \end{cases} \tag{23.9}$$

The algorithmic approach of Hestens and Powell is now generalized as follows. For given $\mathbf{y}^k$ and $r > 0$ we find $\mathbf{x}^k$ minimizing $L_r(\mathbf{x}, \mathbf{y}^k)$, so that

$$0 = \nabla_{\mathbf{x}} L_r(\mathbf{x}^k, y^k) = \nabla f_0(\mathbf{x}^k) + \sum_{i=1}^{m} \theta\left(y_i^k + 2r f_i(\mathbf{x}^k)\right) \nabla f_i(\mathbf{x}^k). \tag{23.10}$$

We then set

$$y_i^{k+1} = \theta\left(y_i^k + 2r f_i(\mathbf{x}^k)\right) \geq 0, \; i = 1, ..., m. \tag{23.11}$$

In other words we have

$$\mathbf{y}^{k+1} = \mathbf{y}^k + 2r \nabla_{\mathbf{y}} L_r(\mathbf{x}^k, \mathbf{y}^k). \tag{23.12}$$

ROCKAFALLAR (1973b) proved a convergence theorem, which will now be formulated.

Let C denoted a convex set in a real vector space V (possibly infinite-dimensional), and let $f_0, f_1, ..., f_m$ be real-valued *convex* functions on C. We are concerned with the following problem

$(\mathcal{P})$ | find min $\{f_0(\mathbf{x}) \mid f_i(\mathbf{x}) \leq 0, \; \mathbf{x} \in C, \; i = 1, ..., m\}$

ROCKAFELLAR (1973a, 1973b), devised the following algorithm for solving Problem $\mathcal{P}$, which depends on the initial choice of $r > 0$, $\mathbf{y}^0 \in R^m$, and a sequence $\{\alpha_k\}, k \in N$ with $0 \leq \alpha_k \to 0$ as $k \to \infty$. We observe that the Lagrangian $L_r(\mathbf{x}, \mathbf{y})$ is convex in $\mathbf{x} \in$ and concave in $\mathbf{y} \in R^m$.

Algorithm. Given $\mathbf{y} \in R^m$, determine $\mathbf{x} \in C$ such that

$$L_r(\mathbf{x}^k, \mathbf{y}^k) \leq \inf_{\mathbf{x} \in C} L_r(\mathbf{x}, \mathbf{y}^k) + \alpha_k. \tag{23.13}$$

Then, define $\mathbf{y}^{k+1}$ by 23.11, or, equivalently, 23.12.

Let m denote the optimal minimum value in $(\mathcal{P})$. A *Kuhn-Tucker vector* for $(\mathcal{P})$ is a vector $\overline{\mathbf{y}} \in R_+^m$ with the property that

$$\inf_{\mathbf{x} \in C} \{f_0 \mathbf{x} + \overline{y}_1 f_1(\mathbf{x}) + ... + \overline{y}_m f_m(\mathbf{x})\} = m > -\infty. \tag{23.14}$$

Now we are in a position to state the convergence result.

Theorem 23.1. *Suppose that* $(\mathcal{P})$ *possesses at least one Kuhn-Tucker vector and that*

$$\sum_{k=1}^{m} \alpha_k^{1/2} < +\infty. \tag{23.15}$$

Then, $\{\mathbf{y}^k\}$ *converges to some Kuhn-Tucker vector, while* $\{\mathbf{x}^k\}$ *satisfies*

$$\lim_{k\to\infty} f_0(\mathbf{x}^k) = m = \min(\mathcal{P}). \tag{23.16}$$

Sketch of the proof. To obtain the desired conclusion about $\{\mathbf{x}^k\}$, it is sufficient to prove that $\{\mathbf{y}^k\}$ is a bounded maximizing sequence for g_r, defined by

$$g_r(\mathbf{y}) = \inf_{\mathbf{x}\in C} L_r(\mathbf{x}, \mathbf{y}).$$

If, in addition, $\{\mathbf{y}^k\}$ is convergent, the whole theorem is established.
Remark 22.2. Further developments of the augmented Lagrangian methods, primarily proposed by HESTENS (1969) and POWELL (1969), can be found in the papers by ROCKAFELLAR (1976,1981,1988), HAGER (1993) and NAKAYAMA et al. (1975), POLYAK and TRETYAKOV (1973). The books by BERTSEKAS (1982) and GOLSHTEIN and TRETYAKOV (1996) summarize and develop many theoretical and more practical aspects of these methods, (see also HIRIART-URRUTY and LEMARÉCHAL (1993b, Chapter XII, Sec. 5.2). GOLSHTEIN and TRETYAKOV (1996) prefer the notion of "modified Lagrangian". BOFFI and LOVADINA (1997) proposed an augmented Lagrangian method for the study of saddle points in the absence of unilateral conditions, cf. also HIPTMAIR et al. (1996).

23.3

Ito and Kunisch augmented Lagrangian methods

ITO and KUNISCH (1990c) generalized the method of HESTENS (1969) and POWELL (1969) to the following nonlinear programming problem:

$$(\mathcal{P}_1) \quad \left| \begin{array}{l} \text{find} \\ \qquad \min\left\{ f(u) | k(u) = 0, g(u) \le 0,\ A(u) \le \hat{z} \right\}, \end{array}\right.$$

where $f : X \to R, k : X \to Y, g : X \to R^m, A : X \to Z, \hat{z} \in Z,$ and with X and Y Hilbert spaces and Z a Banach space. The operator A is assumed to be linear and bounded on X. The ordering on Z is induced by a closed convex cone $\mathcal{K}$ with vertex at the origin.
The setting of Problem $(\mathcal{P}_1)$ generalizes the results available before 1990 for nonlinear programming in infinite dimensional spaces; for a brief discussion see ITO and KUNISCH (1990c). Now the augmented Lagrangian is given by:

$$L_r(u, v; \lambda, \mu, \eta) = L(u; \lambda, \mu, \eta) + \frac{1}{2} r \,||k(u)||_Y^2 + \frac{1}{2} r \,||g(u) + v||_{R^m}^2 \tag{23.17}$$

where L is the standard Lagrangian:

$$L(u;\lambda,\mu,\eta) = f() + <\lambda, k(u)>_{Y\times Y} + <\mu, g(u)>_{R^m\times R^m} + <\eta, A(u)>_{Z^*\times Z} \tag{23.18}$$

Here v plays the role of a slack variable for the constraint $g(u) \leq 0$, i.e., $g(u)+v=0, v\geq 0$. As previously, we have $r \in R, r>0$. Under twice Fréchet differentiability of f, k and g and additional assumptions ITO and KUNISCH (1990c) developed the augmented Lagrangian algorithm and its convergence. We observe that the assumptions for the algorithm to hold are rather strong. In essence, this algorithm is similar to the one which will now be presented for the case where the image space of the inequality constraint is infinite dimensional.

I

n the present section we describe, after ITO and KUNISCH (1990a), the augmented Lagrangian technique for a general class of variational inequalities. This class comprises unilateral frictionless contact problems for geometrically linear solids and structures (beams, plates, shells).
We introduce the following spaces and mappings:
V is a Hilbert space; B is a reflexive Banach space, continuously embedded into V; H is a Hilbert lattice with inner product $(\cdot,\cdot)$; $a(\cdot,\cdot): V\times V \to R$ is a bilinear, continuous, V-eliptic (coercive) form, with $a(v,v)\geq \alpha\|v\|^2$, for some $\alpha>0$; $l: V\to R$ is a continuous linear functional; $g: B\to H$ is a convex, continuous, Gâteaux differentiable mapping. Our framework is slightly more general then that of ITO and KUNISCH (1990a) since we use the Riesz map $j: H\to H^*$. Thus we have $j^{-1}: f\in H \leftrightarrow v_f \in H$, $\|f\|_{H^*} = \|v_f\|_H$. Also, we set: $f(v) = <f, v>_{H^*\times H}$. ITO and KUNISCH (1990a) identify H with its dual.
The problem under investigation reads:

$$(\mathcal{P}_2) \quad \left| \begin{array}{l} \text{find} \\ \qquad \min\left\{\frac{1}{2}a(u,u) - l(u) |\ g(u)\leq 0,\ u\in B\right\}. \end{array}\right.$$

In practice, the order denoted by $\leq$ means pointwise almost everywhere ordering. The following hypothesis is assumed:
(A) There exists $(\overline{u}, j^{-1}\lambda^*) \in B\times H$ such that $\overline{u}$ solves $(\mathcal{P}_2)$, $\lambda^*\geq 0$,
  (i) $<j^{-1}\lambda^*, g(\overline{u})> = 0$,
  (ii) $a(\overline{u},h) - l(h) + <j^{-1}\lambda^*, g'(\overline{u})h> = 0$, for all $h\in B$.
Here $g'(u)$ stands for the Gâteaux derivative of g at $\overline{u}$.
The coercivity of $a(\cdot,\cdot)$ implies uniqueness of the solution $\overline{u}$ of $(\mathcal{P}_2)$. As we already know, λ^* of (A) is referred to as Lagrange multiplier for $(\mathcal{P}_2)$.
The *augmented Lagrangian* is now determined by

$$L_r(u,\lambda) = \frac{1}{2}a(u,u) - l(u) + \frac{1}{2r}\left[\|\sup(0, j^{-1}\lambda + r\,g(u))\|_H^2 - \|j^{-1}\lambda\|_H^2\right]. \tag{23.19}$$

where $\lambda\in H^*, r\in R, r>0$. We recall that $\|\lambda\|_{H^*} = \|j^{-1}\lambda\|_H^2$, by the Riesz representation theorem.

Now we are in a position to define a family of augmented Lagrangian problems:

$$(\mathcal{P}_2)_{r,\lambda} \quad \left| \begin{array}{l} \text{find} \\ \qquad \min \{L_r(u,\lambda) |\ u \in B\}. \end{array} \right.$$

Simple calculation shows that $L_r(u, \lambda)$ can be expressed in alternative manner. Indeed, let $\hat{g} : B \times H \times R^+ \to H$ be defined by

$$\hat{g}(u, \lambda, r) = \sup\left\{g(u), -\frac{j^{-1}\lambda}{r}\right\}. \tag{23.20}$$

Then

$$L_r(u, \lambda) = \frac{1}{2}a(u,u) - l(u) + < j^{-1}\lambda, \hat{g}(u, \lambda, r) > + \frac{r}{2}||\hat{g}(u, \lambda, r)||_H^2. \tag{23.21}$$

As usual, the augmented Lagrangian algorithm for solving Problem $(\mathcal{P}_2)$ consists of a sequence of *unconstrained* optimization problems $(\mathcal{P}_2)_{r_k,\lambda_k}$ whose solutions, provided that they exist, converge to the solution of $(\mathcal{P}_2)$.

The algorithm.

Step 1: Choose $\lambda_1 \in H^*$, $\lambda_1 \geq$ and $r > 0$.

Step 2: Put $k = 1$.

Step 3: Solve $(\mathcal{P}_2)_{r,\lambda_k}$ for u_k.

Step 4: Put $j^{-1}\lambda_{k+1} = j^{-1}\lambda_k + r\hat{g}(u_k, \lambda_k, r) = \sup(0, j^{-1}\lambda_k) + r\, g(u_k)$.

Step 5: Put $k = k + 1$ and reurn to step 3.

We observe that a stopping criterion must be added to the algorithm for actual computations. Let us proceed to the convergence of algorithm just introduced.

Theorem 23.2. *Assume that (A) holds and that $u \to L_r(u, \lambda)$ is radially unbounded (i.e., $L_r(u, \lambda) \to \infty$ when $||u||_V \to \infty$) from B to R for every $\lambda \geq 0$. Then*

$$\alpha||\overline{u} - u_k||_V^2 + \frac{1}{2r}||j^{-1}\lambda_{k+1} - j^{-1}\lambda^*||_H^2 \leq \frac{1}{2r}||j^{-1}\lambda_k - j^{-1}\lambda||_H^2, \tag{23.22}$$

for all $k = 1, 2, \ldots$. We recall that $j^{-1}\lambda_k - j^{-1}\lambda^* = j^{-1}(\lambda_k - \lambda*)$ and $||j^{-1}(\lambda_k - \lambda^*)||_H = ||\lambda_k - \lambda^*||_{H^*}$.

We note that assumptions on the radial unboundedness of $u \to L_r(u, \lambda)$ and on the reflexivity of B are only required in the proof of an audiliary lemma (Lemma 2.1 in Ito and Kunisch, 1990a) and can be replaced by the assumption of existence of a solution of $(\mathcal{P}_2)_{r,\lambda}$. If $B = V$, then the radial unboundedness of $u \to L_r(u, \lambda)$ is implied by the coercivity of $a(\cdot, \cdot)$.

Proposition 1.4 Under the assumption of Theorem ??, $\{||\lambda_k||_{H^*}\}_{k\in N}$ is bounded and $u_k \to \overline{u}$ in V with

$$\alpha \sum_{k=1}^{\infty} ||u_k - \overline{u}||_V^2 \leq \frac{1}{2r}||j^{-1}\lambda_1 - j^{-1}\lambda^*||_H^2. \tag{23.23}$$

Moreover, if $g'(u_k)h \to g'(\overline{u})h$ in H, for every $h \in B$, then every weak cluster point $\overline{\lambda}$ of $\{\lambda_k\}_{k\in N}$ satisfies $(A)(ii)$ with λ^* replaced by $\overline{\lambda}$ and $\overline{\lambda} \geq 0$.

Corollary 23.3.1. *Under the assumptions of Theorem ??, $\overline{u}$ is the unique solution of the constrained Problem $(\mathcal{P}_2)$ if and only if $\overline{u}$ is the solution of unconstrained Problem $(\mathcal{P}_2)_{r,\lambda^*}$,*

Proof. The result follows imediately from Theorem ?? with $\lambda_1 = \lambda^*$. □

The augmented Lagrangian algorithm with variable stepsize
In this case the step 4 is replaced with:
Step 4a: Put $j^{-1}\lambda_{k+1} = j^{-1}\lambda_k + \beta\, r\, \hat{g}(u_k, \lambda_k, r)$, where $\beta \in (0,1]$.
ITO and KUNISCH (1990a) extended Theorem ?? to the algorithm with variable stepsize.

ITO

and KUNISCH (1995, 1996a) generalized the augmented Lagrangian method to the following problem:

$(\mathcal{P}_3)$ | find $\min\{f(u) + \varphi(\Lambda u) | u \in \mathcal{C}\}$

Here $f : X \to R$ is a lower semicontinuous, continuously differentiable, convex function; $\Lambda \in \mathcal{L}(X, H)$ and $\varphi : H \to R$ is a proper, lower semicontinuous convex function. X and H are real Hilbert spaces and $\mathcal{C}$ is a closed convex subset of X. As usual, $\mathcal{L}(X, H)$ denotes the space of linear and bounded operators from X into H. It is assumed that $H = H^*$ and $< \cdot, \cdot >$ denotes now the dual pairing (the dual product) defined on $X^* \times X$. We make the following assumptions:
(A_1) f and φ are bounded below by zero on $\mathcal{C}$.
(A_2) $< f'(v) - f'(u), v - u > \geq \sigma \|v - u\|^2_X$, for all $u, v \in \mathcal{C}$ and $\sigma > 0$.
(A_3) $\varphi(\Lambda u_0 < \infty$ for some $u_0 \in \mathcal{C}$.

Let us pass to the augmented Lagrangian method. Obviously, Problem $(\mathcal{P}_3)$ is eqiuvalent to:

$$\text{find } \min\{f(u) + \varphi(\Lambda u - w) | u \in \mathcal{C}, w = 0 \text{ in} H\} \tag{23.24}$$

The equality constraint $w = 0$ in 23.24 is treated by the augmented Lagrangian method. In fact, consider a sequential minimization over $u \in \mathcal{C}$ and $w \in H$ of the form:

$$\min\{f(u) + \varphi(\Lambda u - w) + (\lambda, w)_H + \frac{r}{2}\|w\|^2_H\}, \tag{23.25}$$

where $\lambda \in H$ is a Lagrange multiplier and $r > 0$ is a penalty parameter. Equivalently, Problem 23.25 is written as follows:

$$\min\{L_r(u, \lambda) = f(u) + \varphi_r(\Lambda u, \lambda) |\ u \in \mathcal{C}\}, \tag{23.26}$$

where φ_r is a smooth approximation of φ and is given by

$$\varphi_r(v,\lambda) = \inf\{\varphi(v-z) + (\lambda, z)_H + \frac{r}{2}\|z\|_H^2 | \ z \in H\} \tag{23.27}$$

for $(r,\lambda) \in R^+ \times H$.

Algorithm.

Step 1: Choose a starting value $\lambda_1 \in H$, a positive number r and set $k=1$.

Step 2: Given $\lambda_k \in H$, find $u_k \in H$ by $L_r(u_k,\lambda_k) = \min L_r(u,\lambda_k)$ over $u \in \mathcal{C}$.

Step 3: Update λ_k by $\lambda_{k+1} = \varphi_r'(\Lambda u_k, \lambda_k)$.

Step 4: If the convergence criterion is not satisfied then set $k=k+1$ and go to Step 2.

ITO and KUNISCH (1996) proved the convergence of the augmented Lagrangian algorithm just introduced.

23.4 CONTACT PROBLEMS

The celebrated contribution by Fichera (1972) opened a route to variational treatment of unilateral contact problems. In the second section of the paper we present our recent result on the application of the augmented Lagrangian method to a beam contact problem. Also, we review recent developments on applications of augmented Lagrangian methods to various contact problems without and with friction. We note that the paper by LANDERS and TAYLOR (1985) was one of the first where the augmented Lagrangian method was applied to a frictionless contact problem, cf. also SIMO at al. (1985).

Application of Ito - Kunisch method to contact beam problem without friction

To illustrate the augmented Lagrangian method presented in Section **??** we consider a simple obstacle problem. Let the functional of the total potential energy of a clamped beam be given by

$$J(v) = \frac{1}{2}\int_0^1 \left[(v''(x))^2 - 2\,f\,v(x)\right]\,dx.$$

We formulate the convex minimization problem:

$$\text{find } w \in K \text{ such that } J(w) = \inf_{v\in K} J(v) \tag{23.28}$$

where

$$K = \{v \in H^2(0,1) \mid v(0) = v(1) = v'(0) = v'(1) = 0, v(x) \geq -h\ , x \in (0,1)\}$$

We assume that $K \neq \emptyset$. Problem **??** is *equivalent* to the following variational inequality:

$$w \in K: \ a(w, v-w) \geq l(v-w), \ \forall v \in K, \tag{23.29}$$

where

$$a(w, v-w) = \int_0^1 w''(x)(v(x) - w(x))''dx, \ l(v) = \int_0^1 f\,v(x)dx.$$

Formally, the variational inequality 23.29 is equivalent to:

$$\begin{aligned} &w^{IV} - f \geq 0, \quad w + h \geq 0, \quad (w^{IV} - f)(w + h) = 0, \text{ in } (0.1) \\ &w(0) = w(1) = 0, \ w'(0) = w'(1) = 0, \end{aligned}$$

To apply the augmented Lagrangian method we set

$$g(w) = -w - h \leq 0, \ (\text{since } w \geq -h)$$

Then

$$\hat{g}\,(v, \lambda, r)) = \sup\left(-v - h, -\frac{\lambda}{r}\right) = \begin{cases} -v - h & \text{if } v \leq -h + \frac{\lambda_1}{r}, \\ -\frac{\lambda_1}{r} & \text{otherwise.} \end{cases}$$

The augmented Lagrangian 23.21 has now the following form

$$L_r(v, \lambda) = \frac{1}{2}\int_0^1 (v'')^2 dx - \int_0^1 f\, v\, dx + \int_0^1 \lambda\, \hat{g}(v, \lambda, r) dx + \frac{r}{2}\int_0^1 (\hat{g}(v, \lambda, c))^2\, dx,$$

To solve Problems $(\mathcal{P}_2)_{r_k, \lambda_k}$ in this specific case we have used the FEM. Figures ?? present the results of calculations.

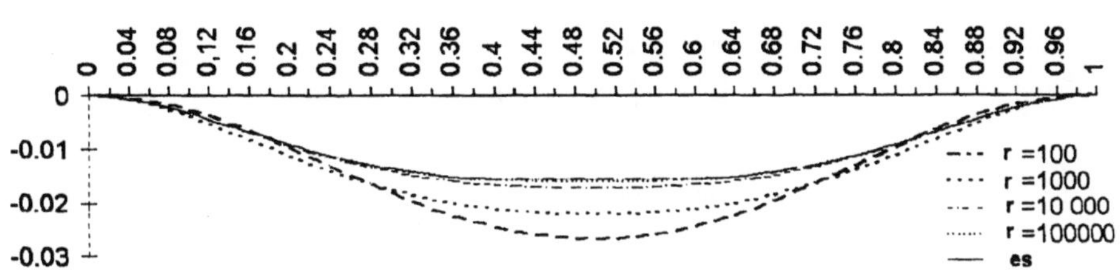

Figure 23.1 Exact solution and the beam displacement in the first step for $\lambda_1 = 1, \ h = 1/64$ and four values of r; es - exact solution.

Brief survey of applications of augmented Lagrangian methods to contact problems

HEEGAARD and CURNIER (1993) developed the augmented Lagrangian approach for the unilateral *frictionless* large-slip contact problem between two-dimensional elastic discretized bodies, cf. also HEEGAARD (1995). The procedure was applied to the analysis of the human kneecap sliding on the femur during flexion.

GWINNER (1989) used a *penalty* approximation method to describe the frictionless unilateral contact of a hyperelastic body with a rigid obstacle. The

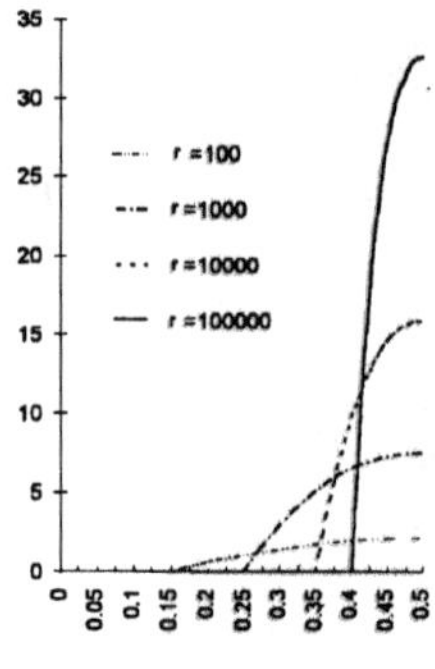

Figure 23.2 λ_2 for four values of r

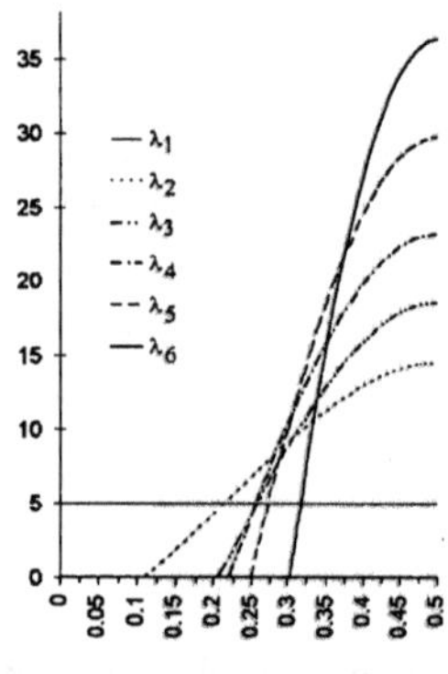

Figure 23.3 λ_i $(i = 1, \ldots, 6)$ for $r = 500, h = 1/64$ in the subsequent iterations.

stored energy function is *polyconvex* while the unilateral boundary condition is assumed in the form

$$j(\psi(x)) \leq 0 \text{ for } x \in \Gamma_2 \tag{23.30}$$

where $\Gamma_2 \subset \Gamma = \partial\Omega$ and ψ stans for a deformation (configuration) field. The function $j : R^3 \to R$ is positively homogeneous and convex, hence continuous and sublinear. It seems possible to develop an augmented Lagrangian also for this case.

CESCOTTO (1995) presented Hu-Washizu type variational functionals with penalties and calls such an approach "the augmented Lagrangian methods".

The Hellinger-Reissner functional combined with the augmented Lagrangian methods is used by CUOMO and VENTURA (1995) for the study of a frictionless unilateral contact problem for a solid made of a masonary - like material. BILLE et al. (1995) used an augmented Lagrangian method based on Uzawa's algorithm to the classical Signorini's problem without friction.

2.3.3 ALDUNCIN (1995) applied decomposition methods and iterative numerical algorithms for mixed finite element approximations of frictionless constrained problems in finite viscoelastcity.

2.2.4 Two unilateral frictionless contact problems for a Reissner- Mindlin plate were studied by AURICCHIO and SACCO (1995). The augmented Lagrangian formulation was used to develop an efficient iterative solution algorithm.

2.2.5 Let us pass to contact problems with friction. It is well known that because the friction law is not associated with the friction condition therefore such contact problems are not amenable to optimization (minimization or maximization) problems. Nevertheless, also for this practically very important class of contact problems one can develop augmented Lagrangian methods. Usually, they are based on the principle of virtual work and its variants, cf. SIMO et al. (1985), CHABRAND and DUBOIS (1995), WRIGGERS and ZAVARISE (1993a, 1993b), ZAVARISE et al. (1995), LAURSEN (1992), SIMO and LAURSEN (1992), LAURSEN and SIMO (1993), ALART and CURNIER (1991), ZAVARISE (1991). For frictional contact problems one still lacks rigorously justified augmented Lagrangian methods and algorithms comparable with those developed by ITO and KUNISCH (1990a, 1990c, 1995, 1996a), cf. also ALDUNCIN (1992), ECK (1996). The same pertains to geometrically nonlinear problems which are, as a rule, nonconvex even in the absence of friction.

ALART (1998) developed the augmented Lagrangian method suitable for numerical treatment of frictional unilateral contact between two primarily discretized bodies. The augmented Lagrangian is sufficiently regular for the applicability of the generalized Newton method to nondifferentiable operators. As far as possible this author exploits convex analysis. We observe that the book in two volumes by HIRIART-URRUTY and LEMAREĆHAL (1993a, 1993b) is confined to finite-dimensional spaces and successfully complements the celebrated book by ROCKAFELLAR (1970).

Inspired by the theory of Generalized Standard Materials DE SAXCÉ and FENG (1991) proposed the augmented Lagrangian method applicable to unilateral contact problems with friction within thw framework of geometrically linear theory. To this end the following functional, called *bifunctional*, is introduced

$$J(\mathbf{u},\boldsymbol{\sigma}) = \int_{\Omega} \mathcal{W}(\mathbf{e}(\mathbf{u}),\boldsymbol{\sigma})dx - L(\mathbf{u}) + \int_{\Gamma_2} b(\mathbf{u},\mathbf{t}(\boldsymbol{\sigma})d\Gamma. \tag{23.31}$$

Here $\mathbf{u} = (u_i)$, $\mathbf{e}(\mathbf{u}) = (u_{(i,j)})$ and $\boldsymbol{\sigma}= (\sigma_{ij})$ denote the the displacement vector, small strain tensor and the stress tensor respectively. The functional $L(\mathbf{u})$ is the loading functional and $\Gamma_2 \subset \partial\Omega$ is the surface of potential contact with a rigid body. For explicit standard materials we have

$$\mathcal{W}(\mathbf{e},\boldsymbol{\sigma}) = \mathcal{V}(\mathbf{e}) + \mathcal{V}^*(\boldsymbol{\sigma}), \tag{23.32}$$

where $\mathcal{V}(\mathbf{e})$ denotes the density of the stored energy function and $\mathcal{V}^*(\boldsymbol{\sigma})$ is the complementary energy.

Contact with friction can be represent by the following ***bipotential***:

$$b(-\Delta u_n, -\Delta \mathbf{u}_t, \Delta t_n, \Delta \mathbf{t}_t) = t_{n_0}\Delta u_n, +\mathbf{t}_{t_0}\cdot \Delta \mathbf{u}_t + \Delta t_n(u_{n_0} + g_0) +$$

$$+\mu(t_{n_0} + \Delta t_n)||\Delta \mathbf{u}_t|| + I_{\Delta K_\mu}(\Delta t_n, \Delta \mathbf{t}_t) + I_{R^+}(\Delta u_n + u_{n_0} + g_0) \tag{23.33}$$

provided that Coulomb's criterion is used, i.e.,

$$||\mathbf{t}_n|| \leq \mu t_n. \tag{23.34}$$

Moreover, ΔK_μ is the convex friction set defined by

$$\Delta K_\mu = \{(\Delta t_n, \Delta \mathbf{t}_t) \text{ such that } ||\mathbf{t}_{t_0} + \Delta \mathbf{t}_t|| \leq \mu(t_{n_0} + \Delta t_n\}, \tag{23.35}$$

where the values at the starting point of the step are noted with a subscript 0.

To extend the augmented Lagrangian method to the bipotential we note that

$$r\, b(\mathbf{z},\mathbf{y}') - r\, b(\mathbf{z},\mathbf{y}) + [\mathbf{y} - (\mathbf{y} + r\,\mathbf{z})]\cdot(\mathbf{y}' - \mathbf{y}) \geq 0 \tag{23.36}$$

for all $\mathbf{y}'$. Consequently $\mathbf{y}$ is the proximal point of the augmented stress associated with the function $r\,b(\mathbf{z},\cdot)$, i.e.,

$$\mathbf{y} = \text{prox}(\mathbf{y} + r\,\mathbf{z}, r\,b(\mathbf{z},\cdot)). \tag{23.37}$$

For the definition and properties of the proximal point the reader is refeared to HIRIART-URRUTY and LAMARÉCHAL (1993b).

23.5 PARAMETER ESTIMATION AND OPTIMAL CONTROL

In this section we shall present applications of the augmented Lagrangian methods to parameter estimation in elliptic and parabolic systems and to optimal control problems.

3.1. ITO and KUNISCH (1990b) considered the problem of determining the unknown coefficient q in the elliptic partial differential equation:

$$-\text{div}(q\,\nabla u) = f \text{ in } \Omega,\ u = 0 \text{ on } \Gamma, \tag{23.38}$$

from an observation z of the solution u, where Ω is a bounded domain in $R^n(n = 1, 2 \text{ or } 3)$ with piecewise smooth boundary Γ and $f \in H^{-1}(\Omega)$. In applications the function z might be constructed by interpolation of poinwise measurements. ITO and KUNISCH (1990b) proposed and analyzed a hybrid method that combines the output least squares and the equation error formulation within the mathematical framework given by the augmented Lagrangian technique. More precisely, the following constrained minimization problem is investigated:

$$(\mathcal{P}) \quad \left| \begin{array}{l} \text{find} \\ \qquad \min\left\{F(q,u) = \frac{1}{2}\|u-z\|^2_{H^1_0} + \frac{\beta}{2}N(q)| \; -\nabla\cdot(q\nabla u) = f \right. \\ \qquad\qquad \left. \text{in } \; H^{-1}(\Omega), \|q\|_{H^2(\Omega)} \leq \gamma, \alpha \leq q \text{ on } \Omega\right\}. \end{array}\right.$$

The second term in the cost functiona represents a regularization term. To solve Problem $(\mathcal{P})$ the augmented Lagrangian algorithm is applied. In essence, it involves minimizing a sequence of functionals of the form:

$$L_{r_k}(q,u;\lambda_k) = F(q,u) + <\lambda_k, h(q,u)>_{H^1_0(\Omega)} + \frac{r_k}{2}\|h(q,u)\|^2_{H^1_0(\Omega)}, \tag{23.39}$$

over

$$q \in Q_{ad} = \{q \in H^2(\Omega) | \; q \geq \alpha \text{ and } \|q\|_{H^2(\Omega)} \leq \gamma\} \tag{23.40}$$

and the multiplier sequence $\{\lambda_k\}$ in $H^1_0(\Omega)$ is given by

$$\lambda_{k+1} = \lambda_k + r_k h(q_k, u_k) \tag{23.41}$$

The function $h : H^2(\Omega) \times H^1_0(\Omega) \to H^1_0(\Omega)$ is defined by

$$h(q,u) = (-\Delta)^{-1}(\nabla\cdot(q\nabla u) + f), \tag{23.42}$$

where $\Delta : H^1_0(\Omega) \to H^1_(\Omega)$ is the Laplacian, and the pair (q_k, u_k) minimizes (3.2). To perform this iterative scheme a (possibly) constant sequence of positive real numbers $\{r_k\}$ and a startup $\lambda_1 \in H^1_0(\Omega)$ for the Lagrange multiplier need to be chosen. We observe that the inequality constraint $\|q\|_{H^2(\Omega)} \leq \gamma$ can be augmented in a manner sugested in Section 1. Detailed study of convergence was also carried out and second-order optimality condition was investigated. Moreover, practical experience with the augmented Lagrangian technique to determine q in (3.1) from data for u was reported.

Remark 3.1. (i) The sensitivity analysis presented by ITO and KUNISCH (1990c) applies also to (3.1). (ii) The augmented Lagrangian method developed in (ITO and KUNISCH, 1990b) for parameter estimation in elliptic equations was extended in KUNISCH and PEICHL (1991) to a parabolic system in one-space dimension. (iii) The equations considered by ITO and KUNISCH (1990b) , KUNISCH and PEICHL (1991) describe only systems made of *isotropic* materials.

3.2. ITO and KUNISCH (1996b) proposed three algorithms, which employ augmented Lagrangian methods, for optimization problems in Hilbert spaces with second order updates of the Lagrangian. More presisely, the technique proposed is the augmented Lagrangian -SQP (sequential quadratic programming) algorithm. Next, in ITO and KUNISCH (1996c) a class of optimal control problems for nonlinear partial differential equations, where the costs is of tracking type has been studied.

The optimal control problem are governed by partial differential equations of the following type:

$$\begin{gathered} -\Delta + f(u) = h \text{ in } \Omega \\ \frac{\partial u}{\partial \mathbf{n}} = g \text{ on } \Gamma_1, \quad \frac{\partial u}{\partial \mathbf{n}} = g_2 \text{ on } \Gamma_2, \end{gathered} \tag{23.43}$$

where $h \in l^2(\Omega)$ and $g_2 \in L^2(\Gamma_2)$ are fixed and $g \in L^2(\Gamma_1)$ is the controle variable. Here $\Gamma = \partial\Omega = \Gamma_1 \cup \Gamma_2$ and Γ_2 may be empty. Eq. (3.6) includes a simplified Ginzburg-Landau model for superconductivity. The optimal control problem is given by:

$$(\mathcal{P}_1) \quad \left| \begin{array}{l} \text{find} \\ \quad \min \frac{1}{2}\|\mathcal{B}u - u_d\|_Z^2 + \frac{\alpha}{2}\|g\|_{L^2(\Gamma_1)}^2 \\ \quad \text{subject to}(u,g) \in H^1(\Omega) \times L^2(\Gamma_1) \text{ a solution of (3.6)} \end{array} \right.$$

Here $\mathcal{B}$ is a bounded linear (observation) operator from $H^1(\Omega)$ to a Hilbert space Z and $u_d \in Z$ and $\alpha > 0$ are fixed.
The second-order convergence rate of the augmented-SQP algorithm for solving $(\mathcal{P}_1)$ is proved. This rate is also shown with ilustrative numerical examples.
3.3. BERGOUNIOUX and KUNISCH (1995) investigated the following optimal control problem:

$$(\mathcal{P}_2) \quad \left| \begin{array}{l} \text{find} \\ \quad \min\{J(y,u) |\ Au = f \text{ in } \Omega,\ y = u \text{ on } \Gamma,\ \Lambda_1 y \in K,\ u \in U\} \end{array} \right.$$

where

$$J(y,u) = \frac{1}{2}\int\limits_\Omega (y - z_d)^2 dx + \frac{\alpha}{2}\int\limits_\Gamma (u - u_d)^2 d\Gamma, \tag{23.44}$$

and
(i) $f, z_d \in L^2(\Omega),\ u, u_d \in L^2(\Gamma)$ and either $\alpha > 0$ or U is bounded in $L^2(\Gamma)$,
(ii) L is a Hilbert space with the dual L^* identified with L and $\Lambda_1 \in \mathcal{L}(W, L)$; moreover

$$W = \left\{y \in L^2(\Omega) | Ay \in L^2(\Omega),\ y_{|\Gamma} \in L^2(\Gamma)\right\}, \tag{23.45}$$

(iii) K and U are nonempty, closed and convex subsets of L and $L^2(\Gamma)$ respectively,
(iv)

$$Ay = -\frac{\partial}{\partial x_i}\left(a_{ij}(x)\frac{\partial y}{\partial x_j}\right) + a_0(x)y \tag{23.46}$$

with

$$a_{ij}, a_0 \in C^2(\overline{\Omega})\ (i,j = 1,\ldots,n), \quad \inf\left\{a_0(x) | x \in \overline{\Omega}\right\} > 0,$$

$$a_{ij}(x)\xi_i\xi_j \geq \delta\|\xi\|_{R^n}^2,\ \forall x \in \overline{\Omega},\ \forall \xi \in R^n, \quad \delta > 0$$

Here we apply the summation convention over repeated indices.
We assume that the admissible domain:

$$\mathcal{D} = \left\{(y,u) \in W \times L^2(\Gamma) | Ay = f \text{ in } \Omega, y = u \text{ on } \Gamma, (\Lambda_1 y, u) \in K \times U\right\} \tag{23.47}$$

is nonempty.
In essence, BERGOUNIOUX and KUNISCH (1995) is based on the augmentation of both the state - equation as well qs the state and control constraints.

Augmented Lagrangian formulation
We set, cf. Sec. 1.3

$$X = W \times L^2(\Gamma),\ H = L \times L^2(\Gamma),\ \mathcal{C} = K \times U.$$

Let Λ be defined from X to H by $\Lambda(y,u) = (\Lambda_1 y, u)$ and let φ be the inicator function of the convex set $\mathcal{C}$. For any $r > 0$ we define the function $\varphi_r : H \times H \to R$ by

$$\varphi_r(z,\lambda) = \inf\left\{\varphi(z-\eta) + (\lambda,\eta)_H + \frac{r}{2}\|\eta\|_H^2 \mid \eta \in H\right\}, \tag{23.48}$$

where $z = (y,u)$. Here $(\cdot,\cdot)_H$ is given by $(\lambda,\eta)_H = (\lambda_1,\eta_1)_\Omega + (\lambda_2,\eta_2)_\Gamma$ with $\lambda = (\lambda_1,\lambda_2)$ and $\eta = (\eta_1,\eta_2)$.
Properties of the function φ_r
(i) For all $z \in H,\ \lambda \in H$, the infimum in (3.11) is attained at a unique point $\eta_r(z,\lambda)$. φ_r is convex, Lipschitz -continuously Fréchet-differentiable in z and

$$\varphi_r'(z,\lambda) = \lambda + r\,\eta_r(z,\lambda). \tag{23.49}$$

Moreover, $\lim\limits_{r\to+\infty} \varphi_r(z,\lambda) = \varphi(z)$.
(ii) For all $z = (y,u) \in H$ and $\lambda = (\lambda_1,\lambda_2) \in H$ we have

$$\begin{aligned}\varphi_r(z,\lambda) &= \tfrac{r}{2}\|z - P_\mathcal{C}(z+\tfrac{\lambda}{r})\|_H^2 + \left(\lambda, z - P_\mathcal{C}(z+\tfrac{\lambda}{r})\right)_H \\ &= \tfrac{r}{2}\|y - P_K(y+\tfrac{\lambda_1}{r})\|_L^2 + \left(\lambda_1, y - P_K(y+\tfrac{\lambda_1}{r})\right)_L + \\ &\quad \tfrac{r}{2}\|u - P_U(u+\tfrac{\lambda_2}{r})\|_\Gamma^2 + \left(\lambda_2, u - P_U(u+\tfrac{\lambda_2}{r})\right)_\Gamma +\end{aligned} \tag{23.50}$$

$$\varphi_r'(z,\lambda) = r\left(y + \frac{\lambda_1}{r} - P_K(y+\frac{\lambda_1}{r}), u + \frac{\lambda_2}{r} - P_U(u+\frac{\lambda_2}{r})\right), \tag{23.51}$$

where P_K (resp. $P_U,\ P_\mathcal{C}$) is the L (resp. $L^2(\Gamma), H$) projection on K (resp. $U,\ \mathcal{C}$).
We now introduce the *augmented Lagrangian function* of $(\mathcal{P}_2)$ associated with the constraints $\Lambda(y,u) \in \mathcal{C}$:

$$F_{r,\lambda}(y,u) = J(y,u) + \varphi_r(\Lambda(y,u),\lambda). \tag{23.52}$$

It can be shown that for all $\lambda \in H$ and $r > 0$ the following minimization problem:
$(\mathcal{P}_3)_{r,\lambda}$ $\quad \min\{F_{r,\lambda}(y,u) \mid Au = f \text{ in}\Omega,\ y = u \text{ on } \Gamma\}$,
possesses a unique solution $(y_{r,\lambda}, u_{r,\lambda})$. Moreover, for every fixed $\lambda \in H$ we have

$\lim\limits_{r\to+\infty} y_{r,\lambda} = \overline{y}$ strongly in W and
$\lim\limits_{r\to+\infty} u_{r,\lambda} = \overline{u}$ strongly in $L^2(\Gamma)$,

where W is equipped with the graph norm and $(\underline{y},\underline{u})$ denotes the solution of $(\mathcal{P}_2)$.

Algorithm 1
Step 1. Initialization: set $k = 0$ and choose $\gamma > 0,\ q_0 \in L^2(\Omega),\ R_0 \in L^2(\Gamma)$.
Step 2. Compute $(y_k, u_k) = \arg\min\{L_\gamma(y, u, q_k, R_k) \mid \Lambda(y, u) \in K \times U\}$, where
$L_\gamma(y,u,q,R) = J(y,u) + (q, Ay)_\Omega + (R, y-u)_\Gamma + \frac{\gamma}{2}\|Ay - f\|^2_\Omega + \frac{\gamma}{2}\|y-u\|^2_\Gamma$,
is the *augmented Lagrangian* with respect to the *state-equation* constraint.
Step 3. Set
$q_{k+1} = q_k + \rho_1(Ay_k - f)$, where $\rho_1 \in (0, 2\gamma]$,
$R_{k+1} = R_k + \rho_2(u_{k|\Gamma} - u_k)$, where $\rho_2 \in (0, 2\gamma]$.
We recall that $\arg\min L_\gamma(y, u, q_k, R_k)$ denotes those functions (y_k, u_k), with $\Lambda(y_k, u_k) \in K \times U$, at which the functional $L_\gamma(\cdot, \cdot, q_k, R_k)$ attains the minimum value. It remains now to solve the auxilary problem of Step 2. To this end the following algorithm is used:
Algorithm 2
Step 1. Initialization: choose $\lambda^0 \in H$ and $r > 0$.
Step 2. Calculate $(y^j, u^j) = \arg\min\{L_\gamma(y,u) + \varphi_r(\Lambda(y,u), \lambda^j) \mid \Lambda(y,u) \in X\}$ where φ_r is defined by (3.11).
Step 3. Set (see (3.14))
$\lambda^{j+1} = \varphi_r'(\Lambda(y^j, u^j), \lambda^j) = r(\Lambda_1 y^j + \frac{\lambda_1^j}{r} - P_K(\Lambda_1 y^j + \frac{\lambda_1^j}{r}), u^j + \frac{\lambda_2^j}{r} - P_U(u^j + \frac{\lambda_2^j}{r}))$.
BERGOUNIOUX and KUNISCH (1995) proposed also the version where Algorithm 2 appears as an inner loop in Algorithm 1. Also, Gauss-Seidel splitting techniques have been used to solve the auxiliary problems. The splitting avoids the minimization of the auxiliary problem with respect to y and u simultaneously. The convergence of the iterates has also been discussed.

23.6 IMAGE RESTORATION

In this section we apply the augmented Lagrangian technique sketched in Section 1.3 to the nonsmooth convex optimization problem that arises in image restoration. This technique allows to substitute the nondifferentiable term by the Lagrangian or dual variable.

Consider the variational problem which means evaluating, see ITO and KUNISCH (1998)

$$\min\left\{\int_\Omega \left(\frac{\mu}{2}|\nabla u|^2 + g|\nabla u|\right) dx + \frac{1}{2}\int_\Omega |u - z|^2 dx \mid u \in H_0^1(\Omega)\right\}. \tag{23.53}$$

This is a problem for restoring the original image u from a noisy image $z \in L^2(\Omega)$. Here $\Omega \subset R^n$ ($n = 1$ or 2) is a bounded domain and $\mu > 0, g > 0$ are fixed but can be adjusted to the statistics of noise. For more details on this type of functional the reader is referred to the book by MOREL and SOLOMINI (1995).

To put the problem (4.1) to the desired form choose

$$X = H_0^1(\Omega), \quad H = L^2(\Omega) \times L^2(\Omega), \quad \Lambda = grad = \nabla$$

and define f and φ by

$$f(u) = \frac{1}{2} \int_{\Omega} (\mu|\nabla|^2 + |u - z|^2 dx,$$

$$\varphi(v_1, v_2) = g \int_{\Omega} \sqrt{v_1^2 + v_2^2} dx.$$

Now the nondifferentiable functional $\varphi(\mathbf{v})$ is finite and continous at each $\mathbf{v} \in H$ and $\boldsymbol{\lambda} \in \partial\varphi(\mathbf{v})$ is given by

$$\boldsymbol{\lambda} \in C \text{ and } \boldsymbol{\lambda}(x) \cdot \mathbf{v}(x) = |\mathbf{v}(x)| \quad \text{a.e. } x \in \Omega \tag{23.54}$$

where

$$C = \{\boldsymbol{\lambda} \in H : |\boldsymbol{\lambda}(x)| \leq 1 \quad \text{a.e. } x \in \Omega\}. \tag{23.55}$$

We observe that the necessary and sufficient optimality condition for (4.1) is given by the existence of $(u, \boldsymbol{\lambda}) \in X \times H$ satisfying

$$-\mu\Delta u + u - g \operatorname{div}\boldsymbol{\lambda} = z, \ \boldsymbol{\lambda}(x) \cdot \nabla u(x) = \ | \nabla u(x)| \quad \text{a.e. } x \in \Omega. \tag{23.56}$$

Since the minimization problem (4.1) is convex therefore the necessary and sufficient condition for $\overline{u} \in X$ to be the minimizer of the problem:

$$\min \{f(u) + \varphi(\Lambda u)| \ u \in X\} \tag{23.57}$$

is given by

$$< f'(\overline{u}), u - \overline{u} > + \varphi(\Lambda u) - \varphi(\Lambda \overline{u}) \geq 0 \quad \text{for all } u \in X \tag{23.58}$$

The last inequality implies the existence of $\boldsymbol{\lambda}^* \in H$ such that

$$\boldsymbol{\lambda}^* \in \partial\varphi(\Lambda\overline{u}) \quad \text{and } f'(\overline{u}) + \Lambda^*\boldsymbol{\lambda}^* = 0 \tag{23.59}$$

The optimality condition (4.7) can equivalently be written as

$$f'(\overline{u}) + \Lambda^*\boldsymbol{\lambda}^* = 0, \quad \text{and} \quad \boldsymbol{\lambda}^* = \varphi_r'(\Lambda\overline{u}, \boldsymbol{\lambda}^*), \tag{23.60}$$

for some and equivalently all $r > 0$; here φ_r is given by 23.27. For the image restoration problem we have

$$\varphi^*(v^*) = gI_C(\mathbf{v}),$$

and

$$\varphi_r(\mathbf{v}, \boldsymbol{\lambda}\ ) = g \int_{\Omega\setminus A} [\frac{1}{r}(|\boldsymbol{\lambda} + r\mathbf{v}| - 1) + \frac{1}{2r}(1 - |\boldsymbol{\lambda}\ |^2)]dx + \frac{g}{2r} \int_{A} [|\boldsymbol{\lambda} + r\mathbf{v}|^2 - |\boldsymbol{\lambda}|^2]dx,$$

where $A = \{x : |\boldsymbol{\lambda}(x) + r\mathbf{v}(x)| \leq 1\}$; moreover

$$\varphi_r'(\mathbf{v}, \boldsymbol{\lambda}\ ) = g \frac{\boldsymbol{\lambda}(x) + r\mathbf{v}(x)}{\max(1, |\boldsymbol{\lambda}(x) + r\mathbf{v}(x)|)}, \quad \text{a.e. } x \in \Omega.$$

Let us pass to the presentation of the first order iterative algorithms for the condition (4.8). The first order augmented Lagrangian method can be written as the fixed point iteration

$$\boldsymbol{\lambda}_{k+1} = \varphi_r'(\Lambda u_{k+1}, \boldsymbol{\lambda}_k) \tag{23.61}$$

where $u_{k+1} \in X$ satisfies

$$f'(u_{k+1}) + \Lambda^* \boldsymbol{\lambda}_{k+1} = 0. \tag{23.62}$$

The first order augmented Lagrangian method is implicit and in contrast to the Uzawa algorithm it converges unconditionally (in $r > 0$), provided that there exists a Lagrange multiplier $\boldsymbol{\lambda}^*$ satisfying (4.7). For the image restoration problem the Uzawa algorithm is given by

$$-\mu\Delta u_k + u_k - g\text{div}\boldsymbol{\lambda}_k = z, \ \boldsymbol{\lambda}_{\ k+1} = \frac{(\boldsymbol{\lambda}_k + r\nabla u_k)(x)}{\max(1, |\boldsymbol{\lambda}_k + r\nabla u_k|(x))}. \tag{23.63}$$

ITO and KUNISCH (1998) considered also the finite difference appeoximation of (4.1) on $\Omega = (0,1) \times (0,1)$.

References

[1969] Hestens, M.R., Multiplier and gradient methods, *J. Optim. Theory Appl.*, **4**, 303-320.

[1969] Powell, M.J.D., A method for nonlinear constraints in minimization problems, in: Optimization, ed. by R. Fletcher, Academic Press, New York, pp. 283-298.

[1970] Rockafellar, R.T., Convex Analysis, Princeton University Press, Princeton.

[1972] Fichera, G., Boundary value problems of elasticity with unilateral consraints, in: Handbuch der Physik, Band 6a/2, Springer-Verlag, Berlin.

[1973a] Rockafellar, R.T., The multiplier method of Hestens and Powell applied to convex programming, *J. Optim. Theory Appl.*, **12**, 555-562.

[1973b] Rockafellar, R.T., A dual approch to solving nonlinear programming problems by unconstrained optimization, *Math. Programming*, **5**, 354-373.

[1973] Polyak, B.T., Tretyakov, N.V., The method of penalty estimates for conditional extremum problems, *Zh. Vychisl. Mat. i Mat. Fiz.*, **13**, 34-46, in Russian

[1975] Nakayama, H., Sayama, H., Sawaragi, Y., A generalized Lagrangian function and multiplier method, *J. Optim. Theory Appl.*, **17**, 211-227.

[1976] Rockafellar, R.T., Augmented Lagrangians and applications of the proximal point algorithm in convex programming, *Math. Res, Oper.*, **12**, 97-116

[1981] Rockafellar, R.T., Proximal subgradients, marginal values and augmented Lagrangians in nonconvex optimization, *Math. Oper. Res.*, **6**, 424-436.

[1982] Bertsekas, D.P., *Constrained Optimization and Lagrangian Multiplier Methods*, Academic Press, New York.

[1983] Le Tallec, P., Augmented Lagrangians in finite elasticity and in plasticity, in: *Computational Aspects of Penetration Mechanics, Lecture Notes in Engineering*, vol. 3, Springer-Verlag, Berlin.

[1985] Landers, J.A., Taylor, R.L., An augmented Lagrangian formulation for the finite element solution of contact problems, Department of Civil Engineering, University of California, Berkeley, California, Report No UCB/SESM-85/09.

[1985] Simo, J., Wriggers, P., Taylor, R., A perturbed Lagrangian formulation for the finite element solution of contact problems, *Comp. Meth. Mech. Eng.*, **50**, 163-180.

[1986] Blanchard, D., Le Tallec, P., Numerical analysis of the equations of small strains quasistatic elastoviscoplasicity, *Numer. Math.*, **50**, 147-169.

[1986] Le Tallec, P., Numerical solution of viscoplastic flow problems by augmented Lagrangians, *IMA J. Numer. Anal.*, **6**, 185-219.

[1986] Kikuchi, N., Oden, J.T., *Contact Problems in Elasticity*, SIAM, Philadelphia.

[1988] Alart, P., Multiplicateurs "augmentés" et méthode de Newton généralisée pour contact avec frottement, Document, Departement de Mécanique, Ecole Polytechnique Fédérale de Lausanne.

[1988] Rockafellar, R.T., First- and second - order epi-differentiability in nonlinear programming, *Trans. Amer. Math. Soc.*, **307**, 75-108.

[1989] Glowinski, R., Le Tallec, P., *Augmented Lagrangian and Operator - Splitting Methods in Nonlinear Mechanics*, SIAM, Philadelphia.

[1989] Gwinner, J., A penalty approximation for unilateral contact problem in non-linear elasticity, *Mat. Meth. in the Appl. Sci.*, **11**, 447-458.

[1990a] Ito, K., Kunisch, K., An augmented Lagrangian technique for variational inequalities, *Appl. Math. Optim.*, **21** 223-241.

[1990b] Ito, K., Kunisch, K., The augmented Lagrangian method for parameter estimation in elliptic systems, *SIAM J. Control Optim.*, **28**, 113-136.

[1990c] Ito, K., Kunisch, K., The augmented Lagrangian Method for equality and inequality constraints in Hilbert spaces, *Math. Programming*, **46** 341-360.

[1991] Alart, P., Curnier, A., A mixed formulation for frictional contact problems prone to Newton like solution methods, *Comp. Meth. Appl. Mech. Eng.*, **92**, 353-375.

[1991] Arora, J.S., Chahande, A.I., Paeng, J.K., Multiplier methods for engineering optimization, *Int. J. Num. Meth. Eng.*, **32**, 1485-1525.

[1991] De Saxcé, G., Feng, Z.Q., New inequality and functional for contact with friction: the implicit standard material approach, *Mech. Struct. & Mach.*, **19**, 301-325.

[1991] Kunisch, K., Peichl, G., Estimation of a temporally and spatially varying diffusion coefficient in a parabolic system by an augmented Lagrangian technique, *Numer. Math.*, **59**, 473-509.

[1991] Zavarise, G., Problemi termomeccanici di contatto-aspetti fizici e computazionali, Ph.D. Thesis, Università di Padova, Facoltá di Ingegneria.

[1992] Alduncin, G., Augmented Lagrangian methods for the quasistatic viscoelastic two-body contact problem with friction, in: Proc. Int. Symp. Contact Mechanics, ed. by A. Curnier, pp. 337-359, Presses Polytechniques et Universitaires Romandes, Lausanne.

[1992] Klarbring, A., Mathematical programming and augmented lagriangian methods for frictional contact oroblems, in: Proc. Int. Symp. Contact Mechanics, ed. by A. Curnier, pp. 409-422 , Presses Polytechniques et Universitaires Romandes, Lausanne.

[1992] Laursen, T.A., Formulation and treatment of frictional contact problems using finite elements, SUDAM Report No 92-6 .

[1992] Simo, J.C., Laursen, T.A., Augmented Lagrangean treatment of contact problems involwing friction, *Comp. and Structures*, **42**, 97-116.

[1993] Hager, W-W., Analysis and implementation of a dual algorithm for constrained optimization, *J. Optim. Theory Applic.*, **79**, 427-462.

[1993a] Hiriart-Urruty, J.-B., Lemaréchal, C., *Convex Analysis and Minimization Algorithms*, vol.I: *Fundamentals*, Springer-Verlag, Berlin.

[1993b] Hiriart-Urruty, J.-B., Lemaréchal, C., *Convex Analysis and Minimization Algorithms*, vol.II: *Advanced Theory and Bundle Methods*, Springer-Verlag, Berlin.

[1993] Laursen, T.A., Simo, J.C., Algorithmic symmetrization of Coulomb frictional problems using augmented Lagrangians, *Comp. Meth. Appl. Mech. Eng.*, **108**, 133-146.

[1993] Heegaard, J.-H., Curnier, A., An augmented Lagrangian method for discrete large-slip contact problems, *Int. J. Num. Meth. Eng.*, **36**, 569-593.

[1993a] Wriggers, P., Zavarise, G., Application of augmented Lagrangian techniques for non-linear consttutive laws in contact interfaces, Comm. appl. numer. methods eng., **9**, 815-824.

[1993b] Wriggers, P., Zavarise, G., Recent trends in computational methods for contact problems, in: Proc. of the XI Polish Conf. on Computer Methods in Mechanics, ed. by W. Gilewski and L. Chodor, vol.3, pp. 127-148, Kielce University of Technology, Poland.

[1995] Alduncin, G., Decomposition methods and mixed finite element approximations of adherence and contact problems in finite viscoelasticity, in:

Contact Mechanics, ed. by M. Raous, M. Jean and J.J. Moreau, pp. 229-236., Plenum Press, New York.

[1995] Auricchia, F., Sacco, E., An effective plate element for contact problems, in: Contact Mechanics, ed. by M. Raous, M. Jean and J.J. Moreau, pp. 237-241., Plenum Press, New York.

[1995] Bergounioux, M., Kunisch, K., Augmented Lagrangian techniques for elliptic state constrained optimal control problems, Université d'Orléans, Département de mathématiques, URA 1803, Report No 95 - 12,

[1995] Bille, J.-P., Cescotto, S., Habraken, A.M., Charlier, R., Numerical approach of contact using an augmented Lagrangian method, in: Contact Mechanics, ed. by M. Raous, M. Jean and J.J. Moreau, pp. 243-246, Plenum Press, New York.

[1995] Cescotto, S., Modelling of frictional contact from mixed variational principles, in: Contact Mechanics, ed. by M. Raous, M. Jean and J.J. Moreau, pp. 217-227., Plenum Press, New York.

[1995] Chabrand, P., Dubois, F., Numerical method for dealing with frictional contact problems in finite elastoplastic deformations, in: Contact Mechanics, ed. by M. Raous, M. Jean and J.J. Moreau, pp. 259-168., Plenum Press, New York.

[1995] Cuomo, M., Ventura, G., Augmented Lagrangian formulation for analysis of no-tension structures with unilateral supports, in: Contact Mechanics, ed. by M. Raous, M. Jean and J.J. Moreau, pp. 247-250., Plenum Press, New York.

[1995] Heegaard, J.H., Geometric properties of unilateral contact constraints, in: Contact Mechanics, ed. by M. Raous, M. Jean and J.J. Moreau, pp. 173-182, Plenum Press, New York.

[1995] Ito, K., Kunisch, K., Augmented Lagrangian formulation of nonsmooth, convex optimization in Hilbert spaces, in: Control of Partial Differential Equations and Applications, ed by E. Casas, Marcel Dekker, pp. 107-117.

[1995] Morel, J.-M., Solomini, S., *Variatioonal Methods in Image Segmentation with seven image processing experiments*, Birkhaüser, Boston.

[1995] Zavarise, G., Wriggers, P., Schrefler, B.A., On Augmented Lagrangian Algorithms for thermomechanical contact problems with friction, *Int. J. Numer. Methods Eng.*, Vol. 38, 2929-2949.

[1996] Eck, C., Existenz und Regularität der Lösungen für Kontaktprobleme mit Reibung, Ph. D. Thesis, Mathematisches Institut der Universität Stuttgart.

[1996] Golshtein, E.G., Tretyakov, N.V., *Modified Lagrangians and Monotone Maps in Optimization*, John Wiley & Sons, New York.

[1996] Hiptmair, R., Shiekofer, T., Wohlmuth, B., Multilevel preconditioned augmented Lagrangian techniques for 2nd order mixed problems, *Computing*, **57**, 25-48

[1996a] Ito, K., Kunisch, K., Augmented Lagrangian methods for nonsmooth, convex optimization in Hilbert spaces, Preprint, Institut für Mathematik, Technische Universität Graz, Austria.

[1996b] Ito, K., Kunisch, K., Augmented Lagrangian- SQP-methods in Hilbert spaces and application to control in the coefficients problems, *SIAM J. Optim.*, **6**, 96-125.

[1996c] Ito, K., Kunisch, K., Augmented Lagrangian- SQP-methods for nonlinear optimal control problems of tracking type, *SIAM J. Control and Optim.*, **34**, 874-891.

[1997] Boffi, D., Lovadina, C., Remarks on augmented Lagrangian formulations for mixed finite element schemes, *Boll. Un. Mat. Ital.*, **11**-A, 41-55.

[1998] Ito, K., Kunisch, K., An active set strategy based on the augmented Lagrangian formulation for image restoration, *Math Modelling Numer. Anal.*, in press.

24 MOUNTAIN PASS THEOREMS, DEFORMATION THEOREMS, AND PALAIS-SMALE CONDITIONS

Stepan A. Tersian

Centre of Applied Mathematics and Informatics,
University of Rousse, Rousse 7017, Bulgaria

tersian@ami.ru.acad.bg

24.1 INTRODUCTION

Let E be a Banach space, $X \subset E$ be an open subset, $f \in C^1(X, \mathbf{R})$ be a functional and

$$\begin{aligned} K &= \{x \in X : f'(x) = 0\}, \\ K_c &= \{x \in X : f(x) = c, f'(x) = 0\} \end{aligned}$$

are the sets of critical points of f.

In thecelebrated paper of Ambrosetti & Rabinowitz [1], critical values f are characterized as numbers $c = \inf_{A \in \mathcal{A}} \sup_{x \in A} f(x)$, where $\mathcal{A}$ is an appropriate class of subsets of X. During last two decades this idea is extensively developed up to Critical point theory for continuous functionals due to Degiovanni & Marzocchi [9].

Our goal in this survey is to systematize various aspects and give a common treatment of existing mountain pass theorems. The paper is divided in two sections. In the first section we consider deformation theorems of Rabinowitz [20], [21], Willem [29], [30], Bartolo, Benci and Fortunato [3] and (PS) type conditions as those of Cerami [7] - $(PSC)_c$ condition, Schechter [23],[24] - $(PS)_{c,\psi}$ condition , Struwe [26], [27] - $(PS)_c^*$ condition in scales of Banach spaces. We introduce a $(PS)_{c,\varphi}$ condition which is between $(PSC)_c$, $(PS)_{c,\psi}$ and generalize some deformation theorems. In the second section we formulate and comment the classical mountain pass theorem of Ambrosetti & Rabinowitz [1]. We note it's generalizations due to Cerami [7], Willem [29], Pucci & Serin [19], Rabinowitz [21], Schechter [23], Brezis & Nirenberg [5], Aubin & Ekeland

R.P. Gilbert et al. (eds.), From Convexity to Nonconvexity, 333–344.

[2], Ghossoub & Preiss [14] . We prove a variant and three critical point theorem with $(PS)_{c,\psi}$ condition which extend results of Willem [29], Rabinowitz [21].

24.2 DEFORMATION THEOREMS AND (PS) CONDITIONS

Let $f \in C^1(X, \mathbf{R})$ be a functional, defined on a open subset X of the Banach space E. We use the following notations:

$$\begin{aligned} f^c &= \{x \in E : f(x) \leq c\}, \\ f_c &= \{x \in E : f(x) \geq c\}, \\ F_\delta &= \{x \in E : d(x, F) < \delta\}, \\ d(x, F) &= \inf\{\|x - y\|, y \in F\}. \end{aligned}$$

A family of continuous mappings $\eta(t, x) : [0,1] \times X \to X$ is said a *homotopy of homeomorphisms* if for every $t \in [0,1]$

1. $\eta(t, x) : X \to X$ is a homeomorphism,
2. $\eta(0, x) = x, \quad \forall x \in X,$

In this section we consider various Palais- Smale conditions ((PS) conditions for short).

Definition 24.1. *We say that the differentiable functional $f : X \to \mathbf{R}$ satisfies the Palais- Smale (PS) condition if every sequence $\{x_j\}$ in X such that $f(x_j)$ is bounded and $f'(x_j) \to 0$ in X^* as $j \to \infty$ has a convergent subsequence.*

From (PS) condition it follows that the set of critical points for a bounded functional is compact. A variant of (PS) condition, introduced by Brezis, Coron & Nirenberg [4] which implies a compactness of critical point at a fixed level is the following

Definition 24.2. *(Brezis, Coron & Nirenberg, 1980) Let $c \in \mathbf{R}$. We say that the differentiable functional $f : X \to \mathbf{R}$ satisfies the $(PS)_c$- condition if every sequence $\{x_j\}$ in X such that $f(x_j) \to c$ and $f'(x_j) \to 0$ in X^* as $j \to \infty$ has a convergent subsequence.*

It is clear that (PS) condition implies the $(PS)_c$- condition for every $c \in \mathbf{R}$. Following deformation theorem is proved by P.Rabinowitz [21].

Theorem 1. *(Rabinowitz, 1978) Let $f \in C^1(X, \mathbf{R})$ and satisfies (PS) conditions. For any $c \in \mathbf{R}$, $\varepsilon_0 > 0$ and any neighborhood N of K_c, there exist $\varepsilon, 0 < \varepsilon < \varepsilon_0$ and a homotopy of homeomorphisms $\eta \in C([0,1] \times E, E)$ such that*
(1) $\eta(0, x) = x, \quad \forall x \in E$.
(2) $\eta(t, x) = x$ if $|f(x) - c| \geq \varepsilon_0, \quad \forall t \in [0,1]$.
(3) $\eta(1, f^{c+\varepsilon} \backslash N) \subset f^{c-\varepsilon}$.
(4) if $K_c = \emptyset, \quad \eta(1, f^{c+\varepsilon}) \subset f^{c-\varepsilon}$.
(5) if f is even, $\eta(t, x)$ is odd in x, $\forall t \in [0,1]$.

Similar results , known as deformation theorems, were in Browder [6], Palais [18]. The proof of Theorem 1 is based on the notion of a *pseudo- gradient vector field* due to Palais [18].

Definition 24.3. *(Palais, 1970) Let $X \subset E$ be an open subset, $f \in C^1(X,\mathbf{R})$ and $0 < \alpha < \beta$ are given. A pseudo- gradient vector field for f on $X \backslash K$ is a locally Lipschitz continuous mapping $V : X\backslash K \to E$ such that*

(1) $\|V(x)\| \le \beta \left\|f'(x)\right\|_*$,

(2) $\alpha \left\|f'(x)\right\|_*^2 \le \left\langle f'(x), V(x)\right\rangle$.

Lemma 24.1. *Under assumptions of Definition 24.3, there exists a pseudo-gradient vector field for f on $X\backslash K$.*

Corollary 24.2.1. *Given $0 < \alpha < \beta$ there exists locally Lipschitz continuous mappings $V_j : X\backslash K \to E$, $j = 1,2$ such that for every $x \in X\backslash K$*

(1) $\|V_1(x)\| \le \beta, \quad \alpha \left\|f'(x)\right\|_* \le \left\langle f'(x), V_1(x)\right\rangle$,

(2) $\alpha \le \left\langle f'(x), V_2(x)\right\rangle \le \left\|f'(x)\right\|_* \|V_2(x)\| \le \beta$.

In the case $\alpha = 1$, $\beta = 2$, Lemma 24.1 is proved in Willem [29], Rabinowitz [21].

We say that a homotopy of homeomorphisms $\eta(t,x) : [0,1] \times X \to X$ is *f-decreasing (f- increasing)* if the function $t \to f(\eta(t,x))$ is decreasing (increacing) for every $x \in X$.

Theorem 1 was generalized by Willem [29], Bartolo, Benci & Fortunato [3], Brezis & Nirenberg [5]. We prove a variant of deformation theorem.

Theorem 2. *Let $f \in C^1(X,\mathbf{R})$, $F, G \subseteq X$ are closed sets, $c \in \mathbf{R}$, $\varepsilon, \delta > 0$ be such that $F_{2\delta} \cap G = \emptyset$ and for every $x \in f^{-1}[c-2\varepsilon, c+2\varepsilon] \cap F_{2\delta}$ it follows $\|f'(x)\|_* \ge \frac{4\varepsilon}{\delta}$. Then there exists a f- decreasing homotopy of homeomorphisms $\eta(t,x) : [0,1] \times X \to X$ such that*

(1) $\eta(t,x) = x$, *if* $x \in G$ *or* $|f(x) - c| \ge 2\varepsilon$,

(2) $\eta(1, f^{c+\varepsilon} \cap F) \subset f^{c-\varepsilon} \cap F_\delta$,

(3) $\|\eta(t,x) - x\| \le \delta t$.

If $G = \emptyset$ we are in the case of Willem [29], Lemma 2. Another weaker form of (PS) condition was introduced by Cerami [7].

Definition 24.4. *(Cerami, 1978) Let $c \in \mathbf{R}$. We say that the differentiable functional $f : X \to \mathbf{R}$ satisfies the $(PSC)_c$- condition if every sequence $\{x_j\}$ in X such that $f(x_j) \to c$ and $(1 + \|x_j\|)\left\|f'(x_j)\right\|_* \to 0$ as $j \to \infty$ has a convergent subsequence.*

Let Φ be the set of positive increasing functions $\varphi(s) : \mathbf{R}^+ \to \mathbf{R}^+$, $\varphi(0) > 0$ such that $\varphi(s) = O(s)$ as $s \to \infty$, i.e. there are $C > 0$, $R > 0$ such that $\varphi(s) \le Cs$ if $s \ge R$.

Definition 24.5. *The functional f satisfies $(PS)_{c,\varphi}$ condition for a given $\varphi \in \Phi$ if every sequence $\{x_j\}$ such that $f(x_j) \to c$, $\varphi(\|x_j\|) \left\|f'(x_j)\right\|_* \to 0$ has a convergent subsequence.*

If $\varphi(s) = 1+s$, $s \geq 0$ we are in the case of $(PSC)_c$ condition. Next Theorem extends a result of Ramos [22], Theorem 2.12.

Theorem 3. *Let $f \in C^1(X, \mathbf{R})$ satisfies $(PS)_{c,\varphi}$ condition, $N = K_{c,\delta}$ is a neighborhood of K_c. There exists $\varepsilon > 0$ and f- decreasing homotopy of homeomorphisms $\eta(t,x) : [0,1] \times X \to X$ such that*
(1) $\eta(t,x) = x$ *if* $|f(x) - c| \geq 2\varepsilon$,
(2) $\eta(1, f^{c+\varepsilon} \backslash N) \subset f^{c-\varepsilon}$.

Another (PS) type condition is introduced by Schechter [23], [24].

Let ψ be the set of positive non-increasing functions $\psi(s) : \mathbf{R}^+ \to \mathbf{R}^+$ such that $\int_1^\infty \psi(s)\, ds = \infty$.

Definition 24.6. *(Schechter, 1991) The function f satisfies $(PS)_{c,\psi}$ condition for a given $c \in \mathbf{R}$, if every sequence $\{x_j\} \subset X$ such that $f'(x_j) \to c$ and $\frac{\left\|f'(x_j)\right\|_*}{\psi(\|x_j\|)} \to 0$ has a convergent subsequence in X.*

If $\psi = C > 0$ we have usual $(PS)_c$ condition, if $\psi(s) = \frac{1}{1+s}$ we are in the case of Cerami $(PSC)_c$ condition. If $\varphi \in \Phi$, then $\frac{1}{\varphi} \in \Psi$. Next result modify a result of Schechter [24], Ramos [22], Theorem 2.16.

Theorem 4. *Let $\psi \in \Psi$, $c \in \mathbf{R}$, $K_c = \emptyset$, and f satisfies $(PS)_{c,\psi}$ condition. Then there exists $\varepsilon > 0$ such that for every $M > 0$ there exists a f- increasing homotopy of homeomorphisms $\eta(t,x) : [0,1] \times X \to X$ such that*
(1) $\eta(t,x) = x$ *,if* $|f(x) - c| \geq 2\varepsilon$.
(2) $\eta(1, f_{c-\varepsilon} \cap B_M(0)) \subset f_{c+\varepsilon}$.

Proof. There exist $\varepsilon_0 > 0$, $\beta > 0$ such that

$$|f(x) - c| \leq \varepsilon_0 \Rightarrow \left\|f'(x)\right\|_* \geq \beta\psi(\|x\|).$$

Otherwise there exists a sequence $\{x_n\}$ such that

$$|f(x_n) - c| \leq \frac{1}{n}, \quad \left\|f'(x_n)\right\|_* < \frac{1}{n}\psi(\|x_n\|).$$

By $(PS)_{c,\psi}$ condition we have $x_n \to x_0$, $f(x_0) = c$, $f'(x_0) = 0$ - a contradiction with $K_c = \emptyset$.

Let $\varepsilon < \min(\varepsilon_0, \beta)$. Then

$$|f(x) - c| \leq \varepsilon \Rightarrow \left\|f'(x)\right\| \geq \varepsilon\psi(\|x\|).$$

Let $T > 0$ be such that

$$2 < \int_M^{M+2T} \psi(s)\, ds$$

and

$$\begin{aligned} A &= \{x : |f(x) - c| \geq 2\varepsilon\} \cup K, \\ B &= \{x : |f(x) - c| \leq \varepsilon\}, \\ \chi(x) &= \frac{d(x, A)}{d(x, A) + d(x, B)}, \end{aligned}$$

$V : X \backslash K \to E$ is a locally Lipschitz mapping such that

$$\left\| f'(x) \right\|_* \leq \left\langle f'(x), V(x) \right\rangle, \qquad \|V(x)\| \leq 2.$$

Let $g(x) = \chi(x) V(x)$ and $\sigma(t, x)$ be the solution of Cauchy problem:

$$\left| \begin{array}{c} \dot{\sigma}(t) = g(\sigma(t)), \\ \sigma(0) = x. \end{array} \right.$$

Since $\|g(u)\| \leq 2$, the solution $\sigma(t, x)$ is defined $\forall t \in [0, +\infty)$. We have

$$\begin{aligned} \|\sigma(t, x) - x\| &\leq \int_0^t \|\dot{\sigma}(s)\| \, ds \leq \int_0^t \|V(\sigma(s))\| \, ds \leq 2t, \\ \frac{d}{dt} f(\sigma(t, x)) &= \left\langle f'(\sigma(t, x)), \chi V(\sigma(t, x)) \right\rangle \\ &= \chi(\sigma(t, x)) \left\langle f'(\sigma(t, x)), V(\sigma(t, x)) \right\rangle \geq 0 \end{aligned}$$

and therefore $t \to \sigma(t, .)$ is f - increasing.

Moreover $\sigma(t, x) = x$ if $|f(x) - c| \geq 2\varepsilon$, because $\chi_{|A} = 0$. Suppose there exists y, $\|y\| \leq M$, $f(y) \geq c - \varepsilon$ such that for every t, $0 \leq t \leq T$, $f(\sigma(t, y)) < c + \varepsilon$.

Then $f(\sigma(t, y)) \in B$, $0 \leq t \leq T$ and

$$\begin{aligned} f(\sigma(T, y)) - f(y) &= \int_0^T \frac{d}{ds} f(\sigma(s, y)) \, ds \\ &= \int_0^T \left\langle f'(\sigma(s, y)), \dot{\sigma}(s, y) \right\rangle ds \\ &= \int_0^T \left\langle f'(\sigma(s, y)), V(\sigma(s, y)) \right\rangle ds \\ &\geq \int_0^T \left\| f'(\sigma(s, y)) \right\| ds \end{aligned}$$

$$\begin{aligned}
&\geq \varepsilon \int_0^T \psi\left(\|\sigma(s,y)\|\right) ds \\
&\geq \varepsilon \int_0^T \psi\left(\|y\| + 2s\right) ds \\
&\geq \varepsilon \int_0^T \psi\left(M + 2s\right) ds \\
&= \varepsilon \int_M^{M+2T} \psi(s)\, ds > 2\varepsilon.
\end{aligned}$$

We have

$$\begin{aligned}
c+\varepsilon &> f(\sigma(T,y)) > f(y) + 2\varepsilon \\
&\geq c - \varepsilon + 2\varepsilon = c + \varepsilon,
\end{aligned}$$

a contradiction.

Therefore for every x, $\|x\| \leq M$, $f(x) \geq c - \varepsilon$ there exists $t_1 \in [0,T]$ such that $f(\sigma(t_1,x)) \geq c + \varepsilon$. Then

$$f(\sigma(T,x)) \geq f(\sigma(t_1,x)) \geq c + \varepsilon$$

and we can take $\eta(t,x) = \sigma(Tt,x)$, $0 \leq t \leq 1$, which satisfies all assertions of the Theorem. □

Another (PS) type condition, which we consider is one in scales of Banach spaces, (cf. Struwe [26], [27], Silva [25], Li & Willem [15]). Let X be a Banach space and $X_1 \subset X_2 \subset \ldots \subset X_n \subset X_{n+1} \subset \ldots \subset X$ be a scale of Banach spaces such that $\|x\|_{n+1} \leq \|x\|_n$, $\forall x \in X_n$, where $\|\cdot\|_n$ is the norm in X_n and $\bigcup_{n=1}^{\infty} X_n$ is dense in X.

Let $f : X \to \mathbf{R}$ be a functional, such that $f \in C^1(X_n, \mathbf{R})$, and $f_n = f_{|X_n}$. Note that $X_1^* \supset X_2^* \supset \ldots \supset X_n^* \supset X_{n+1}^* \supset \ldots \supset X^*$, where X^* is the dual space of X and $\|p\|_{*,n} \leq \|p\|_{*,n+1}$ if $p \in X_{n+1}^*$. Moreover if $x, y \in X_n$, $\langle f_n'(x), y\rangle = \langle f_{n+1}'(x), y\rangle$ and therefore $f'_{n+1|X_n} = f'_n$. If $n \geq n_0$, $x \in X_{n_0}$, $\left\|f_n'(x)\right\|_{*,n} \geq \left\|f_n'(x)\right\|_{*,n_0} = \left\|f_{n_0}'(x)\right\|_{*,n_0}$.

Definition 24.7. *The functional f satisfies $(PS)_c^*$ condition if every sequence $\{x_j\} \subset X$ satisfying:*

1. *there exists a sequence $\{n_j\} \subset \mathbf{N}, n_j \to \infty$ as $j \to \infty$ such that $x_j \in X_{n_j}$ for every $j \in \mathbf{N}$;*

2. *$f(x_j) \to c$, as $j \to \infty$;*

3. *$\|f_{n_j}'(x_j)\|_{*,n_j} \to 0$ as $j \to \infty$, has a strongly convergent subsequence in X.*

Denote

$$K_c^* = \left\{x \in X : f(x) = c, \quad f_n'(x) = 0, \quad \forall n\right\},$$
$$K_c^n = \left\{x \in X : f(x) = c, \quad f_n'(x) = 0\right\}.$$

If $f \in C^1(X, \mathbf{R})$, $K_c \subset K_c^*$. We have if $n \geq n_0$, $\quad K_c^* \subset K_c^n \subset K_c^{n_0} \subset K_c^1$.

Theorem 5. *(Struwe, 1983). Let $f : X \to \mathbf{R}$ be a functional, $f \in C^1(X_n, \mathbf{R})$, f satisfies $(PS)_c^*$ condition and $N = K_{c,4\delta}^*$ be a neighborhood of K_c^*. Then there exist $\varepsilon_0 > 0$, $n_0 \in N$ and f - decreasing homotopy of homeomorphisms $\eta_n(t,x) : [0,1] \times X_n \to X_n$ for $n \geq n_0$ such that if $0 < \varepsilon < \varepsilon_0$*
(1) $\quad$ *$\eta_n(t,x) = x$, if $|f(x) - c| \geq 2\varepsilon$,*
(2) $\quad$ *$\eta_n(1, (f^{c+\varepsilon} \backslash N) \cap X_n) \subset f^{c-\varepsilon}$.*

24.3 MOUNTAIN PASS THEOREMS

The mountain-pass theorem of Ambrosetti & Rabinowitz [1] is as follows

Theorem 6. *(Ambrosetti & Rabinowitz, 1973). Let X be a real Banach space and $f \in C^1(X, \mathbf{R})$. Suppose f satisfies (PS) condition, $f(0) = 0$,*
(i) there exist constants $\rho, \alpha > 0$ such that $f(x) \geq \alpha$, if $\|x\| = \rho$,
(ii) there is an $e \in X$, $\|e\| > \rho$ such that $f(e) \leq 0$.
Then f possesses a critical value $c \geq \alpha$ which can be characterized as

$$c = \inf_{\gamma \in \Gamma} \max_{t \in [0,1]} f(\gamma(t)), \tag{24.1}$$

where

$$\Gamma = \{\gamma \in C([0,1], X) : \gamma(0) = 0, \gamma(1) = e\}. \tag{24.2}$$

Geometrically, when $X = \mathbf{R}^2$ assumptions (i), (ii) mean that the origin lies in a valley surrounded by a mountain $\Gamma_f = \left\{(x, f(x)) \in \mathbf{R}^3 : x \in \mathbf{R}^2\right\}$. Then the level of lowest pass c through the mountain between $(0,0)$ and $(e, f(e))$ is a critical value, i. e. there exists $x_0 \in X$ such that $f(x_0) = c$, $f'(x_0) = 0$.

Note that (PS) condition is essential in Theorem 6 as shows following example. The function $f(x,y) = x^2 + (x+1)^3 y^2$ satisfies assumptions (i), (ii) of Theorem 6 but not satisfies (PS) condition and it's unique critical point is $(0,0)$. Indeed $(0,0)$ is a strict local minima and unique critical point. If (PS) condition is satisfied then $(PS)_c$ with $c > 0$ defined by (24.1) is also satisfied. Let (x_j, y_j) be a sequence such that $x_j^2 + (x_j+1)^3 y_j^2 \to c > 0$, $\quad 2x_j + 3(x_j+1)^2 y_j^2 \to 0$, $\quad 2(x_j+1)^3 y_j \to 0$. Suppose $(x_j, y_j) \to (x_0, y_0)$. Then $x_0^2 + (x_0+1)^3 y_0^2 = c > 0$, $\quad 2x_0 + 3(x_0+1)^2 y_0^2 = 0$, $\quad 2(x_0+1)^3 y_0 = 0$, a contradiction.

The proof of Theorem 6 is based on deformation theorem. Deformation approach is used in various generalizations of Theorem 6 - Rabinowitz [21], Willem [29], Schechter [23], [24], Chang [8], Shi & Chang [28].

We consider now a characterization of critical values b as

$$b = \sup_{N \in \mathcal{N}} \inf_{x \in \partial N} f(x), \tag{24.3}$$

where $\mathcal{N} = \{N \subset X : N \text{ bounded and open}, 0 \in N, e \notin \bar{N}\}$. We prove an extension of Rabinowitz [21], Theorem 3.2 with $(PS)_\psi$- condition using Theorem 4.

Theorem 7. *Let $f \in C^1(X, \mathbf{R})$ and there exists $e \neq 0$, $N_0 \subset \mathcal{N}$ such that*
(j) $f(0) \leq 0, \quad f(x) \geq a > 0, \quad x \in \partial N_0,$
(jj) $f(e) \leq 0,$
and f satisfies $(PS)_{b,\psi}$ condition, where b is defined by (24.3). Then b is a critical value.

Proof. By (j) it follows $b \geq a > 0$. Let $K_b = \emptyset$, $0 < \varepsilon < \frac{a}{2}$. Let $N_1 \in \mathcal{N}$ be such that

$$\inf_{x \in \partial N_1} f(x) \geq b - \varepsilon,$$

and $M > 0$ be such that $N_0 \subset B_M(0)$.

According Theorem 4 there exists a f- increasing homotopy $\eta(t, x) : [0,1] \times X \to X$ such that

$$\eta(1, f_{b-\varepsilon} \cap N_0) \subset f_{b+\varepsilon},$$

because $N_0 \subset B_M(0)$. As $\eta(1, .) : X \to X$ is a homeomorphism $N_2 = \eta(1, N_1)$ is open, $\partial N_2 = \eta(1, \partial N_1)$. Moreover $N_2 \in \mathcal{N}$ because $0 \in N_2$, $e \notin N_2$. Indeed as $b > 0$, $f(0) \leq 0 < b - 2\varepsilon$, $f(e) \leq 0 < b - 2\varepsilon$, it follows $0, e \in \{x : |f(x) - b| \geq 2\varepsilon\}$ and $0 = \eta(1, 0) \in N_2$, $e = \eta(1, e) \notin N_2$. By $f(\eta(1, x)) \geq b + \varepsilon$ if $f(x) \geq b - \varepsilon$, $\partial N_1 \subset f_{b-\varepsilon}$ we have

$$\inf_{x \in \partial N_1} f(\eta(1, x)) \geq b + \varepsilon,$$

s a contradiction with the definition of b. Therefore $K_b \neq \emptyset$. □

Now we generalize Theorem 7 assuming non strict inequality in (jj). We prove

Theorem 8. *Let $f \in C^1(X, \mathbf{R})$ and there exists $e \neq 0$, $N_0 \in \mathcal{N}$ such that*
(j') $f(0) \leq 0$ *and* $f(x) \geq 0, \quad x \in \partial N_0$
(jj') $f(e) \leq 0$

Let f map bounded sets into bounded sets and satisfy $(PS)_{b,\psi}$ condition, where b is defined by (24.3). Then $b \geq 0$ is a critical value. If $b = 0$ there exists a critical point y_0 of f, such that $\|y_0\| = r$, $r = \min(d(0, \partial N_0), d(e, \bar{N}_0))$.

Proof. By assumptions it follows $b \geq 0$. If $b > 0$, $\varepsilon < \frac{b}{2}$, $0 < a < b - 2\varepsilon < b - \varepsilon$ there exists $N_1 \in \mathcal{N}$ such that

$$\inf_{x \in \partial N_1} f(x) > b - \varepsilon > a > 0,$$

and we are in situation of Theorem 7.

Let now $b = 0$ and $r = \min(d(0, \partial N_0), d(e, \bar{N}_0))$, $0 < \delta < r$, $V = \{x : r - \delta \leq \|x\| \leq r + \delta\}$. If ρ, $r - \delta < \rho < r + \delta$, $B(0, \rho) \in \mathcal{N}$, $\partial B(0, \rho) \subset V$ and

$$0 \geq \inf_{\partial B(0,\rho)} \geq \inf_V f.$$

Let $\{x_n\}$ be a minimizing sequence of f in V

$$f(x_n) \leq \inf_V f + \frac{1}{n}, \qquad \|x_n\| = r.$$

By Ekeland variational principle (see [10]) there exists $y_n \in V$ such that

$$f(y_n) \leq f(x_n), \quad \|x_n - y_n\| \leq \frac{1}{n}$$

and

$$f(y_n) \leq f(x) + \frac{1}{n}\|x - y_n\|, \forall x \in V. \tag{24.4}$$

For large n, y_n belongs to the interior of V and if $x = y_n + tv$, where $v \in X$, $\|v\| = 1$ for sufficiently small t by (24.4) letting $t \to 0$ it follows $\left\|f'(y_n)\right\|_* \leq \frac{1}{n}$.

As $r - \delta \leq \|y_n\| \leq r + \delta$ by properties of ψ,

$$\psi(r + \delta) \leq \psi(\|y_n\|) \leq \psi(r - \delta)$$

and $\psi(\|y_n\|)\left\|f'(y_n)\right\|_* \to 0$ as $n \to \infty$. By $(PS)_{b,\psi}$ condition $y_n \to y_\delta$, $y_\delta \in V$ and $f'(y_\delta) = 0$. As δ is arbitrary $0 < \delta < r$ again by $(PS)_{b,\psi}$ condition $y_\delta \to y_0$ as $\delta \to 0$ and $f'(y_0) = 0$, $\|y_0\| = r$. □

As a corollary of Theorem 8 we obtain following three critical point result.

Corollary 24.3.1. *Let $f \in C^1(X, \mathbf{R})$ and satisfy $(PS)_\psi$ condition. Suppose that f has two local minima. Then f has at least one more critical point.*

Similar three critical points theorems with (PS) condition were proved by Mawhin & Willem [16], Figueiredo & Solimini [13], Pucci & Serrin [19].

In above considered variants of mountain pass theorems we use deformation theorems proved in previous section. Another approach to mountain-pass theorems is based on Ekeland variational principle. We use it partially in the proof of Theorem 8. A general result in this direction due to Aubin & Ekeland [2]. To formulate their result we need another form of (PS) condition.

Definition 24.8. *(Aubin & Ekeland, 1984). The Gateaux differentiable functional $f : X \to \mathbf{R}$, satisfies (WPS) condition on subspace $\Omega \subset X$, if for every sequence $\{x_j\}$ in Ω such that*
(1) $|f(x_j)| \leq M$,
(2) $f'(x_j) \neq 0$, $\forall j$ *and* $\left\|f'(x_j)\right\|_* \to 0$,
it follows that there exists $\bar{x} \in X$ such that

$$\liminf f(x_j) \leq f(\bar{x}) \leq \limsup f(x_j), \quad f'(\bar{x}) = 0.$$

Relations between (PS) and (WPS) condition are given in Aubin & Ekeland [2].

Theorem 9. *(Aubin & Ekeland, 1984). Let $f : X \to \mathbf{R}$ be Gateaux-differentiable functional, such that $f' : X \to X^*$ is continuous from X with norm topology to X^* with weak* topology and satisfies following assumptions*
(1) there exists $\alpha > 0$ such that

$$m(\alpha) = \inf \{f(x) : \|x\| = \alpha\} > f(0),$$

(2) there exists $e \in X$ such that

$$\|e\| > \alpha, \quad f(e) < m(\alpha),$$

(3) f satisfies (WPS) condition on $\{x : f(x) \geq m(\alpha)\}$.
Then there exists $\bar{x} \in X$, such that $f(\bar{x}) \geq m(\alpha)$, $f'(\bar{x}) = 0$.

The idea of the proof of last theorem is modified by Ghoussob & Preiss [14] in order to get information about location of critical points. They introduce $(PS)_{G,c}$ condition around a set G at the level c as follows

Definition 24.9. *(Ghoussob & Preiss, 1989) The Gateaux differentiable functional $f : X \to \mathbf{R}$, satisfies $(PS)_{G,c}$ condition around a set G at the level c, if every sequence $\{x_n\}$ in X such that:*
(1) $\lim_{n\to\infty} d(x_n, G) = 0$,
(2) $\lim_{n\to\infty} f(x_n) = c$,
(3) $\lim_{n\to\infty} \left\|f'(x_n)\right\|_* = 0$,
has a convergent subsequence.

The usual (PS) condition corresponds to the case where it verifies $(PS)_{G,c}$ for any $G \subset X$ and any $c \in \mathbf{R}$.

Definition 24.10. *A closed subset G of a Banach space separate two points u and v in X, if u and v belong to disjoint connected components of $X \backslash G$.*

Theorem 10. *(Ghoussob & Preiss, 1989). Let $f : X \to \mathbf{R}$ be Gateaux- differentiable functional, such that $f' : X \to X^*$ is continuous from X with norm topology to X^* with weak* topology. Let $e \neq 0$*

$$c = \inf_{\gamma\in\Gamma} \max_{t\in[0,1]} f(\gamma(t)),$$

where

$$\Gamma = \{\gamma \in C([0,1], X) : \gamma(0) = 0, \gamma(1) = e\},$$

and the closed subset G be such that $G \cap f_c$ separates 0 and e. Assume f satisfies $(PS)_{G,c}$ condition. Then $K_c \neq \emptyset$

Further Ekeland [11] generalizes Theorem 10 assuming $(PS)_{G,c}$ condition of Cerami type. Finally we note that mountain-pass theorem is proved also in scales of Banach spaces by Struwe [26], [27].

Theorem 11. *(Struwe, 1983). Let $f : X \to \mathbf{R}$ be a functional, $f \in C^1(X_n, \mathbf{R})$, where $X_n \subset X_{n+1}$, $\bigcup_{n=1}^{\infty} X_n$ is dense in X. Assume there exist $e \neq 0, \rho > 0, \alpha > 0$ such that:*
(1) $f(0) = 0$, $f(x) \geq \alpha$ *for every* x *with* $||x|| = \rho$,
(2) $f(e) < \alpha$,
(3) f *satisfies* $(PS)_c^*$ *condition with* $c = \inf_{\gamma \in \Gamma} \max_{t \in [0,1]} f(\gamma(t))$.
Then $K_c^* \neq \emptyset$.

References

[1] A. Ambrosetti, P. Rabinowitz. Dual variational methods in critical point theory and applications. J. Funct. Anal. **14** (1973), 349- 381.

[2] J. P. Aubin, I. Ekeland, I. Applied Nonlinear Analysis, John Wiley & Sons, N.Y., 1984.

[3] P. Bartolo, V. Benci, D. Fortunato. Abstract critical point theory and applications to some nonlinear problems with "strong" resonance at infinity. Nonlinear Anal. T.M.A. **7**, 9 (1983), 981- 1012.

[4] H. Brezis, J. M. Coron, L. Nirenberg. Free vibrations for a nonlinear wave equation and a theorem of P. Rabinowitz. Comm. Pure and Appl. Math. **33** (1980), 667–689.

[5] H. Brezis, L. Nirenberg. Remarks on finding critical points. Comm. Pure and Appl. Math. **44** (1991), 939-963.

[6] F. E. Browder. Infinite dimensional manifolds and nonlinear eigenvalue problems. Ann. of Math. 2, **82** (1965), 459- 477.

[7] G. Cerami. Un criterio di esistenza per i punti critci su varieta illimate. Rend. Acad. Sci. Let. Ist. Lombardo **112** (1978), 332-336.

[8] K.C. Chang. On the mountain pass lemma, in Equadiff 6, Brno 1985, Lect. Notes in Math. N 1192, Springer Verlag, Berlin, 1986, 203- 208.

[9] M. Degiovanni, M. Marzocchi. A critical point theory for nonsmooth functionals. Ann. Matem. pura ed appl. (IV), v. **CLXVII** (1994), 73- 100.

[10] I. Ekeland. Nonconvex minimazation problems. Bull. Amer. Math. Soc. (NS) **1** (1979), 443- 474.

[11] I. Ekeland. Convexiy methods in Hamiltoniam Mechanics. Springer- Verlag, NY etc, 1990

[12] D. G. De Figueiredo. Lectures on the Ekeland variational principle with applications and detours. Preliminary Lecture Notes, SISSA, 1988.

[13] D. G. De Figueiredo, S. Solimini. A variational approach to superlinear elliptic problems. Comm. Partial Differential Equations **9** (1984), 699-717.

[14] N. Ghoussoub, D. Preiss. A general mountain pass principle for locating and classifying critical points. Ann. Inst. Henry Poincaré. v. **6,** N 5 (1989), 321-330.

[15] S. Li. , M. Willem. Applications of local linking to critical point theory. Preprint Univ. Catholique de Louvain **N 203**, 1992.

[16] J. Mawhin, M. Willem. Multiple solutions of the periodic boundary value problems for some forced pendulum type equations. J. Diff. Eq. **52** (1984), 264- 287.

[17] J. Mawhin, M. Willem. Critical point theory and Hamiltonian systems. Springer Verlag, N.Y., 1988.

[18] R. S. Palais. Critical point theory and the minimax principle. Proc. Sympos. Pure Math. vol.15, Amer. Math. Soc. Providence, R.I., 1970, 185-212.

[19] P. Pucci, J. Serrin. Extensions of the montain pass lemma. J. Functional Anal. **59** (1984), 185- 210.

[20] P. Rabinowitz . Some minimax theorems and applications to nonlinear partial differential equations. Nonlinear Analysis. A collection of papers in honor of Erich Röthe. Academic Press, N.Y. 1978, 161- 177.

[21] P. Rabinowitz. Minimax methods in Critical Point Theory and Applications to Differential Equations. CBMS Reg. Conf. 65, AMS, Providence, R.I. , 1986.

[22] M.P.N. Ramos. Teoremas de enlace na teoria dos pontos críticos. Iniversidade de Lisboa, Faculdade de Ciências, 1993.

[23] M. Schechter. A bounded Mountain Pass Lemma without the (PS) condition. Trans. AMS **331** (1992), 681-703.

[24] M. Schechter. A variation of the Mountain Pass Theorem and applications. J. London Math. Soc. **44** (1991), 491-502.

[25] E. A. de B. Silva. Linking theorems and applications to semilinear elliptic problems at resonance. Nonlinear Analysis, T.M.A. , v. **16**, N 5 (1991), 455- 477.

[26] M. Struwe. Generalized Palais- Smale condition and applications, Universität Bonn, Preprint N **17,** 1983.

[27] M. Struwe. Variational Methods, Springer Verlag, NY, 1990.

[28] S.Z. Shi., K. C. Chang. A local minimax theorem without compactness. Nonlinear and Convex Analysis, Proc. Confer. in Honor of Ky Fan, Dekker, 1987, 211- 233.

[29] M. Willem. Lecture notes on critical point theory, Fundacao Universidade de Brasilia, **199**, 1983.

[30] M. Willem. Minimax theorems, Birkhäuser, 1997.

25 PROXIMAL METHODS FOR VARIATIONAL INEQUALITIES WITH SET-VALUED MONOTONE OPERATORS

A. Kaplan and R. Tichatschke

University of Trier, D-54286 Trier Germany

tichat@uni-trier.de

Abstract: A general approach to analyse convergence and rate of convergence of the proximal-like methods for variational inequalities with set-valued maximal monotone operators is developed. It is oriented to methods coupling successive approximation of the variational inequality with the proximal point algorithm as well as to related methods using regularization on a subspace and weak regularization. This approach seizes also so-called multi-step regularization methods, in which the number of proximal iterations in the approximated problems is controlled by a criterion characterizing these iterations as to be effective.

Key words: Monotone operators, variational inequalities, convex programming, proximal point methods, weak regularization.

AMS subject classification: 47H05, 47H19, 49M45, 65J20.

25.1 INTRODUCTION

Variational inequalities with maximal monotone operators include convex programs, convex-concave saddle point problems, equations and inclusions with maximal monotone operators etc..

Besides an unification of the study of the problems mentioned, their transformation into variational inequalities provides significant advancements in theoretical and numerical analysis for certain classes of problems. For a series of nonlinear problems in mathematical physics, especially, for so-called free boundary problems, this yields similar advantages as the application of variational formulations of linear problems. In particular that permits to introduce a fruitful concept of a weak solution and to substantiate the application of finite

R.P. Gilbert et al. (eds.), From Convexity to Nonconvexity, 345–361.

element methods.
It is generally accepted to date the beginning of the study of variational inequalities with the paper of FICHERA [6] dedicated to the complete theoretical analysis of Signorini's elastostatics problem. As a generalization of this analysis, in LIONS / STAMPACCIA [27] and FICHERA [7] the existence theorems for abstract variational inequalities with linear semi-coercive operators have been proved, and the structure of the solution set has been described. In the present paper and in [18] the results of G. Fichera are used to construct methods with regularization on a subspace. Concerning the existence theorems for variational inequalities with nonlinear monotone operators we reffer to [26], [35] and [32]. The coupling of finite element methods with convex optimization techniques for solving variational problems in mechanics and physics is investigated extensively in the monographs [9], [10] and [12].
The proximal point method, introduced by MARTINET [29] and later on studied in a more general setting by ROCKAFELLAR [35], has initiated a series of algorithms for solving variational inequality problems. We refer to [2], [14], [36] for convex programming algorithms connected with multiplier methods and to [1], [3], [15], [16] for modifications of the penalty technique. Proximal bundle methods for non-smooth convex optimization problems are given in [22], [25], [30], and for partial inverses of monotone operators as well as for decomposition algorithms, see [4], [8], [38] and [39]. Moreover, it is known (see [16], [24], [36]) that some classical numerical methods can be interpreted as special applications of the proximal point method. Recently, in [5] this fact was established concerning the Douglas-Rachford splitting method for finding a zero of the sum of two monotone operators. That points out new applications of the proximal-based techniques, in particular, to problems in mathematical physics.

The basic results of ROCKAFELLAR [35] on convergence of the proximal point method for solving variational inequalities with maximal monotone operators were generalized in [28] concerning the rate of convergence, and in [11] a similar analysis was made for methods using the proximal procedure on a subspace. In these papers the proximal iterations are assumed to be inexactly performed, however, an approximation of the problem was not considered.

In the present paper the convergence analysis seizes methods which couple a successive approximation of the variational inequality with the proximal point approach as well as related methods using a regularization on a subspace or a weak regularization.

The problem under consideration is the following variational inequality:

$$\textit{find } u \in K \textit{ such that } \exists\, y \in \mathcal{T}u : \quad \langle y, v-u\rangle \geq 0 \quad \forall\, v \in K, \tag{25.1}$$

with K a convex, closed subset of a Hilbert space V, $\mathcal{T} : V \to 2^{V'}$ a monotone operator, $D(\mathcal{T}) \equiv \{v \in V : \mathcal{T}v \neq \emptyset\} \supset K$, V' a dual space of V and $\langle \cdot, \cdot \rangle$ the duality pairing between V and V'. For the terminology of monotone operators we reffer to [34] and [35].

Throughout the whole paper it is supposed that H is a given Hilbert space such that V can be continuously embedded into H, V_1 is a given closed subspace of V and $\mathcal{P} : V \to V_1$ is an orthogonal projection operator (orthoprojector). If V_1 is also closed in H then $\mathcal{P}$ can be defined as the orthoprojector according to the norm of H. In the framework of the approach considered, the solution of (25.1) is obtained by solving approximately the sequence of variational inequalities:

$$u \in K_i \ : \ \langle \mathcal{T}_i u, v-u\rangle + \chi_i(\mathcal{P}u - \mathcal{P}u^{i,s-1}, \mathcal{P}v - \mathcal{P}u)_H \geq 0 \ \ \forall v \in K_i, \quad (25.2)$$

$$s = 1, ..., s(i); \ i = 1, 2, ...,$$

with $\mathcal{T}_i : V \to V'$ and K_i certain approximations for $\mathcal{T}$ and K, respectively, $(\cdot,\cdot)_H$ the inner product in H, $u^{i,s-1}$ a solution of the previous problem $(u^{i,0} \equiv u^{i-1,s(i-1)})$, $0 < \chi_i \leq \bar{\chi}$.

The strong formalization of Method (25.2) is described in Section 2 as *multi-step regularization method* (MSR-method). The notion "multi-step regularization" reflects the presence of an inner cycle (w.r.t. s) of proximal iterations. The case that $\mathcal{T}$ is a subdifferential of a convex functional, $H = V = V_1$, and $s(i) = 1$ for all i, includes several "diagonal" variants of the proximal point method for convex optimization (cf. [16], Sect. 9 and 12, [23] and the references therein).

There is a lot of papers concerning the *diagonal approximation* of the ill-posed Problem (25.1) with the use of the *Browder-Tikhonov regularization*. In this case the auxiliary problems have the form

$$u \in K_i \ : \quad \langle \mathcal{T}_i u + \chi_i \mathcal{M}_0(u - \hat{u}), v - u\rangle \geq 0 \ \ \forall \ v \in K_i,$$

where $\mathcal{M}_0 : V \to V'$ is a strongly monotone operator, $\hat{u} \in V$ is a fixed element, and $\chi_i > 0, lim\ \chi_i = 0$. Fundamental results in this direction were obtained by MOSCO [31], for concrete algorithms see VASIL'EV [40].

Inasmuch $\chi_i \to 0$ is not necessary for the convergence of the proximal methods, they possess a better stability and provide a better efficiency of fast convergent algorithms solving the regularized auxiliary problems.
In methods of type (25.2) the iterations w.r.t. s (for a fixed approximation level i) continue "as long as they remain effective" (see Remark 3 below). The MSR-methods were developed in [16] for convex variational problems. In [18], [13] and [19] they were adapted to some problems in elasticity theory and optimal control with PDE's. In comparison with diagonal proximal processes, the MSR-methods permit to handle with rough approximations of the original problem more efficiently so that the numerical expense can be reduced.
Here we investigate the convergence of such methods in the more general setting that the operator $\mathcal{T}$ may be non-potential. The principal distinction to the convergence analysis made in [16] is caused by the fact that the condition about the uniform approximation of the objective functional used in [16] cannot be adapted to describe the closeness between $\mathcal{T}$ and $\mathcal{T}_i$.

The paper is organized as follows: Section 2 contains the description of a generalized MSR-scheme. The convergence of the MSR-scheme is studied in Section 3, where we also formulate some results about the rate of convergence for the basic variant (with $H = V = V_1$). Due to the limitation of the size of the paper, we cannot include the proofs of the theorems about rate of convergence as well as describe some applications. We refer for that to [21].

25.2 MULTI-STEP PROXIMAL REGULARIZATION SCHEME

We denote by $(\cdot,\cdot)$ the inner product in V, by $\|\cdot\|, \|\cdot\|_{V'}$ and $\|\cdot\|_H$ the norms in V, V', and H, respectively. With z, C, D from V (respectively from V'), let

$$dist(z,D) = inf_{w\in D}\|z-w\|, \quad dist(C,D) = sup_{z\in C} dist(z,D),$$
$$\text{(resp. } dist_{V'}(z,D) = inf_{w\in D}\|z-w\|_{V'}, \ dist_{V'}(C,D) = sup_{z\in C} dist_{V'}(z,D) \text{).}$$

The solution set U^* of Problem (25.1) is supposed to be non-empty.
The following assumption concernes the chosen triple H, V_1 and $\mathcal{P}$.

Assumption 1 *For a given linear, continuous and monotone operator $\mathcal{B} : V \to V'$ with the symmetry property*

$$\langle \mathcal{B}u, v\rangle = \langle \mathcal{B}v, u\rangle \quad \forall u, v \in V$$

and a given $\tilde{\chi} > 0$ there exists $\beta_0 > 0$ such that

$$\tilde{\chi}\langle \mathcal{B}u, u\rangle + \|\mathcal{P}u\|_H^2 \geq \beta_0\|u\|^2 \quad \forall u \in V. \tag{25.3}$$

If this assumption is valid, one can introduce on V a new norm $|||\cdot|||$ by

$$|||u|||^2 = \tilde{\chi}\langle \mathcal{B}u, u\rangle + \|\mathcal{P}u\|_H^2, \tag{25.4}$$

which is equivalent to the original one $\|\cdot\|$ because of

$$\beta_0\|u\|^2 \leq |||u|||^2 \leq (\beta_1\tilde{\chi} + \beta_2^2)\|u\|^2, \tag{25.5}$$

with $\beta_1 = sup_{u\neq 0}\frac{\|\mathcal{B}u\|_{V'}}{\|u\|}$ and $\beta_2 : \|u\|_H \leq \beta_2\|u\| \ \forall u \in V$.

Let $\{\mathcal{T}_i\}, \mathcal{T}_i : V \to V'$ be a sequence of monotone and hemicontinuous operators, approximating $\mathcal{T}$, and $\{K_i\}, K_i \subset D(\mathcal{T}_i) \subset V$, be a sequence of convex closed sets, approximating K. Moreover, let Assumption 1 be fulfilled.
Henceforth, a sphere $S_\tau = \{u \in V : |||u||| \leq \tau\}$, and radii r and r^* are chosen such that

$$U^* \cap S_{r^*/8} \neq \emptyset \text{ and } r > r^*. \tag{25.6}$$

MSR-method (Multi-step regularization)
Let $\{\chi_i\}, \{\varepsilon_i\}, \{\delta_i\}$ be positive controlling sequences with

$$sup_i\ \chi_i \leq \bar{\chi} < \infty, \quad lim_{i\to\infty}\ \varepsilon_i = 0,$$

and $u^0 \in S_{r^*/4}$.

Step i: Given u^{i-1}.

a) Set $u^{i,0} := u^{i-1}, \; s := 1$.

b) Given $u^{i,s-1}$, define

$$u^{i,s} \in D(\mathcal{T}_i) : \|u^{i,s} - \bar{u}^{i,s}\| \le \varepsilon_i, \tag{25.7}$$

where $\bar{u}^{i,s}$ is the exact solution of the variational inequality

$$u \in K_i : \langle \mathcal{T}_i u, v-u\rangle + \chi_i(\mathcal{P}u - \mathcal{P}u^{i,s-1}, \mathcal{P}v - \mathcal{P}u)_H \ge 0 \;\; \forall v \in K_i. \tag{25.8}$$

c) If $\|\mathcal{P}u^{i,s} - \mathcal{P}u^{i,s-1}\|_H > \delta_i$, then set $s := s+1$ and repeat b). Otherwise, set $s(i) := s, u^i := u^{i,s}, i := i+1$ and repeat Step i.

The choice $H = V = V_1$ corresponds to the *basic variant of the MSR-method* [16], and $V_1 \ne V$ reflects the (MSR-)*method with regularization on a subspace* (cf. [16], [13], [18]). If $V_1 = V$ and $\|\cdot\|_H$ is weaker than $\|\cdot\|$, one deals with the (MSR-)*method with weak regularization* (see [16],[18], [37]). For the diagonal method with regularization on a subspace we refer to [11] and to Rockafellar's interpretation for the multiplier method in [36].

25.3 CONVERGENCE ANALYSIS

According to Riesz's representation theorem, the variational inequality (25.8) can be rewritten in the form

$$u \in K_i : \langle \mathcal{T}_i u + \chi_i \mathcal{M}(u - u^{i,s-1}), v - u\rangle \ge 0 \;\; \forall v \in K_i, \tag{25.9}$$

where the linear operator $\mathcal{M} : V \to V'$ is defined by

$$\langle \mathcal{M}u, v\rangle = (\mathcal{P}u, \mathcal{P}v)_H \;\; \forall u, v \in V.$$

In order to prove convergence of the MSR-method, we need some auxiliary statements. The first one can be proved similarly to Lemma 2.3 in [18].

Lemma 1 *Let $G \subset V$ be a closed convex set, and let $\mathcal{A} : V \to V'$ be a single-valued operator with $D(\mathcal{A}) \supset G$. Suppose that the inequality*

$$\langle \mathcal{A}u - \mathcal{A}v, u - v\rangle \ge \langle \mathcal{B}u - \mathcal{B}v, u - v\rangle \;\; \forall u, v \in G \tag{25.10}$$

is valid with $\mathcal{B}$ satisfying Assumption 1.
Moreover, for arbitrary $a^0 \in V$ and $\chi \in (0, 2\bar{\chi}^{-1}]$, let a^1 be a solution of the variational inequality

$$u \in G \; : \; \langle \mathcal{A}u, v - u\rangle + \chi(\mathcal{P}u - \mathcal{P}a^0, \mathcal{P}v - \mathcal{P}u)_H \ge 0 \;\; \forall v \in G. \tag{25.11}$$

Then the inequality

$$|||a^1 - v|||^2 - |||a^0 - v|||^2 \leq -||\mathcal{P}a^1 - \mathcal{P}a^0||_H^2 + \frac{2}{\chi}\langle \mathcal{A}v, v - a^1\rangle \qquad (25.12)$$

is true for each $v \in G$.

Lemma 2 *Let* $G \subset V$ *be a closed convex set and* $\mathcal{A} : V \to 2^{V'}$ *be a maximal monotone operator with* $D(\mathcal{A}) \supset G$, $\mathrm{int}D(\mathcal{A}) \neq \varphi$. *Assume that*

$$sup_{v\in G}\ sup_{y\in \mathcal{A}v} ||y||_{V'} < \infty, \qquad (25.13)$$

and for some $u \in G$ *and each* $v \in G$, *there exists* $y(v) \in \mathcal{A}v$ *satisfying*

$$\langle y(v), v - u\rangle \geq 0. \qquad (25.14)$$

Then, there exists $y \in \mathcal{A}u$ *such that*

$$\langle y, v - u\rangle \geq 0 \textit{ for each } v \in G.$$

Proof Let $\mathcal{A}_1 v = \mathcal{A}v + \mathcal{I}(v-u)$, with $\mathcal{I} : V \to V'$ the operator of the canonical isometry. The operator $\mathcal{A}_1$ is strongly monotone, and due to Theorem 1 in [34], is maximal monotone.
For fixed $q(u) \in \mathcal{A}u$, using the inequality

$$\begin{aligned} \langle q(v) + \mathcal{I}(v-u), v-u\rangle &\geq \langle q(u), v-u\rangle + ||v-u||^2 \\ &\geq ||v-u|| \left(||v-u|| - ||q(u)||_{V'}\right), \end{aligned}$$

which holds for any $q(v) \in \mathcal{A}v$, one can conclude that

$$\langle q(v) + \mathcal{I}(v-u), v-u\rangle \geq 0 \qquad (25.15)$$

if

$$||v|| \geq ||u|| + ||q(u)||_{V'}.$$

Taking into account (25.13) and (25.15), Theorem 5 in [34] guarantees that the variational inequality

$$\textit{find}\quad w \in G \quad \textit{such that}\quad \exists\, y_1(w) \in \mathcal{A}_1 w \;:\; \langle y_1(w), v - w\rangle \geq 0 \quad \forall v \in G$$

is solvable. Now, all we have to do is to use the proof of Lemma 3 in [20]. ◇

Throughout the remaining part of the paper we suppose that Assumption 1 is fulfilled and r, r^* are chosen according to (25.6). Set

$$Q_i = K_i \cap S_r, \quad Q = K \cap S_r, \quad Q^* = U^* \cap S_{r^*}. \qquad (25.16)$$

To investigate the convergence of the MSR-method for Problem (25.1) we make the following general

Assumption 2

(a) $sup_{u \in Q}\ sup_{y \in \mathcal{T}u} \|y\|_{V'} \leq \mu(r) < \infty$;

(b) *with* $\mathcal{B}$ *and* $\tilde{\chi}$ *from Assumption 1, for each* i*, the relation*

$$\langle \mathcal{T}_i u - \mathcal{T}_i v, u - v \rangle \geq \langle \mathcal{B}u - \mathcal{B}v, u - v \rangle \quad \forall u, v \in Q_i$$

is valid, and $\tilde{\chi} \leq 2\bar{\chi}^{-1}$;

(c) *for each* $z \in Q$ *there exist points* $v_i(z) \in Q_i$ *and a compact set* $\Lambda(z) \subset \mathcal{T}z$ *such that*

$$lim_{i\to\infty}\|v_i(z) - z\| = 0, \quad lim_{i\to\infty} dist_{V'}(\mathcal{T}_i v_i(z), \Lambda(z)) = 0,$$

($\Lambda(z)$ *may depend on* $\{v_i(z)\}$*);*

For given sequences $\{\varphi_i\}, \{\sigma_i\}$*, such that* $lim_{i\to\infty}\varphi_i = lim_{i\to\infty}\sigma_i = 0$*, it holds:*

(d) *for each* $u \in Q^*$ *there exist points* $w_i \in Q_i$*, satisfying*

$$\|u - w_i\| \leq \varphi_i, \quad dist_{V'}(\mathcal{T}_i w_i, \hat{\Lambda}(u)) < \sqrt{\beta_0}\,\sigma_i, \ i = 1, 2, ...,$$

where $\hat{\Lambda}(u) = \{y \in \mathcal{T}u : \langle y, v - u\rangle \geq 0 \ \ \forall v \in Q\}$*;*

(e) $dist(Q_i, Q) \leq \varphi_i, \quad i = 1, 2, ...$

Lemma 3 *Let the Assumptions 2(a), (b), (d), (e) be fulfilled and*

$$\frac{1}{4r}\left(\frac{4}{\chi_i}(\mu(r)\varphi_i + r\sigma_i) - \tilde{\varepsilon}_i^2\right) + \beta_3\varepsilon_i < 0, \quad \tilde{\varepsilon}_i = \delta_i - \beta_2\varepsilon_i > 0 \tag{25.17}$$

hold for $i \leq i_0$*, with* $\beta_3 = \left(2\beta_1\bar{\chi}^{-1} + \beta_2^2\right)^{1/2}$*. Moreover, assume that in the MSR-method the relations* $s(i) < \infty$ *for* $i < i_0$*, and* $|||u^{i_0,s}||| < r^*$*,* $|||\bar{u}^{i_0,s}||| < r^*$ *for each* s *are valid. Then* $s(i_0) < \infty$ *is true.*

Proof Fix $u^{**} \in U^* \cap S_{r^*}$. Due to the Assumptions 2(d), (e), one can choose $v^{i_0} \in Q_{i_0}$, $v^{i_0,s} \in Q$ such that

$$\|v^{i_0,s} - \bar{u}^{i_0,s}\| \leq \varphi_{i_0}, \quad \|u^{**} - v^{i_0}\| \leq \varphi_{i_0},$$

and, for some $y \in \hat{\Lambda}(u^{**})$,

$$\left\|\mathcal{T}_{i_0} v^{i_0} - y\right\|_{V'} \leq \sqrt{\beta_0}\,\sigma_{i_0}.$$

Then, regarding Assumption 2(a),

$$\left\langle \mathcal{T}_{i_0} v^{i_0}, v^{i_0} - \bar{u}^{i_0,s}\right\rangle \leq \left\langle y, u^{**} - v^{i_0,s}\right\rangle + 2r\sigma_{i_0} + 2\mu(r)\varphi_{i_0}.$$

But from the definition of $\hat{\Lambda}(u)$ and $v^{i_0,s} \in Q$, one can conclude that

$$\langle y, u^{**} - v^{i_0,s}\rangle \leq 0,$$

hence,

$$\langle \mathcal{T}_{i_0} v^{i_0}, v^{i_0} - \bar{u}^{i_0,s} \rangle \leq 2r\sigma_{i_0} + 2\mu(r)\varphi_{i_0}. \tag{25.18}$$

Now, using Assumption 2(b) and Lemma 1 with $\mathcal{A} = \mathcal{T}_{i_0}$, $G = Q_{i_0}$, $\chi = \chi_{i_0}$, $v = v^{i_0}$ and $a^0 = u^{i_0,s-1}$, in view of $\bar{u}^{i_0,s} \in K_{i_0}$, $|||\bar{u}^{i_0,s}||| < r^*$ and (25.18) we obtain

$$\begin{aligned} &|||\bar{u}^{i_0,s} - v^{i_0}|||^2 - |||u^{i_0,s-1} - v^{i_0}|||^2 \\ &\leq -\|\mathcal{P}\bar{u}^{i_0,s} - \mathcal{P}u^{i_0,s-1}\|_H^2 + \frac{4}{\chi_{i_0}} \left(r\sigma_{i_0} + \mu(r)\varphi_{i_0}\right). \end{aligned} \tag{25.19}$$

With regard to (25.7) and $\|\mathcal{P}u\|_H \leq \beta_2\|u\| \;\; \forall u \in V$, the second inequality in (25.17) together with (25.18) yield

$$|||\bar{u}^{i_0,s} - v^{i_0}|||^2 - |||u^{i_0,s-1} - v^{i_0}|||^2 < -\tilde{\varepsilon}_{i_0}^2 + \frac{4}{\chi_{i_0}} \left(r\sigma_{i_0} + \mu(r)\varphi_{i_0}\right) \tag{25.20}$$

for $1 \leq s < s(i_0)$, and taking into account the first inequality in (25.17), we have

$$|||\bar{u}^{i_0,s} - v^{i_0}||| < |||u^{i_0,s-1} - v^{i_0}|||.$$

Moreover, due to $|||u^{i_0,s}||| < r^*$, $|||\bar{u}^{i_0,s}||| < r^*$, $|||v^{i_0}||| \leq r$, $\tilde{\chi} \leq 2\bar{\chi}^{-1}$ and the relations (25.5), (25.7), and (25.20), we obtain for $1 \leq s < s(i_0)$ that

$$\begin{aligned} &|||u^{i_0,s} - v^{i_0}||| < |||u^{i_0,s-1} - v^{i_0}||| \\ &+ \frac{1}{4r}\left[\frac{4}{\chi_{i_0}} \left(r\sigma_{i_0} + \mu(r)\varphi_{i_0}\right) - \tilde{\varepsilon}_{i_0}^2\right] + \beta_3\varepsilon_{i_0} < 0. \end{aligned} \tag{25.21}$$

Summing up the inequalities (25.20) w.r.t. $s = 1, ..., \bar{s}$, with $\bar{s} < s(i_0)$ arbitrarily chosen, one gets

$$|||u^{i_0,\bar{s}} - v^{i_0}||| < |||u^{i_0,0} - v^{i_0}||| + \bar{s}\left[\frac{1}{4r}\left(\frac{4}{\chi_{i_0}} \left(r\sigma_{i_0} + \mu(r)\varphi_{i_0}\right) - \tilde{\varepsilon}_{i_0}^2\right) + \beta_3\varepsilon_{i_0}\right],$$

therefore,

$$s(i_0) < 2r\left[\frac{1}{4r}\left(\tilde{\varepsilon}_{i_0}^2 - \frac{4}{\chi_{i_0}} \left(r\sigma_{i_0} + \mu(r)\varphi_{i_0}\right)\right) - \beta_3\varepsilon_{i_0}\right]^{-1} + 1. \tag{25.22}$$

◇

Lemma 4 *Let the Assumptions 2 (a), (b), (d), (e) be fulfilled, and let the controlling parameters* ε_i, χ_i, δ_i, σ_i, *and* φ_i *satisfy*

$$\frac{1}{4r}\left(\frac{4}{\chi_i}\left(\mu(r)\varphi_i + r\sigma_i\right) - \left(\delta_i - \beta_2\varepsilon_i\right)^2\right) + \beta_3\varepsilon_i < 0 \tag{25.23}$$

and

$$\sum_{i=1}^{\infty}\left(2\left(\frac{\mu(r)\varphi_i + r\sigma_i}{\chi_i}\right)^{1/2} + \beta_3(\varepsilon_i + 2\varphi_i)\right) < \frac{1}{2}r^*. \tag{25.24}$$

Then, in the MSR-method $s(i) < \infty$ is valid for each i, variational inequality (25.8) is uniquely solvable, and $|||u^{i,s}||| < r^*$, $|||\bar{u}^{i,s}||| < r^*$ are true for all (i, s).

Proof The inequality $\delta_i - \beta_2\varepsilon_i > 0$ follows immediately from (25.23), (25.24) and (25.6). Suppose that i_0 and s_0 are kept fixed, $s(i') < \infty$ for $i' < i_0$, and that $0 \le s_0 < s(i_0)$. Denote

$$\Theta_0 = \{(i', s') \ : \ i' < i_0, \ 0 < s' \le s(i') \text{ and } i' = i_0, \ 0 < s' \le s_0\}.$$

Assume, moreover, that for all $(i, s) \in \Theta_0$ the variational inequality (25.8) is uniquely solvable and the inequalities $|||u^{i,s}||| < r^*$, $|||\bar{u}^{i,s}||| < r^*$ are true.
For fixed $u^{**} \in S_{r^*/8} \cap U^*$ and each i, due to Assumption 2 (d), one can choose $v^i \in Q_i$ and $y^i \in \hat{\Lambda}(u^{**})$ such that

$$\|v^i - u^{**}\| \le \varphi_i, \quad \|\mathcal{T}_i v^i - y^i\| \le \sqrt{\beta_0}\, \sigma_i. \tag{25.25}$$

Then, as in the proof of Lemma 3, we obtain (cf. (25.20)) for $i < i_0, \ 0 \le s < s(i) - 1$ and $i = i_0, \ 0 \le s < s_0$ that

$$|||u^{i,s+1} - v^i||| - |||u^{i,s} - v^i||| \le \frac{1}{4r}\left(\tau_i - \tilde{\varepsilon}_i^2\right) + \beta_3\varepsilon_i < 0, \tag{25.26}$$

with $\tau_i = \frac{4}{\chi_i}\left(\mu(r)\varphi_i + r\sigma_i\right), \ \tilde{\varepsilon}_i = \delta_i - \beta_2\varepsilon_i$.
Also for $i < i_0, \ s = s(i) - 1$, similarly to (25.18), one can conclude that

$$|||\bar{u}^{i,s(i)} - v^i|||^2 - |||u^{i,s(i)-1} - v^i|||^2 \le \tau_i,$$

and regarding (25.5) and (25.7),

$$|||u^{i,s(i)} - v^i||| - |||u^{i,s(i)-1} - v^i||| \le \sqrt{\tau_i} + \beta_3\varepsilon_i. \tag{25.27}$$

Summing up the inequalities (25.26), where $i < i_0$ is fixed and $0 \le s < s(i) - 1$, together with (25.27), we get

$$|||u^{i,s(i)} - v^i||| - |||u^{i,0} - v^i||| \le \sqrt{\tau_i} + \beta_3\varepsilon_i.$$

Due to (25.5) and (25.25), this yields

$$|||u^{i,s(i)} - u^{**}||| - |||u^{i,0} - u^{**}||| \le \sqrt{\tau_i} + \beta_3(\varepsilon_i + 2\varphi_i). \tag{25.28}$$

Assumptions 1 and 2(b) ensure strong monotonicity of the operator $\mathcal{T}_{i_0} + \chi_{i_0} M$ on Q_{i_0}, thus according to Theorem 5 in [34] the variational inequality

$$u \in Q_{i_0} \ : \ \langle \mathcal{T}_{i_0} u, v - u\rangle + \chi_{i_0}\left(\mathcal{P}u - \mathcal{P}u^{i_0,s_0}, \mathcal{P}v - \mathcal{P}u\right)_H \ge 0 \ \ \forall v \in Q_{i_0} \tag{25.29}$$

is uniquely solvable. We denote its solution by $\hat{u}^{i_0,s_0+1}$. The use of Lemma 1 leads to

$$|||\hat{u}^{i_0,s_0+1} - v^{i_0}||| \le |||u^{i_0,s_0} - v^{i_0}||| + \sqrt{\tau_{i_0}}. \tag{25.30}$$

Observing (25.25), (25.26) (taken with $i = i_0$, $0 \leq s < s_0$) and (25.30), the inequality

$$|||\hat{u}^{i_0,s_0+1} - u^{**}||| \leq |||u^{i_0,0} - u^{**}||| + \sqrt{\tau_{i_0}} + 2\beta_3\varphi_{i_0}. \tag{25.31}$$

can be obtained similarly to (25.28). The inequalities (25.28) and (25.31) imply

$$|||\hat{u}^{i_0,s_0+1} - u^{**}||| \leq |||u^{1,0} - u^{**}||| + \sum_{k=1}^{i_0-1} (\sqrt{\tau_k} + \beta_3(\varepsilon_k + 2\varphi_k)) + \sqrt{\tau_{i_0}} + 2\beta_3\varphi_{i_0}. \tag{25.32}$$

Due to (25.24) and the choice of u^{**} and $u^{1,0}$, estimate (25.32) yields

$$|||\hat{u}^{i_0,s_0+1}||| < r^* - \beta_3\varepsilon_{i_0}. \tag{25.33}$$

Now, from Lemma 2 in [20], the variational inequality (25.8) with $i = i_0$ and $s = s_0 + 1$ is uniquely solvable, $\bar{u}^{i_0,s_0+1} = \hat{u}^{i_0,s_0+1}$, and hence

$$|||u^{i_0,s_0+1}||| < r^*. \tag{25.34}$$

This enables to conclude that

$$|||\bar{u}^{i_0,s}||| < r^*, \ |||u^{i_0,s}||| < r^* \text{ for all } s,$$

and Lemma 3 gives $s(i_0) < \infty$. To complete the induction, note that the unique solvability of (25.8) with $i = 1, s = 1$, as well as the inequalities $|||\bar{u}^{1,1}||| < r^*$, $|||u^{1,1}||| < r^*$ and the finiteness of $s(1)$ can be established analogously. ◇

Theorem 1 *Let $\mathcal{T}$ be a maximal monotone operator, Assumption 2 be fulfilled, and suppose that the points $u^{i,s}$, generated by the MSR-method, as well as the points $\bar{u}^{i,s}$ which solve the variational inequalities (25.8), satisfy the relations*

$$|||\bar{u}^{i,s}||| < r^* \text{ and } |||u^{i,s}||| < r^* \text{ for all } i, s.$$

Moreover, assume that the controlling parameters satisfy (25.17) and

$$\sum_{i=1}^{\infty} \sqrt{\frac{\sigma_i}{\chi_i}} < \infty, \quad \sum_{i=1}^{\infty} \sqrt{\frac{\varphi_i}{\chi_i}} < \infty, \quad \sum_{i=1}^{\infty} \varepsilon_i < \infty. \tag{25.35}$$

Then, $s(i) < \infty$ holds for each i and the sequences $\{u^{i,s}\}$, $\{\bar{u}^{i,s}\}$ converge weakly in V to an element $u^* \in U^*$.

Proof The relation $s(i) < \infty$ for each i follows immediately from Lemma 3. Further, with an arbitrary $w \in U^* \cap S_{r^*}$, the inequality

$$|||u^{i+1,0} - w||| - |||u^{i,0} - w||| \leq \sqrt{\tau_i} + \beta_3(\varepsilon_i + 2\varphi_i)$$

can be established similarly to (25.28). Therefore, condition (25.35) and Lemma 2.2.2 in [33] imply the convergence of the sequence $\{|||u^{i,0} - w|||\}$.
Note that the inequalities (25.26), (25.27) remain true with a "new" $v^i \in Q_i$ which is chosen according to Assumption 2 (d) such that

$$\|v^i - w\| \leq \varphi_i, \quad inf_{y\in\hat{\Lambda}(w)}\|\mathcal{T}_i v^i - y\|_{V'} < \sqrt{\beta_0}\,\sigma_i.$$

This leads to the relation

$$-(\sqrt{\tau_i} + \beta_3(\varepsilon_i + 2\varphi_i)) + |||u^{i+1,0} - w||| \leq |||u^{i,s} - w||| < |||u^{i,0} - w||| + 2\beta_3\varphi_i \tag{25.36}$$

if $1 \leq s < s(i)$. Hence, the sequences $\{|||u^{i,s} - w|||\}$ and $\{|||\bar{u}^{i,s} - w|||\}$ converge to the same limit as $\{|||u^{i,0} - w|||\}$.
Now, from inequality (25.18), which can be extended to each i and $1 \leq s \leq s(i)$, one can conclude that

$$lim_{i\to\infty}\; max_{1\leq s\leq s(i)}\; \|\mathcal{P}\bar{u}^{i,s} - \mathcal{P}u^{i,s-1}\|_H = 0. \tag{25.37}$$

Using the definition of $\bar{u}^{i,s}$, we get

$$\langle \mathcal{T}_i\bar{u}^{i,s}, v - \bar{u}^{i,s}\rangle + \chi_i\left(\mathcal{P}\bar{u}^{i,s} - \mathcal{P}u^{i,s-1}, \mathcal{P}v - \mathcal{P}\bar{u}^{i,s}\right)_H \geq 0 \;\; \forall v \in Q_i, \tag{25.38}$$

and, in view of the monotonicity of $\mathcal{T}_i$, this yields

$$\langle \mathcal{T}_i v, v - \bar{u}^{i,s}\rangle + \chi_i\left(\mathcal{P}\bar{u}^{i,s} - \mathcal{P}u^{i,s-1}, \mathcal{P}v - \mathcal{P}\bar{u}^{i,s}\right)_H \geq 0 \;\; \forall v \in Q_i. \tag{25.39}$$

For arbitrary $z \in Q$, let us choose $v^i(z) \in Q_i$ and $y^i \in \Lambda(z)$ such that

$$lim_{i\to\infty}\|v^i(z) - z\| = 0, \quad lim_{i\to\infty}\|\mathcal{T}_i v^i(z) - y^i\|_{V'} = 0. \tag{25.40}$$

This choice is possible due to Assumption 2 (c). Then,

$$\begin{aligned}
&\langle y^i, z - \bar{u}^{i,s}\rangle + \chi_i\left(\mathcal{P}\bar{u}^{i,s} - \mathcal{P}u^{i,s-1}, \mathcal{P}z - \mathcal{P}\bar{u}^{i,s}\right)_H \\
&= \langle y^i, z - v^i(z)\rangle + \chi_i\left(\mathcal{P}\bar{u}^{i,s} - \mathcal{P}u^{i,s-1}, \mathcal{P}z - \mathcal{P}v^i(z)\right)_H \\
&\quad + \langle y^i - \mathcal{T}_i v^i(z), v^i(z) - \bar{u}^{i,s}\rangle \\
&+ \left[\langle \mathcal{T}_i v^i(z), v^i(z) - \bar{u}^{i,s}\rangle + \chi_i\left(\mathcal{P}\bar{u}^{i,s} - \mathcal{P}u^{i,s-1}, \mathcal{P}v^i(z) - \mathcal{P}\bar{u}^{i,s}\right)_H\right],
\end{aligned} \tag{25.41}$$

and in view of (25.39) the term in the square brackets is non-negative.
Let $\{\bar{u}^{i,s}\}_{(i,s)\in\Gamma}$ be a subsequence of $\{\bar{u}^{i,s}\}$, which converges weakly in V to some u^*. One can easily show that $u^* \in K \cap intS_r$. Now, we choose from $\{\bar{u}^{i,s}\}_{(i,s)\in\Gamma}$ an infinite subsequence $\{\bar{u}^{i,s_i}\}_{i\in I}$ and, without loss of generality, suppose that $\{y^i\}_{i\in I}$ converges in V' to an element y (according to Assumption 2 (c), $\{y^i\}$ belongs to the compact set $\Lambda(z)$). Maximal monotonicity of $\mathcal{T}$ yields $y \in \mathcal{T}z$. Taking limit with $s = s_i$, $i \to \infty$, $i \in I$, in inequality (25.41), in view of (25.37), (25.40) and the boundedness of the sequences $\{\chi_i\}$, $\{\bar{u}^{i,s}\}$, we obtain

$$\langle y, z - u^*\rangle \geq 0.$$

Now, on using Assumption 2 (a), Lemma 2 permits to conclude that

$$\langle y^*, z - u^* \rangle \geq 0 \text{ for some } y^* \in \mathcal{T}u^* \text{ and for all } z \in Q.$$

Since $u^* \in int\ S_r$, Lemma 2 in [20] guarantees that $u^* \in U^*$, and due to Opial's Lemma (see Proposition A1.3 in [16]), both sequences $\{\bar{u}^{i,s}\}$ and $\{u^{i,s}\}$ converge weakly in V to u^*. ◇

Remark 1 *Theorem 1 establishes convergence of the MSR-method if the set K is bounded. Indeed, if r^* is chosen such that $K \subset S_\tau$ holds with some $\tau < r^* - \beta_3(sup_i\ \varphi_i + sup_i\ \varepsilon_i)$, then due to Assumption 2 (e) and inequality (25.7), we infer that*

$$|||\bar{u}^{i,s}||| < r^*,\ |||u^{i,s}||| < r^* \text{ for all } i, s.$$

Thus, conditions (25.17) and (25.35) ensure weak convergence of $\{\bar{u}^{i,s}\}$ and $\{u^{i,s}\}$ to an element $u^* \in U^*$.

If boundedness of K is not supposed, compiling Lemma 4 and Theorem 1, we obtained immediately

Theorem 2 *Let $\mathcal{T}$ be a maximal monotone operator and Assumption 2 as well as the conditions of Lemma 4 be fulfilled. Then, the MSR-method started with arbitrary $u^0 \in S_{r^*/4}$ has the following properties:*

(i) $s(i) < \infty$ for all i;

(ii) $|||u^{i,s}||| < r^*$, $|||\bar{u}^{i,s}||| < r^*$ for each (i,s);

(iii) $\{u^{i,s}\}$ and $\{\bar{u}^{i,s}\}$ converge weakly in V to an element $u^* \in U^*$.

Remark 2 *If $V_1 = V = H$, then Assumption 1 is fulfilled with $\mathcal{B} = 0$. Therefore, $|||u||| = ||u||\ \forall u \in V$, and Assumption 2(b) follows from the monotonicity of $\mathcal{T}_i$. In this case the Theorems 1 and 2 are also true (with $\beta_1 = 0$, $\beta_2 = \beta_3 = 1$) if, instead of the Assumptions 2 (d) and (e), the following conditions are valid (cf. [20]):*

(d') $dist(Q^*, Q_i) \leq \varphi_i$, $i = 1, 2, ...,$

(e') for each i and $u_i \in Q_i$ there exists $v_i \in Q$, satisfying

$$||u_i - v_i|| \leq \varphi_i, \quad dist_{V'}(\mathcal{T}_i u_i, \mathcal{T} v_i) < \sigma_i.$$

Remark 3 *Now, we are ready to specify the notion "effective iteration" used in the Introduction. The (i,s)-th iteration is considered to be effective if*

$$||\mathcal{P}u^{i,s} - \mathcal{P}u^{i,s-1}||_H > \delta_i,$$

where δ_i is defined by (25.23), (25.24).

Remark 4 *If δ_i is chosen sufficiently large (for instance, $\delta_i > 2\beta_0^{-1/2}\beta_2 r^*$ for all i), then from Theorem 2 (also from Remark 1) it follows that $s(i) = 1$ for each i. This means that the MSR-method passes over to the usual (one-step) diagonal method. The proximal point method studied in [35] and [28] corresponds formally to the MSR-scheme with $H = V = V_1,\ \mathcal{T}_i = \mathcal{T},\ K_i = K,\ s(i) = 1$.*

The statements of the previous theorems can be improved in the following way.

Theorem 3 *Let the conditions of Theorem 1 or 2 be fulfilled and let*

$$lim_{i\to\infty}\, max_{1\le s\le s(i)}\, \|\mathcal{P}u^{i,s} - \mathcal{P}u^*\|_H = 0, \tag{25.42}$$

with u^* a weak limit of $\{u^{i,s}\}$. Then both sequences $\{u^{i,s}\}$ and $\{\bar{u}^{i,s}\}$ converge to an element $u^* \in U^*$ (strongly) in V.

If $dimV_1 < \infty$, condition (25.42) follows from the weak convergence of $\{u^{i,s}\}$.

Now we formulate some results about the rate of convergence for the basic variant of the MSR-method that corresponds to $H = V = V_1, \mathcal{B} = 0$. We need the following additional

Assumption 3 *There exists a constant $d > 0$ such that, for each $u \in Q$ and each $y \in \mathcal{T}u$, the inequality*

$$inf_{v\in U^*\cap S_{r^*}}\ \langle y, u - v\rangle \ge d\|u - u^*(u)\|^l \tag{25.43}$$

holds with $u^*(u) = argmin_{w\in U^*\cap S_{r^*}} \|u - w\|$ and $l = 1$ or $l = 2$.

Similar conditions are made in [28] (Section 2) and [35] (Section 3) to study the rate of convergence of the proximal point method with exact data.

Let us define by $\bar{s}(i)$ the largest integer for the value $16r^2\delta_i^{-2} + 1$, and let

$$\begin{aligned}
\rho_{i,s} &= dist^2\left(u^{i,s}, U^* \cap S_{r^*}\right), \\
\gamma_i &= 6r\varepsilon_i + 4r\varphi_i + \frac{4}{\chi_i}\left[r\sigma_i + (\mu(r) + 2dr)\varphi_i + 2dr\varepsilon_i\right], \\
c_i &= \rho_{1,0}\prod_{k=1}^{i}\left(1 + \frac{2d}{\chi_k}\right)^{-s(k)+\frac{1}{2}}.
\end{aligned}$$

Theorem 4 *Let the Assumptions 2 (a), (d'), (e') (see Remark 2) and Assumption 3 with $l = 2$ be fulfilled. Moreover, suppose that the controlling parameters of the MSR-method satisfy the relations (25.24),*

$$\gamma_i \le c_{i-1}\left(1 + \frac{2d}{\chi_i}\right)^{-\bar{s}(i)+1}\left[\left(1 + \frac{2d}{\chi_i}\right)^{1/2} - 1\right] \tag{25.44}$$

and

$$-(\delta_i - \varepsilon_i)^2 + \frac{4}{\chi_i}(r\sigma_i + \mu(r)\varphi_i) + 4r\varepsilon_i + max\left\{\frac{\delta_i^2}{2}, \left(2 + \frac{8d}{\chi_i}\right) r\varepsilon_i + \left(4 + \frac{8d}{\chi_i}\right) r\varphi_i\right\} < 0. \qquad (25.45)$$

Then, for each i and $0 \le s < s(i)$, the estimate

$$\rho_{i,s} \le \rho_{1,0}\left(1 + \frac{2d}{\chi_i}\right)^{-s} \prod_{k=1}^{i-1}\left(1 + \frac{2d}{\chi_k}\right)^{-s(k)+\frac{1}{2}} \qquad (25.46)$$

is valid.

Remark 5 *If Problem (25.1) is uniquely solvable, estimate (25.46) shows that $\{u^{i,s}\}$ converges to u^* linearly with the factor $(1 + \frac{2d}{\bar{\chi}})^{-1/4}$. In the general case, analogous estimates can be proved for the distance of $u^{i,s}$ to U^*, and using the proof of Theorem 14.6 in [16], linear convergence of $\{u^{i,s}\}$ to an element $u^* \in U^*$ can be established, too.*

Theorem 5 *Let the Assumptions 2(a), (b), (d'), (e'), Assumption 3 with $l = 1$ and the relations (25.23), (25.24) be fulfilled. Then there exists an index i_0 such that the estimate*

$$\rho_{i,s} \le 3\left(1 + \frac{2r\bar{\chi}}{d}\right)\varepsilon_i + 2\left(1 + \frac{2r\bar{\chi}}{d} + \frac{2\mu(r)}{d}\right)\varphi_i + \frac{4r}{d}\sigma_i \qquad (25.47)$$

holds for $i \ge i_0$ and $1 \le s \le s(i)$.

For the basic variant of the MSR-method this is a generalization of Theorem 3 in [35] about "the finite convergence" of the exact proximal point method.

References

[1] P. ALART AND B. LEMAIRE, "Penalization in non-classical convex programming via variational convergence", *Math. Programming* 51 (1991) 307-331.

[2] A.S. ANTIPIN, "On a method for convex programs using a symmetrical modification of the Lagrange function", *Ekonomika i Mat. Metody* 12 (1976) 1164-1173, (in Russian).

[3] A. AUSLENDER, J.P. CROUZEIX AND P. FEDIT, "Penalty proximal method in convex programming", *JOTA* 55 (1987) 1-21.

[4] G. CHEN AND M. TEBOULLE, "A proximal-based decomposition method for convex minimization problems", *Math. Programming* 64 (1994) 81-101.

[5] J. ECKSTEIN AND D.P. BERTSEKAS, "On the Douglas-Rachford splitting method and the proximal point algorithm for maximal monotone operators", *Math. Programming* 55 (1992) 293-318.

[6] G. FICHERA, "Problemi elastostatici con vincoli unilaterali: il problema di Signorini con ambique condizioni al contorno", *Mem. Accad. Naz. Lincei*, s. VIII, 7 (1964) 91-140.

[7] G. FICHERA, "Boundary value problems of elasticity with unilateral constraints", In: *Encyclopedia of Physics*, (ed. by S. Flügge), vol. VI a/2, Springer, Berlin, 1972.

[8] M. FUKUSHIMA, "The primal Douglas-Rachford splitting algorithm for a class of monotone mappings with applications to the traffic equilibrium problem", *Math. Programming* 72 (1996) 1-15.

[9] R. GLOWINSKI, *Numerical Methods for Nonlinear Variational Problems*, Springer, New York - Berlin - Heidelberg - Tokyo, 1984.

[10] R. GLOWINSKI, J.L. LIONS AND R. TREMOLIERES, *Numerical Analysis of Variational Inequalities*, North-Holland, Amsterdam, 1981.

[11] CU D. HA, "A generalization of the proximal point algorithm", *SIAM J. Control and Optimization* 28 (1990) 503-512.

[12] I. HLAVAČEK, J. HASLINGER, I. NEČAS AND J. LOVIŠEK, *Numerical Solution of Variational Inequalities*, Springer, Berlin, 1988.

[13] R. HETTICH, A. KAPLAN AND R. TICHATSCHKE, "Regularized penalty methods for ill-posed optimal control problems with elliptic equations", Part I and II, *Control and Cybernetics* 26 (1997).

[14] S. IBARAKI, M. FUKUSHIMA AND T. IBARAKI, "Primal-dual proximal point algorithm for linearly constrained convex programming problems", *Comp. Optim. and Applications* 1 (1992) 207-226.

[15] A. KAPLAN, "Algorithm for convex programming using a smoothing for exact penalty functions", *Sibirski Mat. Journal* 23 (1982) 307-316 (in Russian).

[16] A. KAPLAN AND R. TICHATSCHKE, *Stable Methods for Ill-posed Variational Problems - Prox-Regularization of Elliptic Variational Inequalities and Semi-Infinite Problems,* Akademie Verlag, Berlin, 1994.

[17] A. KAPLAN AND R. TICHATSCHKE, "Path-following proximal approach for solving ill-posed convex semi-infinite programming problems", *JOTA* 90 (1996) 113-137.

[18] A. KAPLAN AND R. TICHATSCHKE, "Prox-Regularization and solution of ill-posed variational inequalities", *Applications of Mathematics* 42 (1997), No.2, 111-145.

[19] A. KAPLAN AND R. TICHATSCHKE, "Proximal penalty method for ill-posed parabolic optimal control problems", *Proc. of the Intern. Conf. on Control and Estimation of Distributed Parameter Systems,* Vorau (Oesterreich), July 14-20, 1996.

[20] A. KAPLAN AND R. TICHATSCHKE, "Multi-step proximal method for variational inequalities with monotone operators", in: *Recent Advances in Optimization,* P. Gritzmann, R. Horst, E. Sachs, R. Tichatschke (eds.), Lecture Notes in Economics and Mathematical Systems, Springer 1997, pp. 138-153.

[21] A. KAPLAN AND R. TICHATSCHKE, "Stable methods for variational inequalities with set-valued monotone operators", *Forschungsbericht Mathematik/Informatik* Nr. 97-07, Universität Trier, pp. 26.

[22] K. C. KIWIEL, "Proximal level bundle methods for convex nondifferentiable optimization, saddle-point problems and variational inequalities", *Math. Programming*, 69 (1995) 89-109.

[23] B. LEMAIRE, "Coupling optimization methods and variational convergence", *ISNM* 84 (1988) 163-179.

[24] B. LEMAIRE, "The proximal algorithm", *ISNM* 87 (1989) 73-87.

[25] C. LEMARÉCHAL AND C.A. SAGASTIZÁBAL, "An approach to variable metric bundle methods", In: *Lecture Notes in Control and Information Sciences*, No. 197, Springer, Berlin, 1994, 144-162.

[26] J.L. LIONS, *Quelques Méthodes de Résolution des Problèmes aux Limites non Linéaires,* Dunod, Gauthier-Villars, Paris, 1969.

[27] J.L. LIONS AND G. STAMPACCHIA, "Variational inequalities", *Comm. Pure Appl. Math.* 20 (1967) 493-519.

[28] F. J. LUQUE, "Asymptotic convergence analysis of the proximal point algorithms", *SIAM J. Control and Optimization* 22 (1984) 277-293.

[29] B. MARTINET, "Régularisation d'inéquations variationelles par approximations successives", *RIRO* 4 (1970) 154-159.

[30] R. MIFFLIN, "A quasi-second-order proximal bundle algorithm", *Math. Programming* 73 (1996) 51-72.

[31] U. MOSCO, "Convergence of convex sets and of solutions of variational inequalities", *Advances in Mathematics* 3 (1969) 510-585.

[32] P.D. PANAGIOTOPOULOS, *Inequality Problems in Mechanics and Applications. Convex and Nonconvex Energy Functionals.* Birkhäuser, Boston, 1985.

[33] B. T. POLYAK, *Introduction to Optimization,* Optimization Software, Inc. Publ. Division, New York, 1987.

[34] R. T. ROCKAFELLAR, "On the maximality of sums of nonlinear monotone operators", *Trans. Amer. Math. Soc.* 149 (1970) 75-88.

[35] R. T. ROCKAFELLAR, "Monotone operators and the proximal point algorithm", *SIAM J. Control and Optim.* 14 (1976) 877-898.

[36] R. T. ROCKAFELLAR, "Augmented Lagrange multiplier functions and applications of the proximal point algorithm in convex programming", *Math. Oper. Res.* 1 (1976) 97-116.

[37] H. SCHMITT, "Normschwächere Prox-Regularisierungen", PhD-Thesis, University of Trier, 1996.

[38] J.E. SPINGARN, "Partial inverse of a monotone operator", *Applied Math. and Optim.* 10 (1983) 247-265.

[39] J.E. SPINGARN, "Application of the method of partial inverses to convex programming: Decomposition", *Math. Programming* 32 (1985) 199-223.

[40] F. P. VASIL'JEV, *Methods for Solving Extremal Problems*, Nauka, Moscow, 1981 (in Russian).

26 MAXIMAL MONOTONICITY, SUBDIFFERENTIALS AND GENERALIZATIONS

Mirosław Przeworski

Institute of Mathematics, Technical University of Radom,
Radom 26–600,
ul. Malczewskiego 20a,
Poland

mprzewor@kiux.man.radom.pl

Dariusz Zagrodny

Institute of Mathematics
Technical University of /L/od/z,
/L/od/z 90–924,
ul./Xwirki 36,
Poland

zagrodny@ck-sg.p.lodz.pl

26.1 INTRODUCTION

In 1966 R.T. Rockafellar [14] exploring the operator of subdifferential of convex function on a Banach space revealed that the operator is maximal monotone i.e.

$$\text{if for every } (x, x^*) \in \text{graph } \partial f \text{ it holds } \langle x^* - q^*, x - q \rangle \geq 0, \text{ then } q^* \in \partial f(q).$$

R.P. Gilbert et al. (eds.), From Convexity to Nonconvexity, 363–380.

Since R.T. Rockafellar revealed the maximal monotonicity of subdifferentials several independent proofs of this fact have been provided, see [18] for a detailed discusson. In 1994 S.Simons (see [17, 18]) asked to what extent we can generalize the above implication. Is it still true if we replace q or q^* by sets? Is it true if we take arbitrary maximal monotone operator instead of the subdifferential of convex function? Let us recall one of his questions (see [17, 18]).

Q–Q* QUESTION. Let Q be a nonempty weakly compact convex subset of $\mathbb{E}$, Q^* be a nonempty weak* compact convex subset of $\mathbb{E}^*$ and

$$\text{for every } (z, z^*) \in \operatorname{graph} \partial f \text{ there exists } (q, q^*) \in Q \times Q^* \text{ such that } \langle x^* - q^*, x - q\rangle \geq 0. \quad (26.1)$$

Does it imply that

$$\operatorname{graph} \partial f \cap (Q \times Q^*) \neq \emptyset ? \quad (26.2)$$

In papers [17, 18] S. Simons gave an answer in the affirmative to the question in the case when Q or Q^* is singleton. D. Zagrodny [22] showed that for the real line, $\mathbb{E} = \mathbb{R}$ **Q–Q* QUESTION** has the answer in the affirmative, but for the spaces of higher dimension than one the solution of this problem require an additional assumption (see also [12]), namely

$$\text{there exists } (q_0, q_0^*) \in Q \times Q^* \text{ such that } (Q - q_0) \subset (Q^* - q_0^*)^{\perp}. \quad (26.3)$$

It is an interesting problem whether the question has the answer in the affirmative for larger class of operators than that of subdifferentials of convex functions. In section 4 it is shown that if we replace ∂f by a maximal monotone operator then (26.1) implies (26.2) whenever $\mathbb{E} = \mathbb{R}^n$. Besides finding answers to Simons' questions is of intrest in itself, the implication is a good tool to accomplish revealing answers to basic questions concerning maximal monotone operators. Namely in [7] the following unanswrered problems were refered: Is the sum of two maximal monotone operators T, S with dom $T \cap$ int dom $\neq \emptyset$ still a maximal monotone operator? Is the set cl dom T convex ? These questions have a long history. Many useful tools were worked out while giving partial answers to them. However these questions have been unanswered in nonreflexive Banach spaces up till now [23]. It should be also mentioned at this point that large parts of nonlinear analysis are devoted to the theory of monotone operators and its applications , see [5, 13, 24]. Because of this we should expect that the implication can be handy at dealing with other problems, not necessarily involving maximal monotone operators.

More general class of functions than the convex ones is the class of quasiconvex functions. It is well known, that a subdifferential operator of quasiconvex function can not preserve the properties of the subdifferential operator of convex function (see [2, 10, 11, 13]) or even the maximal monotone one e.g. the function $f : \mathbb{R} \longrightarrow \mathbb{R}$, where $f(x) = 2^{-1}x^2$, is quasiconvex, with $\partial f(x) = x$ for every $x \in \mathbb{R}$, but its subdifferential is not maximally quasimonotone operator . This suggests the question what happens with the implication (1.1) $\Rightarrow$ (1.2), when we addmit f to be quasiconvex. Herein we give a solution to it too (see section 3). We show that **Q–Q* QUESTION** has an answer in the affirmative when $\mathbb{E}$ is a real Banach space and f is a quasiconvex lower semicontinuous function. One should strongly emphasize that maximal quasimonotone operators may differ from maximal monotone one. The essential feature of maximal quasimonotone operators is that for all $x \in \text{dom } T$ the set $T(x)$ is a convex closed cone in $\mathbb{E}^*$. Moreover, if T is a maximal quasimonotone operator, then for every $z \in \mathbb{E}$ we have $0 \in T(z)$. Thus analogous theorem to theorem 4.3 (see section 4) is trivial for maximal quasimonotone operators. On the other hand we can find a quasimonotone operator (even with closed graph) satisfying (1.1), for which (1.2) is not valid. Supposedly subdifferential operators of quasiconvex functions constitutes the largest class of quasimonotone operators (see [10]), for which the exploration of the implication (26.1) $\Longrightarrow$ (26.2) is justyfied.

Researches of nondifferentiable objects have made many attempts to generalize the differential notion in order to desciribe behavior of nondifferentiable functions. Because of many definitions of subdifferential we observe, in the last years, tendency to gather all essential properties of these notions into an axiom system (see [2, 3, 4, 20] and their references). Unfortunatelly, the known systems of axioms are incomplete to explore such basic functions like compositon a convex function with another operator (e.g. linear and continuous one, compare with [2, 3, 4, 20]). This lack does not allow us to deal with problems where such objects are involved. This gap is supplemented herein.

26.2 SOME BASIC FACTS AND DEFINITIONS

Let $\mathbb{E}$ be a real Banach space with the topological dual $\mathbb{E}^*$. For a function $f : \mathbb{E} \longrightarrow \mathbb{R} \cup \{+\infty\}$, the effective domain of f is denoted by $\text{dom } f := \{x \in \mathbb{E} \mid f(x) \in \mathbb{R}\}$. Following [10] we recall that f is quasiconvex if for every

$$x, y \in \text{dom } f,\ w \in [x, y] \text{ we have } f(w) \le \max\{f(x), f(y)\},$$

where $[x, y] := \{z \in \mathbb{E} \mid z = \lambda x + (1 - \lambda)y \text{ for some } \lambda \in [0, 1]\}$. Following [1] we define a lower section of f

$$S(f, \alpha) := \{x \in \mathbb{E} \mid f(x) \le \alpha\}, \text{ for every } \alpha \in \mathbb{R}.$$

These sets are convex and closed whenever f is lsc and quasiconvex. Let Q be a nonempty convex and weakly compact (see [8]) subset of $\mathbb{E}$. Let us put $d_Q^2(x) := \left(\inf_{q \in Q} \| x - q \|\right)^2$, where $\| \cdot \|$ is the norm in $\mathbb{E}$. The function d_Q^2 is continuous convex and locally lipschitzian on $\mathbb{E}$.

Below we give axioms, which should satisfy a subdifferential operator. The system of conditions $(P1)$ – $(P3)$ is a slight modification of that from [3], but those axioms are not enough comprehensive to treat problems with composite functions. Because of that we supplement them with an additional condition $(P4)$.

Definition 1. *Let $f : \mathbb{E} \longrightarrow \mathbb{R} \cup \{+\infty\}$ be a proper lower semicontinuous function. We call subdifferential, denoted by ∂, any operator which associates* *a subset* $\partial f(x)$ *of* $\mathbb{E}^*$ *to any lower semicontinuous* $f : \mathbb{E} \longrightarrow \mathbb{R} \cup \{+\infty\}$ *and any* $x \in \mathbb{E}$, *and satisfies the following properties:*

(P1) $\partial f(x) = \{x^* \in \mathbb{E}^* \mid \langle x^*, y - x\rangle \leq f(y) - f(x),$ *for every* $y \in \mathbb{E}\}$ *whenever* f *is convex and* $x \in \operatorname{dom} f$*;*

(P2) $0 \in \partial f(x)$ *whenever* $x \in \operatorname{dom} f$ *is a local minimum of* f*;*

(P3) $\partial(f + g)(x) \subseteq \partial f(x) + \partial g(x)$ *whenever* g *is a real-valued convex function (is continuous);*

(P4) $\partial(f \circ A)(x) = L^* \partial f(A(x))$ *whenever* A *is any affine operator* $A(x) = L(x) + y$, *with* $y \in \mathbb{E}$, $L \in$ *Isom* $(\mathbb{F}, \mathbb{E})$, $f(A(x)) \in \mathbb{R}$, *where Isom* $(\mathbb{F}, \mathbb{E})$ *stands for the set of all isomorphisms (linear, continuous, injective and surjective operators) from* $\mathbb{F}$ *onto* $\mathbb{E}$.

Many well known subdifferentials satisfy the above properties (see [3]). Axioms $(P1) - (P3)$ ensure the existence of a mean value theorem like in [21] (see also [3, 20]) for example. Thus they are sufficient to get it. However, it is an open problem if they are minimal system of axioms to preserve it i.e *if* ∂ *is an operator such that the mean value theorem holds for every lower semicontinuous function* f, *do* $(P1) - (P3)$ *hold?*

Definition 2. *For any proper quasiconvex and lsc function* $f : \mathbb{E} \longrightarrow \mathbb{R} \cup \{+\infty\}$ *and any operator* ∂ *satisfying conditions* $(P1) - (P4)$ *we define*

$$\operatorname{graph} \partial f := \{(x, x^*) \in \mathbb{E} \times \mathbb{E}^* \mid x^* \in \partial f(x)\}.$$

Below we show that the implication (1.1) $\Rightarrow$ (1.2) holds true even if we translate the sets Q and Q^*. This was proved in [22] for proper convex and lsc functions. The proof in the case of proper quasiconvex lsc functions is similar to that. However for the sake of the convenience of reader we provide one.

Lemma 26.4. *Let* $\mathbb{E}$ *be a real Banach space and* $Q \subset \mathbb{E}, Q^* \subset \mathbb{E}^*$ *be convex nonempty weakly compact and weak* compact, respectively. Let us assume that for every proper quasiconvex lsc function* $\psi : \mathbb{E} \longrightarrow \mathbb{R} \cup \{+\infty\}$ (1.1) *implies* (1.2). *Then for every* $e \in \mathbb{E}$ *and* $e^* \in \mathbb{E}^*$ *and every proper quasiconvex lsc function* $\varphi : \mathbb{E} \longrightarrow \mathbb{R} \cup \{+\infty\}$ (1.1) *implies* (1.2), *if we take* $Q - e$ *and* $Q^* - e^*$ *instead of* Q *and* Q^*, *respectively.*

Proof. Let us fix Q, Q^* and e, e^*. Assume that for every $\psi : \mathbb{E} \longrightarrow \mathbb{R} \cup \{+\infty\}$ which is proper quasiconvex lsc (1.1) implies (1.2). Let us fix a proper quasiconvex lsc function $\varphi : \mathbb{E} \longrightarrow \mathbb{R} \cup \{+\infty\}$ such that

$$\text{for every } (z, z^*) \in \operatorname{graph} \partial\varphi \text{ there exists } (q, q^*) \in Q \times Q^* \text{ such that } \langle z^* - (q^* - e^*), z - (q - e)\rangle \geq 0. \quad (26.5)$$

Let us define

$$\psi(z) := \varphi(z - e) + \langle e^*, z\rangle \text{ for every } z \in \mathbb{E}.$$

Our proof starts with the observation that

$$x^* \in \partial\psi(x) \Leftrightarrow x^* \in \partial\big(\varphi(\cdot - e) + \langle e^*, \cdot\rangle\big)(x) \text{ for every } x \in \mathbb{E}. \tag{26.6}$$

It follows from $(P3)$ that

$$\partial\big(\varphi(\cdot - e) + \langle e^*, \cdot\rangle\big)(x) \subseteq \partial\varphi(\cdot - e) + e^*. \tag{26.7}$$

We have also $\partial\varphi(\cdot - e)(x) = \partial\big(\varphi(\cdot - e) + \langle e^*, \cdot\rangle - \langle e^* . \cdot\rangle\big)(x)$ thus by $(P3)$

$$\partial\varphi(\cdot - e)(x) \subseteq \partial\big(\varphi(\cdot - e) + \langle e^*, \cdot\rangle\big)(x) - e^*. \tag{26.8}$$

It follows from (26.7) and (26.8) that

$$\partial\varphi(\cdot - e) + e^* \subseteq \partial\big(\varphi(\cdot - e) + \langle e^*, \cdot\rangle\big)(x) \subseteq \partial\varphi(\cdot - e)(x) + e^*,$$

thus

$$\partial\varphi(\cdot - e)(x) + e^* = \partial\big(\varphi(\cdot - e) + \langle e^*, \cdot\rangle\big)(x) \text{ for every } x \in \mathbb{E}.$$

By (26.6) we get

$$x^* \in \partial\psi(x) \Leftrightarrow x^* - e^* \in \partial\varphi(\cdot - e)(x) \text{ for every } x \in \mathbb{E}.$$

By $(P4)$ we have $\partial\varphi(\cdot - e)(x) = \partial\varphi(x - e)$. Hence for $z := x - e$ and $z^* := x^* - e^*$ by (26.5) we have $< x^* - q^*, x - q > \geq 0$ for some $q \in Q$ and $q^* \in Q^*$, so (26.1) is satisfied for ψ. Thus $\partial\psi(\bar{q}) \cap Q^* \neq \emptyset$ for some $\bar{q} \in Q$. Since $\partial\varphi(\cdot - e)(\bar{q}) = \partial\varphi(\bar{q} - e)$, so $\big(\partial\varphi(\bar{q} - e) + e^*\big) \cap Q^* \neq \emptyset$. It follows from the above that

$$\text{graph } \partial\varphi \cap \big((Q - e) \times (Q^* - e^*)\big) \neq \emptyset,$$

which completes the proof. □

The next two celebrated theorems concern variational inequalities. We provide them following [1].

THEOREM 26.2.1 (Ky Fan's inequality). *Suppose that E is a convex, compact subset of $\mathbb{R}^n$ and that f is a function from $\mathbb{R}^n \times \mathbb{R}^n$ to $\mathbb{R}$ satisfying*

(i) $\forall y \in E$, $x \longrightarrow f(x, y)$ is lower semicontinuous

(ii) $\forall x \in E$, $y \longrightarrow f(x, y)$ is concave.

Then there exists $\bar{x} \in E$ such that

$$\sup_{y \in E} f(\bar{x}, y) \leq \sup_{x \in E} f(x, x). \qquad \square$$

THEOREM 26.2.2 (von Neumann's inequality). *Suppose that $f : \mathbb{R}^n \times \mathbb{R}^n \longrightarrow \mathbb{R}$, $E, F \subset \mathbb{R}^n$ and*

(i) E is convex and compact

(ii) $\forall y \in F$, $x \longrightarrow f(x, y)$ is convex and l.s.c. and that

(iii) F is convex and compact

(iv) $\forall x \in E$, $y \longrightarrow f(x,y)$ *is concave and u.s.c.*

Then there exists a saddle point $(\bar{x},\bar{y}) \in E \times F$ *i.e.*

$$\forall x \in E,\ y \in F \quad f(\bar{x},y) \leq f(\bar{x},\bar{y}) \leq f(x,\bar{y}). \qquad \square$$

Ending this section we recall the Ekeland Variational Principle (see [6]).

THEOREM 26.2.3. *Assume that* f *is a proper lsc function on a real Banach space* $\mathbb{E}$ *which is bounded from below. Suppose that* $\varepsilon > 0$ *and that*

$$f(x_0) \leq \inf_{\mathbb{E}} f + \varepsilon$$

for some $x_0 \in \mathbb{E}$. *Then for any* $0 < \lambda$ *there exists a point* $x_\varepsilon \in \mathrm{dom}\ f$ *such that*

$$f(x_\varepsilon) + \frac{\varepsilon}{\lambda} \parallel x_\varepsilon - x_0 \parallel \leq f(x_0), \tag{26.9}$$

$$f(x_\varepsilon) < f(x) + \frac{\varepsilon}{\lambda} \parallel x_\varepsilon - x \parallel \text{ for every } x \in \mathbb{E} \setminus \{x_\varepsilon\}. \tag{26.10}$$

26.3 A SOLUTION OF THE GENERALIZED SIMONS' PROBLEM

In this section we present a solution to Simons' problem in the case, when ∂ is any subdifferential operator satisfying conditions $(P1) - (P4)$ and $f : \mathbb{E} \longrightarrow \mathbb{R} \cup \{+\infty\}$ is a proper quasiconvex lsc function with $Q^* = \{0\}$. The case when $Q = \{0\}$ and Q^* is a nonempty convex weak* compact subset of $\mathbb{E}^*$ can be done similarly. Without loss of generality we can assume that $Q = \{0\}$ or $Q^* = \{0\}$, if not we apply Lemma 26.4 to reduce our cosiderations to this case. Let us recall that graph $\partial f = \{(x,x^*) \in \mathbb{E} \times \mathbb{E}^* \mid x^* \in \partial f(x)\}$.

THEOREM 26.3.1. *Let* Q *be a nonempty convex weakly compact subset of a real Banach space* $\mathbb{E}$. *Let* $f : \mathbb{E} \longrightarrow \mathbb{R} \cup \{+\infty\}$ *be a proper quasiconvex lsc function, which satisfies the following condition*

$$\text{for every } (x,x^*) \in \text{graph } \partial f \quad \text{there exists } q \in Q \text{ such that } < x^*, x - q > \geq 0. \tag{26.11}$$

Then

$$\text{graph } \partial f \cap (Q \times \{0\}) \neq \emptyset. \tag{26.12}$$

Proof. Let us fix $\bar{x} \in \mathrm{dom}\ f$ such that $\bar{x} \in \mathrm{dom}\ f \cap Q$ if $\mathrm{dom}\ f \cap Q \neq \emptyset$ otherwise take any $\bar{x}$ for which f is finite. Put $Q_1 := \mathrm{conv}\,(Q \cup \{(\bar{x}\})$. This set is weakly compact convex nonempty and $\mathrm{dom}\ f \cap Q_1 \neq \emptyset$. Let us define $r := \inf_{Q_1} f$. Of course $r < +\infty$. Let $\{r_n\}_{n=1}^{\infty} \subset \mathbb{R}$, $r_n \searrow r$ and $B(n) := S(f, r_n) \cap Q_1$ for every $n \in \mathbb{N}$. Because of the lower semicontinuity and quasiconvexity of f the sets $B(n)$ are convex and closed for every $n \in \mathbb{N}$. They forming a nested family of weakly closed subsets of the weakly compact set Q_1. Hence $\bigcap_{n=1}^{\infty} B(n) \neq \emptyset$.

Thus there exists $\bar{q} \in Q_1$ such that $\bar{q} \in \bigcap_{n=1}^{\infty} B(n)$, which implies that $f(\bar{q}) = r$ and $r > -\infty$. Thus

$$\bar{q} \in \arg\min_{Q_1} f. \tag{26.13}$$

If $S(f, \alpha) = \emptyset$ for every $\alpha \in (-\infty, r)$, then $\inf_{\mathbb{E}} f = \inf_{Q_1} f$. By (26.13) there is $\bar{q} \in \arg\min f$, so by $(P2)$ we get

$$0 \in \partial f(\bar{q}). \tag{26.14}$$

If $\bar{q} \notin Q$, so by the definitions of Q_1 and $\bar{x}$ we have $f_{|Q} = \infty$. The function $f + d_Q^2$ is bounded from below on $\mathbb{E}$ since f is. Hence the Ekeland Variational Principle ensures that for every $n \in \mathbb{N}$ there exists $\bar{u_n} \in \mathbb{E} \setminus Q$ (since $f_{|Q} = \infty$) such that $f(\bar{u_n}) + d_Q^2(\bar{u_n}) \leq f(\bar{q}) + d_Q^2(\bar{q})$ and

$$f(\bar{u_n}) + d_Q^2(\bar{u_n}) \leq f(x) + d_Q^2(x) + n^{-1} \| x - \bar{u_n} \| \text{ for every } x \in \mathbb{E}.$$

By $(P1) - (P3)$ we obtain

$$0 \in \partial f(\bar{u_n}) + \partial d_Q^2(\bar{u_n}) + n^{-1} B(0,1).$$

Hence there are $\bar{u_n}^* \in \partial f(\bar{u_n})$, $\bar{v_n}^* \in 2d_Q(\bar{u_n})\partial d_Q(\bar{u_n})$ and $x_n^* \in B^*(0,1)$ such that

$$0 = \bar{u_n}^* + \bar{v_n}^* + n^{-1} x_n^*.$$

By (26.11) for some $\bar{q_n} \in Q$ we have $0 \leq \langle \bar{u_n}^*, \bar{u_n} - \bar{q_n} \rangle$. On the other hand $(P1)$ yields $\langle \bar{v_n}^*, \bar{u_n} - \bar{q_n} \rangle \geq 2d_Q^2(\bar{u_n})$. We have also

$$0 = \langle \bar{u_n}^*, \bar{u_n} - \bar{q_n} \rangle + \langle \bar{v_n}^*, \bar{u_n} - \bar{q_n} \rangle + n^{-1} \langle x_n^*, \bar{u_n} - \bar{q_n} \rangle.$$

Because of the inequalities $f(\bar{u_n}) + d_Q^2(\bar{u_n}) \leq f(\bar{q}) + d_Q^2(\bar{q})$ we get $d_Q^2(\bar{u_n}) \leq d_Q^2(\bar{q})$ for every $n \in \mathbb{N}$ (keep in mind that a global minimum of f is attainned at $\bar{q}$). Thus we have

$$n^{-1}(\text{diam } Q + d_Q(\bar{q})) \geq n^{-1} \| \bar{u_n} - \bar{q_n} \| \geq 2d_Q^2(\bar{u_n}) \text{ for every } n \in \mathbb{N}.$$

Hence $d_Q^2(\bar{u_n}) \to 0$ whenever $n \to \infty$. On the other hand each $\bar{u_n}$ belongs to $S(f, f(\bar{q}) + d_Q^2(\bar{q}))$. Of course $S(f, f(\bar{q}) + d_Q^2(\bar{q}))$ is nonempty convex closed, so weakly closed. We have also

$$S(f, f(\bar{q}) + d_Q^2(\bar{q})) \cap Q = \emptyset.$$

We can strongly separate them (see [8]). There are $\delta > 0$ and a functional $p^* \in B(0,1) \subset \mathbb{E}^*$ such that $\langle p^*, \bar{u_n} - \bar{q_n} \rangle \geq \delta > 0$ for every $n \in \mathbb{N}$, which implies $\| \bar{u_n} - \bar{q_n} \| \geq \delta$ for every $n \in \mathbb{N}$, a contradiction. Thus (26.14) holds with $\bar{q} \in Q$. We are done.

Now consider only these $\alpha \in (-\infty, r)$ for which $S(f, \alpha) \neq \emptyset$. By (26.13) we have $S(f, \alpha) \cap Q_1 = \emptyset$. Because of the closedness and convexity of $S(f, \alpha)$ and, weak compactness and convexity of Q_1 we can apply the strong separation theorem (see [8]) to these sets. Similarly as above we can find $0 < \gamma$ such that

$$0 < \gamma \leq \inf\{\| x - q \| \mid q \in Q_1 \text{ and } x \in S(f, \alpha)\},$$

from which we infer that

$$\inf_{Q_1} d(q, S(f,\alpha)) > 0 \text{ for every } \alpha \in (-\infty, r), \tag{26.15}$$

where $d(\cdot, S(f,\alpha))$ is the distance function from the set $S(f,\alpha)$. Now we prove that

$$\text{there exists } \varepsilon_0 > 0 \text{ such that } f \text{ is bounded from below on the set } Q' := B(Q_1, \varepsilon_0), \tag{26.16}$$

where $B(Q_1, \varepsilon_0) := \{x \in \mathbb{E} \mid d(x, Q_1) \leq \varepsilon_0\}$. For a time being suppose that (26.16) is false. Then for every $\varepsilon > 0$ the function f is unbounded from below on the sets $B(Q_1, \varepsilon)$. Let $\{\varepsilon_n\}_{n=1}^{\infty} \subset \mathbb{R}^+$, $\varepsilon_n \searrow 0$ and $Q(n) := B(Q_1, \varepsilon_n)$. There exists a sequences $\{e_n\}_{n=1}^{\infty} \subset \mathbb{E}$ such that $e_n \in Q(n)$ and $f(e_n) < -n$ for all $n \in \mathbb{N}$. We are able to choose $k \in \mathbb{N}$ such that $\inf_{Q_1} f = r > -k$. Hence $\{e_n\}_{n=k+1}^{\infty} \subset S(f, -k)$. We observe that $d(e_n, Q_1) \leq \varepsilon_n$, hence for some sequence $\{q_n\}_{n=1}^{\infty} \subset Q_1$ we have $\| e_n - q_n \| \leq 2\varepsilon_n$ for all $n >$k. Thus

$$d(q_n, S(f,-k)) \leq \| e_n - q_n \| + d(e_n, S(f,-k)) \leq 2\varepsilon_n$$

for every $n > k$, because $e_n \in S(f, -k)$. It follows from the above that

$$\inf_{q \in Q_1} d(q, S(f,-k)) = 0$$

which contradicts (26.15). Thus (26.16) is true. Now we define $f_n : \mathbb{E} \longrightarrow \mathbb{R} \cup \{+\infty\}$ by

$$f_n(x) = f(x) + n d_{Q_1}^2(x) \text{ for every } n \in \mathbb{N} \text{ and } x \in \mathbb{E}.$$

These functions are lsc for every $n \in \mathbb{N}$. Let us apply the Ekeland Variational Principle to the functions f_n on the set Q' the existence of which is guaranteed by (26.16). For every $n \in \mathbb{N}$ and $\varepsilon > 0$ with $x_0 := \bar{q} \in \arg\min_{Q_1} f$ and $\lambda := 1$ there is $x_\varepsilon^n \in Q'$ such that

$$f_n(x_\varepsilon^n) + \varepsilon \| \bar{q} - x_\varepsilon^n \| \leq f_n(\bar{q}) \tag{26.17}$$

and

$$f_n(x_\varepsilon^n) < f_n(x) + \varepsilon \| x - x_\varepsilon^n \| \text{ for every } x \in Q' \setminus \{x_\varepsilon^n\}. \tag{26.18}$$

It follows from (26.17) that $\inf_{Q'} f + n d_{Q_1}^2(x_\varepsilon^n) \leq f(\bar{q})$. Thus

$$n d_{Q_1}^2(x_\varepsilon^n) \leq \inf_{Q_1} f - \inf_{Q'} f,$$

so $d_{Q_1}^2(x_\varepsilon^n) \to 0$ whenever $n \to \infty$, and $x_\varepsilon^n \in \operatorname{int} Q'$ for n large enough and every ε. It follows from (26.18) that for some n_0 and every $\varepsilon > 0$

$$x_\varepsilon^n \in \arg\min \left(f(\cdot) + n d_{Q_1}^2(\cdot) + \varepsilon \| \cdot - x_\varepsilon^n \|\right) \text{ and } x_\varepsilon^n \in \operatorname{int} Q' \quad \text{for every } \varepsilon > 0 \text{ and } n \geq n_0. \tag{26.19}$$

By $(P1)-(P3)$ we obtain

$$0 \in \partial f(x_\varepsilon^n) + \partial(nd_{Q_1}^2)(x_\varepsilon^n) + \partial(\varepsilon \parallel \cdot - x_\varepsilon^n \parallel)(x_\varepsilon^n) \quad \text{for every } n \geq n_0 \text{ and } \varepsilon > 0.$$

Let $x_\varepsilon^* \in \partial f(x_\varepsilon^{n_0})$, $n_0 y^* \in \partial(n_0 d_{Q_1}^2)(x_\varepsilon^{n_0})$ and $\varepsilon z^* \in \partial(\varepsilon \parallel \cdot - x_\varepsilon^{n_0} \parallel)(x_\varepsilon^{n_0})$ be such that

$$0 = x_\varepsilon^* + n_0 y^* + \varepsilon z^*. \tag{26.20}$$

Then $0 = \langle x_\varepsilon^* + n_0 y^* + \varepsilon z^*, q - x_\varepsilon^{n_0} \rangle$ for every $q \in Q$. On the other hand for $q_\varepsilon \in Q$ the existence of which is ensured by (26.11) we obtain $0 = \langle x_\varepsilon^* + n_0 y^* + \varepsilon z^*, q_\varepsilon - x_\varepsilon^{n_0} \rangle$. Since $\langle x_\varepsilon^*, q_\varepsilon - x_\varepsilon^{n_0} \rangle \leq 0$, so

$$\langle n_0 y^* + \varepsilon z^*, q_\varepsilon - x_\varepsilon^{n_0} \rangle \geq 0. \tag{26.21}$$

By $(P1)$ applied to the function $d_{Q_1}^2(\cdot)$ we get

$$\langle n_0 y^*, x_\varepsilon^{n_0} - q_\varepsilon \rangle \geq n_0 d_{Q_1}^2(x_\varepsilon^{n_0}).$$

Plugging the last inequality into (26.21) we obtain

$$n_0 {d_{Q_1}}^2(x_\varepsilon^{n_0}) \leq \varepsilon \langle z^*, q_\varepsilon - x_\varepsilon^{n_0} \rangle \leq \varepsilon \parallel z^* \parallel \parallel q_\varepsilon - x_\varepsilon^{n_0} \parallel \leq \varepsilon(\text{diam } Q_1 + \varepsilon_0), \tag{26.22}$$

keep in mind that by $(P1)$, $(P4)$ $\partial(\varepsilon \parallel \cdot - x_\varepsilon^{n_0} \parallel)(x_\varepsilon^{n_0}) = \varepsilon B^*(0,1)$, with $B^*(0,1) = \{x^* \in \mathbb{E}^* \mid \parallel x^* \parallel \leq 1\}$. It follows from (26.22) that

$$d_{Q_1}^2(x_\varepsilon^{n_0}) \leq \frac{\varepsilon(\text{diam } Q_1 + \varepsilon_0)}{n_0}$$

hence

$$d_{Q_1}^2(x_\varepsilon^{n_0}) \to 0 \text{ as } \varepsilon \to 0. \tag{26.23}$$

We shall show that for $\bar{q}$, defined in (26.17), we have

$$f(\bar{q}) \leq f(x) + n_0 d_{Q_1}^2(x) \text{ for every } x \in Q'. \tag{26.24}$$

If not then for some $\bar{x}$ we get $\delta := f(\bar{x}) + n_0 d_{Q_1}^2(\bar{x}) < f(\bar{q})$. It follows from (26.18) that $f(x_\varepsilon^{n_0}) < \frac{\delta + f(\bar{q})}{2}$ for $\varepsilon > 0$ small enough. Hence by (26.23) and (26.15) we get a contradiction. Hence (26.24) holds and by $(P1)-(P3)$ we get that $0 \in \partial f(\bar{q})$. Our aim is to show that $\bar{q} \in Q$. If not by the definitions of Q_1 and $\bar{q}$ we have that $f_{|Q} = \infty$. We remind that $f_{|Q_1}$ is bounded from below by the number $r := \inf_{Q_1} f$. Let us take j_0 such that $\frac{n_0 d_Q^2(\bar{q})}{j_0} < \varepsilon_0^2$. Similarly to (26.15) we can prove that

$$\inf_{q \in Q} d(q, S(f, f(\bar{q}) + n_0 d_Q^2(\bar{q}))) \geq \gamma > 0 \text{ for some } \gamma > 0. \tag{26.25}$$

Now we apply the Ekeland Variational Principle to the function $f + n_0 d_Q^2 + j_0 d_{Q_1}^2$ on the set Q' taking $x_0 := \bar{q}$ $(f(\bar{q}) = r = \inf_{Q_1} f)$ and $\lambda := 1$. We obtain

$$f(x_\varepsilon) + n_0 d_Q^2(x_\varepsilon) + j_0 d_{Q_1}^2(x_\varepsilon) + \varepsilon \parallel x_\varepsilon - \bar{q} \parallel \leq f(\bar{q}) + n_0 d_Q^2(\bar{q}) \tag{26.26}$$

and

$$x_\varepsilon \in \arg\min{}_{Q'} \left(f(\cdot) + n_0 d_Q^2(\cdot) + j_0 d_{Q_1}^2(\cdot) + \varepsilon \parallel \cdot - x_\varepsilon \parallel\right). \tag{26.27}$$

By (26.24) and (26.26) we get $j_0 d_{Q_1}^2(x_\varepsilon) \leq n_0 d_Q^2(\bar{q})$, hence $d_{Q_1}^2(x_\varepsilon) < \varepsilon_0^2$. Thus $x_\varepsilon \in$intQ'. Again by (26.26) we get

$$f(x_\varepsilon) \leq f(\bar{q}) + n_0 d_Q^2(\bar{q}),$$

so $x_\varepsilon \varepsilon S(f, f(\bar{q}) + n_0 d_Q^2(\bar{q}))$. By (26.25) we have $d_Q^2(x_\varepsilon) \geq \gamma^2$. It follows from (26.27) and $(P2)$ that there exist $x^* \in \partial f(x_\varepsilon)$, $z^* \in B^*(0,1)$ and $y^* \in \partial d_Q^2(x_\varepsilon)$, $v^* \in \partial d_{Q_1}^2(x_\varepsilon)$ such that $0 = x^* + n_0 y^* + \varepsilon z^* + j_0 v^*$. Let us choose $q_\varepsilon \in Q$ satisfying (26.11). We have

$$\langle x^*, x_\varepsilon - q_\varepsilon\rangle \geq 0 \text{ and } \langle x^*, x_\varepsilon - q_\varepsilon\rangle = n_0\langle y^*, q_\varepsilon - x_\varepsilon\rangle + \varepsilon\langle z^*, q_\varepsilon - x_\varepsilon\rangle + j_0\langle v^*, q_\varepsilon - x_\varepsilon\rangle,$$

thus $0 \leq -n_0 d_Q^2(x_\varepsilon) + \varepsilon \parallel q_\varepsilon - x_\varepsilon \parallel - j_0 d_{Q_1}^2(x_\varepsilon)$. Hence by (26.25)

$$0 < n_0\gamma^2 \leq \varepsilon \,\mathrm{diam}\, Q' \quad \text{for every } \varepsilon > 0,$$

a contradiction. Thus $\bar{q} \in Q$ and $0 \in \partial f(\bar{q})$. □

26.4 LOCATING ELEMENTS OF GRAPHS OF MAXIMAL MONOTONE OPERATORS

In this section we provide an answer to Simons' question, concerning the maximal monotone operators. Our method of solving the problem strongly relies on the fact that cl (dom T) is a convex set for any maximal monotone T, which by [16, Proposition 12.37 (horizon cones and limits)] is a virtue of $\mathbb{R}^n$, more generally of any reflexive Banach space (see [19, Thorem 1 (b)]). There are results, which quarantee the convexity of cl (dom T) in many cases, we refer to [9, 15, 16, 19] for more on that.

The point of the below lemma is that it allows one to link the domain of a maximal monotone operator with its values. Namely, if $|u_n^*| \to \infty$, $x_n \to x$ with $u_n^* \in T(x_n)$, then any cluster point $(|u_n^*|^{-1}u_n^*)$ is in the normal cone to cl dom T at x.

Lemma 26.28. *Let* $T : \mathbb{R}^n \rightrightarrows \mathbb{R}^n$ *be a maximal monotone operator with* dom $T := \{x \in \mathbb{R}^n \,|\, T(x) \neq \emptyset\} \neq \emptyset$. *Assume that* $(x_k) \subseteq$ dom T *and* $(u_k^*) \subset \mathbb{R}^n$ *are such that*

$$|u_k^*| \to \infty,\ |u_k^*|^{-1}u_k^* \to u^*,\ x_k \to x \text{ and } u_k^* \in T(x_k) \text{ for every } k \in \mathbb{N}.$$

Then we have:

(i) dom $T \cap (x + \{u \in \mathbb{R}^n | \langle u^*, u\rangle > 0\}) = \emptyset$, *which amounts to* $u^* \in N_{\mathrm{cl\ dom}\ T}(x)$, *where* $\langle\cdot,\cdot\rangle$ *is the scalar product, and* $N_{\mathrm{cl\ dom}\ T}(x)$ *is the normal cone to the closure of* dom T *at* x.

(ii) if $x \in \operatorname{dom} T$ then $T(x) + [0,\infty)u^* \subseteq T(x)$.

Proof. Let $x_k \to x$, $|u_k^*| \to \infty$, $|u_k^*|^{-1}u_k^* \to u^*$ and $u_k^* \in T(x_k)$ for every $k \in \mathbb{N}$. If $(y, y^*) \in \operatorname{graph} T := \{(z, z^*) \in \mathbb{R}^n \times \mathbb{R}^n \,|\, z^* \in T(z)\}$, then by the monotonicity we have

$$\langle u_k^* - y^*, x_k - y\rangle \geq 0 \quad \text{for every } k \in \mathbb{N}.$$

Dividing by $|u_k^*|$ and passing to the limit we get (i). If $x \in \operatorname{dom} T$ then for any $x^* \in T(x)$ and $(y, y^*) \in \operatorname{graph} T$ we have

$$\langle x^* - y^* + su^*, x - y\rangle \geq \langle x^* - y^*, x - y\rangle \geq 0 \quad \text{for any } s \geq 0.$$

□

Assuming that (26.2) holds true, it becomes obvious that we can take $Q \cap \operatorname{aff} \operatorname{dom} T$ instead of Q, and (26.1) is still true. However, it is a question if the condition

$$\text{for every } (x, x^*) \in \operatorname{graph} T \text{ there is } (q, q^*) \in Q \times Q^* \text{ such that } \langle x^* - q^*, x - q\rangle \geq 0$$

implies that

$$\text{for every } (x, x^*) \in \operatorname{graph} T \text{ there is } (q, q^*) \in \Big(Q \cap (\operatorname{aff} \operatorname{dom} T)\Big) \times Q^*.$$

Below we answer the question whenever $Q^* = \{0\}$.

Lemma 26.29. *Let Q be a compact, convex, nonempty subset of $\mathbb{R}^n$, $n \geq 2$, and $T : \mathbb{R}^n \rightrightarrows \mathbb{R}^n$ be a maximal monotone operator. If*

for every $(x, x^*) \in \operatorname{graph} T$ there is $q \in Q$ such that
$\langle x^*, x - q\rangle \geq 0$

then

for every $(x, x^*) \in \operatorname{graph} T$ there is $q \in Q \cap (\operatorname{aff} \operatorname{dom} T)$ such that
$\langle x^*, x - q\rangle \geq 0$.

Proof. If $\operatorname{aff} \operatorname{dom} T = \mathbb{R}^n$ then there is nothing to prove. Assume that $\operatorname{aff} \operatorname{dom} T \neq \mathbb{R}^n$. Let $X_0 \subset \mathbb{R}^n$ be a linear subspace and $x_0 \in \mathbb{R}^n$ be such that

$$\operatorname{aff} \operatorname{dom} T = x_0 + X_0 \text{ and } \dim X_0 + k = n \quad \text{for some } k \in \mathbb{N}.$$

We are able to find $u_1^*, \ldots, u_k^* \in \mathbb{R}^n$ such that $X_0 = \bigcap_{i=1}^k \ker u_i^*$. It is obvious that for every $s \in \mathbb{R}^k$ and every $(x, x^*) \in \operatorname{graph} T$ we get $x^* + \sum_{i=1}^k s_i u_i^* \in T(x)$, which imply that for every $(s_1, \ldots, s_k) \in \mathbb{R}^k$ there is $q \in Q$ such that

$$\langle x^* - \sum_{i=1}^k s_i u_i^*, x - q\rangle \geq 0.$$

Let us define $h_j : Q \times [-j,j]^k \to \mathbb{R}$, $j \in \mathbb{N}$ as follows

$$h_j\big(q,(s_1,\dots,s_k)\big) := -\langle x^* - \sum_{i=1}^{k} s_i u_i^*, x - q\rangle,$$

where $[-j,j]^k := \underbrace{[-j,j] \times \dots \times [-j,j]}_{k \text{ times}}$. By von Neumann's inequality there are $q_j \in Q$, we may assume that $q_j \to q_x \in Q$, if not take a converging subsequence, such that $\max_{s\in[-j,j]^k} h_j(q_j,s) \le 0$. Hence there is $M > 0$ such that for every $j \in \mathbb{N}$ we have

$$\sum_{i=1}^{k} s_i \langle u_i^*, x - q_j\rangle \le M \text{ for every } s \in [-j,j]^k.$$

Thus there is $q_x \in Q$ such that $\langle x^*, x - q_x\rangle \ge 0$ and $x - q_x \in \bigcap_{i=1}^k \ker u_i^*$, so $x - q_x \in X_0$. Now it is easy to see that $q_x \in \text{aff dom } T$ and the proof is complete. □

The below theorem corresponds to theorem 26.3.1. It preserves the implication whenever T is a maximal monotone operator.

THEOREM 26.4.1. *Let $T : \mathbb{R}^n \rightrightarrows \mathbb{R}^n$ be a maximal monotone operator with* $\text{dom } T := \{x \in \mathbb{R}^n \,|\, T(x) \neq \emptyset\} \neq \emptyset$ *and $n \ge 2$. If Q is a convex, compact, nonempty subset of $\mathbb{R}^n$ such that*

$$\textit{for every } (x,x^*) \in \text{graph } T, \textit{ there is } q \in Q \textit{ such that } \langle x^*, x - q\rangle \ge 0, \tag{26.30}$$

then

$$\text{graph } T \cap Q \times \{0\} \neq \emptyset. \tag{26.31}$$

Proof. Choose $r > 0$ such that $Q \subset B(0,r-1) := \{x \in \mathbb{R}^n \,|\, |x| \le r-1\}$ and $B(0,r-1) \cap \text{rint dom } T \neq \emptyset$, and put $S := B(0,r)$. We are able to find $x_0 \in \text{dom } T$ and subspaces $X_0, Y_0 \subset \mathbb{R}^n$ such that

$$\text{aff dom } T = x_0 + X_0, \quad X_0 + Y_0 = \mathbb{R}^n \quad \text{and} \quad X_0^\perp = Y_0, \tag{26.32}$$

where

$$X_0^\perp := \{y \in \mathbb{R}^n \,|\, \forall x \in X_0,\ \langle y, x\rangle = 0\}$$

with $\langle\cdot,\cdot\rangle$ being the scalar product. Suppose that $x_0 = 0$. Of course, the maximal monotonicity entails the inclusion $T(x) + Y_0 \subseteq T(x)$ for every $x \in \text{dom } T$, thus

$$T(x) \cap X_0 \neq \emptyset \text{ for every } x \in \text{dom } T. \tag{26.33}$$

Let us define two auxiliary functions $f : \mathbb{R}^n \to \mathbb{R}^n$ and $g : \mathbb{R}^n \to \mathbb{R}$ as follows

$$f(x^*) := \big(\max\{(1 + \text{diam } S), |x^*|\}\big)^{-1} x^*,$$

for every $x^* \in \mathbb{R}^n$ and

$$g(x) := d_Q^2(x) \text{ for every } x \in \mathbb{R}^n,$$

where $d_Q(x)$ is the distance of x from Q. Let us put $A := \text{cl dom } T \cap S$ and for every $x \in A$ define

$$F(x) := \begin{cases} f\big(T(x) \cap X_0 + \partial g(x)\big) + N_A(x) \cap B(0,1) & \text{if } x \in \text{dom } T, \\ \\ \limsup\limits_{\substack{x' \to x \\ x' \in \text{dom } T \cap S}} \; f\big(T(x') \cap X_0 + \partial g(x')\big) & \\ \qquad\qquad + N_A(x) \cap B(0,1) & \text{if } x \in \text{cl dom } T \setminus \text{dom } T, \end{cases}$$

where $N_A(x)$ stands for the normal cone to A at x and lim sup corresponds to the set convergence, see [16]. We want to ensure the existence of a point $\bar{x} \in Q \cap \text{dom } T$ such that

$$0 \in F(\bar{x}) \tag{26.34}$$

using the Ky Fan variational inequality. The function which we employ $h : S \times S \to \mathbb{R}$ has the form

$$h(x,y) = \inf_{x^* \in F(\text{pr}_A(x))} \langle x^*, \text{pr}_A(x) - y \rangle,$$

where

$$\text{pr}_A(x) := \{a \in A \,|\, |x - a| = d_A(x)\}.$$

We first observe that $h(x,x) \leq 0$ for every $x \in S$. In fact, if $x \in A$ then $x = \text{pr}_A(x)$. If $x \notin A$ then

$$\frac{x - \text{pr}_A(x)}{|x - \text{pr}_A(x)|} \in N_A\big(\text{pr}_A(x)\big) \cap B(0,1),$$

hence on account of that any term from the upper limit has the norm less than 1, by the definition of F we get $h(x,x) \leq 0$. It is evident that the function $y \to h(x,y)$ is concave for all $x \in S$. We claim that the function $x \to h(x,y)$ is lower semicontinuous for every $y \in S$ as well. Let us fix y and x in S. We will show that

$$\liminf_{k \to \infty} h(x_k, y) \geq h(x,y) \tag{26.35}$$

for every sequence $(x_k) \subset S$, $x_k \to x$. Let $x_k^* \in F(\text{pr}_A(x_k))$ for every $k \in \mathbb{N}$ and $x_k^* \to x^*$. In order to get (26.35) it is enough to have

$$h(x,y) \leq \lim_{k \to \infty} \langle x_k^*, \text{pr}_A(x_k) - y \rangle. \tag{26.36}$$

Let $f(v_k^* + w_k^*) + z_k^* \to x^*$, where

$z_k^* \in N_A\left(\text{pr}_A(x_k)\right) \cap B(0,1)$, $v_k^* \in T(x_k') \cap X_0$ and $w_k^* \in \partial g(x_k')$ for every $k \in \mathbb{N}$, and some $(x_k') \subset \text{dom } T \cap S$, $x_k' \to \text{pr}_A(x)$.

If (v_k^*) contains a bounded subsequence then by the maximal monotonicity of T any cluster point of (v_k^*) is in $T(\mathrm{pr}_A(x)) \cap X_0$, so $\mathrm{pr}_A(x) \in \mathrm{dom}\, T$. The upper semicontinuity of $\partial g(\cdot)$ and of $N_A(\cdot)$ and the continuity of f ensure that

$$x^* \in F(x). \tag{26.37}$$

Of course, (26.37) implies (26.36), thus (26.35) holds true.

Assume that $|v_k^*| \to \infty$ and

$$v^* \in \limsup_{n\to\infty} f(v_k^* + w_k^*) = \limsup_{k\to\infty} f(v_k^*),$$

where lim sup is in the set convergence sens. If $\mathrm{pr}_A(x) \in A \setminus \mathrm{dom}\, T$ we get (26.37) as well, choosing a proper subsequence of (z_k^*). In case $\mathrm{pr}_A(x) \in \mathrm{dom}\, T$, by Lemma 26.28 (ii) we obtain

$$[0,\infty)v^* + T(\mathrm{pr}_A(x)) \subseteq T(\mathrm{pr}_A(x)),$$

so $v^* \in \mathrm{cl}\Big(f\left(T\left(\mathrm{pr}_A(x)\right) \cap X_0\right) + \partial g\left(\mathrm{pr}_A(x)\right)\Big)$, and (26.37) still holds. By the Ky Fan theorem we can assert that there is $\bar{x} \in S$ such that

$$h(\bar{x}, y) \leq 0 \text{ for every } y \in S. \tag{26.38}$$

It follows from the definition of h that (26.38) is still valid even if we take $\mathrm{pr}_A(\bar{x})$ instead of $\bar{x}$. Hence, we may assume that $\bar{x} = \mathrm{pr}_A(\bar{x}) \in A$. We first exclude the case $\bar{x} \notin \mathrm{dom}\, T$. For this reason let us assume that $\bar{x} \notin \mathrm{dom}\, T$ and take $\bar{z} \in \mathrm{rint\,dom}\, T \cap \mathrm{int}\, S$, and $\varepsilon > 0$ such that

$$B(\bar{z}, \varepsilon) \cap X_0 \subset \mathrm{rint\,dom}\, T \cap \mathrm{int}\, S.$$

By Lemma 26.28 (i) if $u^* \in \limsup\limits_{\substack{x'\to\bar{x}\\ x'\in \mathrm{dom}\, T}} f\big(T(x') \cap X_0 + \partial g(x')\big)$ then

$$|u^*| = 1 \text{ and } \langle u^*, z - \bar{x}\rangle \leq 0 \text{ for every } z \in B(\bar{z}, \varepsilon) \cap X_0 = B(\bar{z}, \varepsilon) \cap (\bar{z} + X_0),$$

hence $\langle u^*, \bar{x} - \bar{z}\rangle \geq \varepsilon$. Thus, if $x^* = u^* + v^*$, $u^* \in \limsup\limits_{\substack{x'\to\bar{x}\\ x'\in \mathrm{dom}\, T}} f\big(T(x') \cap X_0 + \partial g(x')\big)$, $v^* \in N_A(\bar{x})$, then $\langle x^*, \bar{x} - \bar{z}\rangle \geq \varepsilon$, which contradicts (26.38), so $\bar{x} \in \mathrm{dom}\, T$. Hence

$$T(\bar{x}) \cap X_0 + N_{\mathrm{cl\,dom}\, T}(\bar{x}) \cap X_0 \subset T(\bar{x}) \cap X_0. \tag{26.39}$$

Let us take any $y \in S$ and observe that if $\langle z^*, \bar{x} - y\rangle < 0$ for some $z^* \in N_{\mathrm{cl\,dom}\, T}(\bar{x}) \cap X_0$, then by (26.39) we get

$$p(y) := \sup\{\langle x^*, y - \bar{x}\rangle \,|\, x^* \in T(\bar{x}) \cap X_0 + \partial g(\bar{x})\} > 0. \tag{26.40}$$

Let us also observe that the function p defined in (26.40) is convex lower semicontinuous. As an easy consequence of the monotonicity we get that

$$p(y) < +\infty, \text{ whenever } \mathrm{pr}_{X_0}(y) \in S \cap \mathrm{dom}\, T. \tag{26.41}$$

Hence, p is continuous on int $S \cap (\text{rint dom } T + Y_0)$, where $Y_0 = X_0^{\perp}$. Thus by the convexity we get

$$\inf_{y \in \text{int } S \cap \text{rint dom } T} p(y) = \inf_{y \in A} p(y). \tag{26.42}$$

Let us fix any $y \in \text{int } S \cap \text{rint dom } T$. Since $y - \bar{x} \in \text{rint dom } T - \bar{x}$, by the definition of h there are sequences $(u_k^*) \subset T(\bar{x}) \cap X_0$ and $(w_k^*) \subset \partial g(\bar{x})$; $(\alpha_k) \subset (0, \infty)$ such that

$$\langle \alpha_k (u_k^* + w_k^*), \bar{x} - y \rangle \to h(\bar{x}, y) \leq 0. \tag{26.43}$$

If $\alpha_k \to 0$, then $\alpha_k = |u_k^*|^{-1}$ for k large enough and $\alpha_k w_k^* \to 0$. We can choose converging subsequence of $\alpha_k u_k^*$ to u^* with $|u^*| = 1$. By Lemma 26.28 (i) $u^* \in N_{\text{cl dom } T}(\bar{x}) \cap X_0$, thus $\langle u^*, \bar{x} - y \rangle > 0$ since $y - \bar{x} \in \text{rint dom } T - \bar{x}$, but it contradicts to (26.43). In case $\liminf_{k \to \infty} \alpha_k > 0$, we can choose simultaneously converging subsequences of (α_k), (u_k^*), (w_k^*) to $\alpha > 0$, $u^* \in T(\bar{x}) \cap X_0$ and $w^* \in \partial g(\bar{x})$, respectively. By (26.43) we have $\langle u^* + w^*, \bar{x} - y \rangle \leq 0$, thus by (26.41)

$$+\infty > p(y) \geq 0 \quad \text{for every} \quad y \in \text{int } S \cap \text{rint dom } T,$$

hence by (26.42)

$$p(y) \geq 0 \quad \text{for} \quad y \in A. \tag{26.44}$$

By the definition of p we get

$$p\big((1-t)\bar{x} + ty\big) = tp(y) \quad \text{for every } t > 0 \text{ and } y \in \mathbb{R}^n, \tag{26.45}$$

thus by (26.40) and (26.44) we obtain

$$p(y) \geq 0 \quad \text{for every} \quad y \in S \cap X_0. \tag{26.46}$$

In fact, by (26.40) we have

$$p(y) \geq 0 \quad \text{whenever} \quad y - \bar{x} \notin T_{\text{cl dom } T}(\bar{x}) + Y_0,$$

and since p is finite on int $S \cap$ rint dom T, so (26.44) and (26.45) imply that

$$p(y) \geq 0 \quad \text{whenever} \quad y - \bar{x} \in T_A(\bar{x}) = \text{cl} \bigcup_{\lambda > 0} \lambda(A - \bar{x}),$$

where $T_C(\bar{x})$ is the tangent cone to C at $\bar{x}$ (see [16], convex case). Because of the equality

$$T_{\text{cl dom } T}(\bar{x}) \cap T_S(\bar{x}) = T_A(\bar{x}),$$

we infer that

$$p(y) \geq 0 \quad \text{whenever} \quad y - \bar{x} \notin T_{\text{cl dom } T}(\bar{x}) + Y_0 \text{ or } y - \bar{x} \in (S - \bar{x}) \cap T_{\text{cl dom } T}(\bar{x}),$$

which gives (26.46). Let us consider the convex minimizing problem

$$p(y) \to \min$$
$$y \in S \cap X_0.$$

We know that $p(\bar{x}) = 0$, so by (26.46) and the sufficient condition for minimum we get

$$0 \in T(\bar{x}) \cap X_0 + \partial d_Q^2(\bar{x}) + N_{S\cap X_0}(\bar{x}),$$

hence, there are $u^* \in T(\bar{x}) \cap X_0$, $w^* \in \partial d_Q^2(\bar{x})$ and $z^* \in N_{S\cap X_0}(\bar{x})$ such that

$$0 = u^* + w^* + z^*.$$

By (26.30) and Lemma 26.29 there is $\bar{q} \in Q \cap X_0 \subset S$ such that

$$\langle u^*, \bar{x} - \bar{q}\rangle \geq 0$$

hence by property $(P1)$ $\langle z^*, \bar{x} - q\rangle < 0$, whenever $\bar{x} \notin Q$, which contradicts the choice of $z^* \in N_{S\cap X_0}(\bar{x})$. Thus $\bar{x} \in Q$ and $\partial d_Q^2(\bar{x}) = \{0\}$, so $z^* \in N_{S\cap X_0}(\bar{x}) \cap X_0$. As $\bar{x} \in Q \subset \text{int } S$ we have $z^* = 0$, which implies $u^* = 0$.

In order to complete the proof we have to consider the case $x_0 \neq 0$ in (26.32). For this reason let us define

$$\tilde{T}(x) := T(x + x_0) \quad \text{for every } x \in \mathbb{R}^n \quad \text{and } \tilde{Q} := Q - x_0$$

and observe that

$$\text{dom } \tilde{T} = \text{dom } T - x_0 \subseteq X_0.$$

Thus (26.32) is satisfied for $\tilde{T}$ (with $x_0 = 0$). Let us also observe that on account of the equivalence

$$x^* \in \tilde{T}(x) \Longleftrightarrow x^* \in T(x + x_0) \quad \text{for every } x \in \mathbb{R}^n \tag{26.47}$$

condition (26.30) is also satisfied with T and Q replaced by $\tilde{T}$ and $\tilde{Q}$, respectively. Hence, the above reasoning can be applied to $\tilde{T}$, $\tilde{Q}$ in order to ensure the existence of $\tilde{q} \in \tilde{Q}$ such that

$$0 \in \tilde{T}(\tilde{q}),$$

which by (26.47) implies (26.31). □

The above theorem has been proved for $Q^* = \{0\}$. Applying a similar lemma to lemma 2.3 and using properties of maximal monotone operators, similarly as in [22], we can generalize this theorem to the case when Q^* is greater than singleton.

References

[1] J.-P. Aubin, *Optima and equilibria*, Springer-Verlag, 1993.

[2] D. Aussel, J.-N. Corvellec, M. Lassonde, *Subdifferential characterization of quasiconvexity and convexity*, J. Convex Analysis 1:195-201, 1994.

[3] D. Aussel, J.-N. Corvellec, M. Lassonde, *Mean value property and subdifferential criteria for lower semicontinuous function*, Trans. Amer. Math. Soc. 147: 4147-4161, 1995.

[4] R. Correa, A. Jofre, L. Thibault, *Subdifferential monotonicity as characterization of convex function*, Numer. Funct. Anal. and Optimiz. 15:531-535, 1994.

[5] K. Deimling, *Nonlinear functional analysis*, Spriger-Verlag, Berlin,1985.

[6] I. Ekeland *Nonconvex minimization problems*, Bull. of the AMS 1:443–474, 1979.

[7] S.P. Fitzpatrick and R.R. Phelps, *Some properties of maximal monotone operators on nonreflexive Banach spaces*, Set–Valued Analysis 3:51–69,1995.

[8] G. Kothe, *Topological Vector Space I*, Springer-Verlag, 1985.

[9] E. Krauss, *Maximal monotone operators and saddle functions I.* Zeitschrift für Analysis und ihre Anwendungen, 5:333–346, 1986.

[10] D. T. Luc, *Characterization of quasiconvex function* Bull. Austral. Math. Soc. 48:393-406, 1993.

[11] D. T. Luc, *On generalized convex nonsmooth functions*, Bull. Ausrtal. Math. Soc. 49:139-149, 1994.

[12] D. T. Luc, *A resolution of Simons' maximal monotonicity problem*, Journal of Convex Analysis 3:367–370,1996.

[13] R. R. Phelps, *Convex functions, monotone operators and differentiability*, Lecture Notes in Math. 1364, Springer-Verlag, 1993 (second edition).

[14] R.T. Rockafellar, *Characterization of the subdifferentials of convex functions*, Pacific J. Math 17:497–509,1966.

[15] R.T. Rockafellar, *Local boundedness of nonlinear, monotone operators.* Michigan Math. J., 16:397–407, 1969.

[16] R.T. Rockafellar and R.J.-B. Wets, *Variational Analysis*, (book to be published by Springer–Verlag), 1995.

[17] S. Simons, *Subtangents with controlled slope*, Nonlinear Anal. Theory Meth. App. 22:1373-1389, 1994.

[18] S. Simons, *Swimming below icebergs*, Set-Valued Analysis 2:327-337, 1994.

[19] S. Simons, *The range of a monotone operator.* J. Math. Anal. Appl., 196:176–201, 1996.

[20] L. Thibault, D. Zagrodny, *Integration of subdifferentials of lower semicontinuous functions on Banach spaces*, J. Math. Anal. Appl. 189:33-58, 1995.

[21] D. Zagrodny *Approximate mean value theorem for upper subderivatives,* Nonlinear Anal. Theory Meth. Appl 12:1413–1428, 1988.

[22] D. Zagrodny, *The maximal monotonicity of the subdifferentials of convex functions: Simons' problem*, Set-Valued Anal. 4:301-314, 1996.

[23] D. Zagrodny, *Answers to the basic questions on maximal monotone operators on nonreflexive Banach spaces* (in preparation).

[24] E. Zeidler, *Nonlinear functional analysis and its applications II: Monotone operators; IV: Applications to mathematical physics*, Springer- Verlag, New-York, Berlin, 1986.

27 DENSITY ESTIMATES AND FURTHER PROPERTIES OF BLAKE & ZISSERMAN FUNCTIONAL

Michele Carriero*, Antonio Leaci*, Franco Tomarelli**

*Dipartimento di Matematica " Ennio De Giorgi"
Via Arnesano
73100 Lecce
Italia
**Dipartimento di Matematica " Francesco Brioschi"
Politecnico
Piazza Leonardo da Vinci 32
20133 Milano
Italia

carriero@ingle01.unile.it
leaci@ingle01.unile.it
fratom@mate.polimi.it

Abstract: We prove some properties of strong minimizers for functionals depending on free discontinuities, free gradient discontinuities and second derivatives, which are related to image segmentation.

27.1 INTRODUCTION

In previous papers ([CLT3],[CLT4]) we proved the existence of minimizers for the following functional

$$F(K_0, K_1, u) :=$$
$$\int_{\Omega\setminus(K_0\cup K_1)} \left(|D^2u|^2 + \mu|u-g|^q\right) dy + \alpha\mathcal{H}^1(K_0\cap\Omega) + \beta\mathcal{H}^1((K_1\setminus K_0)\cap\Omega) \,, \tag{27.1}$$

R.P. Gilbert et al. (eds.), From Convexity to Nonconvexity, 381–392.

where $\Omega \subset \mathbf{R}^2$ is an open set, $\mathcal{H}^1$ denotes the length (in the sense of the one dimensional Hausdorff measure), and $\alpha, \beta, \mu, q \in \mathbf{R}$, with

$$q \geq 1 \ , \ \mu > 0 \ , \ 0 < \beta \leq \alpha \leq 2\beta \ , \ g \in L^{2q}_{loc}(\Omega) \cap L^q(\Omega) \ , \tag{27.2}$$

are given; while $K_0, \ K_1 \subset \mathbf{R}^2$ are Borel sets (a priori unknown) with $K_0 \cup K_1$ closed, $u \in C^2(\Omega \setminus (K_0 \cup K_1))$ and it is approximately continuous on $\Omega \setminus K_0$.

Functional (27.1) (called *thin plate surface under tension* in computer vision modelling) was proposed by Blake & Zisserman as an energy to be minimized in order to achieve a segmentation of a monochromatic picture ([BZ]). In this context g describes the light intensity level of the image Ω, μ is a scale parameter, α is a contrast parameter and a measure of immunity to noise, β is a gradient-contrast parameter.
The elements of a minimizing triplet (K_0, K_1, u) play respectively the role of edges, creases and smoothly varying intensity in the region $\Omega \setminus (K_0 \cup K_1)$ for the segmented image. This second-order model (27.1) was introduced to overcome the over-segmentation of steep gradients (ramp effect) and other inconvenients which occur in lower order models as in case of Mumford & Shah functional ([MS],[DGCL]).

The energy (27.1) is a functional depending on bulk energy and a surfacic discontinuity energy; their coupling is rather intriguing, since there is dependence on second derivatives, while there is no bound on the first derivatives. Moreover the discontinuities take place on the sets $K_0, \ K_1$ which are "a priori" unknown, hence the associated minimization problem turns out to be essentially nonconvex, and non uniqueness of minimizers may develop for suitable data g.

The existence of minimizers of (27.1) was proved by a suitable definition of weak solutions ([CLT3]), and hence by showing regularity properties for them ([CLT4]).

Our technique was based on the application of a new Poincaré type inequality ([CLT4]) which entails a suitable decay outside the singular set of weak solutions.

In this paper we prove additional properties of the solutions: *upper and lower energy density* of the *essential minimizing triplet* (see Definition in Section 3); an *elimination property* and a precise estimate of the *Minkowski content* for the segmentation set $K_0 \cup K_1$ (see Theorems in Section 4).

In particular the elimination property states that, when an optimal segmentation has length, in a small ball, less than an absolute constant times the radius of the ball, then such segmentation does not intersect the ball with half the radius. This is also a useful information for numerical analysis of the problem, in the sense that a suitable algorithm can eliminate such essential isolated parts of $K_0 \cup K_1$ because they are "needless energy" for the segmentation.

The result about the Minkowski content expresses the agreement between the Hausdorff one dimensional measure and the Minkowski content of the segmentation $K_0 \cup K_1$. Roughly speaking, the theorem says that a uniform fattening of an optimal segmentation is a reasonable approximation of the segmentation itself.

The assumption $g \in L^{2q}_{loc}(\Omega)$ is sharp in the sense that for every $s < 2q$ there exist $g \in L^s(\Omega)$ such that functional (27.1) has no minimizing triplet. This is proved here by a counterexample in section 5.

The outline of the paper is the following.

1. Introduction.
2. Notation and preliminary results.
3. Local weak minimizers and essential minimizing triplets.
4. Density estimates for essential minimizing triplets.
5. A counterexample.

27.2 NOTATION AND PRELIMINARY RESULTS

From now on we denote by Ω an open set in $\mathbf{R}^2$. Given two vectors a, b, we set $a \cdot b = \sum_i a_i b_i$.

For a given set $U \subset \mathbf{R}^2$ we denote by ∂U its topological boundary, by $\mathcal{H}^1(U)$ its one–dimensional Hausdorff measure and by $|U|$ its Lebesgue outer measure; χ_U is the characteristic function of U. We indicate by $B_\rho(x)$ the open ball $\{y \in \mathbf{R}^2; |y-x| < \rho\}$. If Ω, Ω' are open subsets in $\mathbf{R}^2$, by $\Omega \subset\subset \Omega'$ we mean that $\overline{\Omega}$ is compact and $\overline{\Omega} \subset \Omega'$.

We say that a subset E of $\mathbf{R}^2$ is *countably* $(\mathcal{H}^1, 1)$ *rectifiable* if it is $\mathcal{H}^1$ measurable and E (up to a set of vanishing $\mathcal{H}^1$ measure) is the countable union of C^1 images of bounded subsets of $\mathbf{R}$; if in addition $\mathcal{H}^1(E) < +\infty$ then we say that E is $(\mathcal{H}^1, 1)$ *rectifiable.*

For any Borel function $v : \Omega \to \mathbf{R}$ the approximate upper and lower limits of v are the Borel functions $v^+, v^- : \Omega \to \overline{\mathbf{R}} = \mathbf{R} \cup \{\pm\infty\}$ defined for any $x \in \Omega$ by

$$\begin{aligned} v^+(x) &= \inf\{t \in \overline{\mathbf{R}} : \lim_{\rho\to 0} \rho^{-2}|\{v > t\} \cap B_\rho(x)| = 0\}, \\ v^-(x) &= \sup\{t \in \overline{\mathbf{R}} : \lim_{\rho\to 0} \rho^{-2}|\{v < t\} \cap B_\rho(x)| = 0\}. \end{aligned}$$

The set

$$S_v = \{x \in \Omega;\ v^-(x) < v^+(x)\ \}$$

is a Borel set, of negligible Lebesgue measure (see e.g. [F], 2.9.13); we say that v is approximately continuous on $\Omega \setminus S_v$ and we denote by $\tilde{v} : \Omega \setminus S_v \to \bar{\mathbf{R}}$ the function

$$\tilde{v}(x) = \text{ap}\lim_{y\to x} v(y) = v^+(x) = v^-(x).$$

Let $x \in \Omega \setminus S_v$ be such that $\tilde{v}(x) \in \mathbf{R}$; we say that v is approximately differentiable at x if there exists a vector $\nabla v(x) \in \mathbf{R}^2$ (the approximate gradient of v at x) such that

$$\text{ap}\lim_{y\to x} \frac{|v(y) - \tilde{v}(x) - \nabla v(x)\cdot(y-x)|}{|y-x|} = 0.$$

If v is a smooth function then ∇v is the classical gradient. In the following with the notation $|\nabla v|$ we mean the euclidean norm of ∇v and we set $\nabla_i v = (\mathbf{e}_i \cdot \nabla)v$, $\{\mathbf{e}_i\}$ denoting the canonical base of $\mathbf{R}^2$. We recall the definition of the space of real valued functions with bounded variation in Ω :

$$BV(\Omega) = \{v \in L^1(\Omega); Dv \in \mathcal{M}(\Omega)\}$$

where $Dv = (D_1 v, D_2 v)$ denotes the distributional gradient of v and $\mathcal{M}(\Omega)$ denotes the space of vector–valued Radon measure with finite total variation. We denote by $\int_\Omega |Dv|$ the total variation of the measure Dv in Ω.

For every $v \in BV(\Omega)$ the following properties hold ([F]):

1. $v^+(x), v^-(x) \in \mathbf{R}$ for $\mathcal{H}^1$–almost all $x \in \Omega$;
2. S_v is countably $(\mathcal{H}^1, 1)$ rectifiable;
3. ∇v exists a.e. in Ω and coincides with the Radon–Nikodym derivative of Dv with respect to the Lebesgue measure;
4. for $\mathcal{H}^1$ almost all $x \in S_v$ there exists a unique $\nu = \nu_v(x) \in \partial B_1(0)$ such that, setting $B_\rho^+ = \{y \in B_\rho(x) : (y - x) \cdot \nu > 0\}$ and $B_\rho^- = \{y \in B_\rho(x) : (y - x) \cdot \nu < 0\}$, then

$$\lim_{\rho \to 0} \left(\fint_{B_\rho^+} |v(y) - v^+(x)|^2 dy + \fint_{B_\rho^-} |v(y) - v^-(x)|^2 dy \right) = 0.$$

Moreover $\nu_v(x)$ is an approximate normal vector to S_v at x, and also

$$\int_\Omega |Dv| \geq \int_\Omega |\nabla v| \, dy + \int_{S_v} |v^+ - v^-| \, d\mathcal{H}^1.$$

We recall the definitions of some function spaces related to first derivatives which are De Giorgi special measures, and we refer to [DGA] and [A] for their properties.

Definition 27.1. *$SBV(\Omega)$ denotes the class of functions $v \in BV(\Omega)$ such that*

$$\int_\Omega |Dv| = \int_\Omega |\nabla v| \, dy + \int_{S_v} |v^+ - v^-| \, d\mathcal{H}^1.$$

Moreover we set

$$SBV_{loc}(\Omega) := \{v \in SBV(\Omega') : \forall \Omega' \subset\subset \Omega\},$$

$$GSBV(\Omega) := \{v : \Omega \to \mathbf{R} \text{ Borel function}; -k \vee v \wedge k \in SBV_{loc}(\Omega) \ \forall k \in \mathbf{N}\}.$$

Notice that, if $v \in GSBV(\Omega)$, then S_v is countably $(\mathcal{H}^1, 1)$ rectifiable and ∇v exists a.e. in Ω.

We recall also a function space related to second derivatives, that allows the definition of a finite energy set of competing functions (see [CLT1,2,3,4]).

Definition 27.2. *We set*

$$GSBV^2(\Omega) := \{v \in GSBV(\Omega),\ (\nabla_1 v, \nabla_2 v) \in [GSBV(\Omega)]^2\}.$$

Notice that $Dv \neq \nabla v$ in $GSBV^2(\Omega)$; moreover we set

$$S_{\nabla v} = S_{\nabla_1 v} \cup S_{\nabla_2 v}.$$

Eventually we define the strong and weak energy functionals.

Definition 27.3 (Strong formulation of Blake & Zisserman functional). *For $\Omega \subset \mathbf{R}^2$ open set, $A \subset \Omega$ Borel set , $K_0, K_1 \subset \mathbf{R}^2$ Borel sets with $K_0 \cup K_1$ closed, $v \in C^2(\Omega \setminus (K_0 \cup K_1))$ and approximately continuous on $\Omega \setminus K_0$, under the assumptions (1.2), we set*

$$F(K_0, K_1, v, A) := \tag{27.3}$$
$$\int_{A\setminus(K_0\cup K_1)} \left(|D^2 v|^2 + \mu|v-g|^q\right) dy + \alpha\mathcal{H}^1(K_0 \cap A) + \beta\mathcal{H}^1((K_1 \setminus K_0) \cap A).$$

We shortly write $F(K_0, K_1, v)$ when $A = \Omega$.

The space $X(\Omega) := GSBV^2(\Omega) \cap L^q(\Omega)$ is the natural space for a weak formulation of functional (1.3).

Definition 27.4 (Weak formulation of Blake & Zisserman functional). *For $\Omega \subset \mathbf{R}^2$ open set, under the assumptions (1.2), we define $\mathcal{F} : X(\Omega) \to [0, +\infty]$ by*

$$\mathcal{F}(v) := \int_\Omega (|\nabla^2 v|^2 + \mu|v-g|^q)\, dy + \alpha\mathcal{H}^1(S_v) + \beta\mathcal{H}^1(S_{\nabla v} \setminus S_v). \tag{27.4}$$

We proved the following results in [CLT3],[CLT4].

Theorem 27.1 (Existence of weak solutions). *Let $\Omega \subset \mathbf{R}^2$ be an open set and assume (1.2). Then there is $v_0 \in X(\Omega)$ such that*

$$\mathcal{F}(v_0) \leq \mathcal{F}(v) \qquad \forall v \in X(\Omega).$$

We recall that assumption $\beta \leq \alpha \leq 2\beta$ is necessary for lower semicontinuity of $\mathcal{F}$.

Theorem 27.2 (Existence of strong solutions). *Let $\Omega \subset \mathbf{R}^2$ be an open set. Assume (1.2). Then there is at least one triplet among $K_0, K_1 \subset \mathbf{R}^2$ Borel sets with $K_0 \cup K_1$ closed and $u \in C^2(\Omega \setminus (K_0 \cup K_1))$ approximately continuous on $\Omega \setminus K_0$ minimizing the functional (27.1) with finite energy. Moreover the sets $K_0 \cap \Omega$ and $K_1 \cap \Omega$ are $(\mathcal{H}^1, 1)$ rectifiable.*

Theorem 27.3. *Let $\Omega \subset \mathbf{R}^2$ be an open set. Assume (1.2) and $\alpha = \beta$. Then there is at least one pair among $K \subset \mathbf{R}^2$ closed set and $u \in C^2(\Omega \setminus K)$ minimizing the functional*

$$\int_{\Omega \setminus K} \left(|D^2 u|^2 + \mu |u - g|^q\right) dy + \alpha \mathcal{H}^1(K \cap \Omega)$$

with finite energy. Moreover the set $K \cap \Omega$ is $(\mathcal{H}^1, 1)$ rectifiable.

27.3 LOCAL WEAK MINIMIZERS AND ESSENTIAL MINIMIZING TRIPLETS.

We refine the definition of solution by the concept of essential minimizing triplet, in order to eliminate unimportant points in the segmentation set.

We start with a localization of the functional $\mathcal{F}$ and of the notion of weak minimizers.

Definition 27.5. *For every Borel subset $A \subset \Omega$ and $v \in X(\Omega)$ we define*

$$\mathcal{F}(v, A) := \int_A (|\nabla^2 v|^2 + \mu |v - g|^q)\, dy + \alpha \mathcal{H}^1(S_v \cap A) + \beta \mathcal{H}^1((S_{\nabla v} \setminus S_v) \cap A)$$

Definition 27.6 (Local weak minimizers). *A function $v \in X(\Omega)$ is a local weak minimizer of $\mathcal{F}(\cdot, \Omega)$ if, for every compact set $H \subset \Omega$,*

$$\mathcal{F}(v, H) = \min_{w \in X(\Omega)} \{\mathcal{F}(w, H) : \ w = v \text{ a.e. in } \Omega \setminus H\} < +\infty.$$

From [CLT4] we get a decay property for local weak minimizers.

Theorem 27.4 (Decay Theorem). *Let $\Omega \subset \mathbf{R}^2$ be an open set and assume (1.2). Then there is an absolute positive constant $c_{2,q}$ (depending only on the dimension and the exponent) such that for every $k > 2$, $\eta, \sigma \in (0,1)$ with $\eta^\sigma < \frac{1}{c_{2,q}}$, there is $\varepsilon_0 > 0$ such that for every $\varepsilon \in (0, \varepsilon_0]$ and $\overline{B}_\rho(x) \subset \Omega$, if $u \in GSBV^2(\Omega)$ is a local weak minimizer of $\mathcal{F}(\cdot, \Omega)$ with*

$$\rho \le \varepsilon^k, \qquad \int_{B_\rho(x)} |g|^{2q} dy \le \varepsilon^k$$

and

$$\alpha \mathcal{H}^1(S_u \cap B_\rho(x)) + \beta \mathcal{H}^1((S_{\nabla u} \setminus S_u) \cap B_\rho(x)) \le \varepsilon \rho,$$

then

$$\mathcal{F}(u, B_{\eta\rho}(x)) \le \eta^{2-\sigma} \mathcal{F}(u, B_\rho(x)).$$

Definition 27.7 (Strong minimizing triplet). *A triplet (T_0, T_1, v) such that $T_0, T_1 \subset \mathbf{R}^2$ are Borel sets, $T_0 \cup T_1$ is a closed set, $v \in C^2(\Omega \setminus (T_0 \cup T_1))$ and approximately continuous in $\Omega \setminus T_0$, is a strong minimizing triplet of the functional (27.1) if*

$$(T_0, T_1, v) \ \in \ \operatorname{argmin} F \ .$$

Remark 27.1 - Notice that, if (T_0, T_1, v) is a strong minimizing triplet then v is a weak minimizer and $F(T_0, T_1, v) = \mathcal{F}(v) = \min \mathcal{F}$ (see [CLT4], Lemma 3.2 and section 6).

Definition 27.8 (Essential minimizing triplet). *Given a strong minimizing triplet (T_0, T_1, v) of the functional (27.1), there is another triplet (K_0, K_1, u), called essential minimizing triplet, uniquely defined by*

$$\begin{aligned} K_0 &= \overline{T_0 \cap K} \setminus (S_{\nabla v} \setminus S_v) \\ K_1 &= \overline{T_1 \cap K} \setminus S_v \\ u &= \tilde{v} \end{aligned}$$

where K is the smallest closed subset of $T_0 \cup T_1$ such that $\tilde{v} \in C^2(\Omega \setminus K)$.

Remark 27.2 - For every $v \in X(\Omega)$ that is a weak minimizer of $\mathcal{F}$, we set

$$\Omega_0 = \left\{ x \in \Omega : \lim_{\varrho \to 0_+} \varrho^{-1} \mathcal{F}(v, B_\varrho(x)) = 0 \right\}.$$

By the argument of section 6 in [CLT4], we get that Ω_0 is an open set. Moreover

$$\Omega_0 \cap (\overline{S_v \cup S_{\nabla v}}) = \emptyset,$$

$$\mathcal{H}^1 \left(\Omega \cap \left((\overline{S_v \cup S_{\nabla v}}) \setminus (S_v \cup S_{\nabla v}) \right) \right) = 0.$$

27.4 DENSITY ESTIMATES FOR ESSENTIAL MINIMIZING TRIPLETS.

In this section we state and prove the main results.

We emphasize that in all the statements of this section it is assumed that the open set Ω is contained in $\mathbf{R}^2$ and the assumptions (1.2) are always understood.

Theorem 27.5 (Density upper bound for the functional F). *Let (K_0, K_1, u) be a strong minimizing triplet for the functional (27.1) with $g \in L^{2q}_{loc}(\Omega)$. Then for every $0 < \rho \le 1$ and for every $x \in \Omega$ such that $\overline{B}_\rho(x) \subset \Omega$ we have*

$$F(K_0, K_1, u, \overline{B}_\rho(x)) \le c_0 \rho$$

where $c_0 = \pi^{\frac{1}{2}} \mu \|g\|^q_{L^{2q}(B_\rho(x))} + 2\pi\alpha$.
If $q = 2$ and $g \in L^\infty(\Omega)$, then $c_0 = \pi\mu\|g\|^2_{L^\infty(\Omega)} + 2\pi\alpha$.

Proof - By minimality of (K_0, K_1, u) for F we get

$$F(K_0, K_1, u) \le F(Q_0, Q_1, w),$$

where

$$w = u\chi_{\Omega \setminus B_\rho(x)}, \quad Q_0 = \left(K_0 \setminus B_\rho(x)\right) \cup \partial B_\rho(x), \quad Q_1 = K_1.$$

Taking into account $\beta \leq \alpha$, by subtraction we obtain

$$\begin{aligned}\int_{B_\rho(x)\setminus(K_0\cup K_1)} & \left(|D^2u|^2\mu|u-g|^q\right) dy \\ & + \alpha\mathcal{H}^1(K_0\cap \overline{B}_\rho(x)) + \beta\mathcal{H}^1\left((K_1\setminus K_0)\cap \overline{B}_\rho(x)\right) \\ & \leq \mu\int_{B_\rho(x)} |g|^q\,dy + \alpha\mathcal{H}^1(\partial B_\rho(x)) \\ & \leq \mu\|g\|^q_{L^{2q}(B_\rho(x))}(\pi\rho^2)^{\frac{1}{2}} + 2\pi\alpha\rho,\end{aligned}$$

hence we achieve the proof. ■

Theorem 27.6 (Density lower bound for the functional F). *Let (K_0, K_1, u) be an essential minimizing triplet for the functional (27.1) with $g \in L^{2q}(\Omega)$. Then there exist $\varepsilon_0 > 0$, $\varrho_0 > 0$ such that*

$$F(K_0, K_1, u, B_\varrho(x)) \geq \varepsilon_0\varrho \qquad \forall x \in K_0\cup K_1, \quad \forall \varrho \leq \varrho_0.$$

Proof - By Remark 27.1 u is a weak minimizer and $F(K_0, K_1, u, B_\rho(x)) = \mathcal{F}(u, B_\rho(x))$ for every $B_\rho(x) \subset \Omega$. Let k, η, σ and ε_0 be as in the Decay Theorem 27.4. Let $\rho_0 > 0$ such that $\rho_0 \leq \varepsilon_0^k$, $\int_{B_{\rho_0}(x)} |g|^{2q}dy \leq \varepsilon_0^k$ for every $x \in \Omega$. Assume, by contradiction, that the thesis is false. In such case, there exist $x \in K_0 \cup K_1$ and $0 < \rho \leq \rho_0$ such that

$$F(K_0, K_1, u, B_\rho(x)) < \varepsilon_0\rho.$$

Then, by Theorem 27.4, for every $h \in \mathbf{N}$,

$$F(K_0, K_1, u, B_{\eta^h\rho}(x)) < \eta^{h(2-\sigma)}\varepsilon_0\rho$$

so that

$$\lim_{\varrho\to 0_+} \varrho^{-1} F\left(K_0, K_1, u, B_\varrho(x)\right) = 0,$$

hence $x \in \Omega_0$. Since Ω_0 is open, then u is a C^2 function in a neighbourhood of x and this contradicts the assumption that the minimizing triplet is an essential one. ■

Theorem 27.7. (**Density lower bound for the segmentation length**)
Let (K_0, K_1, u) be an essential minimizing triplet for the functional (27.1) with $g \in L^{2q}(\Omega)$. Then there exist $\varepsilon_1 > 0, \varrho_1 > 0$ such that

$$\mathcal{H}^1((K_0 \cup K_1) \cap B_\varrho(x)) \geq \varepsilon_1 \varrho \qquad \forall x \in K_0 \cup K_1, \ \forall \varrho \leq \varrho_1.$$

Proof - Let k, η, σ and ε_0, ρ_0 be as in Theorem 27.6. We can fix $h_0 \in \mathrm{N}$ such that $\eta^{h_0(1-\sigma)} c_0 < \varepsilon_0$, where c_0 is given in Theorem 27.5. Define $\varepsilon_1 = \frac{\varepsilon_0}{\alpha}\eta^{h_0-1}$ and $\rho_1 = \min\{\rho_0, 1\}$. If we assume, by contradiction, that there exist $x \in K_0 \cup K_1$ and $\rho \leq \rho_1$ such that

$$\mathcal{H}^1((K_0 \cup K_1) \cap B_\rho(x)) < \varepsilon_1 \rho,$$

then we can use Theorem 27.4 h_0 times until we get

$$\begin{aligned} F(K_0, K_1, u, B_{\eta^{h_0}\rho}(x)) &\leq \eta^{h_0(2-\sigma)} F(K_0, K_1, u, B_\rho(x)) \\ &\leq \eta^{h_0(1-\sigma)} c_0 \eta^{h_0} \rho \\ &< \varepsilon_0(\eta^{h_0}\rho) \end{aligned}$$

which contradicts Theorem 27.6. ■

Theorem 27.8. (**Elimination Property**)
Let (K_0, K_1, u) be an essential minimizing triple for the functional (27.1) with $g \in L^{2q}(\Omega)$ and let $\varepsilon_1 > 0, \varrho_1 > 0$ as in Theorem 27.7 and $\rho \leq \rho_1$. If $x \in \Omega$ and

$$\mathcal{H}^1((K_0 \cup K_1) \cap B_\varrho(x)) < \frac{\varepsilon_1}{2}\rho$$

then

$$(K_0 \cup K_1) \cap B_{\rho/2}(x) = \emptyset.$$

Proof - Assume, by contradiction, that there exists $y \in (K_0 \cup K_1) \cap B_{\rho/2}(x)$. Then $B_{\rho/2}(y) \subset B_\rho(x)$, hence

$$\mathcal{H}^1((K_0 \cup K_1) \cap B_{\varrho/2}(y)) \leq \mathcal{H}^1((K_0 \cup K_1) \cap B_\varrho(x)) < \varepsilon_1 \left(\frac{\rho}{2}\right),$$

therefore $y \notin K_0 \cup K_1$ by Theorem 27.7. ■

Theorem 27.9. (**Minkowski content of the segmentation**)
Let (K_0, K_1, u) be an essential minimizing triplet for the functional (27.1) with $g \in L^{2q}(\Omega)$. Then

(i) $K_0 \cup K_1$ is $(\mathcal{H}^1, 1)$ rectifiable;

(ii) for every $\Omega' \subset\subset \Omega$ the following equality holds

$$\lim_{\rho \to 0} \frac{|\{x \in \Omega;\ dist(x, (K_0 \cup K_1) \cap \Omega') < \rho\}|}{2\rho} = \mathcal{H}^1((K_0 \cup K_1) \cap \Omega').$$

Proof - By Remark 27.1 the function u is a weak minimizer of $\mathcal{F}$, so that $S_u \cup S_{\nabla u}$ is $(\mathcal{H}^1, 1)$ rectifiable and $\mathcal{H}^1\big((K_0 \cup K_1) \cap \Omega \setminus (S_u \cup S_{\nabla u})\big) = 0$. Hence (i) follows. The result (ii) is achieved with the same argument as in [AT] and in [CL], by using the previous Theorem 27.7. ■

We enphasize that the various constants $c_0, \varepsilon_0, \varepsilon_1, \rho_0, \rho_1$ depend on the data α, β, μ, g.

27.5 A COUNTEREXAMPLE

In this section we show that the functional (27.1) does not achieve the infimum when g has not enough summability. This fact holds true in any dimension: in this section Ω denotes an open set in $\mathbf{R}^n$, for any integer $n \geq 2$.

We use in the n dimensional case the notations introduced in section 2 and we set $\omega_n = |B_1(0)|$.

We show that for any $s < nq$ there is a function in $g \in L^s(\Omega) \cap L^q(\Omega)$ such that there are no minimizing triplets of the following functional

$$\begin{aligned} G(K_0, K_1, u) &= \int_{\Omega\setminus(K_0\cup K_1)} \left(|D^2u|^p + \mu|u-g|^q\right) dy \\ &\quad +\alpha\mathcal{H}^{n-1}(K_0 \cap \Omega) + \beta\mathcal{H}^{n-1}((K_1 \setminus K_0) \cap \Omega) \end{aligned} \tag{27.5}$$

where $\Omega \subset \mathbf{R}^n$ is an open set, $n \in \mathbf{N}$, $n \geq 2$, $\mathcal{H}^{n-1}$ denotes the $(n-1)$ dimensional Hausdorff measure and $\alpha, \beta, \mu, p, q \in \mathbf{R}$, with

$$p > 1,\ q \geq 1\ ,\ \mu > 0\ ,\ 0 < \beta \leq \alpha \leq 2\beta\ , \tag{27.6}$$

are given; while K_0, $K_1 \subset \mathbf{R}^n$ are Borel sets (a priori unknown) with $K_0 \cup K_1$ closed, $u \in C^2(\Omega \setminus (K_0 \cup K_1))$ and it is approximately continuous on $\Omega \setminus K_0$.

Notice that (27.5) reduces to (27.1) if $n = p = 2$.

Let $b > \left(\frac{n\alpha}{\mu}\right)^{\frac{1}{q}}$, $a > 1$, $r_h = a^{-h}$ for every $h \in \mathbf{N}$, and $\{x_n\}_{n\in\mathbf{N}}$ a dense sequence in Ω. Setting $A_h = B_{r_h}(x_h)$, we define

$$g = b\sum_{h=1}^{\infty} {r_h}^{-\frac{1}{q}} \chi_{A_h}.$$

Then for every $s < nq$ we have $g \in L^s(\Omega)$, but $g \notin L^{nq}(\Omega)$. In fact, if $s < nq$

$$\sum_{h=1}^{\infty} \|{r_h}^{-\frac{1}{q}}\chi_{A_h}\|_{L^s(\Omega)} = \sum_{h=1}^{\infty} {r_h}^{-\frac{1}{q}}|A_h|^{\frac{1}{s}} = {\omega_n}^{\frac{1}{s}} \sum_{h=1}^{\infty} {r_h}^{\frac{n}{s}-\frac{1}{q}} < +\infty,$$

while

$$\int_\Omega \left|\sum_{h=1}^{\infty} {r_h}^{-\frac{1}{q}}\chi_{A_h}\right|^{nq} dy \geq \int_\Omega \sum_{h=1}^{\infty} \left|{r_h}^{-\frac{1}{q}}\chi_{A_h}\right|^{nq} dy = \omega_n \sum_{h=1}^{\infty} 1 = +\infty\ .$$

Assume by contradiction that (K_0, K_1, u) is a minimizing triplet for G.
Let $\bar{B} \subset \Omega \setminus (K_0 \cup K_1)$ a closed ball. Taking into account that $u \in C^2(\bar{B})$ and $r_h \to 0$, we choose $k_0 \in \mathbf{N}$ such that $A_{k_0} \subset \bar{B}$ and

$$b{r_{k_0}}^{-\frac{1}{q}} \geq \max_{\bar{B}} u.$$

Let $k_1 > k_0$ be an index such that $A_{k_1} \subset A_{k_0}$ and set

$$Q_0 = (K_0 \setminus A_{k_1}) \cup \partial A_{k_1} \qquad Q_1 = K_1$$
$$v = \max\left\{ u\,,\ b\left({r_{k_0}}^{-\frac{1}{q}} + {r_{k_1}}^{-\frac{1}{q}}\right)\chi_{A_{k_1}} \right\}.$$

Now (Q_0, Q_1, v) is an admissible triplet for the functional G. Since $g \geq v \geq u + b\,(r_{k_1})^{-\frac{1}{q}}$ in A_{k_1} we get $0 \leq g - v \leq g - u - b\,(r_{k_1})^{-\frac{1}{q}}$ in A_{k_1} and

$$\begin{aligned}\int_{\Omega\setminus(Q_0\cup Q_1)} |v-g|^q\,dy &= \int_{\Omega\setminus(K_0\cup K_1\cup A_{k_1})} |u-g|^q\,dy + \int_{A_{k_1}} |v-g|^q\,dy \\ &\leq \int_{\Omega\setminus(K_0\cup K_1)} |u-g|^q\,dy - \int_{A_{k_1}} b^q\,(r_{k_1})^{-1}\,dy \\ &< \int_{\Omega\setminus(K_0\cup K_1)} |u-g|^q\,dy - \frac{\alpha}{\mu} n\omega_n\,(r_{k_1})^{n-1}.\end{aligned}$$

Moreover we get easily

$$\int_{\Omega\setminus(Q_0\cup Q_1)} |D^2 v|^p\,dy \leq \int_{\Omega\setminus(K_0\cup K_1)} |D^2 u|^p\,dy,$$

$$\mathcal{H}^{n-1}(Q_0\cap\Omega) \leq \mathcal{H}^{n-1}(K_0\cap\Omega) + n\,\omega_n\,(r_{k_1})^{n-1},$$

$$\mathcal{H}^{n-1}((Q_1\setminus Q_0)\cap\Omega) \leq \mathcal{H}^{n-1}((K_1\setminus K_0)\cap\Omega).$$

Hence $G(Q_0, Q_1, v) < G(K_0, K_1, u)$, which contradicts the minimality of (K_0, K_1, u).

We remark that also in the lower order model (Mumford & Shah functional) the infimum is not achieved for a poorly summable datum g (see [L]).

Acknowledgments

Work partially supported by the National Research Projects "Equazioni Differenziali e Calcolo delle Variazioni" and "Analisi Nonlineare e Calcolo delle Variazioni" of the MURST.

References

[A] L. Ambrosio: *Existence theory for a new class of variational problems*, Arch. Rational Mech. Anal., **111** (1990), 291–322.

[AT] L.Ambrosio & V.M.Tortorelli: *Approximation of functionals depending on jumps by elliptic functionals via Γ-convergence*, Comm. Pure Appl. Math., **43** (1990), 999–1036.

[BZ] A. Blake & A. Zisserman: *Visual Reconstruction*, The MIT Press, Cambridge, 1987.

[CL] M. CARRIERO & A. LEACI: *S^k-valued maps minimizing the L^p norm of the gradient with free discontinuities*, Ann. Sc. Norm. Sup. Pisa, (4) **18** (1991), 321-352.

[CLT1] M. CARRIERO, A. LEACI & F. TOMARELLI: *Special Bounded Hessian and elastic-plastic plate*, Rend. Accad. Naz. delle Scienze (dei XL), (109) **XV** (1992), 223–258.

[CLT2] M. CARRIERO, A. LEACI & F. TOMARELLI: *Free gradient discontinuities*, in "Calculus of Variations, Homogeneization and Continuum Mechanics", Buttazzo, Bouchitté, Suquet Eds, World Scientific, Singapore, 1994, 131-147.

[CLT3] M. CARRIERO, A. LEACI & F. TOMARELLI: *A second order model in image segmentation: Blake & Zisserman functional*, in Variational Methods for Discontinuous Structures, R.Serapioni, F.Tomarelli Eds., Birkäuser, 1996, 57-72.

[CLT4] M. CARRIERO, A. LEACI & F. TOMARELLI: *Strong minimizers of Blake & Zisserman functional*, Ann. Sc. Norm. Sup. Pisa, (4) **25** (1997), 257-285.

[DGA] E. DE GIORGI & L. AMBROSIO: *Un nuovo tipo di funzionale del Calcolo delle Variazioni*, Atti Accad. Naz. Lincei, Rend. Cl. Sci. Fis. Mat. Natur., **82** (1988), 199–210.

[DGCL] E. DE GIORGI, M. CARRIERO & A. LEACI: *Existence theorem for a minimum problem with free discontinuity set*, Arch. Rational Mech. Anal., **108** (1989), 195–218.

[F] H. FEDERER: *Geometric Measure Theory*, Springer, Berlin, 1969.

[L] A. LEACI: *Free discontinuity problems with unbounded data: the two dimensional case*, manuscripta math., **75** (1992), 429–441.

[MS] D. MUMFORD & J. SHAH: *Optimal approximation by piecewise smooth functions and associated variational problems*, Comm. Pure Appl. Math., **XLII** (1989), 577–685.

Nonconvex Optimization and Its Applications

1. D.-Z. Du and J. Sun (eds.): *Advances in Optimization and Approximation.* 1994 ISBN 0-7923-2785-3
2. R. Horst and P.M. Pardalos (eds.): *Handbook of Global Optimization.* 1995 ISBN 0-7923-3120-6
3. R. Horst, P.M. Pardalos and N.V. Thoai: *Introduction to Global Optimization* 1995 ISBN 0-7923-3556-2; Pb 0-7923-3557-0
4. D.-Z. Du and P.M. Pardalos (eds.): *Minimax and Applications.* 1995 ISBN 0-7923-3615-1
5. P.M. Pardalos, Y. Siskos and C. Zopounidis (eds.): *Advances in Multicriteria Analysis.* 1995 ISBN 0-7923-3671-2
6. J.D. Pintér: *Global Optimization in Action.* Continuous and Lipschitz Optimization: Algorithms, Implementations and Applications. 1996 ISBN 0-7923-3757-3
7. C.A. Floudas and P.M. Pardalos (eds.): *State of the Art in Global Optimization.* Computational Methods and Applications. 1996 ISBN 0-7923-3838-3
8. J.L. Higle and S. Sen: *Stochastic Decomposition.* A Statistical Method for Large Scale Stochastic Linear Programming. 1996 ISBN 0-7923-3840-5
9. I.E. Grossmann (ed.): *Global Optimization in Engineering Design.* 1996 ISBN 0-7923-3881-2
10. V.F. Dem'yanov, G.E. Stavroulakis, L.N. Polyakova and P.D. Panagiotopoulos: *Quasidifferentiability and Nonsmooth Modelling in Mechanics, Engineering and Economics.* 1996 ISBN 0-7923-4093-0
11. B. Mirkin: *Mathematical Classification and Clustering.* 1996 ISBN 0-7923-4159-7
12. B. Roy: *Multicriteria Methodology for Decision Aiding.* 1996 ISBN 0-7923-4166-X
13. R.B. Kearfott: *Rigorous Global Search: Continuous Problems.* 1996 ISBN 0-7923-4238-0
14. P. Kouvelis and G. Yu: *Robust Discrete Optimization and Its Applications.* 1997 ISBN 0-7923-4291-7
15. H. Konno, P.T. Thach and H. Tuy: *Optimization on Low Rank Nonconvex Structures.* 1997 ISBN 0-7923-4308-5
16. M. Hajdu: *Network Scheduling Techniques for Construction Project Management.* 1997 ISBN 0-7923-4309-3
17. J. Mockus, W. Eddy, A. Mockus, L. Mockus and G. Reklaitis: *Bayesian Heuristic Approach to Discrete and Global Optimization.* Algorithms, Visualization, Software, and Applications. 1997 ISBN 0-7923-4327-1
18. I.M. Bomze, T. Csendes, R. Horst and P.M. Pardalos (eds.): *Developments in Global Optimization.* 1997 ISBN 0-7923-4351-4
19. T. Rapcsák: Smooth Nonlinear Optimization in R^n. 1997 ISBN 0-7923-4680-7
20. A. Migdalas, P.M. Pardalos and P. Värbrand (eds.): *Multilevel Optimization: Algorithms and Applications.* 1998 ISBN 0-7923-4693-9
21. E.S. Mistakidis and G.E. Stavroulakis: *Nonconvex Optimization in Mechanics.* Algorithms, Heuristics and Engineering Applications by the F.E.M. 1998 ISBN 0-7923-4812-5

Nonconvex Optimization and Its Applications

22. H. Tuy: *Convex Analysis and Global Optimization.* 1998 ISBN 0-7923-4818-4
23. D. Cieslik: *Steiner Minimal Trees.* 1998 ISBN 0-7923-4983-0
24. N.Z. Shor: *Nondifferentiable Optimization and Polynomial Problems.* 1998 ISBN 0-7923-4997-0
25. R. Reemtsen and J.-J. Rückmann (eds.): *Semi-Infinite Programming.* 1998 ISBN 0-7923-5054-5
26. B. Ricceri and S. Simons (eds.): *Minimax Theory and Applications.* 1998 ISBN 0-7923-5064-2
27. J.-P. Crouzeix, J.-E. Martinez-Legaz and M. Volle (eds.): *Generalized Convexitiy, Generalized Monotonicity: Recent Results.* 1998 ISBN 0-7923-5088-X
28. J. Outrata, M. Kočvara and J. Zowe: *Nonsmooth Approach to Optimization Problems with Equilibrium Constraints.* 1998 ISBN 0-7923-5170-3
29. D. Motreanu and P.D. Panagiotopoulos: *Minimax Theorems and Qualitative Properties of the Solutions of Hemivariational Inequalities.* 1999 ISBN 0-7923-5456-7
30. J.F. Bard: *Practical Bilevel Optimization.* Algorithms and Applications. 1999 ISBN 0-7923-5458-3
31. H.D. Sherali and W.P. Adams: *A Reformulation-Linearization Technique for Solving Discrete and Continuous Nonconvex Problems.* 1999 ISBN 0-7923-5487-7
32. F. Forgó, J. Szép and F. Szidarovszky: *Introduction to the Theory of Games.* Concepts, Methods, Applications. 1999 ISBN 0-7923-5775-2
33. C.A. Floudas and P.M. Pardalos (eds.): *Handbook of Test Problems in Local and Global Optimization.* 1999 ISBN 0-7923-5801-5
34. T. Stoilov and K. Stoilova: *Noniterative Coordination in Multilevel Systems.* 1999 ISBN 0-7923-5879-1
35. J. Haslinger, M. Miettinen and P.D. Panagiotopoulos: *Finite Element Method for Hemivariational Inequalities.* Theory, Methods and Applications. 1999 ISBN 0-7923-5951-8
36. V. Korotkich: *A Mathematical Structure of Emergent Computation.* 1999 ISBN 0-7923-6010-9
37. C.A. Floudas: *Deterministic Global Optimization: Theory, Methods and Applications.* 2000 ISBN 0-7923-6014-1
38. F. Giannessi (ed.): *Vector Variational Inequalities and Vector Equilibria.* Mathematical Theories. 1999 ISBN 0-7923-6026-5
39. D.Y. Gao: *Duality Principles in Nonconvex Systems.* Theory, Methods and Applications. 2000 ISBN 0-7923-6145-3
40. C.A. Floudas and P.M. Pardalos (eds.): *Optimization in Computational Chemistry and Molecular Biology.* Local and Global Approaches. 2000 ISBN 0-7923-6155-5
41. G. Isac: *Topological Methods in Complementarity Theory.* 2000 ISBN 0-7923-6274-8
42. P.M. Pardalos (ed.): *Approximation and Complexity in Numerical Optimization: Concrete and Discrete Problems.* 2000 ISBN 0-7923-6275-6
43. V. Demyanov and A. Rubinov (eds.): *Quasidifferentiability and Related Topics.* 2000 ISBN 0-7923-6284-5

Nonconvex Optimization and Its Applications

44. A. Rubinov: *Abstract Convexity and Global Optimization.* 2000 ISBN 0-7923-6323-X
45. R.G. Strongin and Y.D. Sergeyev: *Global Optimization with Non-Convex Constraints.* 2000 ISBN 0-7923-6490-2
46. X.-S. Zhang: *Neural Networks in Optimization.* 2000 ISBN 0-7923-6515-1
47. H. Jongen, P. Jonker and F. Twilt: *Nonlinear Optimization in Finite Dimensions.* Morse Theory, Chebyshev Approximation, Transversability, Flows, Parametric Aspects. 2000 ISBN 0-7923-6561-5
48. R. Horst, P.M. Pardalos and N.V. Thoai: *Introduction to Global Optimization.* 2nd Edition. 2000 ISBN 0-7923-6574-7
49. S.P. Uryasev (ed.): *Probabilistic Constrained Optimization.* Methodology and Applications. 2000 ISBN 0-7923-6644-1
50. D.Y. Gao, R.W. Ogden and G.E. Stavroulakis (eds.): *Nonsmooth/Nonconvex Mechanics.* Modeling, Analysis and Numerical Methods. 2001 ISBN 0-7923-6786-3
51. A. Atkinson, B. Bogacka and A. Zhigljavsky (eds.): *Optimum Design 2000.* 2001 ISBN 0-7923-6798-7
52. M. do Rosário Grossinho and S.A. Tersian: *An Introduction to Minimax Theorems and Their Applications to Differential Equations.* 2001 ISBN 0-7923-6832-0
53. A. Migdalas, P.M. Pardalos and P. Värbrand (eds.): *From Local to Global Optimization.* 2001 ISBN 0-7923-6883-5
54. N. Hadjisavvas and P.M. Pardalos (eds.): *Advances in Convex Analysis and Global Optimization.* Honoring the Memory of C. Caratheodory (1873-1950). 2001 ISBN 0-7923-6942-4
55. R.P. Gilbert, P.D. Panagiotopoulos† and P.M. Pardalos (eds.): *From Convexity to Nonconvexity.* 2001 ISBN 0-7923-7144-5
56. D.-Z. Du, P.M. Pardalos and W. Wu: *Mathematical Theory of Optimization.* 2001 ISBN 1-4020-0015-4

KLUWER ACADEMIC PUBLISHERS – DORDRECHT / BOSTON / LONDON

MIX
Papier aus verantwortungsvollen Quellen
Paper from responsible sources
FSC® C105338

If you have any concerns about our products,
you can contact us on
ProductSafety@springernature.com

In case Publisher is established outside the EU,
the EU authorized representative is:
Springer Nature Customer Service Center GmbH
Europaplatz 3, 69115 Heidelberg, Germany

Printed by Libri Plureos GmbH
in Hamburg, Germany